Signal Integrity Characterization Techniques

Executive Editors

Mike Resso
Signal Integrity Application Scientist
Component Test Division
Agilent Technologies

Eric Bogatin
Signal Integrity Evangelist
Bogatin Enterprises

IEC
Chicago, Illinois

ISBN: 978-1-7325670-2-3

International Engineering Consortium
300 West Adams Street, Suite 1210
Chicago, Illinois 60606-5114, USA
+1-312-559-4100 voice • +1-312-559-4111 fax
publications@iec.org • *www.iec.org*

About the Executive Editors

Mike Resso is the signal integrity application scientist in the component test division at Agilent Technologies and has over twenty-five years of experience in the test and measurement industry. His professional background includes the design and development of electro-optic test instrumentation for aerospace and commercial applications. His most recent activity has focused on the complete multiport characterization of high-speed digital interconnects using time domain reflectometry and vector network analysis. Mike has authored over 30 professional publications in such diverse fields as infrared detector probe systems, linearly variable optical filters and electrically conductive antireflection coatings. He has been awarded one U.S. patent and has twice received the Agilent "Spark of Insight" Award for his contribution to the company. Mike received a B.S. degree in electrical and computer engineering from University of California in 1983.

Eric Bogatin is signal integrity evangelist at Bogatin Enterprises, which specializes in training for signal integrity and interconnect design. His company offers a complete curriculum in short courses and training materials to help accelerate engineers and managers up the learning curve to be more effective in fields related to signal integrity. He has held senior engineering and management positions at such companies as AT&T Bell Labs, Raychem Corporation, Advanced Packaging Systems, and Sun Microsystems. For 20 years, he has been involved in various aspects of signal integrity and inter-connect design, from the materials side, manufacturing, product design, measurements and, most recently, education and consulting. Eric has written four books on signal integrity and inter-connect design, over 200 papers, and most recently wrote a book entitled *Signal Integrity-Simplified* published in 2004. Eric received his Ph.D. in physics from the University of Arizona in Tucson in 1980 and his B.S. in physics from the Massachusetts Institute of Technology in 1976.

About the Publisher

The International Engineering Consortium (IEC) is a non-profit organization dedicated to catalyzing technology and business progress worldwide in a range of high technology industries and their university communities. Since 1944, the IEC has provided high-quality educational opportunities for industry professionals, academics, and students. In conjunction with industry-leading companies, the IEC has developed an extensive, free on-line educational program. The IEC conducts industry-university programs that have substantial impact on curricula. It also conducts research and develops publications, conferences, and technological exhibits that address major opportunities and challenges of the information age. More than 70 leading high-technology universities are IEC affiliates, and the IEC handles the affairs of the Electrical and Computer Engineering Department Heads Association and Eta Kappa Nu, the honor society for electrical and computer engineers. The IEC also manages the activities of the Enterprise Communications Consortium.

IEC Corporate Members / University Program Sponsors

The IEC's University Program, which provides grants for full-time faculty members and their students to attend IEC Forums, is made possible through the generous contributions of its Corporate Members. For more information on Corporate Membership or the University Program, please call +1-312-559-3309 or send an e-mail to *cmp@iec.org*.

IEC Corporate Members

accenture High performance. Delivered.
CableLabs
HOSTWAY
NEC
sandvine
tmforum
Actelis Networks
CNC 中国网通
HUAWEI
NETadmin
SZ Schmid
UTEL
ADVA Optical Networking
CloudShield
IBM
NetCracker
SPIRENT
verizon
Alcatel-Lucent
DETECON Consulting
IET
Nokia Siemens Networks
Tech Mahindra
xeround
Allied Telesis
Deutsche Telekom
infineon
orange
TDS Telephone and Data Systems, Inc.
XTERA
ARTHUR J. SCHMITT FOUNDATION
ERICSSON TAKING YOU FORWARD
ITALTEL
Pacific Telecommunications Council
Telcordia.
ZTE中兴
AttivaCorp Achievement Through Best Practice
ethernet alliance
Juniper Networks
PacketFront
TELECOM ITALIA
belgacom
freescale semiconductor
KEYMILE
PCCW
TELUS
broadband forum
FUJITSU
KT
PeerApp
THOMSON images & beyond
BSF Broadband Services Forum
Genesis
LeadCom
PIRELLI BROADBAND SOLUTIONS
TIA
BT
hp invent
MOTOROLA
Redback Networks AN ERICSSON COMPANY
tilgin

These are some of the universities that participate in the University Grant Program:

The University of Arizona
Arizona State University
Auburn University
University of California at Berkeley
University of California, Davis
University of California, Santa Barbara
Carnegie Mellon University
Case Western Reserve University
Clemson University
University of Colorado at Boulder
Columbia University
Cornell University
Drexel University
École Nationale Supérieure des Télécommunications de Bretagne
École Nationale Supérieure des Télécommunications de Paris
École Supérieure d'Électricité
University of Edinburgh
University of Florida
Georgia Institute of Technology
University of Glasgow
Howard University
Illinois Institute of Technology
University of Illinois at Chicago
University of Illinois at Urbana/Champaign
Imperial College of Science, Technology and Medicine
Institut National Polytechnique de Grenoble
Instituto Tecnológico y de Estudios Superiores de Monterrey
Iowa State University
KAIST
The University of Kansas
University of Kentucky
Lehigh University
University College London
Marquette University
University of Maryland at College Park
Massachusetts Institute of Technology
University of Massachusetts
McGill University
Michigan State University
The University of Michigan
University of Minnesota
Mississippi State University
The University of Mississippi
University of Missouri-Columbia
University of Missouri-Rolla
Technische Universität München
Universidad Nacional Autónoma de México
North Carolina State University at Raleigh
Northwestern University
University of Notre Dame
The Ohio State University
Oklahoma State University
The University of Oklahoma
Oregon State University
Université d'Ottawa
The Pennsylvania State University
University of Pennsylvania
University of Pittsburgh
Polytechnic University
Purdue University
The Queen's University of Belfast
Rensselaer Polytechnic Institute
University of Southampton
University of Southern California
Stanford University
Syracuse University
University of Tennessee, Knoxville
Texas A&M University
The University of Texas at Austin
University of Toronto
VA Polytechnic Institute and State University
University of Virginia
University of Washington
University of Wisconsin-Madison
Worcester Polytechnic Institute

Preface

Today's emerging high-speed digital applications require a special kind of design engineer who understands the subtle signal integrity issues at hand. Although a classical electrical and computer engineering education is helpful, it is the high-frequency microwave effects that normally cause the most problems within telecommunications and computer systems, channels, and components. Reflections from impedance discontinuities, crosstalk, intra-line skew, and a multitude of other problems can immediately stop a system from working properly.

Clearly, there is a gap in the college electrical engineering courses between the traditional digital and microwave curricula. This is why learning never stops for signal integrity engineers. This book addresses this gap with a focus on a practical and intuitive understanding of signal integrity effects within the data transmission channel. High-speed interconnects such as connectors, printed circuit boards (PCBs), cables, integrated circuit (IC) packages, and backplanes are critical elements of differential channels that must be designed using today's most powerful analysis and characterization tools. Both measurements and simulation must be done on the device under test, and both activities must yield data that correlates with each other. Most of this book focuses on real-world applications of signal integrity measurements.

Since the most intuitive measurements for digital engineers are usually done in the time domain, this book starts with a fundamental understanding of single-ended and differential time domain reflectometry (TDR) measurements in chapters 1 and 2. Chapters 3, 4, and 5 complete the first major section of this book by describing vector network analyzers (VNAs) and S-parameters (including 12-port S-parameters).

Section 2 of this book delves into the longest, densest, and highest-bandwidth application for interconnects: the backplane. While many high-speed PCBs exhibit difficult signal integrity problems, none can compare with the typical backplane for design challenges.

The sheer number of layers and signal channels in a backplane create a design challenge most microwave engineers would find daunting. In section 2, we describe a methodology for characterizing even the most complex backplanes.

Section 3 of this book advances the discussion of linear passive device characterization with the implementation of sophisticated error correction techniques. One of the main advantages of using a VNA is the ultra-precise calibration and de-embedding capabilities. Hence, we cover this information in great detail in this section of the book.

Section 4 covers various jitter measurements, including such diverse topics as laser transmitter driver circuitry jitter and a novel statistical jitter measurement methodology called STATEYE.

New measurement technologies co-evolve with design and technology innovation within the engineering communities, so the authors wanted to inspire some new thinking in sections 5 and 6. Analyzing some of these new technologies can provide valuable insight into the future direction that our fast-paced world takes us. Solving today's problems can teach us valuable lessons indeed, but investigating the future trends is sometimes akin to viewing a crystal ball. Our hope is to trigger inspiration to learn more about signal integrity and the high-speed technology around us.

Authors

Mike Resso, Signal Integrity Application Scientist, Component Test Division, Agilent Technologies

Eric Bogatin, Signal Integrity Evangelist, Bogatin Enterprises

Acknowledgements

We would like to acknowledge the professional help from colleagues who have assisted in the creation of this signal integrity book. The list of authors reads like a "who's who" of signal integrity, so it is with much thanks that we publish these individuals' names:

Heidi Barnes
Michael Baxter
Shelley Begley
Orlando Bell
Osvaldo Buccafusca
William Burns
Jack Carrel
Jinhua Chen
Antonio Ciccomancini
Michael Comai
Steven Corey
John D'Ambrosia
Bill Dempsey
Jay Diepenbrock
Dave Dunham
Vince Duperron
John Goldie
Thomas Gneiting
Kevin Grundy
Crescencio Gutierrez
Abraham Islas
Haw-Jyh Liaw
Jim Maynard
Tom McCarthy
Will Miller
Roland Modinger
José Moreira
Gary Otonari
Gautam Patel
Stephen Reddy
Jason Roe
Anthony Sanders
Edward Sayre
Hong Shi
Dina Smolyansky
Laurie Taira-Griffin
Francisco Tamayo-Broes
Ming Tsai

A special thanks to Bob Schaefer for his extensive work on chapter 5 that entailed many months of experiments for comparing detailed measurement data of TDRs and VNAs.

Mike Resso
Agilent Technologies

Eric Bogatin
Bogatin Enterprises

January 2009

Table of Contents

Part IV: Jitter and Active Signal Analysis

Part V: Analysis of New Technologies

Part I

Getting Started: Introducing TDR and VNA Techniques and the Power of S-Parameters

Chapter 1

Single-Port TDR, TDR/TDT, and Two-Port TDR: Interconnect Analysis Is Simplified with Physical Layer Tools

1.1 Introduction

The time domain reflectometer (TDR) has come a long way since the early days when it was used to locate faults in cables. Time domain reflectometry can be used for more than 40 characterization, modeling, and emulation applications, many of which are illustrated in this application note series.

If your applications involve signals with rise times shorter than one nanosecond, transmission line properties of the interconnects are important. TDR is a versatile tool to provide a window into the performance of your interconnects to quickly and routinely answer the three important questions: does my interconnect meet specifications, will it work in my application, and where do I look to improve its performance?

The TDR is not just a simple radar station for transmission lines, sending pulses down the line and looking at the reflections from impedance discontinuities. It is also an instrument that can directly provide first-order topology models, S-parameter behavioral models, and with up to four channels, characterize rise time degradation, interconnect bandwidth, near- and far-end crosstalk, odd mode, even mode, differential and common impedance, mode conversion, and the complete differential channel characterization.

To provide a little order to the wide variety of applications explored in this signal integrity book, the series is divided into the following three parts covering four general areas:

- Part 1 (Chapter 1)—Those that use a single-port TDR, those that use TDR/time domain transmission (TDT), and

those that use two-port TDR.
- Part 2 (Chapter 2)—Those that use four-port TDR or four-port vector network analyzer (VNA) with physical layer test system (PLTS).
- Part 3 (Chapter 10)—Those that use advanced signal integrity measurements and calibration.

The principles of TDR and VNA operation are detailed in other chapters in this book and references listed in the bibliography. This application note series concentrates on the valuable information that can be quickly obtained with simple techniques that can be used to help get the design right the first time.

1.2 Single-Port TDR

Overview

This section will look at the seven most important applications of one-port TDR. The first two refer to the complete characterization of a uniform transmission line, extracting the characteristic impedance and time delay.

But we can get more than this with specially designed test structures. We can also get a fundamental, intrinsic property of the transmission line, the velocity of a signal, and from this, the intrinsic bulk dielectric constant of the laminate.

When the line is not uniform and has discontinuities, we can build first-order, topology-based models right from the front screen. If this is not high-bandwidth enough, we can bring the measured data into a simulation tool such as Agilent's Advanced Design System (ADS) and build very–high-bandwidth models, which can then be used in simulations to evaluate whether this interconnect might be acceptable in a specific application.

Finally, we can emulate the final application system's rise time with the TDR to directly measure the reflection noise generated by physical structures in the interconnect and whether they might pose a potential problem or, equally of value, might be ignored.

Measuring Characteristic Impedance and Uniformity of Transmission Lines

Historically, the most common use of the TDR has been to characterize the electrical properties of a transmission line. For an ideal, lossless transmission line, there are only two parameters that fully characterize the interconnect: its characteristic impedance and its time delay. This is the easiest and most common application for TDR.

The TDR sends a calibrated step edge of roughly 200 mV into the device under test (DUT). Any changes in the instantaneous impedance the edge encounters along its path will cause some of this signal to reflect back, depending on the change in impedance it sees. The constant incident voltage of 200 mV, plus any reflected voltage, is what is displayed on the screen of the TDR.

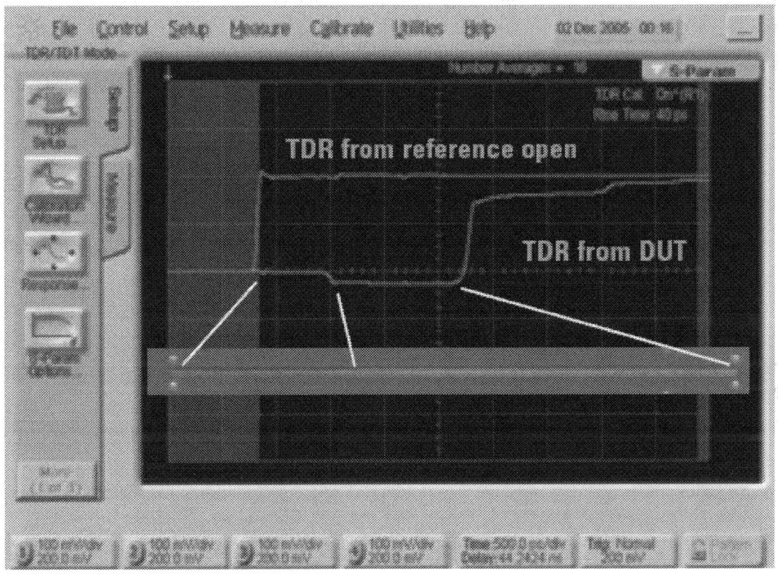

Figure 1.1: *Measured TDR Response from a Microstrip Transmission Line. Top Trace Is the Reflection from the End of the Cable; Bottom Trace Is the Reflected Signal from the DUT*

In *Figure 1.1*, the bottom line is the measured TDR response when the DUT is the microstrip trace shown. The first two inches of the transmission line has a characteristic impedance of roughly 50 Ohms, while the next four inches of the transmission line has a characteristic impedance of roughly 40 Ohms. The far end of the line is open.

The voltage displayed on the screen is the total voltage: the incident, constant 200 mV, plus the reflected voltage. Note on the bottom of the screen, the vertical voltage scale is 100 mV/div. The top line is the TDR response for the cable not connected to the transmission line. This defines the beginning of the cable, which is an open. On the bottom line, at this instant of time, is the small reflected voltage from the surface-mounted assembly (SMA) launch, followed by the roughly 50 Ohm section of the line, and about one division later, the small drop in voltage from the lower-impedance second half of the transmission line.

Contained in this reflected signal is the information about the impedance profile of the transmission line. We could read the voltages off the front screen and use pencil and paper to back out the impedance of the line, or we can take advantage of some of the built-in features of this TDR.

We can use the two markers, which will automatically perform the calculations to back out the instantaneous impedance from the measured data. There are clearly two regions of relatively uniform impedance on this transmission line. We move the markers so that one is in each region, as shown in *Figure 1.2*, and then we can read the impedance of each region from the screen.

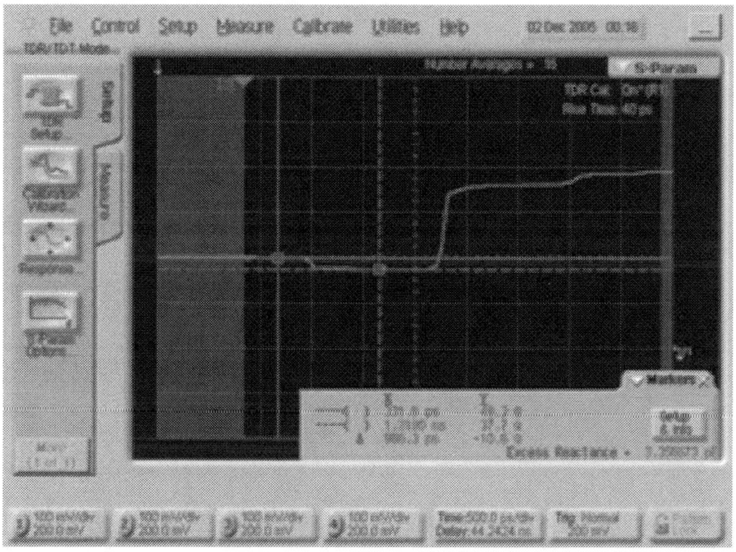

Figure 1.2: *Using Markers to Measure the Characteristic Impedance of a Transmission Line*

The impedance of the first region, read from the solid-line marker, is 48.3 Ohms. The impedance in the second region, read from the dotted-line marker, is 37.7 Ohms. The nominal design impedances were 50 Ohms and 40 Ohms, so we see that actual and fabricated impedances are off by about 3.5 and 6 percent, respectively.

The one caveat when using markers is to watch out for masking effects. The impedance read by the marker can be interpreted as the instantaneous impedance of the transmission line at the location of the marker, as long as it is the first interface, or there have been only small impedance discontinuities up to the location of the marker. This feature makes extracting the instantaneous impedance of a uniform transmission line almost trivial. In addition, we can see that the impedance in each region is relatively uniform, as there is little deviation in the reflected voltage up and down the line segments.

In addition to using the marker to identify the specific instantaneous impedance of the transmission line, we can also convert the vertical voltage scale into an impedance scale.

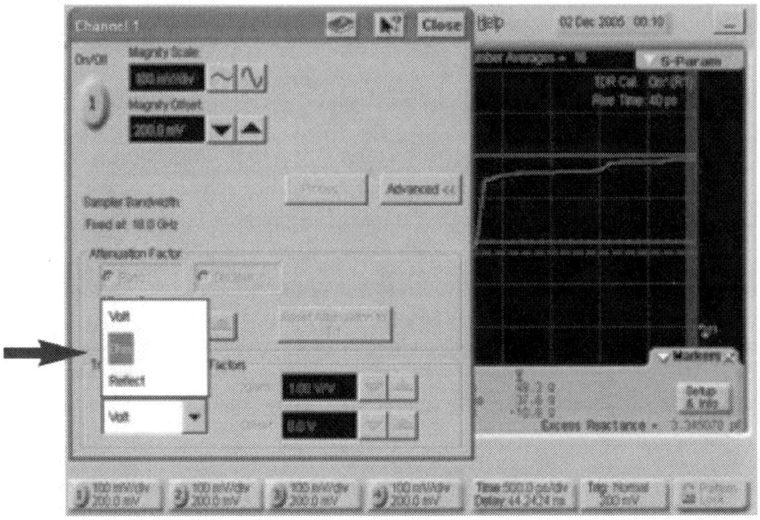

Figure 1.3: *The Advanced Settings Function Can Adjust the Vertical Scale to Display the Impedance Directly*

By selecting vertical scale 1, 100 mV/div: 200.0 mV and then clicking Advanced, we can access the scale settings command functions, as shown in *Figure 1.3*. When we select the Ohms scale, the TDR will convert every point of the reflected voltage into an equivalent instantaneous impedance.

Effectively, the TDR takes each measured voltage point, subtracts 200 mV to get the reflected voltage, then takes the ratio of this voltage to the 200 mV of incident voltage to get the reflection coefficient, and from the reflection coefficient, uses the simple relationship: $Z = 50\ \Omega \times (1 + \text{rho})/(1 - \text{rho})$ to calculate the instantaneous impedance of each point. Finally, this extracted instantaneous impedance is plotted on the screen.

The offset and scale settings, now calibrated in Ohms, can be used to adjust the scale for our application.

Figure 1.4 is the same TDR data for this two-segment transmission line but now with the instantaneous impedance displayed directly on the vertical scale. In this case, the scale is 10 Ohms/div with the center location set to 50 Ohms. On this scale we can literally read off the screen the impedance of the first section as about 48 Ohms and the impedance of the second section as about 38 Ohms.

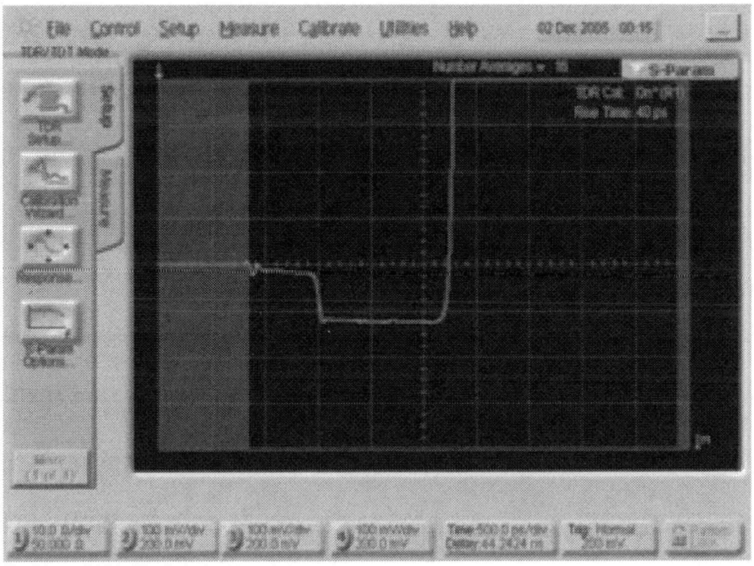

Figure 1.4: *The Same Transmission Line Displayed on the Impedance Scale at 10 Ohms/div, with 50 Ohms in the Center*

This scale setting allows a direct and effortless graphical display of the impedance profile of a transmission line, with the one caveat that we are assuming all the measured voltage coming back to the TDR is due to reflections from impedance discontinuities. This is a good assumption as long as the impedance changes up to each point are small.

It looks like, for this transmission line, the impedance of the first section is decreasing slightly down the line, while the impedance profile of the second section is mostly constant. We can use this technique to evaluate how uniform the impedance of a transmission line is.

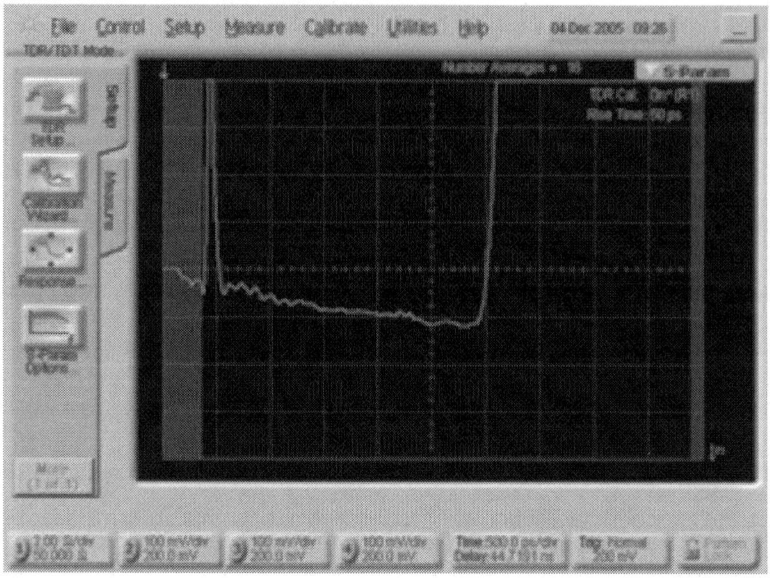

Figure 1.5: *High-Resolution TDR Profile of a Nominally Uniform Transmission Line, at 2 Ohms/div and 50 Ohms in the Center of the Screen*

Figure 1.5 shows the measured TDR response of a nominally uniform transmission line on an expanded scale of 2 Ohms/div. The impedance at the center of the screen is set at 50 Ohms. This scale information can be read next to the channel 1 button of the screen.

The large peak at the beginning of the line is the inductive discontinuity of the SMA launch that, on this high-resolution scale, looks huge. At 2 Ohms/div on the vertical scale, it looks like this uniform transmission line is not so uniform. It appears to have a variation of as much as 1 Ohm from the beginning to the end of the line. This is roughly 2 percent.

Is this variation real, or could it be some sort of artifact? There are two important artifacts that might give rise to this sort of behavior. It could be there is rise-time degradation in the incident signal. It may not be perfectly flat, like an ideal Gaussian step edge. After all, the reflected signal displayed on the TDR is really the reflection of the incident signal. If the incident signal has a long tail, we will see this long tail in the TDR response and may mistakenly interpret this as an impedance profile variation.

One way around this problem is to use the calibrated response feature of the DCA 86100C TDR, which is being done in this case.

Other sources of artifact can be either distributed series resistance in the trace or distributed shunt conductance in the trace due to the lossy nature of the line. The series resistance will cause the reflected voltage to increase down the line, while the shunt conductance will cause the reflected TDR response to decrease down the line, as in this case.

One way to evaluate whether an impedance profile is really showing a variation in the instantaneous impedance of the transmission line or an artifact is to measure the TDR response of the line from both ends. If it is real, we should see the slope of the response change, depending on which end of the line we launch from. If it is one of the two artifacts, the response will look the same on the screen, independent of which end we launch from.

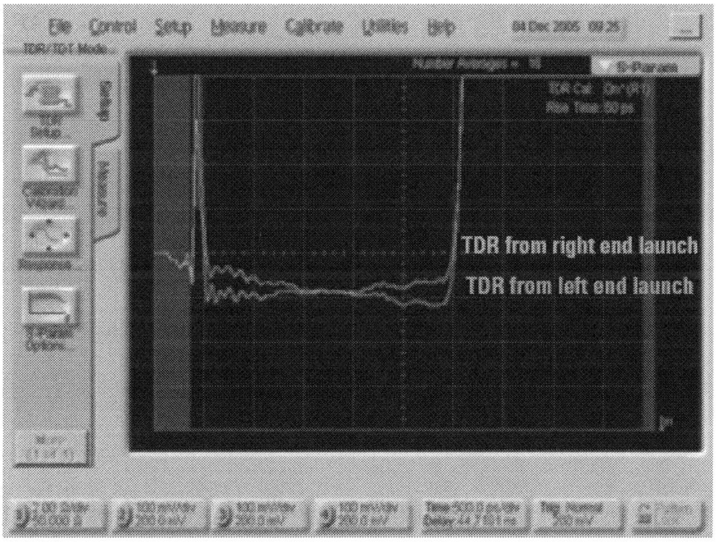

Figure 1.6: *High-Resolution TDR Response from Each End of the Same Uniform Transmission Line, Verifying the Impedance Variation Is Real*

Figure 1.6 shows the measured TDR response launching from each end of the line where the scale is 2 Ohms/div in both cases.

The *TDR from left end launch* line shows the left side of the line is the higher impedance. While the *TDR from right end launch* line also confirms that the impedance of the trace is higher on the left side. This variation in the instantaneous impedance is confirmed to be real and is not due to the series resistance, shunt conductance, or non-ideal step edge. Using the technique of comparing the launches from both ends, we can unambiguously identify real, non-uniform impedance effects in a transmission line.

In this example, the microstrip is showing a variation of about half a division, or 1 Ohm out of 50 Ohms, or 2 percent, from one end to the other. This could be due to a variation in the laminate thickness, the slight drift in the alignment of the trace width over a fiber bundle in the glass weave, or a variation in the etching of the line due to photo resist developer variation across the board.

By measuring the variation in other lines across the board or inspecting the dimensions of the board, the root cause might be identified and the process stabilized.

Measuring Time Delay of a Transmission Line

The second important property of a transmission line is the time delay from one end to the other. This can also be measured directly from the screen of the TDR using markers. However, to get an accurate measure of the time delay, we need to know the starting point of the transmission line.

By removing the DUT and recording the TDR response from the open end of the cable, we can use this as a reference to define the beginning of the line. This is the top line in *Figure 1.7*. When we reconnect the DUT and record the TDR response, we see the reflection from the open at the far end of the transmission line, just visible at the far right edge of the screen.

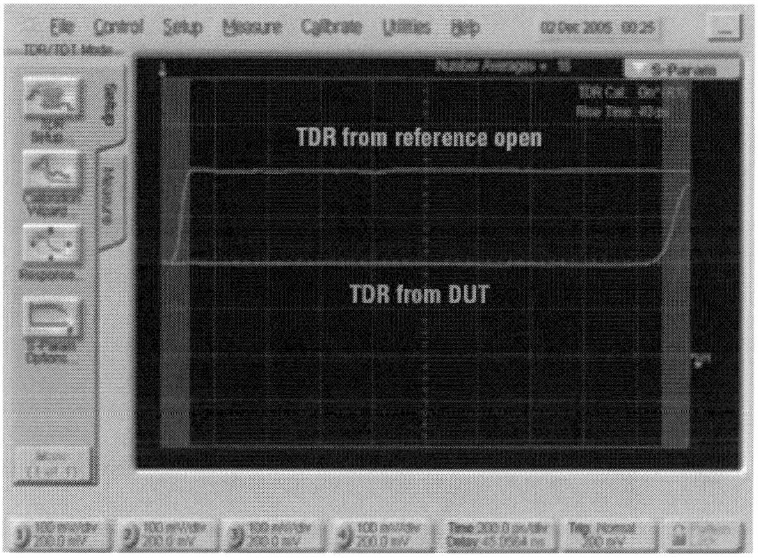

Figure 1.7: *TDR Response of a Uniform Six-Inch Transmission Line Open at the Far End*

The total round-trip time delay is the time interval from the beginning of the reflection from the open end of the cable to the reflection from the open far end of the DUT. To increase the accuracy, we use the time from the midpoint of the two open responses. This can be measured simply and easily using the vertical markers directly from the screen.

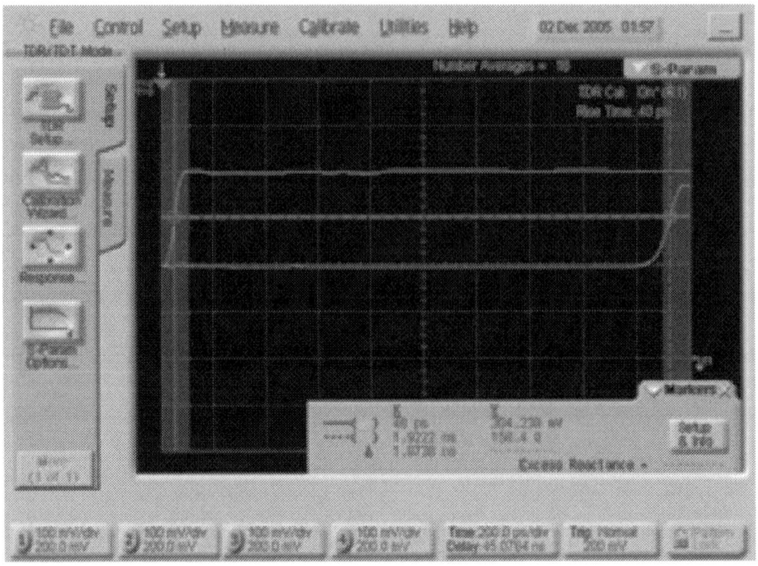

Figure 1.8: *TDR Response of the Reference Open and Uniform Six-Inch Transmission Line, with Markers Showing the Beginning and End of the Traces*

Using the marker buttons below the screen, we can position the markers in *Figure 1.8* so that they define the midpoint-to-midpoint distances. We can read off the screen that the total time delay is 1.87 ns. This is the round-trip time delay. The one-way time delay is half of this, or .935 ns. This is the time delay (TD) of the transmission line.

From the physical length of the transmission line, six inches, and the time delay, 0.935 ns, we can also calculate the speed of the signal down this transmission line. The speed is 6 in/0.935 ns = 6.42 in/ns. This is an intrinsic property of the transmission line and would be true for any transmission line of the same width built on this layer of the board, independent of the length of the line.

One of the artifacts in this measurement is the uncertainty of how much of the total TD is due to the connector at the front of the line. Is the open reference really the beginning of the line, or is there some contribution to the launch into the transmission line of the circuit board? We can take advantage of a simple test feature to get around this artifact and extract a more accurate value for the speed of a signal on the trace.

This trick is useful only if we have the option of designing the test line to aid in the characterization of the circuit board and each particular layer. The secret is to add small imperfections to the line such as reference pads at two locations with a known separation.

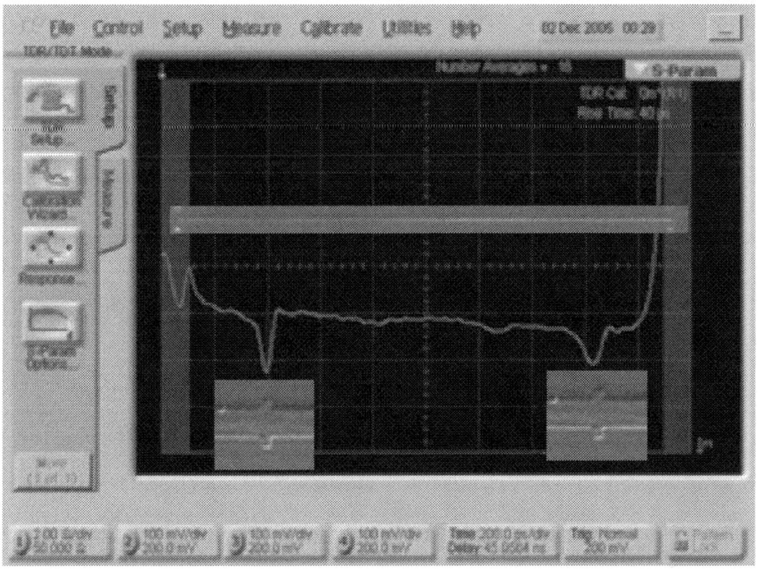

Figure 1.9: *TDR Response of a Uniform Transmission Line with Two Small Reference Pads Located on Four-Inch Centers*

Figure 1.9 shows an example of a six-inch-long transmission line with two reference pads (in close-up), located with a center spacing of four inches. These pads can be easily detected with the TDR. The TDR response is displayed on a scale with 2 Ohms/div. This is a high-sensitivity scale. On the far left, it shows the beginning of the line with a few ripples from the SMA launch. About two divisions from the beginning is the dip from the first pad, which acts as a small capacitive load, a lower impedance. Some time later, the TDR signal shows the response from the second reference pad.

Accurate Measurement of Signal Speed in a Transmission Line

The time difference between these two negative dips is the round-trip time difference between the pads separated by four inches. By measuring this time delay from the screen, we can get an accurate measure of the speed of the signal, independent of the nature of the launch into the transmission line.

We can measure the time delay between the dips using the on-screen markers. By aligning each marker with the center of the dip, we can measure this location within a few picoseconds' accuracy. We can see from the screen in *Figure 1.10* that the round-trip time delay is 1.238 ns. From this round-trip delay, we can calculate the one-way time delay as half this, or 0.619 ns.

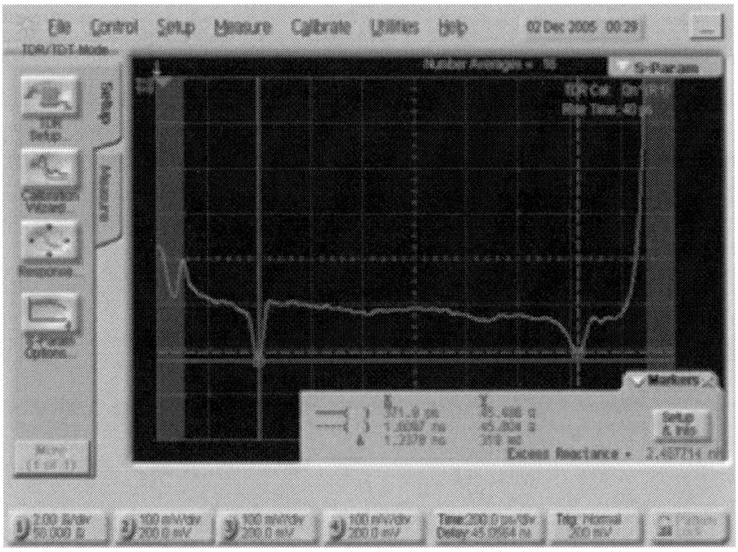

Figure 1.10: *TDR Response from a Microstrip with Two Reference Pads Using Markers to Measure the Round-Trip Time Delay*

Given the physical distance between the two reference pads as four inches, the speed of the signal down the microstrip can be calculated as 4 in/0.619 ns = 6.46 in/ns. This is very close to the 6.42 in/ns calculated as the speed of the signal using the end-to-end method.

Using this value of the speed of the signal, we can extract the laminate's dielectric properties.

Extracting the Bulk Dielectric Constant of the Laminate

The speed of the signal down a transmission line is directly related to the dielectric constant the signal sees. In a stripline structure, such as shown in *Figure 1.11*, the signal sees a uniform, homogeneous material with a composite dielectric constant that is made up of a combination of the resin dielectric constant and the glass weave dielectric constant. Small variations in the local relative combination can affect the local dielectric constant, which is an important source of skew between adjacent lines in a differential pair.

$$Dk = \left(\frac{11.803}{\text{velocity}}\right)^2$$

Figure 1.11: *A Stripline Construction and Extracting the Bulk Dielectric Constant*

From a measurement of the speed of a signal down a stripline transmission line, the effective dielectric constant the signal sees, Dk, can be extracted using the simple relationship shown. The 11.803 number is the speed of light in air, in the units of inches/ns.

However, in a microstrip, the effective dielectric constant the signal sees is not the bulk value of the laminate.

In a microstrip, some of the electric field lines are in the bulk laminate, and see the laminate composite dielectric constant, but some of the field lines, as shown in *Figure 1.12*, are in the air, with a dielectric constant of one. The signal sees a composite of these two materials, which creates an effective dielectric constant, Dk_{eff}. It is this value that affects the signal speed and can be extracted from the measured speed of the signal.

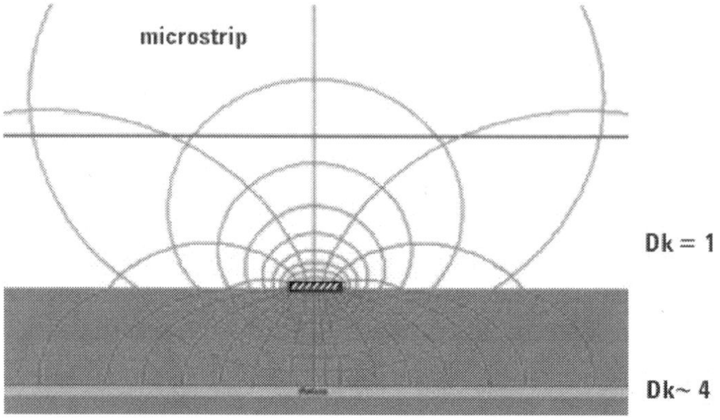

$$Dk_{eff} = \left(\frac{11.803}{velocity} \right)^2$$

Figure 1.12: *Effective Dielectric Constant in Microstrip*

In this example, the speed is 6.46 in/ns. The extracted effective dielectric constant would be 3.34. This is unfortunately not a very useful number. It is not the bulk dielectric constant of the laminate. We cannot use this value of the effective dielectric constant in a field solver or approximation to help us calculate the impedance of any other geometries, for example. We really need to convert the effective dielectric constant into the actual bulk dielectric constant.

This conversion is related to the precise nature of the electric field lines, and what fraction is in the air and the bulk laminate. It also depends very much on the cross-section geometry of the microstrip. The only way to convert the extracted, effective dielectric constant into the bulk laminate dielectric constant is to use a 2D field solver.

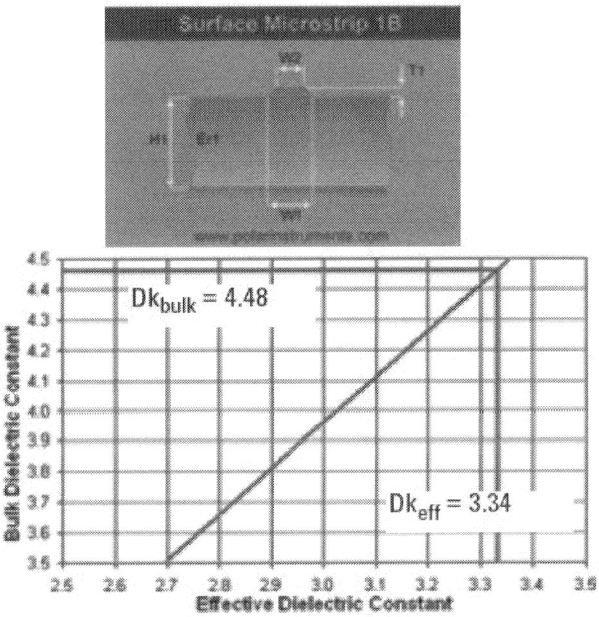

Figure 1.13: *Using a Field Solver to Back out the Bulk Dielectric Constant from the Effective Dielectric Constant*

In *Figure 1.13*, a 2D field solver is used to calculate the effective dielectric constant for different bulk values, using the same geometry as the trace that is measured. We set up the field solver with the cross-section information about the specific microstrip that was measured and use the field solver to calculate the effective dielectric constant for different bulk dielectric constant values.

When we plot up the bulk dielectric constant versus the effective dielectric constant, we get a relatively straight line, as shown in *Figure 1.13*. We use this curve to back out the bulk dielectric constant, given the effective value of 3.34 that was measured. This analysis gives a bulk dielectric constant for this laminate of 4.48.

The TDR enables the measurement of the effective dielectric constant, while the 2D field solver enables the conversion of the effective dielectric constant into the bulk dielectric constant.

Building a Model of a Discontinuity Such as a Corner, Test Pad, Gap in the Return Path, SMA Launch, or Terminating Resistor

Extracting a Model for Capacitive Discontinuities

Not all interconnect structures are uniform transmission lines. As much as we might try to eliminate them, there will often be discontinuities that are unavoidable. For example, test pads, component leads, 90-degree corners, gaps in the return path, or even engineering change wires will all create discontinuities. These structures, by their nature, are non-uniform and often difficult to calculate other than with a 3D field solver. Sometimes, the quickest way to evaluate their impedance is to build a structure and measure it.

From the measured response, we can empirically evaluate the impact on the signal if we match the TDR's rise time to the rise time of the application. We could then directly measure off the screen of the TDR the amount of reflected voltage noise we might see in the system. Alternatively, we could use the TDR to extract a simple, first-order model for the structure and use this model in a system-level simulation to evaluate the impact of the discontinuity. Finally, if we need more accuracy or a higher-bandwidth model than what we can get directly from the screen, we can take the measured data from the TDR and bring it into a modeling and simulation tool such as SPICE or ADS to fit a more accurate model. These processes are illustrated in this section.

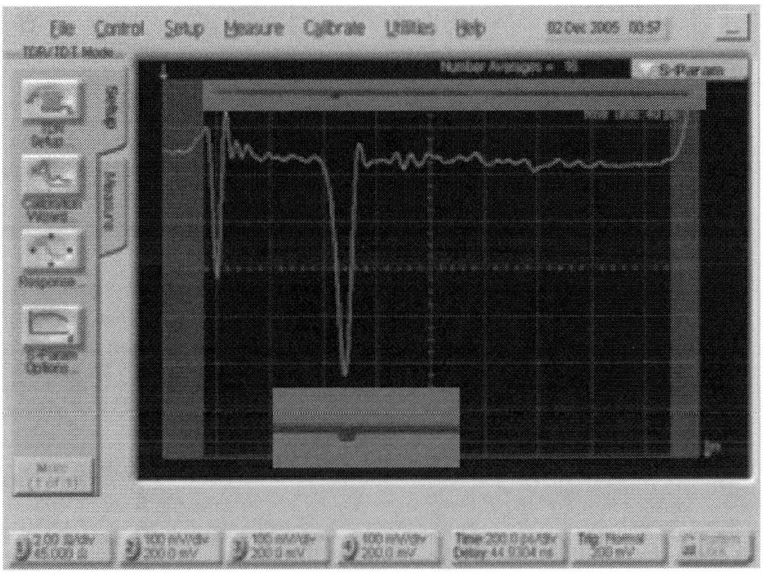

Figure 1.14: *TDR Response from a Uniform Transmission Line Having a Small Test Pad*

Let us start with a simple test pad on an otherwise uniform line, as show in *Figure 1.14*. The TDR response is shown as the yellow line on the screen, displayed in an Ohms scale, with 2 Ohms/div. The small dip near the beginning of the line is due to the SMA launch. The large dip about three divisions from the left edge is from the test pad.

On this scale, the reflected signal from the small test pad looks huge, but is a discontinuity of only 4.5 divisions or about nine Ohms. This can be interpreted as the instantaneous impedance a signal would see, if it had the rise time of the TDR, in this case, about 40 ps. Since this test pad is not a uniform transmission line, the instantaneous impedance is not related to a characteristic impedance, and the impedance a signal would see is going to depend on the rise time of the signal. We can use the TDR to directly emulate any rise time from as fast as 20 ps up to longer than one nanosecond, to directly evaluate the impact of the discontinuity on the system rise time.

Using the built in calibration feature of the DCA 86100C, we can change the effective rise time of the stimulus and directly display the response from this small discontinuity. The structure is the same, and the scale is the same for each of the four rise times of 40, 100, 200, and 500 ps.

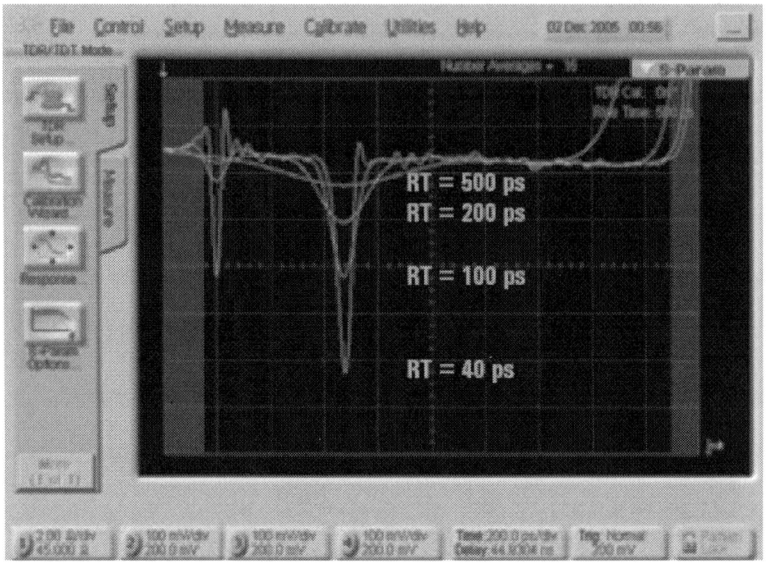

Figure 1.15: *TDR Response for a Uniform Transmission Line with a Small Test Pad, at Four Rise Times of 40, 100, 200, and 500 ps*

Figure 1.15 clearly shows that the instantaneous impedance a signal would see encountering this test pad is strongly dependent on the rise time of the signal. If the rise time were 40 ps, the signal would see an impedance discontinuity of about nine Ohms. At 100 ps, this is only about five Ohms; at 200 ps, it is 2.5 Ohms; and at 500 ps, it is less than one Ohm, hardly noticeable to the signal. Based on the noise budget allocated for discontinuities, we could determine the shortest rise time at which this discontinuity would begin to cause problems or could be ignored.

For example, if three Ohm discontinuities were allowable, this particular test pad could be used for rise times as short as 250 ps. Much below this, and the impact might be felt. The way to know for sure would be to build a model for the discontinuity and use it in a simulation.

By inspection, the simplest model for this discontinuity is a single lumped capacitor. At 40 ps rise time, the TDR response is close to that from an ideal lumped capacitor. We can use the built-in "excess reactance" feature to build a model and extract the parameter values directly from the screen using markers.

The excess reactance feature built into the DCA 86100C will model the DUT as a uniform transmission line having a single discontinuity—either a lumped inductor or lumped capacitor. The software will use the position of the two vertical markers to define the region of the response where the capacitance or inductance will be extracted.

To use this feature, position the markers on either side of the discontinuity and read the amount of capacitance or inductance from the "excess reactance" value on the screen. One hint in using this feature is to position the markers so that they have roughly the same impedance value on either side of the discontinuity. It does not matter what the vertical scale is when the excess reactance function is used.

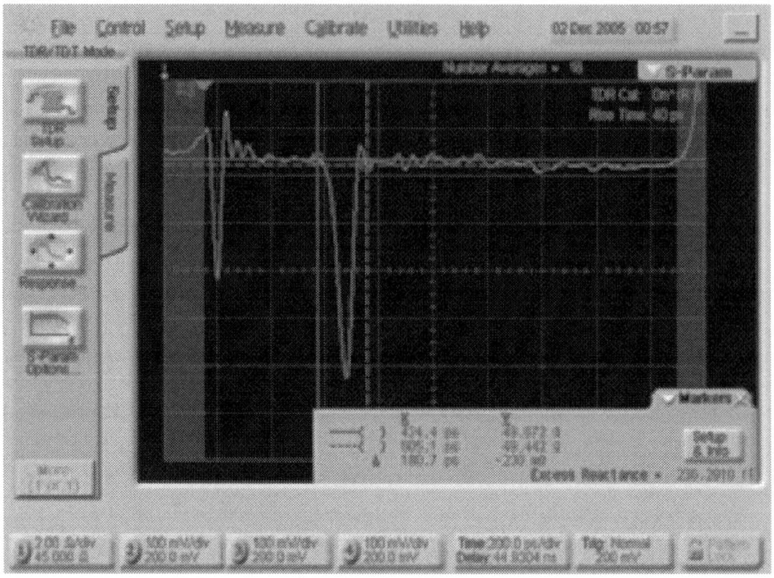

Figure 1.16: *Using the Excess Resistance Feature to Extract the Capacitance of a Test Pad*

In *Figure 1.16*, the markers are used to extract the capacitance of the test pad. The model we are assuming is a single lumped capacitor. The value of this capacitance is read off the screen as 236 fF. This capacitance, plus the impedance of the uniform part of the line, 49 Ohms, provides a complete model for this transmission line structure.

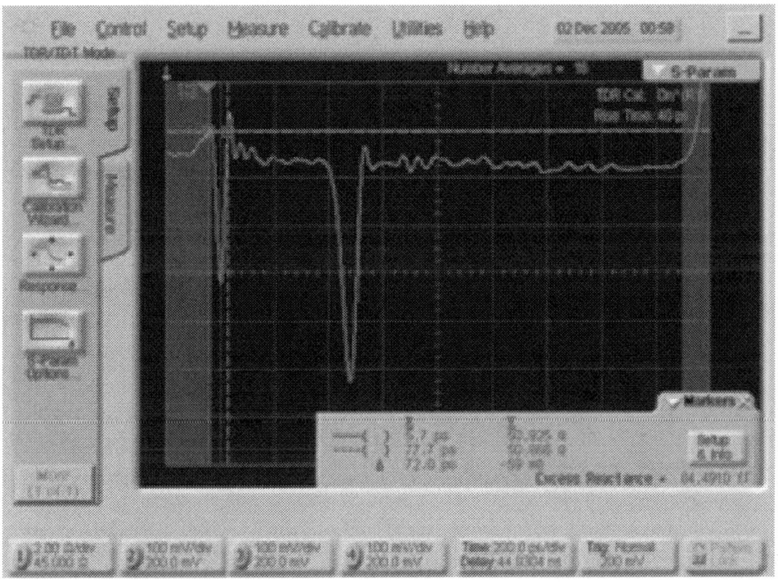

Figure 1.17: *Using Markers to Extract the Excess Capacitance of the SMA Launch*

While we are at it, we can also extract the capacitance associated with the pads used on the end of the transmission line for the SMA launch at the beginning of the line. Using the markers in *Figure 1.17*, we get 84 fF of capacitance. It is clear that the TDR has a very high sensitivity for extracting discontinuity values. On this scale, 84 fF of capacitance is a very large and easily measurable effect.

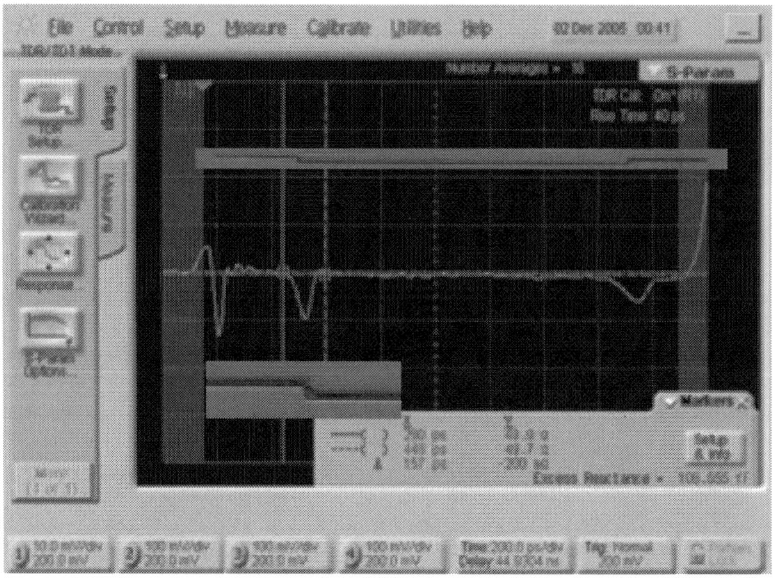

Figure 1.18: *Using Markers to Extract the Excess Capacitance of Two Corners*

We can apply this technique to measure the capacitance associated with a corner. Corners, or 90-degree bends on traces, have stimulated a lot of discussion and concern over the years to signal integrity engineers. *Figure 1.18* shows a simple, uniform transmission line structure built as a microstrip that has two small jags in it. Each jag is a combination of two 90-degree bends.

On this scale of 10 mV/div, it is absolutely clear that a corner causes an impedance discontinuity that is easily measured. On this scale, each division is a reflection coefficient of 10 mV out of 200 mV incident signal, or 5 percent. The two corners in each jag together create a reflected signal on the order of 4 percent, so each corner creates about a 2 percent reflection at a rise time of 40 ps. This is in a 50 Ohm line with a line width of about 60 mils.

In many analog radio frequency (RF) circuits where a flat response over a narrow frequency range is important or a return loss below -40 dB is sometimes required, a corner can introduce big problems. Historically, many RF and microwave circuits were built on thick ceramic substrates where line widths were 100 mils or wider. This would almost double the impact from a corner. This is one of the reasons why corners have developed a reputation as a potential problem and should be avoided.

Using the TDR measurement, we can build a model for a corner and use this model in a system simulation to evaluate whether a corner might pose a potential problem or can be ignored. Clearly, from the TDR response, we can see that the impact of the two corners in this jag looks like the response from a single lumped capacitor. Using the two markers, we can measure the excess capacitance from the two corners as 107 fF. Since this is from two corners, this corresponds to about 53 fF of capacitance per corner. This value can be put into a circuit simulation tool such as SPICE or ADS to simulate the impact from a 53 fF capacitor.

It is also important to note that the amount of capacitance in a corner will scale with the width of the line, for the same impedance traces. If a 60 mil wide line has a capacitance in one corner of roughly 60 fF, then a 5 mil wide line will have a capacitance of roughly 5 fF. This is a good rule of thumb to remember: the capacitance of a corner is about 1 fF per mil of line width for a 50 Ohm line.

Extracting a Model for Inductive Discontinuities

The second type of discontinuity is an inductive discontinuity. These arise, for example, when the line width of a trace necks down, as when passing through a via field; if the return path is disturbed, such as when the trace crosses a gap; or when there is an engineering change wire. An inductive discontinuity will look like a higher impedance and give a peak reflection as the TDR response.

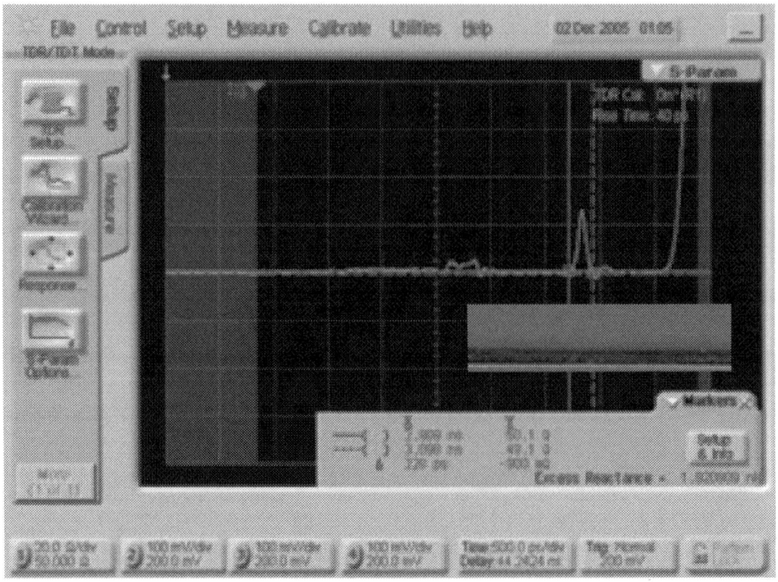

Figure 1.19: *Using Markers to Extract the Excess Inductance of a Short Gap in the Return Path*

Figure 1.19 is an example of the TDR response from a uniform transmission line with a signal line that passes over a very short gap in the return path. This commonly occurs when passing through a field where the clearance or antipads are large enough or on tight enough centers to overlap, inadvertently creating a gap.

The TDR response is a positive peak, just as we would expect from a lumped inductor. We can position the markers on either side of the discontinuity and read the lumped inductance right off the screen as 1.8 nH. For this gap, roughly 100 mils long, the loop inductance created is about 1.8 nH.

If the gap length were increased, the inductive discontinuity created would increase as well. *Figure 1.20* is an example of the TDR response from two return path gap structures. The smaller line trace is the response described previously, from a 100 mil long gap. The inductance was about 1.8 nH.

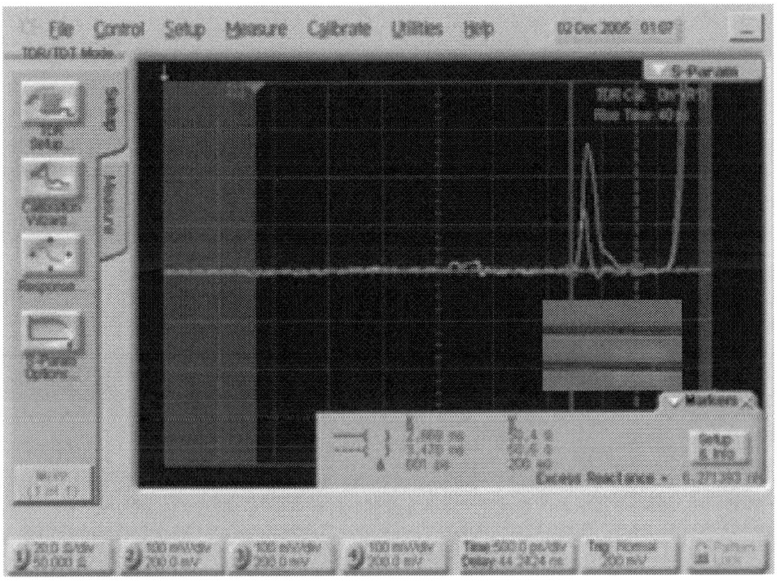

Figure 1.20: *Using Markers to Extract the Excess Inductance of a Large Gap in the Return Path*

The longer trace in *Figure 1.20* is the TDR response from a longer gap of 500 mils. Using the markers, we can measure this lumped inductive discontinuity as 6.3 nH. If we were concerned about the impact from these features, we could easily use the model of the uniform transmission line with these lumped series inductors in a circuit simulation to determine if the additional noise or impact on timing was sufficient to warrant attention.

Modeling Termination Resistors

It is not just interconnects that can be characterized and modeled with a TDR. We can also use a one-port TDR to build a model for discrete components such as termination resistors. *Figure 1.21* is an example of the TDR response from a 50 Ohm axial lead termination resistor connected to the end of a 50 Ohm transmission line.

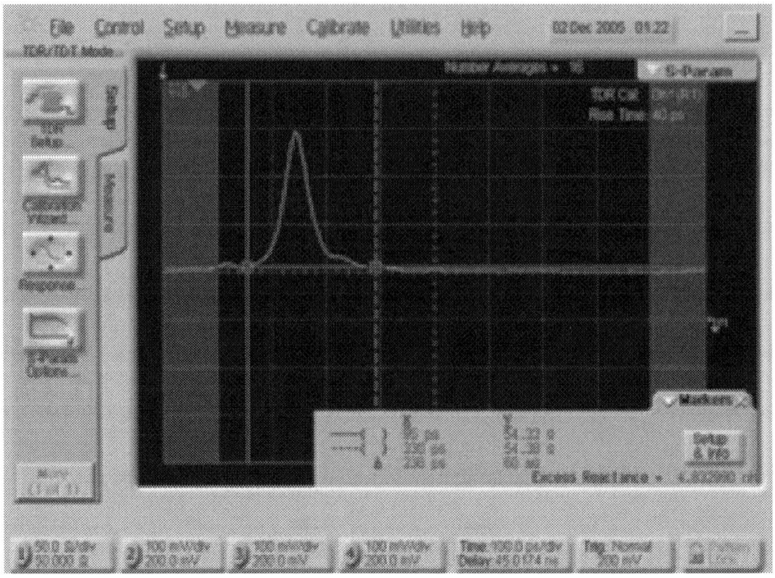

Figure 1.21: *Using Markers to Extract the Excess Inductance of an Axial Lead Termination Resistor*

On the left side of the peak is the transmission line going to the resistor, which, on this scale of 50 Ohms per division, has an impedance of about 50 Ohms. The resistor itself also has an impedance on the order of 50 Ohms. This is seen on the right side of the peak. It is just that it also seems to have some lumped inductance associated with it. This series inductance arises from the long body of the resistor and the leads connecting the signal to the return path.

Using the excess reactance function, we can read the excess lumped inductance of this resistor by positioning the markers on either side of the discontinuity and reading the series loop inductance off the screen as about 4.8 nH. The equivalent circuit model we are assuming is a uniform transmission line with an ideal 4.8 nH series inductor, followed by an ideal resistor of 50 Ohms.

With a rise time of 40 ps, the signal sees a peak impedance of about 200 Ohms. This is the 150 Ohms discontinuity in addition to the 50 Ohms of the line. Of course, as we saw earlier, the impedance a signal would see when it interacts with this inductance will depend on the rise time of the signal. A longer rise time will see a lower impedance; however, the excess inductance of this resistor will not change with the rise time. It is only a function of the geometry of the device.

This is a huge amount of inductance and would probably limit the operation of any circuit it was used in to rise times greater than about 1 ns. At 1 ns rise time, the roughly 5 nH inductance would create a noise level of about 10 percent.

Of course, for high-speed circuits, axial lead resistors are out of the question. Surface mount technology (SMT) resistors, which are physically smaller and can be mounted with much less equivalent series inductance, are required. *Figure 1.22* is an example of an 0603 SMT resistor soldered between the signal and return path on a signal integrity test board, available from BeTheSignal.com. On the top side of the board is an SMT SMA connector, which is connected to the TDR.

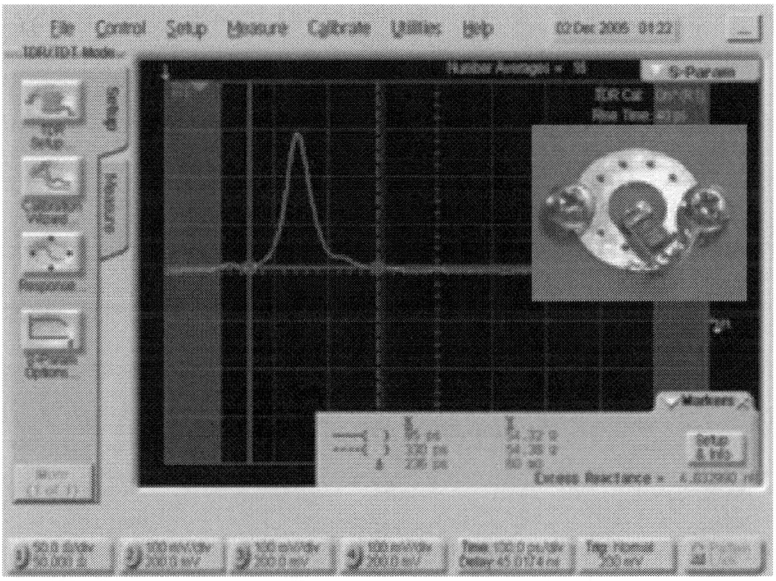

Figure 1.22: *Using Markers to Extract the Excess Inductance of an SMT Termination Resistor Mounted to the MCW620 Test Board*

The TDR response from this component is also shown in *Figure 1.22*, using a scale of 10 Ohms per division, much more sensitive than for the axial lead example.

On the left side of the peak is the transmission line and connector going to the resistor, which has an impedance of about 50 Ohms. The resistor also has an impedance on the order of 50 Ohms, within about 1 percent. This is seen on the right side of the peak. The small peak is the reflection from the series inductance arising from the resistor body, the surface trace, and the vias going to the top layer. The design for the attach of this particular SMT resistor has been optimized for low mounting inductance.

Using the marker function, we can read the excess lumped inductance of this resistor as about 480 pH. This is about an order of magnitude lower series loop inductance than an axial lead resistor and is typical of what can be obtained with an optimized mounting design for a body size of 0603. This applies to resistor components as well as capacitor components.

With a rise time of 50 ps, it looks like it has a series impedance of about 11 Ohms. Of course, as we saw earlier, the impedance a signal would see when it interacts with this inductance will depend on the rise time of the signal. A shorter rise time will see a higher impedance. However, the excess inductance of this resistor will not change with the rise time. It is only a function of the geometry of the device.

Using the marker function to read the excess inductance off the front screen assumes a simple model for the DUT. In the case of this terminating resistor, we assume the model is a single, series lumped inductor. The excess inductance we read off the screen is then the series loop inductance of this resistor. However, we do not have any clear sense from looking at the screen how high the bandwidth is for this simple model, nor can we build more sophisticated models of components easily from just the front screen. This is a case where switching to the frequency domain can get us to the answer faster.

In TDR measurements, a time domain, fast-rising step edge is sent into the DUT and the reflected signal measured. In addition, we can look at the signal that is transmitted through the DUT. This is the time domain transmitted signal, TDT.

If we look at the signal incident into the DUT, it can be thought of as being composed of a series of sine waves, each with a different frequency, amplitude, and phase. Each sine wave component will interact with the DUT independently. When a sine wave reflects from the DUT, the amplitude and phase may change a different amount for each frequency. This variation gives rise to the particular reflected pattern. Likewise, the transmitted signal will have each incident frequency component with a different amplitude and phase.

There is no difference in the information content between the time domain view of the TDR or TDT signal, or the frequency domain view. Using Fourier transform techniques, the time domain response can be mathematically transformed into the frequency domain response and back again without changing or losing any information. These two domains tell the same story, they just emphasize different parts of the story.

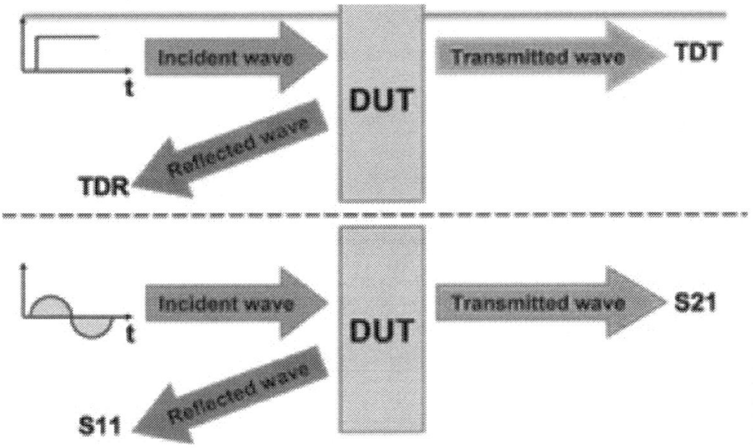

Figure 1.23: *TDR and VNA Techniques*

TDR measurements are more sensitive to the instantaneous impedance profile of the interconnect, while frequency domain measurements are more sensitive to the total input impedance looking into the DUT. To distinguish these two domains, we also use different words to refer to them. Time domain measurements are TDR and TDT responses, while frequency domain reflection and transmission responses are referred to as S (scattering) parameters. S11 is the reflected signal, S21 is the transmitted signal. They are also often called the return loss and the insertion loss. This is illustrated in *Figure 1.23*.

Depending on the question asked, the answer may be obtained faster from one domain or the other. If the question is, what is the characteristic impedance of the uniform transmission line, the time domain display of the information will get us the answer faster. If the question is what is the bandwidth of the model, the frequency domain display of the information will get us the answer faster.

In the DCA 86100C, the frequency domain response of the TDR or TDT waveform can be instantly displayed by pulling down the S-Param window, located in the upper right-hand corner of the screen.

Building a High-Bandwidth Model of a Component

To evaluate the bandwidth of the model for this SMT resistor, we can bring the measured S11 data of the DUT into a simulation tool such as Agilent's ADS and perform more sophisticated modeling. The corresponding frequency domain response is S11, the return loss. This is one of the S-parameter matrix elements.

By selecting the S-param tab on the upper right corner of the screen, we can bring down the converted time domain response as a frequency domain response, S11. This is still the reflection coefficient, but now it is displayed in the frequency domain and is related to reflections, not from instantaneous impedances, but from the total, integrated impedance of the entire DUT, looking into its input.

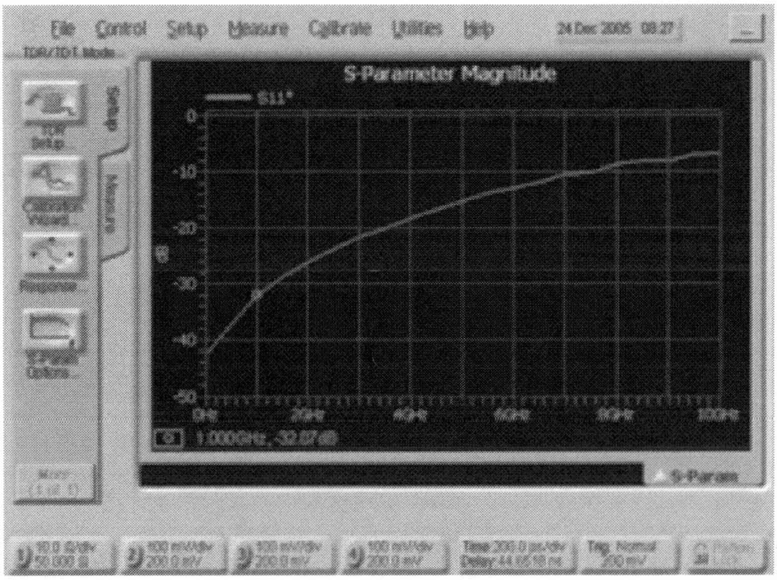

Figure 1.24: *Converted S11 of the SMT Termination Resistor*

Displayed in *Figure 1.24* is the measured S11 of this terminating resistor component, up to about 10 GHz. At low frequency, the magnitude of S11 is very low; not much of the signal comes back because the impedance the TDR sees is a pretty good match to 50 Ohms. As we go up in frequency, the impedance of the terminating resistor increases due to the higher impedance of the series inductance, causing more signal to reflect, which we see as a decrease in the negative dB values.

We can save this data in a .s1p Touchstone formatted file and bring it into ADS for further analysis.

ADS is a powerful analysis and modeling tool. In this example, we are interested in how well this simple model of the SMT resistor, being a series resistor and inductor, fits the actual, measured response and up to what bandwidth it is accurate.

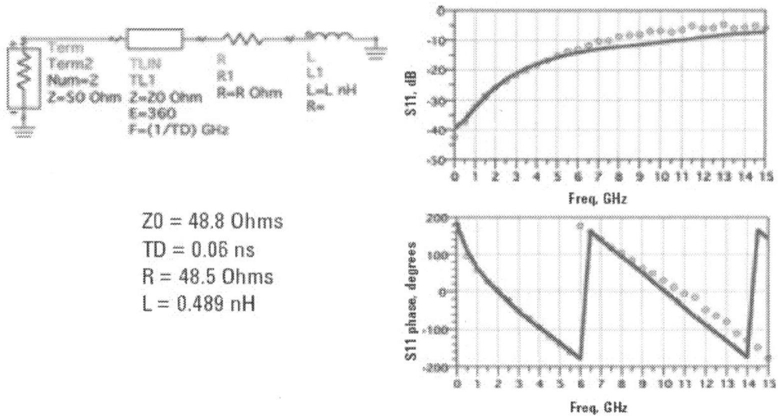

Figure 1.25: *ADS Model of Resistor and the Measured and Simulated S-Parameters*

In ADS, we build a circuit topology to match what we think this DUT is. This simple circuit is shown in *Figure 1.25*. We start with an ideal, lossless, uniform transmission line with an ideal resistor in series with an ideal inductor, which terminate the end of the line. There are four parameters that fully characterize this circuit—the characteristic impedance and TD of the line, the resistance, and the inductance. We just do not know what their parameter values are.

However, we can take advantage of the built-in, powerful optimization features in ADS to find the best set of parameter values that gives the closest agreement between the actual measured S11 values from the TDR to the simulated S11 values of this circuit model. We perform the optimization for the measured data at 4 GHz and below and find the best set of parameter values is a characteristic impedance of 48.8 Ohms, a time delay of 0.06 ns, a resistance of 48.5 Ohms, and an inductance of 0.489 nH.

Also shown in *Figure 1.25* is the comparison of the measured return loss, as red circles, and the simulated return loss, as solid lines, up to 15 GHz. We see that the simulated return loss of this simple model matches the actual, measured return loss of the component very accurately up to about 7 GHz. The bandwidth of this model is 7 GHz.

We also find that the extracted value of the equivalent series inductance of this SMT resistor is 0.489 nH. Using the markers on the screen, we estimated it to be 0.481 nH. This estimate from the excess reactance is within 2 percent of what we get using higher-bandwidth modeling techniques.

It is also important to note that the actual frequency values that are measured by the TDR are rather sparse. In the time domain response, the scale was 200 ps per division. This scale gave a comfortable time resolution to see the inductive spike, using the 50 ps rise time. With 10 horizontal divisions full-scale, the entire time sweep was 2 ns. When we convert this measured data from the time domain to the frequency domain, the 2 ns window converts to a first harmonic frequency value of 1/(2 ns) or 500 MHz. This is the step size used for all the frequency values.

We see in the display of the measured data in *Figure 1.25* that there is a circle at every 500 MHz. This is the frequency resolution of the TDR. If we wanted finer frequency resolution, we would have had to use a longer time window and a larger number of picoseconds per division. For example, if we use 1 ns per division for a total of 10 ns full scale, the frequency resolution would have been 1/(10 ns) or 100 MHz.

Directly Emulating the Impact on a Signal with the System Rise Time from a Discontinuity

A short discontinuity will look like either a lumped capacitor or inductor. The impact on the reflected signal will depend on the rise time of the incident signal. A shorter rise time will create a larger reflected signal. This means that the instantaneous impedance of a discontinuity, as measured off the screen of a TDR, is rise-time–dependent. It really does not mean anything to describe the impedance of a discontinuity unless we also specify at what rise time. Even then, we cannot do much with this information.

One way of evaluating the impact of this discontinuity in a particular application is to build a model for the structure, using excess reactance, and bring it into a circuit simulator. We could then use the driver models to simulate signals at the system's rise time and calculate the amount of reflection noise based on this interconnect model.

Another way of evaluating the expected reflection noise is to use the TDR to emulate the system rise time. After the TDR stimulus is calibrated, we can change the rise time of the stimulus to match any rise time from as low as 20 ps to well above 1 ns. We can then directly measure the reflected noise for different rise times and measure the impact on the signal a specific interconnect discontinuity would have.

In *Figure 1.26* is an example of a uniform 50 Ohm transmission line that has a region about 200 mils long that necks down. This is what typically occurs when a line has to pass through a via field associated with a connector footprint. The impedance of this line goes from 50 Ohms up to about 70 Ohms and then back to 50 Ohms.

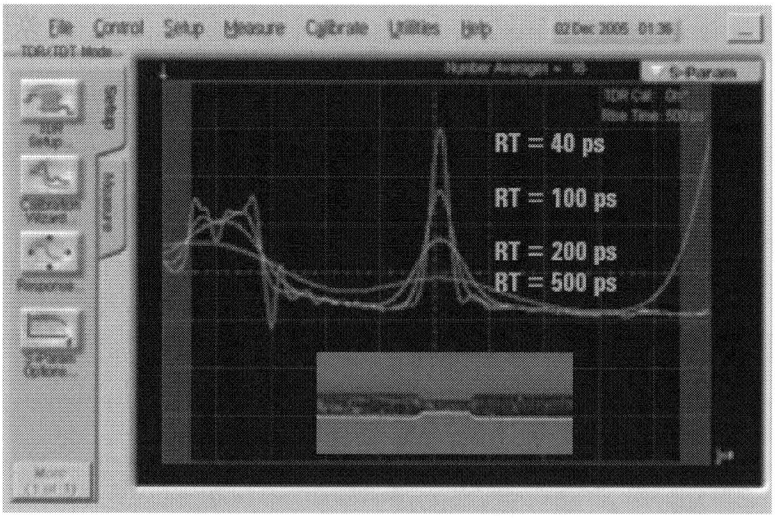

Figure 1.26: *Emulating System Rise Time Responses for a 200 mil Long Neck Down Region with RT = 40, 100, 200, and 500 ps*

By changing the rise time of the stimulus, we can directly measure the reflected noise at rise times of 40, 100, 200, and 500 ps. The vertical scale is 10 mV/div. In the region of the discontinuity, the peak reflected noise is about 37, 22, 11, and 5 mV, respectively.

This is with an incident signal of 200 mV or a reflection coefficient of 18, 11, 5.5, and 2.5 percent. The decrease in reflected amplitude does not scale directly with rise time because of the complication of the finite size of the discontinuity and the rise time of the signal. This complexity is automatically taken into account in the TDR measurement.

For example, if a 500 ps signal were to encounter this 200 mil long neck down, it would see a reflection, but it would only be 2.5 percent, which might be acceptable. This would demonstrate the advantage of necking down the line to get through the via field is a reasonable compromise, compared with possibly adding more layers in the board to keep the line width uniform.

Alternatively, if the system rise time were 100 ps, we would see that the reflection noise of 11 percent might exceed the typical 5 percent noise budget allocated to refection noise, and it might not be acceptable to neck down the trace, but may require a rerouting around the connector field.

Using the built-in adjustment of the rise time of the stimulus, we can emulate the actual system's rise time for a specific application and directly measure the performance of an interconnect in the specific application without having to first build a model and run a simulation. This can save a lot of time and help us to get to an acceptable answer faster.

1.3 Two-Port TDR/TDT

Overview

As seen in the previous section, a TDR generates a stimulus source that interacts with an interconnect. With one port, we were able to measure the response from one connection to the interconnect. This limited us to just looking at the signal that reflects right back into the source. From this type of measurement, we got information about the impedance profile and properties of the interconnect and extracted parameter values for uniform transmission lines, with discrete discontinuities.

By adding a second port to the TDR, we can dramatically expand the sort of measurements possible and the information we can extract about an interconnect. There are three important new measurements that can be performed with an additional port: the transmitted signal, coupled noise, and the differential or common signal response of a differential pair. The most important applications that can be addressed with these techniques and examples of each are illustrated in this chapter.

Introduction to TDR/TDT

When the second port is connected to the far end of the same transmission line and is a receiver, we call this TDT. A schematic of this configuration is shown in *Figure 1.27*. The combination of measuring the TDR response and TDT response of an interconnect allows accurate characterization of the impedance profile of the interconnect, the speed of the signal, the attenuation of the signal, the dielectric constant, the dissipation factor of the laminate material, and the bandwidth of the interconnect.

Figure 1.27: *Configuration for TDR/TDT Measurements*

The TDR can be set up for TDR/TDT operation by selecting the stimulus response as single-ended and identifying channel two as the TDT channel. This is shown in *Figure 1.28*.

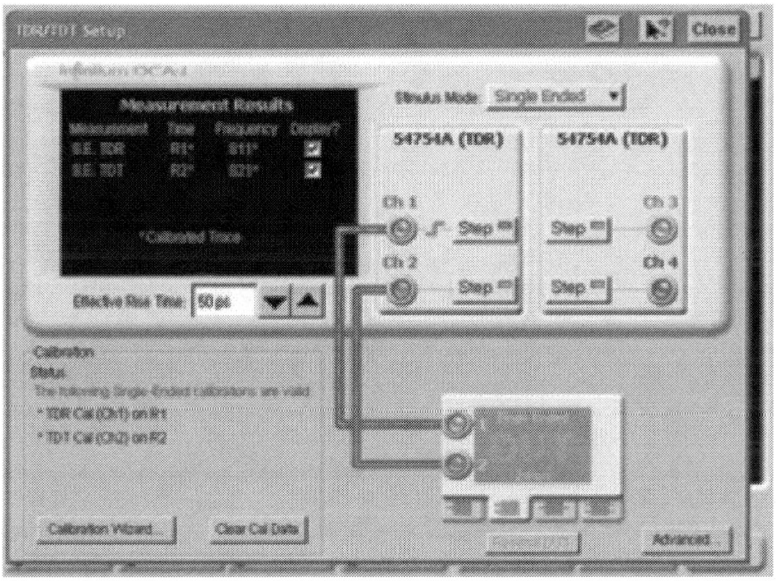

Figure 1.28: *TDR Setup Screen for TDR/TDT Operation*

Measuring Insertion Loss and Return Loss

In the simplest application, the ports of the TDR are connected to each end of the single-ended transmission line. Port 1 is the TDR response we are familiar with, while channel 2 is the transmitted signal. In the TDR response of a uniform, eight-inch microstrip transmission line, as shown in *Figure 1.29*, the end of the line is seen as having an impedance of 50 Ohms. This is the cable connected to the end of the DUT, and then ultimately, the source termination inside the TDR's second channel.

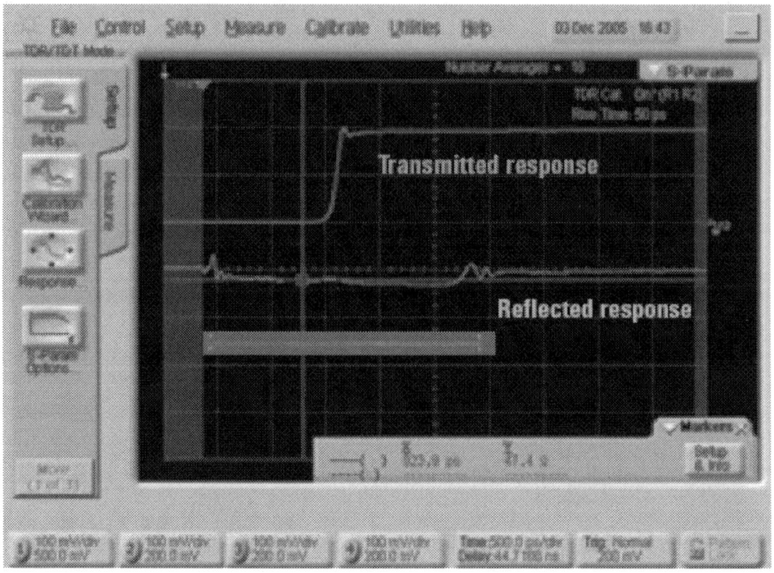

Figure 1.29: *Example of TDR/TDT Response from Eight-Inch-Long Microstrip Transmission Line on 20 mV/div and 500 ps/div Scales*

The time base in this application is 500 ps/div, with the vertical scale at 20 mV/div. The marker is being used to extract the impedance of the line as 47.4 Ohms. Note that the top trace, the signal transmitted through the interconnect, on 100 mV/div scale, shows the signal coming out exactly halfway between the time the signal goes into the front of the line, reflects off the back end, and is received at the source.

The TDR signal looks at the round-trip time of the flight down the interconnect then back to the front, while the TDT signal sees the one-way path through the interconnect. In the time domain display, we can see the impedance discontinuities of the SMA launches on the two ends of the line, and see that the line is not a perfectly uniform transmission line. On this scale of 20 mV/div or a reflection coefficient of 10 percent/div, the variation in impedance is about 1 Ohm down the line.

The transmitted signal is a relatively fast edge, but it is difficult to get much information off the screen from this received signal.

Though we could measure the 10/90 or 20/80 rise time directly off the screen, it is not clear what we would do with this information, as the interconnect distorts the edge into a not really Gaussian edge. This is a case where we can take the same information content but change how it is displayed to interpret it more quickly and easily.

Figure 1.30 shows the same measured response as shown in the time domain, but now transformed into the frequency domain. This screen is accessed by clicking on the S-Param tab in the upper right-hand corner of the TDR response screen. In the frequency domain, we call the TDR signal S11 and the TDT signal S21. These are two of the S-parameters that describe scattered waveforms in the frequency domain. S11 is also called return loss and S21 is called insertion loss. The vertical scale is the magnitude of the S-parameter, in dB.

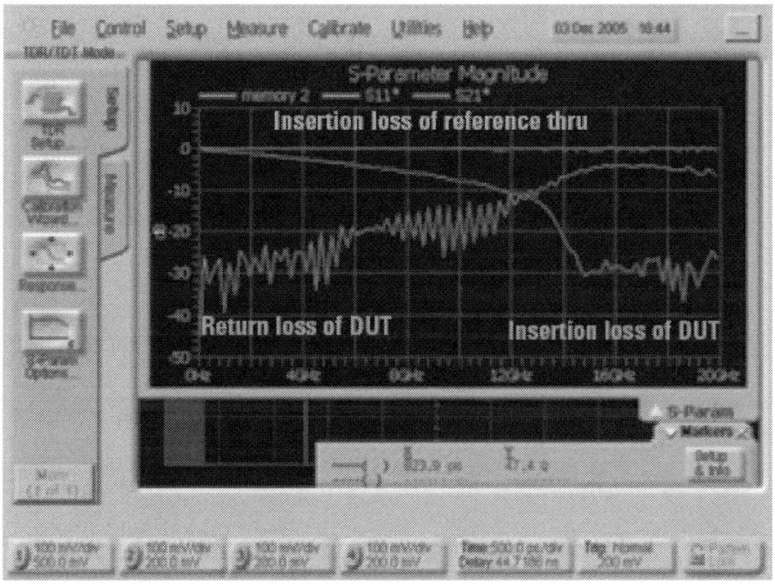

Figure 1.30: *TDR/TDT Response Converted into Frequency Domain: Return Loss/Insertion Loss*

The top trace is the insertion loss for a reference thru. Of course, if we have a perfect thru, every frequency component will be transmitted with no attenuation, and the amplitude of the received signal is the same as the incident signal. The magnitude of the insertion loss is always 1, and in dB, is 0 dB. This is flat across the entire 20 GHz frequency range.

The *return loss of DUT* trace, starting at about -30 dB at low frequency, is the return loss for this same transmission line, which is really S11 in the frequency domain. The *insertion loss of DUT* trace is the insertion loss of this transmission line, or S21. On this display, we are only showing the magnitude of the S-parameters; the phase information is there, it is just less important to display.

The return loss starts at relatively low values, near –30 dB, and then creeps up, eventually reaching the –10 dB range, above about 12 GHz. This is a measure of the impedance mismatch of this transmission line and the 50 Ohm connections on either end.

The insertion loss has immediately useful information. In a high-speed serial link, the transmitters and receivers work together to enable high–bit-rate signals to be transmitted and then received. In simple CMOS drivers, an insertion loss of –3 dB might be acceptable, before a significant bit-error rate. With simple serializer/deserializer (SerDes) chips, an insertion loss of -10 dB might be acceptable, while for state-of-the-art, high-end SerDes chips, -20 dB might be acceptable. If we know the acceptable insertion loss for a particular SerDes technology, we can directly measure off the screen the maximum bit rate an interconnect is capable of.

As a rough rule of thumb, if the bit rate in Gbps is BR and the bandwidth of the signal is BW, the highest sine wave frequency component is roughly BW = 0.5 x BR, or BR = 2 x BW. The BW is defined by the highest frequency signal that can be transmitted through the interconnect and still have less attenuation than the SerDes can compensate for. Using low-end SerDes, the acceptable insertion loss might be –10 dB, and the bandwidth for this eight-inch-long microstrip, we can read right off the screen in *Figure 1.30*, would be about 12 GHz. This would allow operation well above 20 Gbps bit rates. But, this is for a wide conductor, only eight inches long. In a longer backplane or motherboard with connectors, daughtercards, and vias, the transmission properties are not as clean.

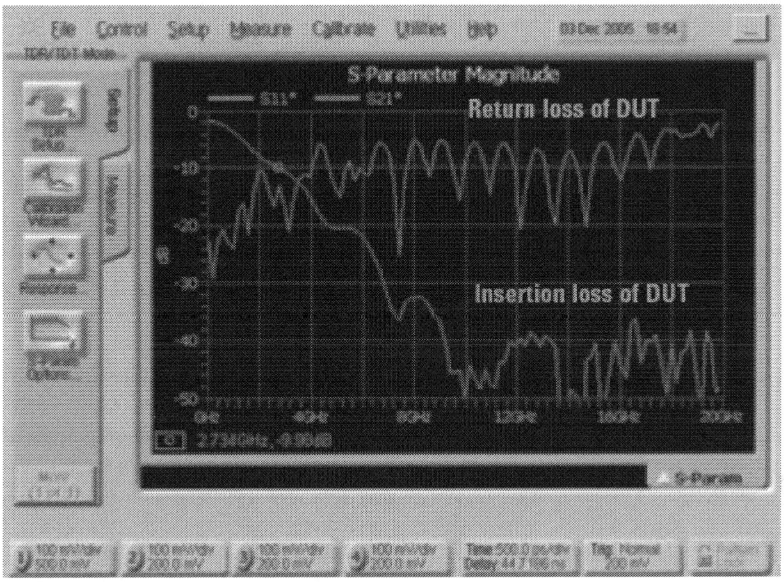

Figure 1.31: *Return and Insertion Loss of a 24-Inch Interconnect on a Motherboard with Two Daughtercards*

The TDR/TDT response of a 24-inch-long stripline interconnect in a typical motherboard is shown in *Figure 1.31*. In this example, an SMA launch connects the TDR cable to a small card, through a connector, a via field, back through a connector, and into the second channel of the TDR.

The *insertion loss of DUT* trace is the insertion loss displayed as S21. For this interconnect, the -10 dB insertion loss bandwidth is 2.7 GHz. For this interconnect, the maximum transmitted bit rate would be about 5 Gbps, using low-end SerDes drivers and receivers.

Interconnect Modeling to Extract Interconnect Properties
The ability to take the measured data and display it as either time domain responses or frequency domain responses means we can easily extract more information than if we were limited to just one

domain. Further, by exporting the frequency domain insertion and return loss measurements as a Touchstone formatted file, we can use sophisticated modeling tools such as Agilent's ADS to extract more information than we could get off the screen.

In this example, we will look at the uniform, eight-inch-long microstrip and how we can use modeling and simulation tools to extract material properties. The simplest model to describe this physical interconnect is an ideal transmission line. We can use the built-in multilayer interconnect library (MIL) of ADS to build a physical model of this very microstrip, with the material properties parameterized, and extract their values from the measurement.

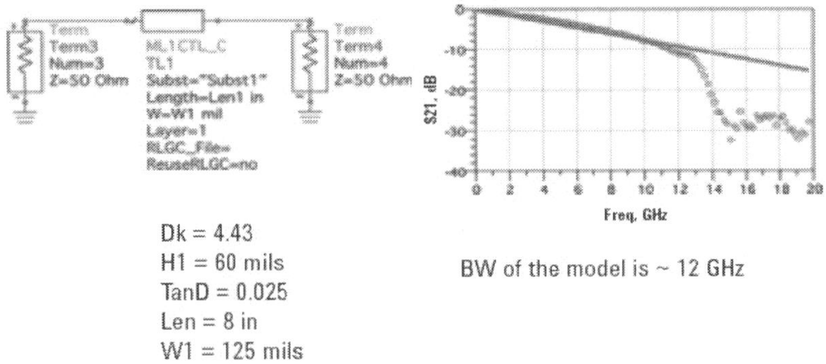

Figure 1.32: *ADS Modeling of a Uniform Eight-Inch-Long Microstrip Showing the Bandwidth of the Simple Model to be ~12 GHz*

Figure 1.32 shows the simplest model to describe this transmission line, as a single trace on a substrate, with a length of eight inches, a dielectric thickness of 60 mils, and a line width of 125 mils. These are parameters measured directly from the physical interconnect. What we do not know initially are the laminate's bulk dielectric constant and its bulk dissipation factor. However, we have the measured insertion loss. *Figure 1.32* displays the measured insertion loss of the interconnect as red circles. This is exactly the same data as displayed previously, from the screen of the TDR. The phase response is also used in the analysis, it is just not displayed in this figure.

Given this simple model, with the two unknown parameters, the dielectric constant, and dissipation factor, we use the built-in optimizer in ADS to search all parameter space for the best-fit values of these two terms to match the measured insertion loss response to the simulated insertion loss response. The diagonal line in *Figure 1.32* is the final value of the simulated insertion loss using a value of 4.43 as the dielectric constant and 0.025 as the dissipation factor. We can see from this display, the agreement between the measured and simulated insertion loss is excellent, up to about 12 GHz. This is the bandwidth of the model. There is even better agreement in the phase not shown in this figure.

By building a simple model, fitting parameter values to the model, and taking advantage of the built-in 2D boundary element field solver and optimization tools of ADS, we are able to extract very accurate values for the material properties of the laminate from the TDR/TDT measurements. We are also able to convince ourselves that, in fact, this interconnect is very well behaved. There are no unusual, unexplained properties of this transmission line. There are no surprises, at least up to 12 GHz.

Identifying Design Features That Contribute to Excessive Loss

Being able to quickly and easily bring measured TDR/TDT data from the TDR instrument directly into a modeling tool can sometimes cut the debug time from days to minutes by helping us unravel the root cause of surprising or anomalous behavior. *Figure 1.33* is an example of the measured TDT response from three structures. The top horizontal line is the insertion loss measured from a reference thru, showing the very flat response when the interconnect is basically transparent. This is a direct measurement of the capabilities of the instrument.

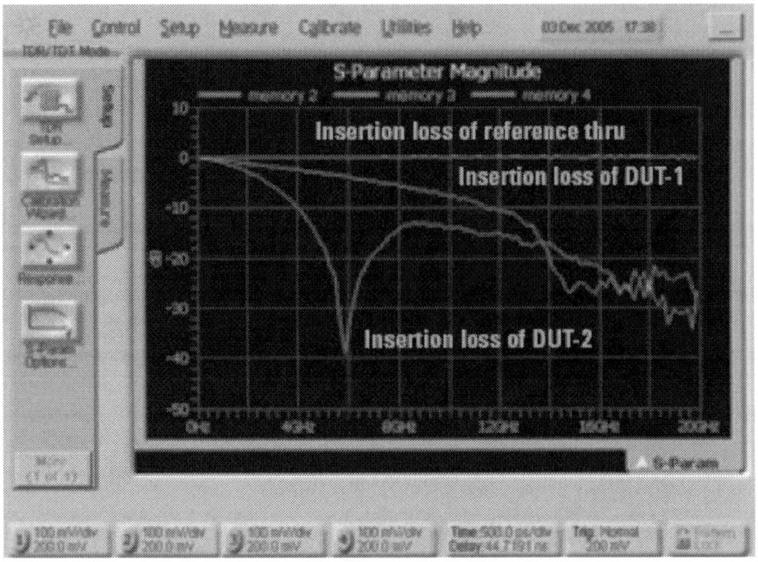

Figure 1.33: *Measured Insertion Loss of a Reference through a Uniform Line (DUT–1) and a Uniform Line That Is Part of a Differentiated Pair (DUT–2)*

The second line from the top is the insertion loss of the eight-inch, single-ended microstrip we saw before. The third line is the measured insertion loss of another nine-inch-long, uniform microstrip transmission line. However, this transmission line has a very large dip in the insertion loss at about 6 GHz. This dip would dramatically limit the usable bandwidth of this interconnect. The -10 dB bandwidth of the first transmission line is about 12 GHz, while the -10 dB bandwidth of the second line is about 4 GHz. This is a reduction of a factor of three in usable bandwidth. Understanding the origin of this dip would be the first step in optimizing the design of the interconnect. What could cause this very large dip?

In this second transmission line, there are no vias. It is a uniform microstrip. The SMA launch is identical as in the first transmission line. It happens that, though this is a single-ended measurement, there is another transmission line physically adjacent and parallel to this measured transmission line, with a spacing about equal to the line width. However, this adjacent line is also terminated with 50 Ohm resistors on its ends. Is it possible that the proximity of this other trace could somehow cause this large dip? If so, what feature of this other line influences the dip frequency?

One way we can answer this question is by building a parameterized model for the physical structure of the two coupled lines, verifying that its simulated insertion loss matches the measured insertion loss, and then tweaking terms in the model to explore design space.

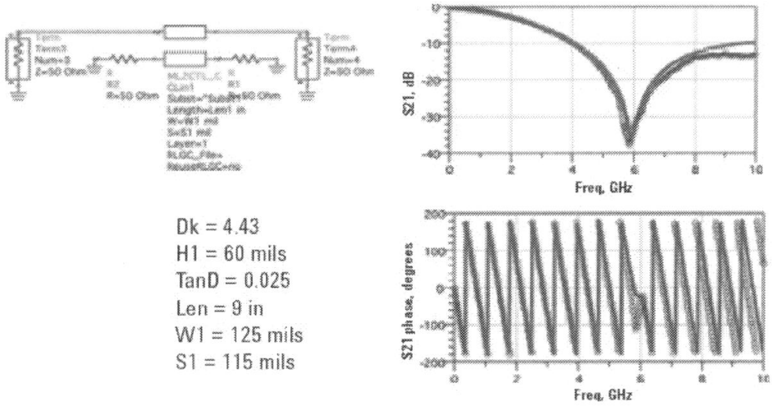

Figure 1.34: *ADS Model of the Nine-Inch-Long Trace, Modeling the Coupling to the Adjacent Quiet Line, Showing the Bandwidth of the Model to be ~8 GHz*

Figure 1.34 shows the simple model of two coupled transmission lines using the MIL structures in ADS. All the physical and material properties are parameterized so that we can vary them later. We assume a simple model of two uniform, equal width lines, with a spacing, length, dielectric thickness, dielectric constant, and dissipation factor. We use all the geometry terms as measured with a micrometer from the structure and use the same dielectric constant and dissipation factor as measured from the uniform transmission line.

The integrated 2D field solver in ADS will automatically take these geometry values and calculate the complex impedance and transmission properties of the line and simulate the frequency domain insertion and return loss performance, configured exactly as in the actual measurement.

We bring the measured insertion loss data in Touchstone format from the TDR into ADS and compare the measured response and the simulated response. Shown in *Figure 1.34* is the magnitude of the insertion loss in dB and the phase of the insertion loss. The red circles are the measured data, same as that displayed on the screen of the TDR instrument. The lines are the simulated response based on this simple model, with no parameter fitting.

The agreement is astonishingly good up to about 8 GHz. This indicates there is nothing anomalous going on. There is nothing that is not expected from the normal behavior of two coupled, lossy lines. In this case, the second line, which is not being driven, is terminated by 50 Ohm resistors at the ends. The model is set up to match this same behavior. That we see this anomalous dip in the insertion loss in a single line when it is part of a pair of lines, but not when a similar line is isolated, and that we confirm this with the field solver, suggests there is something about the proximity of the adjacent line that causes this dip.

The effect that gives rise to this disastrous behavior is not anomalous, it is just very subtle. We could spend weeks spinning new boards to test for one effect after another, searching for the knob that influences this behavior. For example, we could vary the coupled length, line width, spacing, dielectric thickness, and even dielectric constant and dissipation factor, looking for what influences the resonance frequency. Or we could perform these same experiments as virtual experiments using a simulation tool such as ADS. It is only after we have confidence the tool accurately predicts this behavior that we can use it to explore design space.

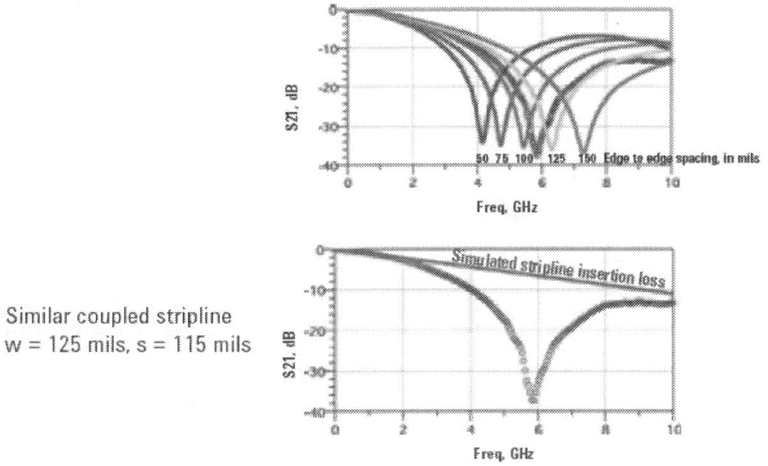

***Figure 1.35:** Changing Separation between the Two Transmission Lines Showing the Impact on the Insertion Loss Dip*

One obvious virtual experiment to try is varying the line spacing. What happens to the resonant absorption dip in the insertion loss of one line as the traces are moved closer or farther apart? *Figure 1.35* is the simulated insertion loss of one line in the simple two-coupled line model, when we use separations of 50, 75, 100, 125, and 150 mils. The red circles are the measured insertion loss for the single-ended trace. Each line is the simulated response of the insertion loss with a different separation. The trace with the lowest frequency resonance has a separation of 50 mil, followed by 75 mil, and finally 150 mil.

As the separation distance increases, the resonance frequency increases. This is almost counter-intuitive. Most resonance effects decrease in frequency as a dimension is increased. Yet, in this effect, the resonance frequency increases, as the dimension—the spacing—increases. If we did not have the confirmation of the very close agreement between the simulation and measurement in the previous figure, we would possibly begin to doubt the results from the simulation.

The explanation of the dip is clearly not a resonant effect. Its origin is very subtle, but is intimately related to far-end crosstalk. In the frequency domain, when the sine wave enters the front of the first line, it will couple into the second line. As it propagates along, there is a frequency where all of the energy couples from the first line into the adjacent line, leaving none in the first line, and hence, the large dip.

As we increase frequency more, the energy will couple back to the first line. This process will repeat. This is a fundamental property of modes and tightly coupled systems. It is ultimately related to the fact that the two modes, which propagate down the pair of lines, the odd and even modes, travel at different speeds in microstrip. If this were the true explanation, and if the two coupled lines were constructed in stripline, where the even and odd modes travel at the same speed, there would be no dip.

Also shown in *Figure 1.35* is the simulated insertion loss of a single stripline transmission line, with the same line width, having an adjacent, terminated trace with a spacing of 115 mils. There is no dip at 6 GHz and the insertion loss is smoothly decreasing with frequency, all due to the dielectric loss of the laminate.

This suggests an important design rule: To get the absolute highest bandwidth in a single-ended transmission line, you want to avoid having a closely spaced adjacent line, however terminated it may be.

1.4 Two-Port TDR/Crosstalk

Overview

So far, we have evaluated the electrical performance of single transmission lines. When an adjacent transmission line is present, some of the energy from one line can couple into the second line, creating noise in the second line. To distinguish the two lines, we sometimes call the driven line the active line or the aggressor line. The second line is called the quiet line or victim line. This is illustrated in *Figure 1.36*.

One end of the active line is driven by the TDR stimulus. We get the TDR response of the active line for free. If we connect the second port to the far end of the active line, we can measure the

TDT response. If we connect the second port to the end of the quiet line adjacent to the stimulus, we can measure the noise induced on the quiet line. To distinguish the two ends of the quiet line, we refer to the end near the stimulus as the near end, and the end far from the stimulus as the far end.

The ratio of the voltage noise measured on the near end of the quiet line to the incident stimulus voltage going into the active line is defined as the near-end crosstalk (NEXT). The ratio of the far-end voltage noise on the quiet line to the incident stimulus voltage going into the active line is defined as the far-end crosstalk (FEXT). These two terms are figures of merit in describing the amount of crosstalk between two parallel, uniform transmission lines. They can be measured directly by a two-port TDR.

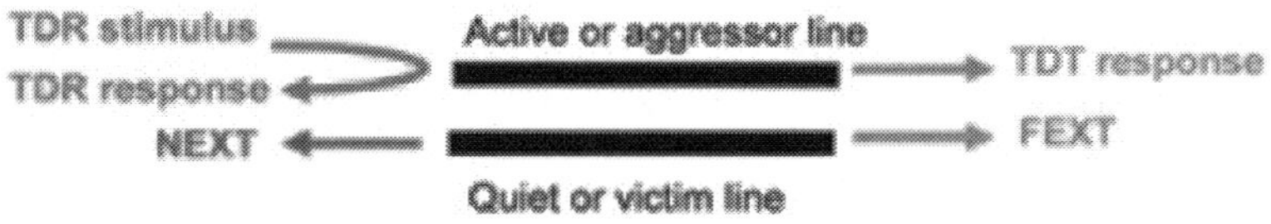

Figure 1.36: *Configuration for Two-Port TDR Measurements*

Measuring NEXT

As a simple example, the NEXT in a pair of tightly coupled microstrips was measured and is displayed in *Figure 1.37*. These are two, roughly 50 Ohm microstrips, nine inches long, with a spacing about equal to their line widths. The bottom line is the measured TDR response of one line in the pair. The vertical scale is 5 Ohms per division. The large peak on the left edge of the trace is the high impedance of the SMA edge launch, while the far end shows a smaller discontinuity at the launch.

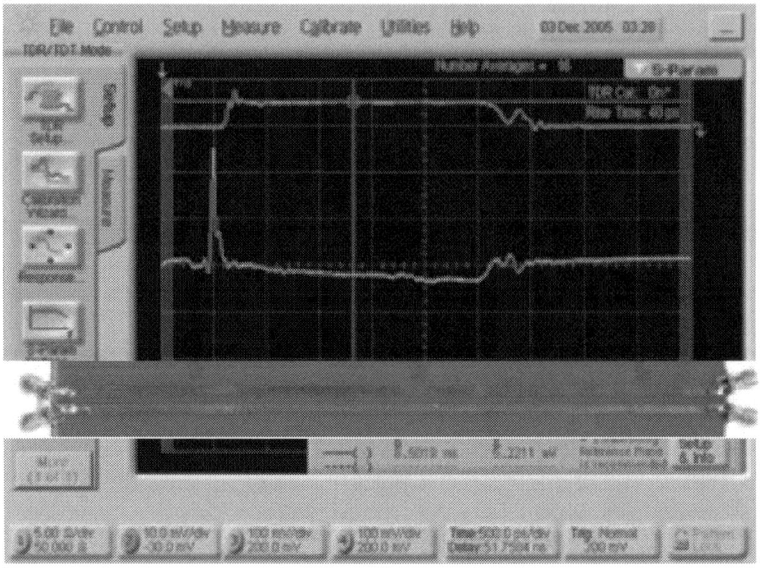

Figure 1.37: *Measurement of the NEXT on a Quiet Line Using the Marker*

The top trace is the measured voltage picked up on the near end, while the far end is terminated into 50 Ohms. We can use the markers to read the near-end noise directly from the screen as 5.22 mV. This is with an incident signal going into the active line of 200 mV. The NEXT is 5.22 mV/200 mV = 2.6 percent. It turns on with the rise time of the signal and lasts for the same amount of time as the TDR response, a round-trip time of flight.

Measuring FEXT

By connecting the second channel of the TDR to the far end of the quiet line, the far-end noise can be measured. At the same time, a 50 Ohm terminating resistor is added to the near end of the quiet line. The measured far-end and near-end noise on the quiet line is shown in *Figure 1.38*. Both are on the same scale of 20 mV/div. This corresponds to 10 percent crosstalk per division. The white line is the near-end noise, while the bottom line is the measured far-end noise.

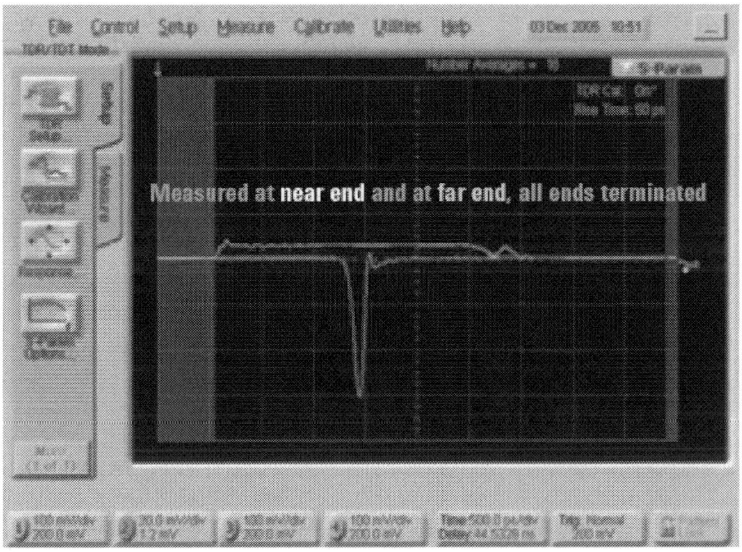

Figure 1.38: *Measuring the NEXT and FEXT with the Second Channel in the TDR*

In this example, while the NEXT is only 2.6 percent, the FEXT is seen to be about 30 percent, a huge amount. It appears to be coming out of the quiet line at a time equal to half the round-trip time of flight, which is the one-way time of flight, the time it takes the signal to propagate from the input to the output.

The width of the far-end noise is the rise time of the signal. In fact, the shape of the far-end noise is roughly the derivative of the rising edge of the signal.

These values of NEXT and FEXT are defined for the special case of all ends of the two lines terminated, so there are no reflections of the signals or noise. Normally, performing these measurements of NEXT and FEXT requires connecting and disconnecting the second port of the TDR to each of the two ends of the quiet line, while connecting a termination to the unused end. By taking advantage of reflections, we can actually perform both measurements of near- and far-end noise from one end only. We just need to understand that changing the terminations will change the noise voltages picked up.

Emulating FEXT for Different System Rise Times

If we measure the noise voltage coming out of the near end, we will obviously pick up the near-end voltage. However, if we keep the far end of the quiet line open terminated, the far-end noise propagating down the quiet line to the far end will reflect at the open and head back to the near end, where it can be measured from the near-end side of the quiet line.

In addition, if we keep the far end of the active line open terminated, the signal will also reflect. As it heads back down to the source end of the active line, it will be generating additional far-end noise in the quiet line, but it will be heading in the direction back to the near end. This will increase the amount of far-end noise picked up in the quiet line, at the near end.

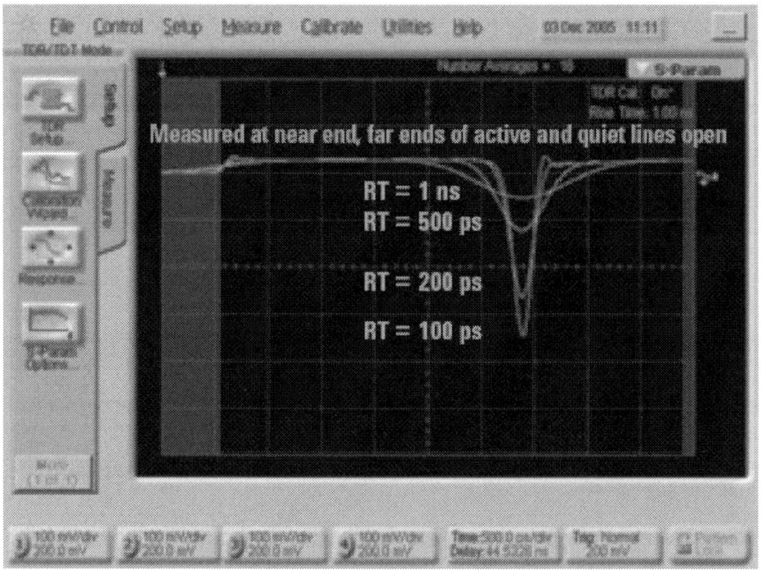

Figure 1.39: *Emulating the FEXT with Different System Rise-Time Responses with RT= 100, 200, 500 ps and 1 ns*

In *Figure 1.39* is the measured noise at the near end of the quiet line, showing the initial near-end noise, followed, one round-trip time of flight later, with the reflected far-end noise.

The signature of the far-end noise is the derivative of the rising edge of the signal. This means that as the rise time changes, the peak value of the far-end noise will change. However, the area under the curve of the far-end noise will always be constant, as long as the signal voltage transition levels are the same.

In *Figure 1.39* is an example of the combination of near- and far-end noise, measured at the near end, as we change the rise time of the signal. We see two important features. The magnitude of the near-end noise is independent of the rise time. Second, the peak value of the far-end noise decreases as the rise time increases, but the area under the far-end noise voltage is constant. As the rise time decreases, the peak decreases, but the far-end noise spreads out, since the derivative of the rising edge spreads out with longer rise time.

Understanding the impact of terminations on the measured noise at one end or the other of a quiet line can often help resolve the interpretation of the noise picked up.

Identifying Design Features That Contribute to NEXT

For this example, we will look at the measured noise between two parallel interconnects in a motherboard with two plug-in cards. In this case, the lines are stripline, but the dielectric distribution is not uniform, so there will be some far-end noise. The total parallel run length is about 24 inches, including the path on the motherboard.

The TDR response of one of the single-ended lines is shown in *Figure 1.40*. The scale is 10 Ohms/div. We see the very first peak at the SMA launch. The constant impedance region, with an impedance value we can read off the first, solid marker as 56.8 Ohms, is the interconnect on the daughtercard.

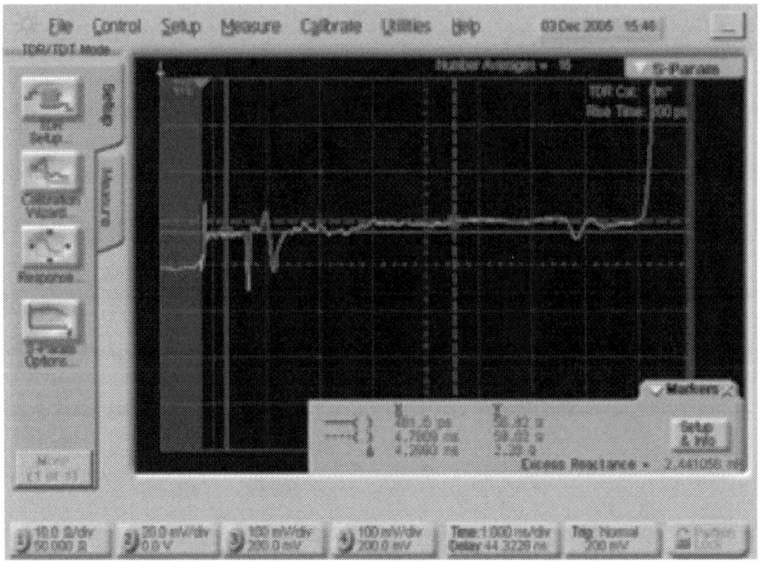

Figure 1.40: *Measured TDR Response of 24-Inch-Long Trace in a Motherboard Using Markers to Measure the Impedance in the Daughtercard and Motherboard*

The first dip is the via field on the daughtercard where the connector is. This short uniform region is inside the connector, followed by the second dip of the via field in the motherboard. The long region of uniform impedance, with a value read from the second, dashed marker, about 59 Ohms, is the interconnect in the motherboard. At the end of the motherboard trace is the second daughtercard. Because of the lossy interconnects, the initial rise time of the TDR stimulus has increased by the time it gets to the far end of the active line, and the spatial resolution has decreased.

This incident signal from the TDR will be used as the active signal, while we measure the near- and far-end noise on the adjacent quiet line.

The near-end noise on the quiet line is shown in *Figure 1.41* as the *NEXT* trace, on a scale of 10 mV/div. This is with a signal magnitude of 200 mV, so the scale is really 5 percent per division. We see that in this case, the near-end noise has a peak of about 11 percent. With a typical crosstalk noise allocation in the noise budget of about 5 percent, the 11 percent near-end noise can be considered a lot. In order to even consider how to reduce it, the first step is to identify where it is coming from.

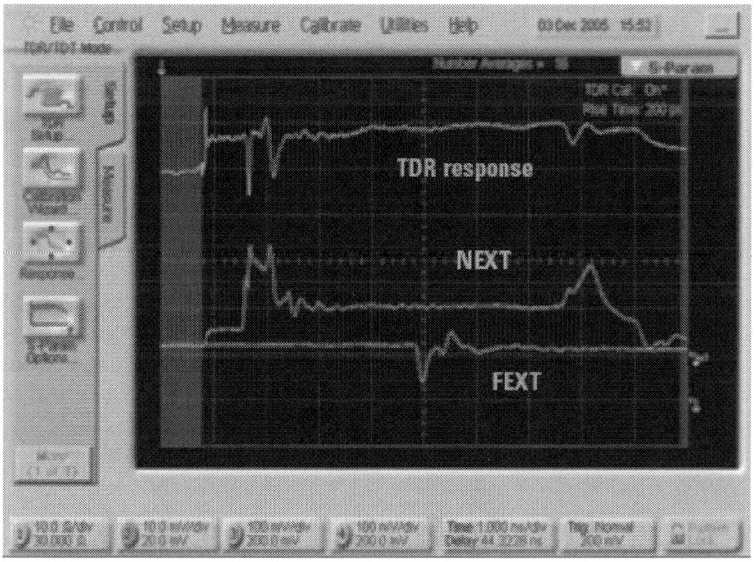

Figure 1.41: *Measured NEXT and FEXT in a 24-Inch-Long Trace on a Motherboard with All Ends Terminated*

By comparing the location of the near-end noise with the TDR response, we can quickly identify where in the interconnect path this noise is created. As the active signal propagates down the active line, the TDR response picks up reflections from impedance changes. The time at which these reflections are picked up at the near end of the active line is the round-trip time from the source to the discontinuity.

Likewise, the time value when we pick up the near-end noise at the near end of the quiet line corresponds to the time it takes for the signal to hit that region of the interconnect, plus the time it takes for the generated noise to propagate back to the near end of the quiet line. This is the round-trip time of flight. This means that by comparing the time response of the near-end noise to the TDR response, we can identify which specific interconnect features in the active line might have generated the near-end noise.

Looking at the *NEXT* line in *Figure 1.41* and comparing it to the *TDR response* line of the TDR response, we can see that the near-end noise in the daughtercard trace is small, only about 4 percent. The very large peak corresponds to the connector between the daughtercard and motherboard. We see the double peak corresponding to the via field in the two boards, and a large contribution from the connector itself.

The near-end noise from the traces in the motherboard is also high, on the order of 5 percent, followed by a peak in near-end noise from the connector on the other end of the interconnect. This says, to minimize the near-end noise in this motherboard application, the place to look is in the design or selection of the connector. For single-ended signal applications, we would want to use a connector with lower coupling than this particular one. The coupling between signal lines in the motherboard interconnects is high but probably would meet a typical noise budget.

Also shown, as the *FEXT* line in *Figure 1.41*, is the measured far-end noise on the quiet line. We see on this scale, the far-end noise is only about 4 percent. Even though the interconnects are stripline, any inhomogeneities in the dielectric distribution will generate far-end noise. This is low enough to not really cause a problem.

Exploring the Impact of Terminations on NEXT and FEXT

Given these noise levels on the quiet line when all the ends of the lines are terminated, we can also evaluate what would be the impact if the ends were not terminated, as would be the case if the receivers were tri-stated. Because of the high impedance, any voltages hitting these open ends will reflect. There is one situation in particular that can cause significant problems.

If the far end of the active line is tri-stated and open and the far end of the quiet line is a receiver, while the near end of the quiet line is tri-stated and open, as illustrated in *Figure 1.42*, the noise picked up at the far end of the quiet line can exceed 15 percent, a very large amount. In this configuration, the signal propagating down the active line will generate near-end noise in the quiet line, which will propagate back to the source end of the quiet line. But, if this end of the quiet line is tri-state and open, the near-end noise will be reflected and head back to the far end of the quiet line.

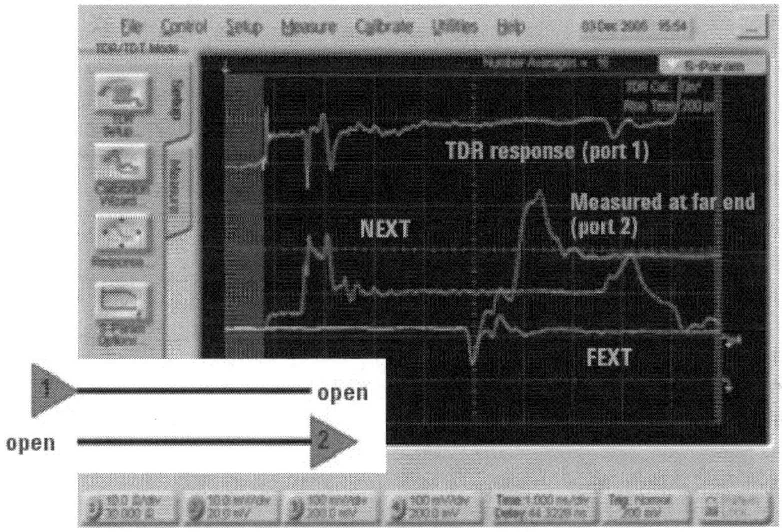

Figure 1.42: *Measured Crosstalk in Quiet Line with Worst-Case Termination*

The front edge of the reflected, near-end noise will be coincident with the front edge of the signal heading down to the receiver on the active line. When this signal hits the far end of the active line and the receiver there is still tri-stated, the signal will see an open and reflect. This reflected signal will be heading back to the driver and generate in the quiet-line, near-end noise that is heading to the far-end receiver.

The near-end noise generated on the quiet line that reflected hits the far end of the quiet line at the same time the active signal hits the far end and reflects, generating another round of near-end noise. This means there will be almost twice the near-end noise picked up by the receiver on the far end of the quiet line.

In *Figure 1.42*, the *measured at far end (port 2)* line is the measured noise picked up on the far end of the quiet line when the far end of the active line is open and the near end of the quiet line is open. This is a noise value with a peak of 15 percent. It is not the 20 percent double of the NEXT, mostly because of the smearing out of the near-end noise across the short length of the connector, compared to the rise time of the signal.

This example points out that the actual measured noise is related to not just the coupling between the lines, but also to how the ends of the lines are terminated. When measuring crosstalk, care should be taken to consider the termination configuration for worst-case coupling.

Measuring Ground Bounce

Crosstalk is created by capacitive and inductive coupling. In uniform, adjacent transmission lines, the capacitive and inductive coupling is uniformly distributed along their length. In addition, coupling can be localized. In particular, any discontinuities in the return path under signal conductors can dramatically increase the inductive coupling. We call the noise generated by inductive coupling switching noise. A special case of switching noise is when the return currents of multiple signal paths share the same return paths. We call the noise generated in this case ground bounce.

Ground bounce occurs mostly in connectors and packages, where multiple signal lines typically share the same return pin. Ground bounce can also occur in board-level interconnects if there are any discontinuities in the return path that force return currents to overlap. This happens if there is a gap in the return plane and return currents are confined to narrow paths.

Figure 1.43 is an example of two gaps in the return path underneath two coupled microstrip lines in FR4. Each transmission line is about 50 Ohms, and the spacing is about equal to the line width. The dark color adjacent to the copper traces is the circuit board color when there is a solid ground plane underneath. In the two regions marked by arrows, the copper plane has been removed, making the area light in color.

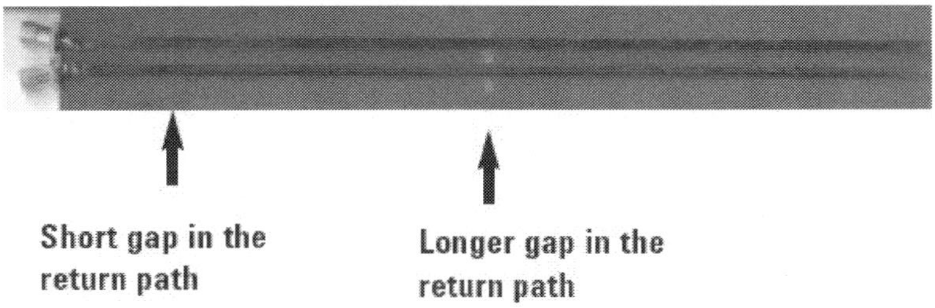

Figure 1.43: *Tightly Coupled Pair of Transmission Lines with Small Gaps in the Return Path That Will Generate Ground Bounce*

In this configuration, the near-end noise is about 2.5 percent, an acceptable level in most applications. However, where the gap is under the traces, the return currents will have to meander around the gap. This will increase the loop inductance in this section of the transmission line and increase the mutual inductance between the two lines. The noise generated on the quiet line due to the higher mutual inductance in this region is called ground bounce and can be measured at the near end of the quiet line.

One of the lines in the pair is used as the aggressor. The TDR response of this line is shown in *Figure 1.44*. We see an initial peak from the SMA launch, another reflection peak from the first gap, a uniform region, then another peak from the second gap, a uniform region, and then the open at the far end.

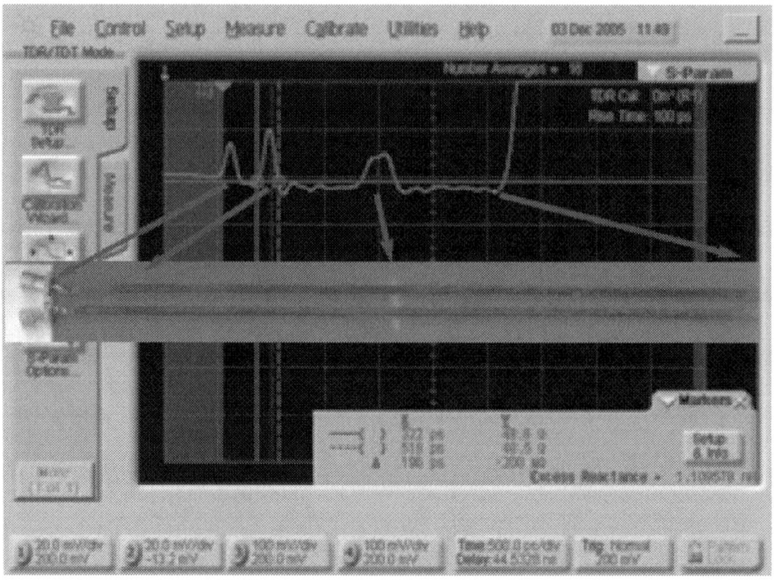

Figure 1.44: *TDR of a Single-Ended Transmission Line Crossing Gaps in the Return Path, Showing the Inductive Discontinuities*

Identifying Design Features That Contribute to Ground Bounce

While the signal is propagating down the active line, noise is being generated on the quiet line from the capacitive and inductive coupling. In *Figure 1.45*, we show the measured noise on the near end of the quiet line, with the far end of the quiet line open terminated. The near-end noise is on a scale of 20 mV/div or 10 percent noise per division.

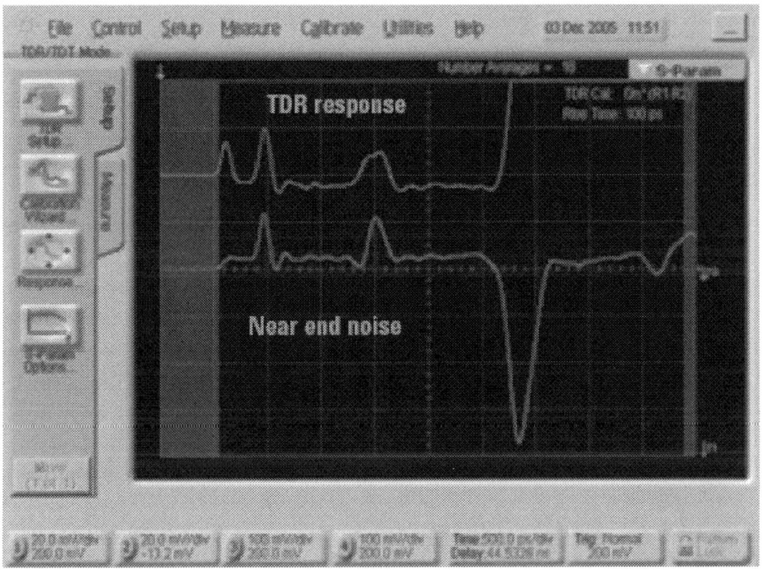

Figure 1.45: *Measured Ground Bounce on the Quiet Line from Gaps in the Return Path*

Initially, the small near-end noise of roughly 2.5 percent is from the uniform section of the transmission lines. The first peak in the near-end noise of roughly 11 percent is due to the small gap in the return path. This is a direct measure of the ground bounce voltage generated across the gap that is picked up in the adjacent, quiet line.

On the other side of this gap is the 2.5 percent near-end noise from the uniform region, followed by another 11 percent of near-end noise from the second gap. Then comes the near-end noise from the uniform section and finally, we see the reflected far-end noise at the end of the quiet line. The ground bounce also contributes to an increase in the far-end noise.

Emulating Ground Bounce Noise for Different System Rise Times

All switching noise is generated because of a switching current through some mutual inductance. The peak value of the switching noise is related to the mutual inductance times the dI/dt. This magnitude will depend on the rise time of the signal. If we can slow the edge down, we can reduce the magnitude of the switching noise. If the rise time can be increased enough using slew rate control in the driver and not impact timing, the switching noise might be reduced below a problem level.

Using the rise time control feature of the DCA 86100C, we can change the rise time of the stimulus and measure the resulting ground bounce when the rise time is longer. *Figure 1.46* shows the TDR and near-end crosstalk response for these two coupled lines with the two gaps in the return path. The bottom line details the response we saw before, with a rise time of 100 ps. The top trace details it for a rise time of 500 ps, while the *near end noise* trace shows the noise measured on the near end with a rise time of 500 ps.

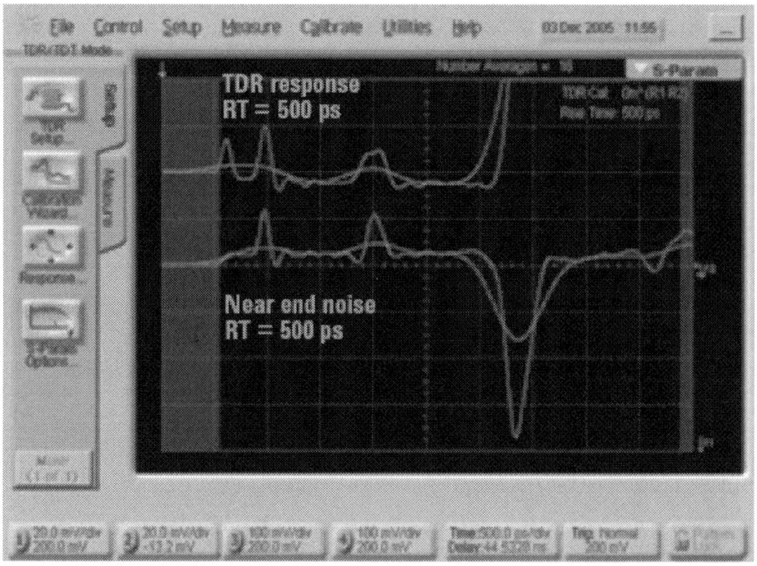

Figure 1.46: *Emulating Impact of Rise Time on the Ground Bounce Noise in a Pair of Coupled Lines with a Rise Time of 500 ps*

We see that in each case, the magnitude of the noise peak has been dramatically reduced. The ground bounce has been spread out over a larger area, to a level that could be perfectly acceptable. The far-end noise has been significantly reduced by the increase in system rise time but is still large, approximately 15 percent. However, we are measuring the far-end noise from the near-end side. In this configuration, the magnitude of the measured far-end noise is actually twice what would appear at the far end if there were a receiver present, so in fact, the far-end noise may also be acceptable with a 500 ps rise time. If we know the final application system's rise time, we could emulate the system's signal and empirically determine if the switching noise generated in the interconnect was acceptable or if it had to be reduced.

In the previous example, the gap in the return path was very slight, and the ground bounce noise generated was small. We could probably find this level of ground bounce acceptable. But often, the gap in the return path is large. In the next example, the gap has been increased to be a large, wide slot. *Figure 1.47* shows the top view of a pair of 50 Ohm microstrip transmission lines with a solid plane as the dark tan region. In the middle of the board, the copper return plane has been removed in a region about an inch long and an inch wide. The region with no copper plane is the top line.

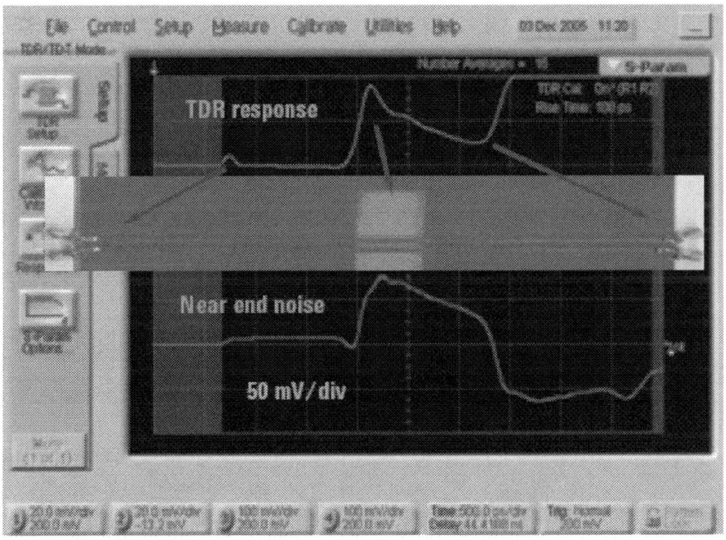

Figure 1.47: *Measured TDR Response of a Single Line Crossing a Large Gap in the Return Path and the Ground Bounce Noise in the Quiet Line*

The TDR stimulus launches into one of the lines and the second channel is used to measure the noise on the near end of the trace. The two far ends of the transmission lines are left terminated in an open. As the TDR signal propagates down the line, it is sensitive to impedance discontinuities. The top trace on a 50 mV/div scale in *Figure 1.47*, shows the small inductive peak at the beginning of the line from the SMA launch. The very large peak midway down the line is from the gap in the return path. The gap dramatically increases the loop inductance of the signal path, as the return current must make a large detour around the gap to reach the source. This extra path length increases the series loop inductance of the signal path. Just before the reflected voltage settles down, the open end of the line is reached.

At the same time the TDR stimulus is propagating down the active line, the second channel is measuring the near-end noise on the quiet line, the bottom trace, also on a 50 mV/div scale. We see the very slight near-end noise initially due to the tight coupling between the uniform transmission line segments. On this scale it is barely at the detectable level. However, as soon as the TDR signal hits the inductive discontinuity and generates the ground bounce voltage across the two regions of the circuit board, this voltage is picked up in the quiet line. In fact, we see that the ground bounce voltage in the quiet line is just about the same magnitude as the reflected voltage in the active line. All the reflected voltage was really ground bounce voltage, shared by the quiet line.

This amount of noise in the quiet line, about 75 mV out of 200 mV, or 37 percent of the incident signal, is far higher than any reasonable noise budget and would be a disaster. In fact, every trace in a bus that shared this return path, meandering around the gap, would see the same ground bounce. The more lines that switched simultaneously, the more dI/dt ground bounce would be generated, and the larger the switching noise would be on the quiet lines.

One way to identify switching noise is to look for narrow, isolated regions where the near-end noise dramatically increases. The TDR response of the active line can be used to guide us to the physical location where the near-end noise is being generated.

While increasing the rise time will decrease the magnitude of the switching noise, sometimes it can still be too large. *Figure 1.48* is an example of comparing a 100 ps rise time and a 1 ns rise time as the system rise time for this large gap.

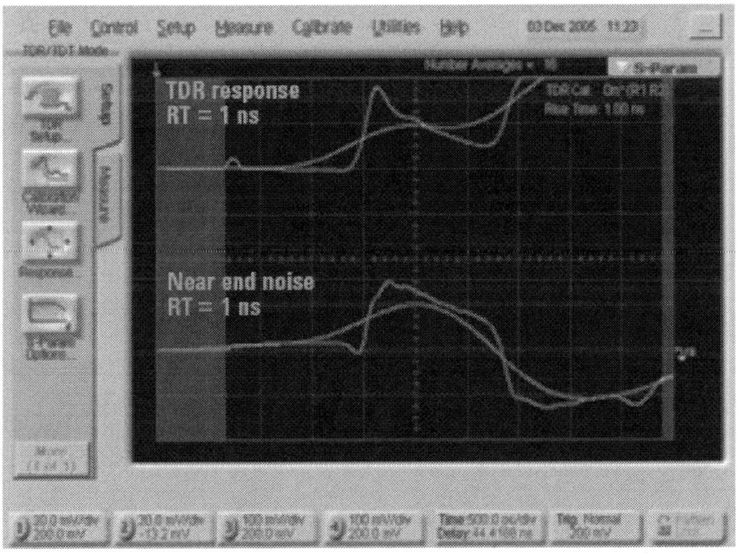

Figure 1.48: *Emulating Ground Bounce Noise from Large Gap at Rise Times of 100 ps and 1 ns*

Though the ground bounce voltage does decrease a little when we increase the rise time by an order of magnitude, the amount of ground bounce is still too large. This is an example of using the TDR to emulate the system rise time and evaluate the impact of a discontinuity on the amount of crosstalk generated. This suggests that no amount of slew rate control would have a hope of getting around this problem. Instead, it would be necessary to identify the source of the ground bounce and either remove the gap or route the signals around the gap, rather than letting them cross it.

1.5 Two-Port Differential TDR (DTDR)

Overview

Previously, we explored two single-ended lines with coupling. Each line had its properties of an impedance profile and TD and there was near- and far-end noise on one line from the signal on the other line. This is one way of describing these two individual lines.

An equivalent way of describing these same two lines is as a single differential pair. Two types of signals can propagate on a differential pair: a differential signal and a common signal. In a differential signal, the voltage on one line is the negative of the other. The differential signal component on a differential pair is the difference in voltage between the two lines. This means in a differential signal, the voltage on one line, measured with respect to the return plane, is the negative of the other. Most high-speed serial links use a differential signal to transmit information. Because of the nature of the receivers, a differential signal can have much better signal-to-noise ratio and noise immunity than a single-ended signal.

The common signal component is the average of the two signals on each line of a differential pair. This means a common signal is really a measure of how much voltage the lines in a pair have in common. While a common signal is rarely used to carry information, it can sometimes cause complications if it becomes so large as to saturate the differential receivers or if it gets out of the product on external cables, as it would contribute to electromagnetic interference (EMI).

When a differential signal propagates on an interconnect, it drives the odd mode of the differential pair and the differential signal sees the differential impedance of the interconnect. When the common signal propagates on an interconnect, it drives the even mode of the differential pair and the common signal sees the common impedance of the differential pair.

To characterize a differential pair, the TDR must drive either a differential signal or a common signal, and measure the response as the reflected differential signal or common signal. This requires two channels to be connected to the same end of the diff pair and have the equivalent of two simultaneous stimuli—either launching a differential signal or launching a common signal into the DUT. This is done with a differential TDR (DTDR).

When set for differential stimulus, as shown in *Figure 1.49*, the stimulus from the two channels is exactly opposite, while, when it is set for common stimulus, the output voltages are exactly the same.

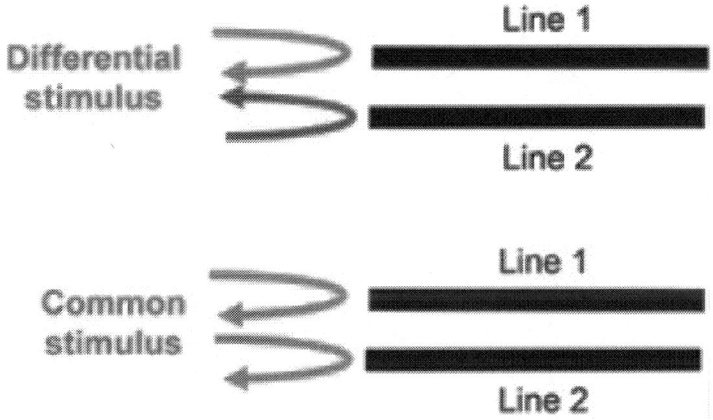

Figure 1.49: *Configuration for Differential Pair Characterization*

In application, the DTDR is set up for one operating mode or the other. To adjust the DTDR for the differential stimulus operating mode, the TDR setup window is opened by clicking the soft bottom on the left side. The stimulus pull-down list allows selecting the single-ended operating mode, the differential operating mode, or the common operating mode.

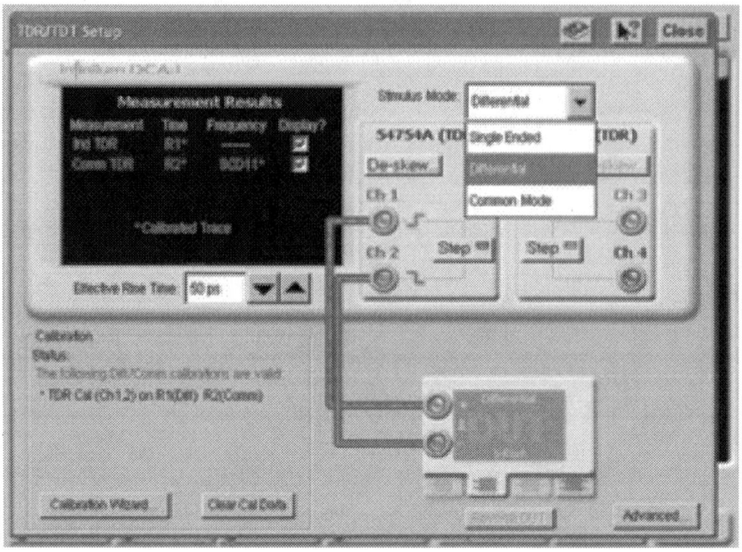

Figure 1.50: *DTDR Setup Screen for Differential Measurements*

Figure 1.50 shows the setup screen when the operating mode is adjusted for differential operating mode. As a side note, though it is sometimes confusing, do not mix up the common mode of operation with the even mode in which the differential pair can be driven. The mode in the screen label with common refers to the mode of operation, not a mode in which the differential pair is driven.

Measuring Each of the Five Impedances Associated with a Differential Pair

If we have a single-ended transmission line that is part of a differential pair, it really has three different impedances that characterize it. It has a single-ended impedance, its instantaneous impedance when the other line in the pair has a constant voltage on it; an odd-mode impedance, the instantaneous impedance of the line when the pair is driven in the odd mode; and an even-mode impedance, the instantaneous impedance of the line when the pair is driven in the even mode.

In *Figure 1.51* is an example of the measured TDR response from a single line in a differential pair. The lines are nine inches long, roughly 50 Ohm microstrip traces, with a spacing about equal to their line width. On this scale of 20 mV/div, one division corresponds to a reflected voltage of 10 percent.

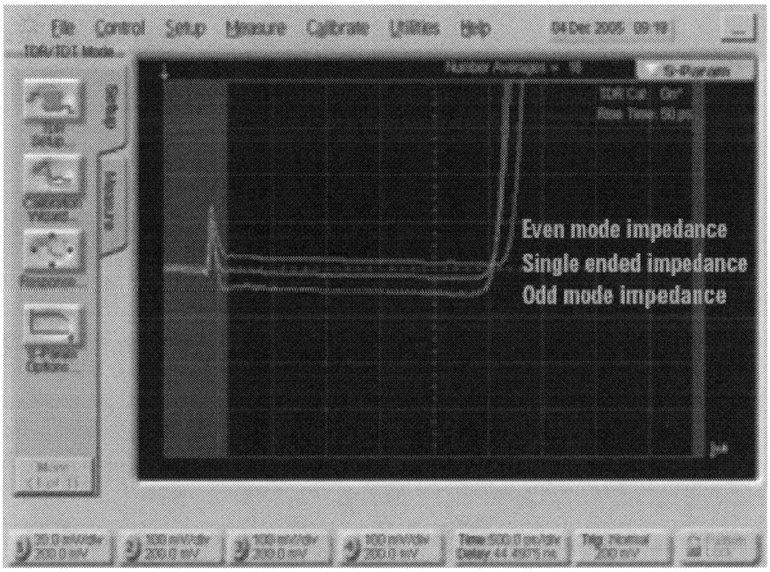

Figure 1.51: *Measured TDR Response of a Single Transmission Line Configured for the Even Mode*

When the DTDR is set up as single-ended, the response is the single-ended impedance. When it is set up as differential, the TDR response from each channel is the odd-mode impedance of the line, and when the DTDR is set up in the common mode of operation, the TDR response from either channel is the even-mode impedance of the line. While we could take the measured reflected voltages and calculate the corresponding impedances, it is much easier to let the TDR do it for us.

The vertical scale can be changed to Ohms so that the first-order impedance is directly displayed. *Figure 1.52* is the same measured response as previously, but the reflected voltage has been converted into the instantaneous impedance. The scale has been expanded to 2 Ohms/div with 50 Ohms right at the center.

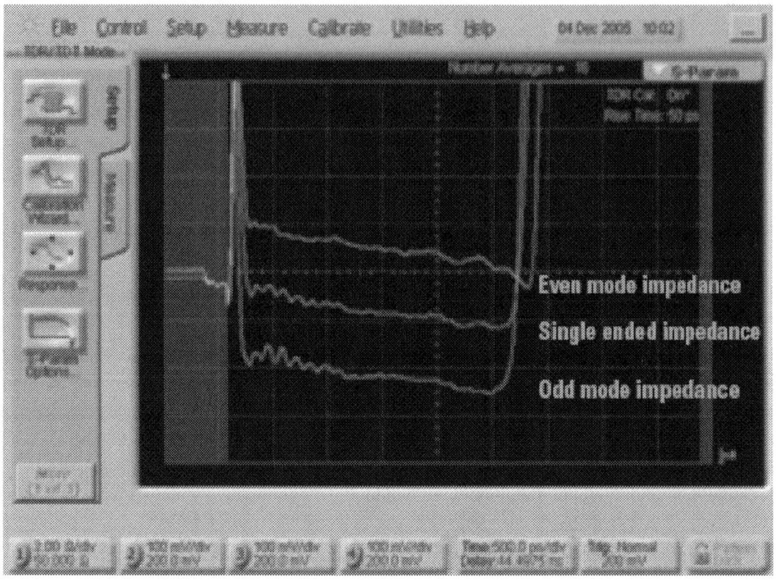

Figure 1.52: *The Three Impedances of a Single Line Displayed Directly on an Impedance Scale*

We can now read the three impedances of this line directly off the screen. In each case, the impedances start high on one end and drop about 1 to 2 Ohms by the other end. The even mode drops 2 Ohms, while the odd-mode impedance changes by only 1 Ohm. This suggests it is probably a dielectric thickness variation that causes the small change in impedance across the length of the board, as the even-mode impedance is more sensitive to dielectric thickness than the odd-mode impedance.

Everything we ever wanted to know about the impedance properties of the differential pair is contained in these three impedance values of each line.

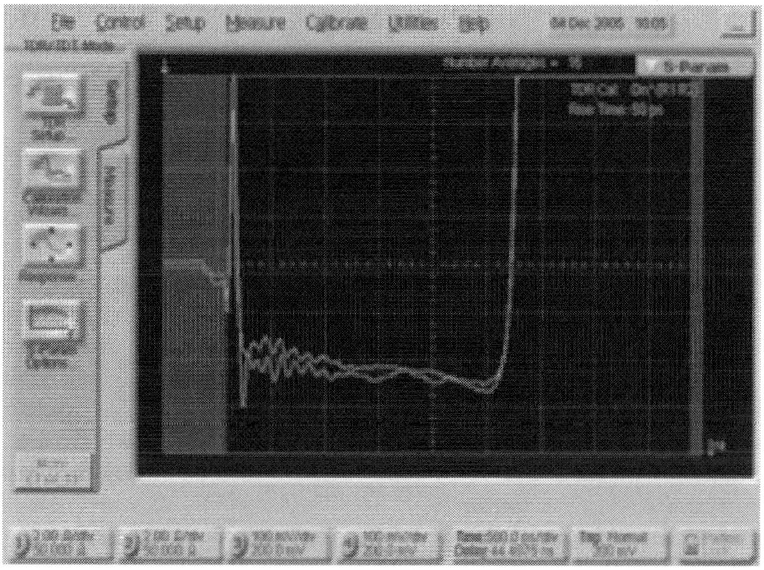

Figure 1.53: *Measured Odd-Mode Impedance of Each Line in a Differential Pair, Displayed Directly on an Impedance Scale*

In DTDR measurements on a differential pair, the odd mode impedances of both lines are measured simultaneously. Previously, we displayed the impedance of just one line. *Figure 1.53* is the odd-mode response of both lines on a 2 Ohms per division scale. This is the individual response from each channel. In this case, we see the differential impedances of the two lines are matched to within a small fraction of an Ohm.

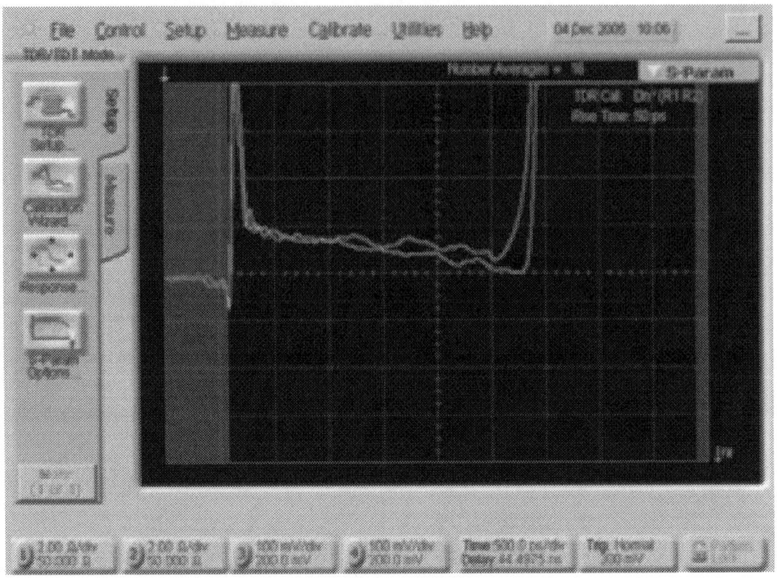

Figure 1.54: *Measured Even-Mode Impedance of Each Line in a Differential Pair, Displayed Directly on an Impedance Scale*

Likewise, when the stimulus is set for the common mode of operation, the even mode of each line can be measured as the responses from the two channels. Again, we see from *Figure 1.54* that for this differential pair, the even-mode impedances are matched to within a small fraction of an Ohm.

The odd- and even-mode impedances are only part of the story. Though each line may have an odd-mode impedance when a differential signal propagates down the differential pair, the differential signal itself sees a differential impedance. It is numerically equal to the sum of the odd-mode impedances of both lines. When the odd-mode impedances of the two lines are the same, the differential impedance of the pair is just twice the odd-mode impedance of either line.

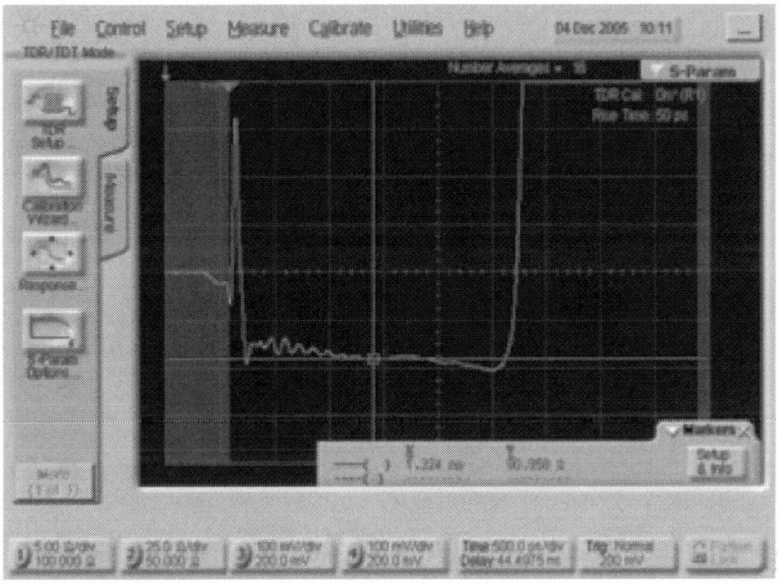

Figure 1.55: Measured Differential Impedance of a Pair of Microstrip Traces, Displayed Directly on an Impedance Scale

The DTDR can simply and easily display the differential impedance profile of the pair of lines. With the stimulus set for differential mode of operation, the differential impedance is selected in the Response 2 setting. As shown in *Figure 1.55*, the differential impedance profile can be plotted directly from the screen. In this case, it is on a 5 Ohms per division scale with 100 Ohms at the very center. The marker can be used to read the differential impedance as about 91 Ohms.

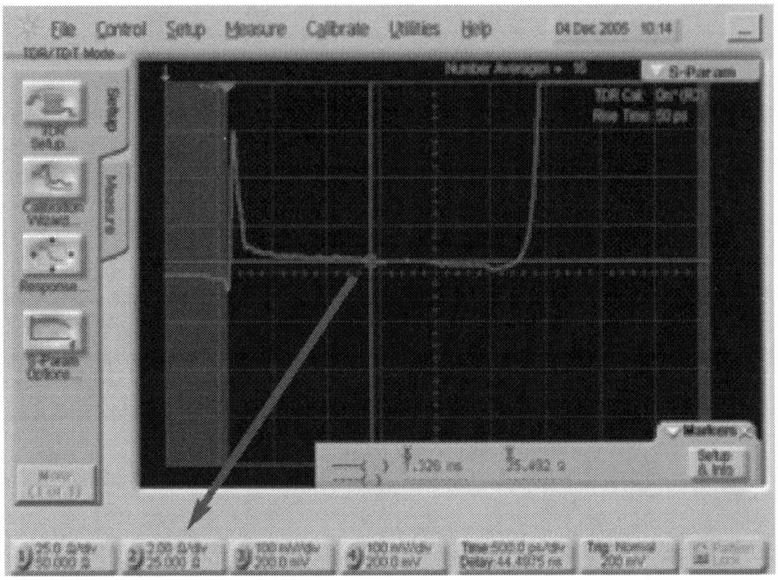

Figure 1.56: *Measured Common Impedance of a Pair of Microstrip Traces, Displayed Directly on an Impedance Scale*

In the same way, the common impedance profile can be displayed directly on the screen. The stimulus mode is set for common mode of operation and the common impedance is selected for Response 2. *Figure 1.56* shows the common impedance profile on a scale of 2.00 Ω/div. The marker is set to read the common impedance off the front screen as about 25.5 Ohms. In this way, we can extract the complete impedance profile characterization of either line or both lines in a differential pair.

Measuring the Degree of Coupling between Lines in a Differential Pair

The differential impedance of a pair of lines is twice the odd-mode impedance. When there is very little coupling, the single-ended impedance of one line is the same as the odd-mode impedance of that line, and the differential impedance is really twice the single-ended impedance of the line.

However, if there is any coupling, the single-ended impedance is not the same as the odd-mode impedance. The odd-mode impedance of that line will be reduced by the coupling. We cannot easily and accurately measure the odd-mode impedance or differential impedance of a pair of lines unless we drive the pair in the odd mode with a differential signal.

The difference between the odd-mode impedance and the single-ended impedance for a typical trace on a motherboard is shown in *Figure 1.57*. The top line is the single-ended TDR response of one line on the motherboard. On the scale of 5 Ohms/div, the single-ended impedance of the trace on the daughterboard and motherboard is seen to be about 58 Ohms.

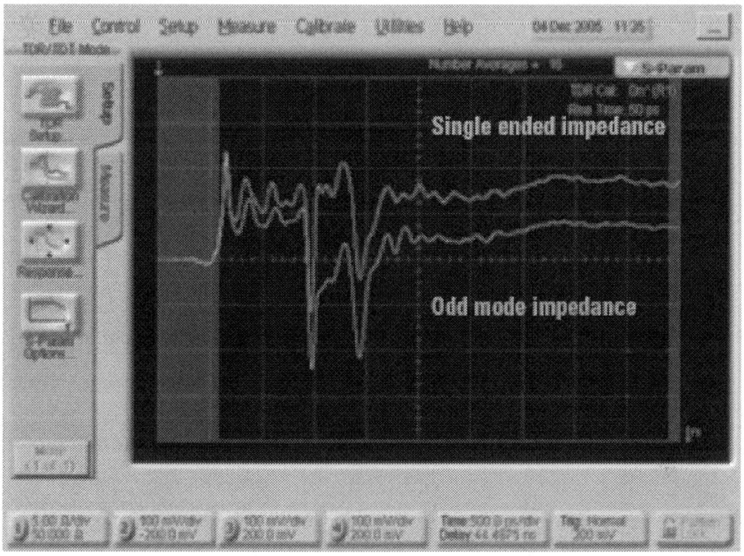

Figure 1.57: *Comparison of the Measured Single-Ended Impedance and Odd-Mode Impedance of a Single Line in a Long Motherboard Trace*

When the pair is driven with a differential signal, the odd-mode impedance of the same line is seen to have dropped in some cases by as much as 5 Ohms. The daughtercard trace is still a little high, at about 55 Ohms, while the connector is seen to be very close to 50 Ohms, though still with large capacitive dips from the via field. The long line on the motherboard is about 53 Ohms.

If the odd-mode impedance is 53 Ohms, the differential impedance of the pair would be 106 Ohms, close to the target value of 100 Ohms.

Measuring the Differential Impedance of a Twisted-Pair Cable

Many applications require a differential signal to be transported from one board to another through twisted-pair cables. The DTDR can be used to measure not only the differential impedance of a differential pair on a circuit board, but also the differential impedance of a twisted-pair cable.

In this case, there is no return plane in proximity. However, as long as the coupling between the two lines in the pair is much tighter than either line to an adjacent plane, the return currents for each line, when driven with a differential signal, will exactly overlap in the adjacent plane and the presence of the plane will be irrelevant. In a twisted pair, any plane, or literally the ground that could act as a return path, carries no current and will not play a role in determining the differential impedance of the pair or in a measurement of the differential impedance.

To measure the differential impedance of the twisted pair, we have to connect each of the lines in the twisted pair to the signal lines in the cable. This establishes a 100 Ohm launch into the twisted pair. In this example, we look at the measured differential impedance of two types of twisted-pair cable. The first case is a two-foot length of twisted-pair cable taken from a low-cost, plain old telephone service (POTS) cable. The second case is a two-foot length of twisted pair taken from a CatV Ethernet cable.

In *Figure 1.58*, the DTDR responses from these two twisted pairs are shown. The top line is the DTDR response from a twisted pair of wires as found in a low-cost POTS hook-up cable. On this scale of 20 Ohms per division, the differential impedance of the cable can be seen to be relatively constant, but on the order of 125 Ohms. This impedance is related to the precise wire diameter and dielectric thickness of the insulation. This cable is typically specified for 120 Ohms and is not rated for high bit rate. As we can see, it is a relatively controlled impedance.

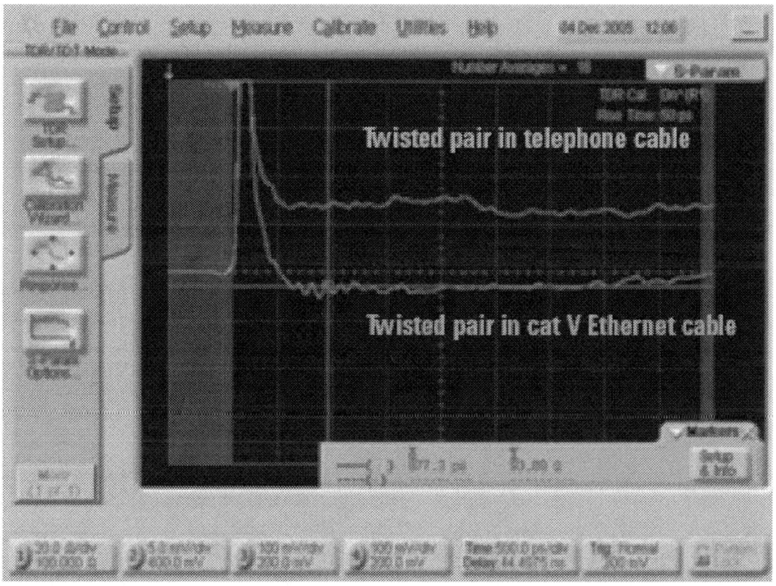

Figure 1.58: *Measured Differential Impedance of Two Twisted-Pair Cables Connected to a Coax Launch*

The bottom line is the measured differential impedance of the CatV twisted pair. It is specified for 100 Ohms, and we can read its impedance with the marker as 94 Ohms. It is also seen to be a very constant, controlled impedance.

The large peak at the beginning of the DTDR response is due to the poor launch into the twisted pair. In both cases, the wires were pulled apart in order to solder two separate SMA connectors, which connected to the coax cables from the DTDR. Part of the twisted-pair connector design is optimizing this launch to minimize the discontinuity.

In this example, we are measuring the reflected differential signal. Once it gets through the connector, the differential impedance of the twisted-pair cable is very close to the differential impedance of the two coax cables. What about the common impedance? While the signal is in the two coax cables, the common impedance is half the even-mode impedance of either cable, which is about 25 Ohms. What is the common impedance of the twisted pair?

The common impedance is the impedance between the two signal lines, with respect to the return path, which in the case of a twisted pair, is literally the floor. As we might imagine, when the return path is far away, this common impedance can be pretty high, easily a few hundred Ohms. To the TDR, it will look like an open.

In addition to measuring the differential impedance profile of the twisted pair, we can also measure the common impedance profile as the common signal travels from the coax cable to the twisted pair. In *Figure 1.59*, the measured differential impedance profile on a scale of 25 Ohms/div is shown as the bottom trace.

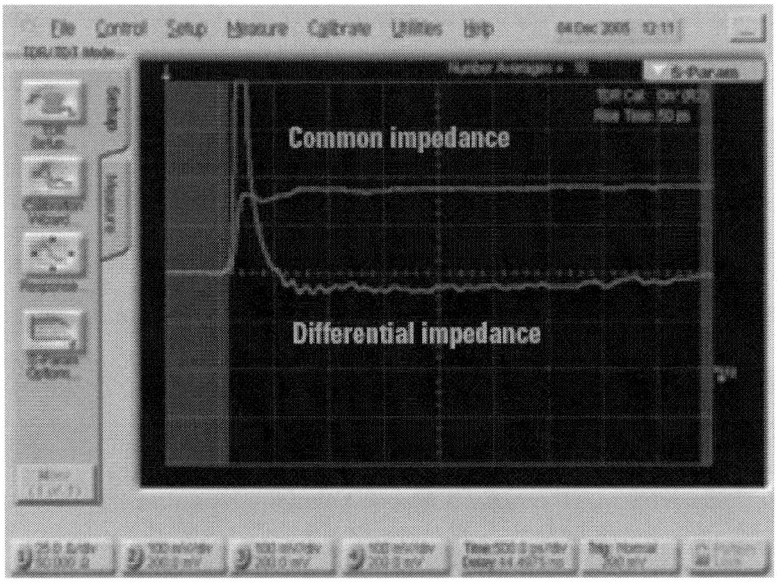

Figure 1.59: *Measured Reflected Common Signal from a Coax to Twisted-Pair Transition with an Incident Common Signal*

The DTDR was set up to use a common signal as the stimulus and then the reflected common signal is measured. The impedance was so high, we changed the scale format to voltage scale and recorded the reflected voltage of the common signal on a scale of 100 mV/div. The incident common signal is 200 mV. We can see in this plot that the reflected common signal is almost 200 mV. In the transitions from the coax cable to the twisted pair, other than the discontinuity of the connector, the differential signal is able to transition to the twisted pair and propagate down the twisted pair, but virtually all of the common signal is reflected due to the very high common impedance of the twisted pair. The little bit of common signal that does get out on the cable will contribute to radiate emissions. This is why it is important in the design of twisted-pair connections to make the impedance the common signal sees as high as possible, so there is little common signal on the external cable to radiate.

Measuring the Reflected Noise of a Differential Signal Crossing a Gap

When a single-ended signal encounters a large gap in the return path, it will see a large inductive discontinuity and generate ground bounce in the plane, which will be picked up by any adjacent signal traces. Crossing a gap in a single-ended transmission line can be a disaster.

However, the same gap can be crossed with less of a problem by a differential signal. *Figure 1.60* shows an example of a microstrip differential pair crossing a gap in the return path. The gap is the top trace in the board where the copper plane has been removed. The bottom trace represents the single-ended TDR response for the signal on one of the lines crossing the gap. There is a huge reflected signal, which affects the reflected signal for the duration of the TD down the line.

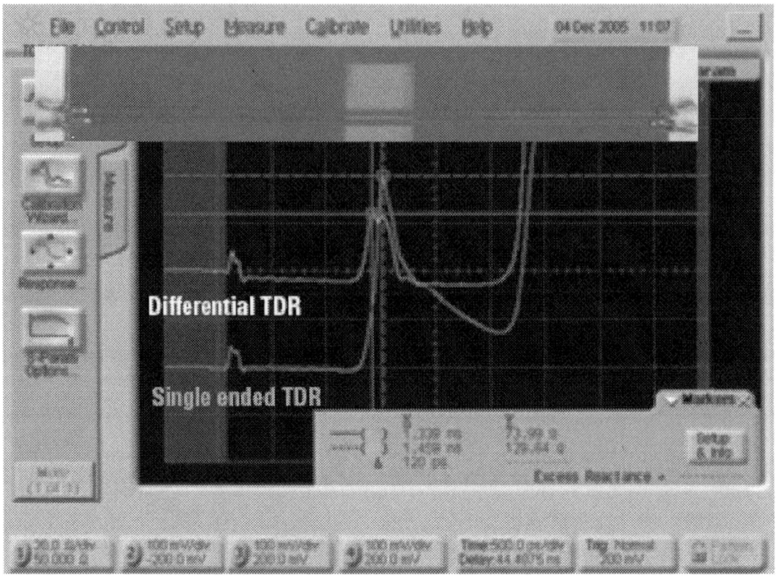

Figure 1.60: *Measured Differential Impedance Profile of a Differential Pair Crossing a Wide Gap in the Return Path*

The white trace is the DTDR response for a differential signal on this same pair of traces, crossing the large gap. The differential signal sees a differential impedance of about 100 Ohms in the region where the plane is continuous. In the region where the plane is removed, the differential impedance of the pair is about 130 Ohms, as read by the dotted marker. Where the plane is removed, the differential impedance is uniform, it is just high. This 130 Ohm discontinuity lasts for the time of flight of the gap, and then the differential impedance the signal sees comes back down to roughly 100 Ohms.

Like all discontinuities, if we keep the length of the discontinuity short compared to the rise time of the signal, the impact of the discontinuity can be reduced.

The impact of this gap on the system's rise time can be emulated by changing the DTDR rise time. For this same differential pair with the one-inch-long gap, the DTDR response was measured for rise times of 100, 200, and 500 ps and 1 ns. *Figure 1.61* shows the

DTDR response of this discontinuity for these rise times, on a scale of 10 Ohms/div.

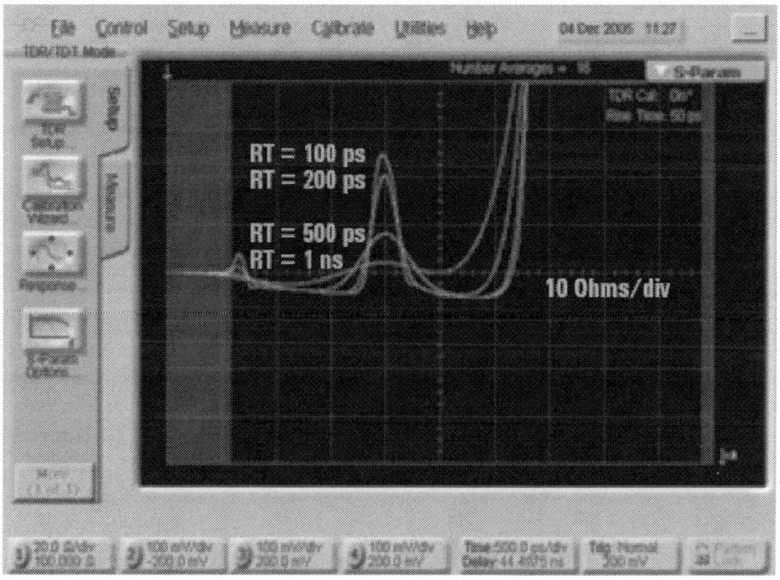

Figure 1.61: *Emulating the Differential Impedance Profile of a Differential Signal Crossing a Large Gap at Four Rise Times*

The longer the rise time, the lower the effective impedance the gap appears as. When the rise time is 1 ns, the impact of the gap has almost disappeared. This suggests an important design rule: if a signal must cross a gap, keep the length of the gap short, keep the rise time of the signal as long as possible, and use a tightly coupled differential pair to cross the gap.

Using the DTDR adjusted to the system rise time would allow a quick and simple evaluation of the impact on the signal's reflected noise from this gap.

Measuring the Mode Conversion in a Differential Pair

In addition to the impedance profiles outlined so far, there is another problem a DTDR can assist in debugging. When a differential signal enters a differential pair, some of the differential signal can reflect back to the source, due to discontinuities in the differential impedance of the interconnect. Of course, these reflected differential voltages are detected by the receiver in the TDR, and we use this received differential voltage to extract information about the differential impedance profile of the interconnect. Under some situations, not only will differential signals reflect and head back to the source, but the incident differential signal can be converted into a common signal and head back to the source as well.

The generation of a common signal, sent back to the source, when a pure differential signal is incident, is called mode conversion. In mode conversion, some of the differential signal is converted into common signal. The presence of the converted common signal is only an issue if the common signal level is large enough to saturate a receiver or if any of the common signal gets out on an external twisted pair, where it can radiate and cause the product to fail the Federal Communications Commission's (FCC's) electromagnetic compatibility (EMC) certification testing.

Otherwise, it is not the converted common signal that causes a problem but the distortion of the differential signal, because of the mode conversion. After some of the differential signal is converted into a common signal, what is left of the differential signal will have a distorted rise time that can cause inter-symbol interference, deterministic jitter, and collapse of the eye diagram. All these factors will limit the maximum bit rate through the interconnect.

Identifying Specific Physical Features That Contribute to Mode Conversion in a Differential Pair

Identifying the physical sources that cause mode conversion will be the first step in eliminating them and enabling higher bit rates and lower radiated emissions.

The fundamental cause of mode conversion is an asymmetry between either the individual signal launches into each line of the differential pair or an asymmetry between the two lines that make up the pair. When the differential signal source is the DTDR stimulus, the asymmetries in the signals can be reduced to below -40 dB. Using the Agilent DCA 86100C DTDR, we are mostly sensitive to asymmetries in the interconnect.

By using the timing information of when the converted common signal returns, we can identify physically where, down the line, the asymmetry might be located. As an example, *Figure 1.62* shows in the bottom line the differential TDR response from a symmetric microstrip differential pair, with a differential impedance of about 90 Ohms. We see a small peak at the beginning of the measurement, corresponding to the SMA launch into the differential pair, and a small dip about midway down the line and then the open of the line. In this example, the signal was a + and – 200 mV differential voltage launched into the differential pair.

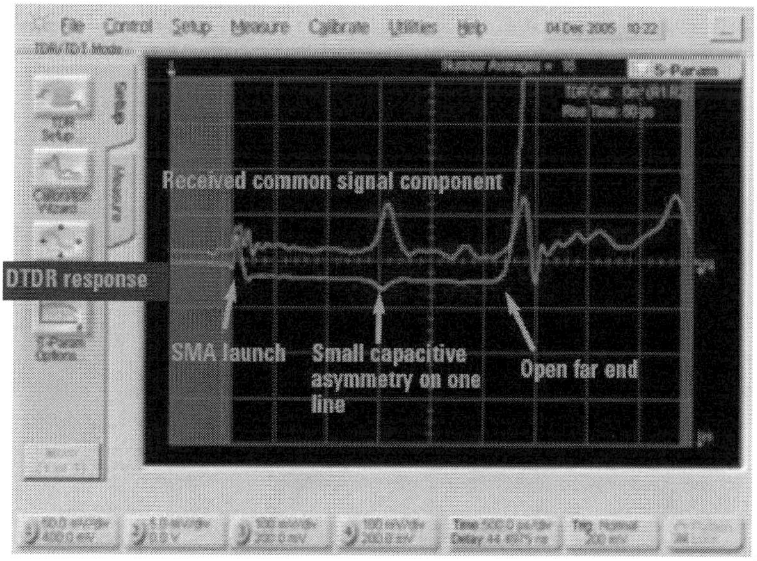

Figure 1.62: *Measured Mode Conversion from Differential to Common Signals Due to an Asymmetry on One Line in a Pair*

The dip in the middle was caused by adding a small capacitive load to one of the lines. This caused the differential impedance of the pair to decrease a small amount and reflect some of the differential signal. In DTDR operation, the receivers are sensitive to the reflected differential signal. In addition, while the stimulus is set to the differential mode of operation, we can adjust the receivers to measure the common signal by selecting the Response 2 and setting it for the common mode of operation so it measures the common signal, which reflects back.

The detected, received common signal voltage is displayed in *Figure 1.62* as the top line. If there was no common signal reflecting back, the top trace would have registered as zero common voltage. The scale for the common signal is 5 mV per division.

Instead, we see that near the beginning of the line, there is a small common signal generated where the SMA launches are and very little common signal except in the middle, where the asymmetric capacitive load is. Finally, we see an additional common signal detected, coincident with when the signal has hit the end of the line and reflected back. This last peak in the common signal is the common signal generated by the asymmetry, moving in the forward direction that hit the end of the differential pair, where the common signal saw an open and reflected back to the source.

The sign of the reflected converted common signal depends on whether the discontinuity occurred on the + or the – line of the pair. If we move the asymmetry to the other line, we change the sign of the converted common signal.

In *Figure 1.63* is the measured common signal at the receivers, for the same capacitive discontinuity, first on line 1 and then taken off line 1 and placed on line 2. We see that the time at which the converted common signal is detected is the same, which means the physical location of the discontinuity is the same.

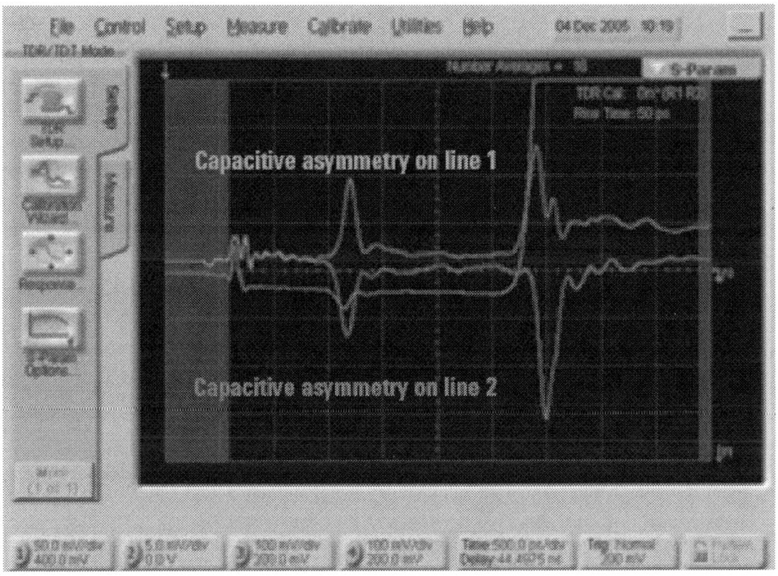

Figure 1.63: *Measured Mode Conversion on a Differential Pair When the Capacitive Asymmetry Is Moved from One Line to the Other*

At the discontinuity, the common signal is converted and scatters backward, back to the source and the detectors in the DTDR, and forward. The forward-traveling common signal propagates down the differential pair and hits the open at the far end, where it reflects with a reflection coefficient of 1.

This reflected wave heads back to the receiver, where it is detected as a common signal, one round-trip time of flight later. We see that both the backward- and forward-scattered common signals have the same sign when we place a capacitive discontinuity alternatively on one line and then the other.

References

[1] Many of the principles described in this application note are introduced in detail in the book, *Signal Integrity-Simplified* by Eric Bogatin, published by Prentice Hall, 2003.

[2] Additional application notes can be found at www.BeTheSignal.com and are available for free

download.

[3] Many of the examples of transmission line structures are available in the circuit boards provided with the Master Class Workshops listed on www.BeTheSignal.com and reviewed in the on-line lectures that can be found on this Web site.

[4] *Signal Integrity Solutions,* Brochure, Literature Number 5988-5405EN, August 29, 2005.

[5] Limitations and Accuracies of Time and Frequency Domain Analysis of Physical Layer Devices, Application Note, Literature Number 5989-2421EN, November 1, 2005.

Authors

Eric Bogatin, Signal Integrity Evangelist, Bogatin Enterprises

Mike Resso, Signal Integrity Application Scientist, Component Test Division, Agilent Technologies

Chapter 2

4-Port TDR/VNA/PLTS – Interconnect Analysis Is Simplified with Physical Layer Test Tools

2.1 Introduction

The vector network analyzer (VNA) has come a long way since it was used to test antenna arrays for military applications. VNA can be used to perform more than one-hundred critical characterization, modeling, and emulation applications for high-speed digital design, many of which are illustrated in this signal integrity book.

If your application requires the measurement of very low-level signals such as near-end crosstalk (NEXT) and far-end crosstalk (FEXT), using a test system with high dynamic range becomes very important. Unlike a wide-bandwidth time domain reflectometer (TDR), a VNA allows the user to set a narrow receiver bandwidth (known as intermediate frequency [IF] bandwidth). This highly accurate tool provides a window into the performance of high-speed digital interconnects that propagate signals with rise times of 500 ps or shorter. The VNA will expand the design validation capability of all signal integrity laboratories around the world and answer important questions such as what is the limitation of my current design, where do I need to focus my attention to increase my data rate, and will my interconnect survive the next-generation application?

The VNA is no longer limited to microwave applications traditionally utilized for aerospace and defense work. Today's commercial electronic designs push the limit of what can be achieved on copper, and the power of scattering parameters (S-parameters) are critical to assure proper performance of these components and systems. Interconnect analysis using a VNA is now simplified with a popular software application called physical layer test system (PLTS). Utilizing a graphical user interface designed for digital designers enables the power of the VNA with the ease of use of a TDR. The ultimate in test accuracy can be provided for topology models, S-parameter behavioral models, characterization of rise-time degradation, interconnect bandwidth, NEXT and FEXT, odd mode, even mode, differential and common impedance, mode conversion, and the complete differential channel characterization.

To provide a little order to the wide variety of applications explored in this signal integrity book, the series is divided into three parts: part 1 (Chapter 1): those that use a single-port TDR, those that use TDR/time domain TDT, and those that use two-port TDR; part 2 (Chapter 2): those that use four-port TDR; and part 3 (Chapter 10) : those that use advanced signal integrity measurements and calibration. The principles of TDR, VNA, and PLTS operation are detailed in other chapters of this book and references listed in the bibliography. We concentrate this application note series on the valuable information we can quickly obtain with simple techniques that can be used to help us get the design right the first time.

2.2 Four-Port Techniques

Complete Differential Pair Characterization
How a TDR can provide valuable signal integrity characterization information about interconnects was reviewed in Part 1. *Figure 2.1* summarizes the various applications for one-port and two-port TDR configurations. Though we are able to obtain some information about a differential pair from two-port measurements, the complete characterization of a differential pair requires four ports.

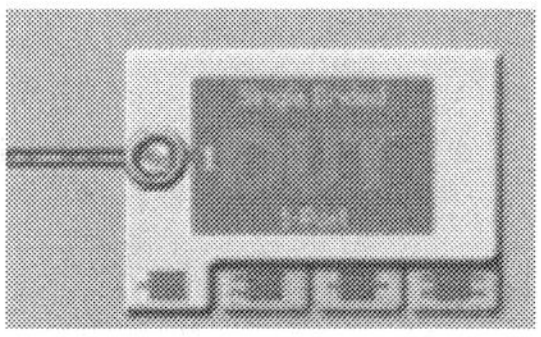

Applications: single ended transmission line characteristic impedance, time delay and discontinuity characterization

Applications: single ended transmission line insertion loss, return loss, and materials characterization

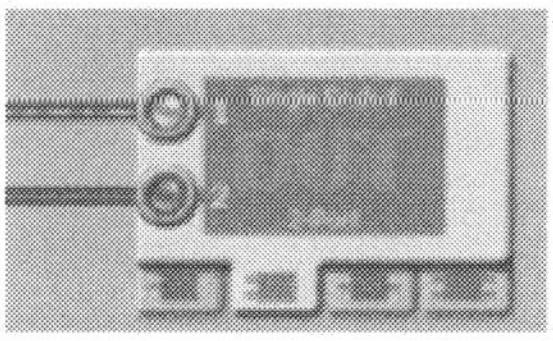

Applications: cross talk between two single ended transmission lines, ground bounce, differential and common impedance characterization

Figure 2.1: *TDR Configurations and Applications*

In a four-port configuration, each of the instrument's ports is connected to each end of the transmission lines in the differential pair. In practice, it is a coaxial connection made at each end of each transmission line, so that each port has a signal and return connection to the transmission line.

Application: complete characterization of differential pair: differential return loss, differential insertion loss, common return loss, common insertion loss, mode conversion

Figure 2.2: *Differential Pair Characterization*

In the configuration shown in *Figure 2.2*, everything important about a differential pair can be extracted. This includes the differential and common return and insertion loss and all forms of mode conversion. From these measurements, details of the differential or common impedance profiles, material properties, and asymmetries can be extracted.

A four-port measurement can be performed in the time domain using a four-port TDR or in the frequency domain using a four-port VNA. There is exactly the same information content in both measurements. There are differences in the dynamic range, or the noise floor of the measurement, so if higher-accuracy measurements are required, you should use a VNA.

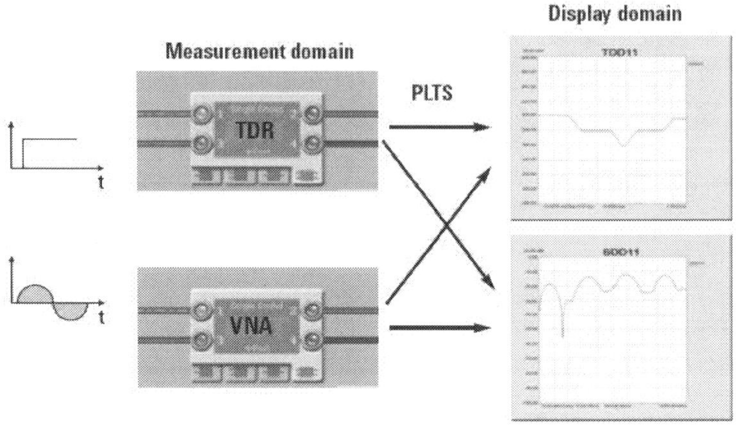

Figure 2.3: *Time and Frequency Domains*

As illustrated in *Figure 2.3*, we can take the measured data from either domain and, using Agilent's N1930 PLTS, translate it mathematically using Fourier transform techniques to display the same data in the time or the frequency domain. These two domains tell the same story. They just emphasize different parts of the story. With PLTS, the display and analysis of the information is completely independent of the instrument used to collect the data. What is important is the information we extract. The flexibility of moving back and forth between the time and frequency domains gives us the flexibility of extracting the most information as quickly and easily as possible. The proliferation of high-speed serial links has driven the widespread use of differential pairs. A differential pair is nothing more than two single-ended transmission lines, with some coupling, used together to carry a differential signal from a transmitter to a receiver. Every single backplane produced today, and in the foreseeable future, is composed of multiple channels of differential pairs.

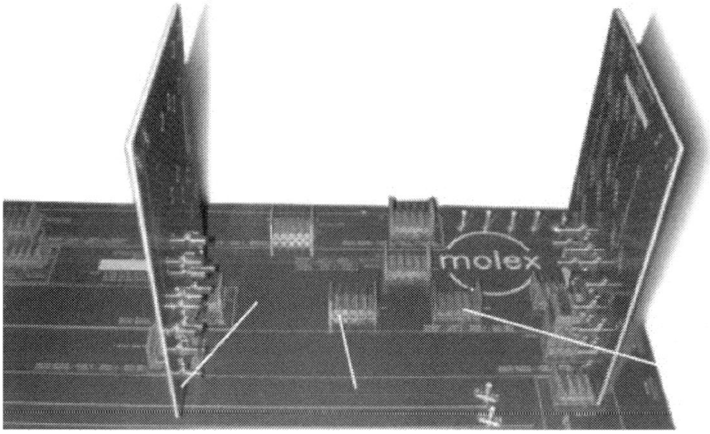

Figure 2.4: *Typical Small Backplane*

Figure 2.4 shows an example of a section of a 46-layer backplane, 18 inches wide by 48 inches long, designed to test out many differential pair cross-sections and via designs. This backplane, designed with Molex GBX connectors, is similar to many state-of-the-art backplanes in use. All the important properties of differential pairs in backplanes such as these can be measured with four ports and analyzed with PLTS.

Four-Port Single-Ended S-Parameters

There are two commonly used types of four-port S parameters: single-ended and differential. The four-port single-ended S-parameters are an extension of one- and two-port S-parameters. In a differential pair, which is really an example of a four-port device, we conventionally label the ends of the device as shown in *Figure 2.5*, with port 1 connected through to port 2 and port 3 connected through to port 4. This is also shown schematically.

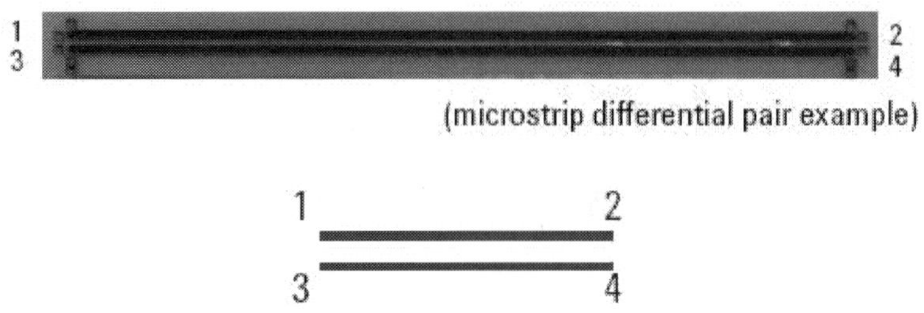

$$S_{out,in} = \frac{V_{out}}{V_{in}}$$

Figure 2.5: *Four-Port, Single-Ended S-Parameters Definition*

By definition, each S-parameter is the ratio of the voltage wave coming out of some port to the voltage wave that was going in. For example, S21 is the ratio of the wave coming out from port 2 to the wave going in at port 1. This is also called the insertion loss. S31 is the ratio of the wave coming out of port 3 to the wave going in at port 1. This is a measure of the NEXT. As the ratio of two sine waves, each S-parameter is complex with a magnitude and a phase. The four-port, single-ended S-parameters have become a de facto standard for describing the electrical properties of any four-port interconnect. There are 16 possible combinations of waves going in and waves coming out. Conventionally, these 16 terms are described in a matrix representation. The S-parameter formalism is not complicated. It is just confusing and somewhat anti-intuitive.

We would expect that the order of the indices that define each term would have the first index being the going-in port and the second index the coming-out port. For mathematical reasons, the definition is the exact opposite. The first index is the coming-out port while the second index is the going-in port. The first index of each element, the rows, represent the response sources—where the wave is coming out. The second index, the columns, is the stimulus, where the wave is going in. *Figure 2.6* shows an example of a generic, four-port S-parameter matrix, which includes all 16 elements.

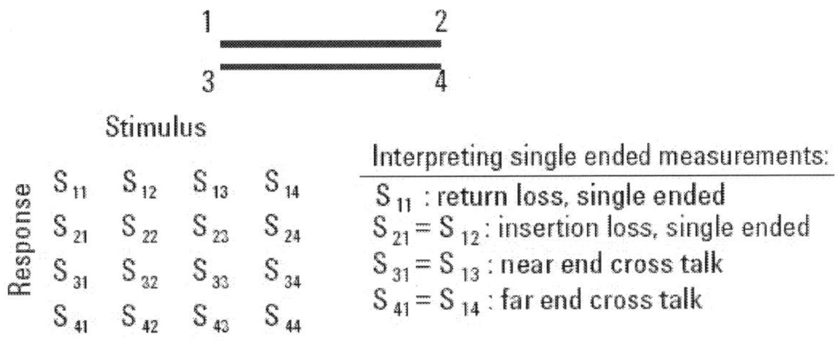

Figure 2.6: *Four-Port Single-Ended S-Parameter Matrix*

When the interconnect is two transmission lines, each element has special meaning. The diagonal elements are the return loss or reflection coefficients. S11 is the return loss of one line from the left end, while S22 is the return loss for the same line from the right end. Since all ends of the device are terminated into the measuring instrument, they are all effectively terminated into 50 Ohms. The S21 and S43 terms are the insertion loss of each line. This is the ratio of what gets transmitted through the transmission line from one end to the other. S31 is the near-end noise and S41 is the far-end noise.

Though there is no industry standard for labeling the ends of the lines, there is a commonly adopted practice. When we describe the interconnect as two separate single-ended lines, it is conventional to use the labeling as shown in *Figure 2.6.* A signal travels from port 1 to port 2 and from port 3 to port 4. In this way S21 is the transmitted signal coming out of port 2 from port 1. As long as we always use this format, S21 will always refer to a transmitted signal and S31 will be the NEXT term.

S-Parameters in the Time Domain

In the complete matrix of measured four-port S-parameters, there is a lot of data. *Figure 2.7* shows an example of the measured single-ended S-parameters of two traces in a small backplane. There are 16 elements, each with magnitude and phase information, for each frequency value. Not shown in any individual plot is the phase information for each element.

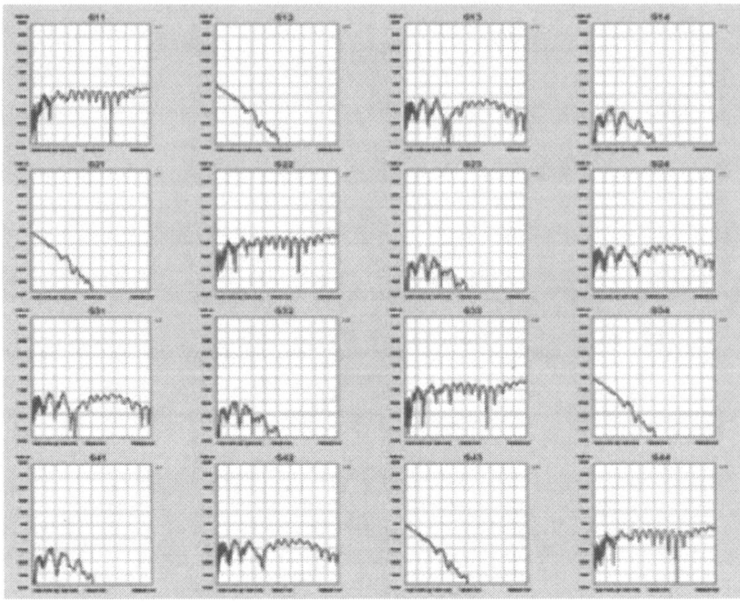

***Figure 2.7:** Measured, Complete, Four-Port Single-Ended S-Parameter Matrix*

Keeping track of all the information can be a difficult task without a tool such as PLTS. This tool will allow us to collect the data from the instrument and display each or all of the elements in precisely the format that will get us to the answer quickly. In this example, we could read the insertion loss of the interconnect at any frequency from the S21 and S43 terms and the near-end noise from the S31 and S42 terms.

The same information contained in the frequency domain S-parameters can be transformed into the time domain form of the S-parameters. The frequency data in each element can be converted into the time domain response of the same element. A return loss becomes a reflected signal. An insertion loss becomes a transmitted signal.

By convention, when the S-parameters are displayed in the frequency domain, they are called S-parameters, but when displayed in the time domain, they are referred to as T-parameters. S11 in the frequency domain becomes T11 in the time domain.

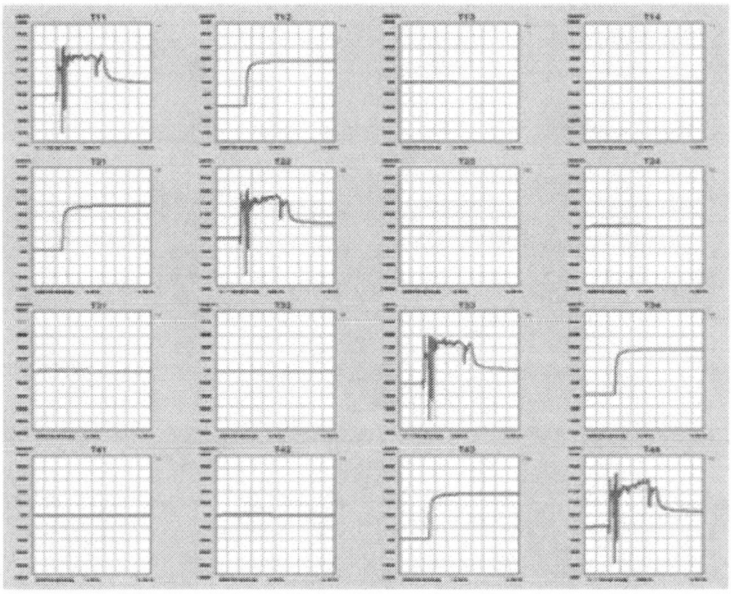

Figure 2.8: *Time Domain S-Parameters*

With the 16 single-ended T-parameters displayed in a matrix, as shown in *Figure 2.8*, many of the important performance questions can be answered at a glance. In its simplest form, T11, when displayed as the step response with a 200 mV incident voltage applied, is identical to the TDR response we are used to seeing on the front screen of a TDR instrument. *Figure 2.9* shows an example of the T11 step response for one trace in a short backplane, on a time base of 1 ns/div, with 10 mV/div on the vertical scale. While the general features can be seen, it is difficult to quantify the impedance profile on this scale.

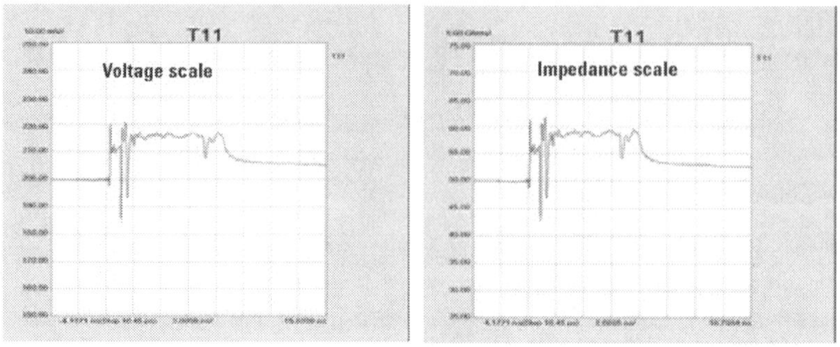

Figure 2.9: *T11 as Reflected Voltage and Impedance Profile*

However, the reflected voltage can be converted into a reflection coefficient, and from this, the first order impedance profile can be directly displayed so that the general features of the interconnect can be extracted right from the screen. In the example shown in *Figure 2.9*, the scale is 5 Ohms per division. We can see the impedance of the daughtercard trace is about 55 Ohms, while the single-ended impedance of the backplane trace is about 58 Ohms.

Everything we ever wanted to know about the behavior of these two single-ended lines can be found in one form or another in the S-parameter matrix or the T-parameter matrix.

Four-Port Differential S-Parameters

As demonstrated in *Figure 2.10*, two individual transmission lines with coupling are also, at the same time, a single, differential pair. As two single-ended transmission lines, we described their electrical properties in terms of their single-ended characteristic impedance and time delay and their NEXT and FEXT.

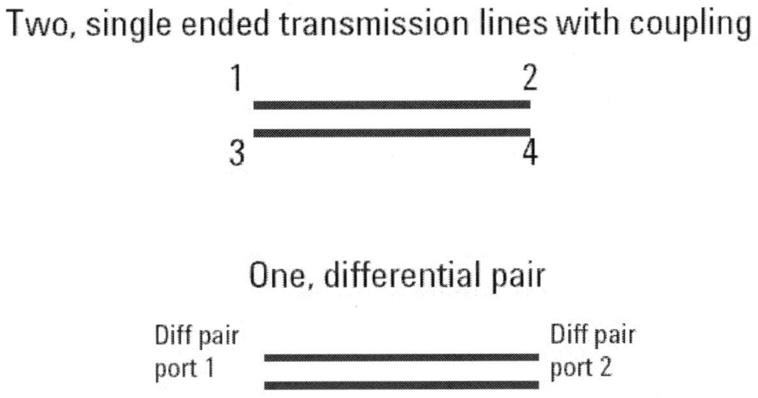

Figure 2.10: *Two Equivalent Views*

As a differential pair, we are interested in how differential and common signals interact with the pair. When we look at these same two lines as a single, differential pair, we describe the differential pair in terms of differential impedance, common impedance, and differential or common time delay. Since we are describing the exact same interconnects, these two views must be exactly equivalent. For linear, passive devices, which includes all interconnects except ferrites, the single-ended S-parameters can be mathematically transformed into differential S-parameters.

When describing a single differential pair, the stimulus and response can only be a differential signal or a common signal. There are four possible outcomes. A differential signal enters the differential pair and a differential signal comes out, a differential signal enters and a common signal comes out, a common signal enters and a common signal comes out or a common signal enters and a differential signal comes out.

			Stimulus			
			Differential signal		Common signal	
			port 1	port 2	port 1	port 2
Response	Differential signal	port 1	SDD11	SDD12	SDC11	SDC12
		port 2	SDD21	SDD22	SDC21	SDC22
	Common signal	port 1	SCD11	SCD12	SCC11	SCC12
		port 2	SCD21	SCD22	SCC21	SCC22

Figure 2.11: *Differential S-Parameters*

Each of these four outcomes is partitioned into a different quadrant of the differential S-parameter matrix, as shown in *Figure 2.11*. To distinguish these quadrants, we use the same index format as the single-ended S-parameters, using a D or C to designate differential or common, stimulus or response. The first index is the coming-out index, while the second index is the going-in index.

The SDD quadrant describes differential signals going in and coming out, and the SCC quadrant describes common signals going in and coming out. The SCD quadrant in the lower left corner of the matrix describes differential signals going in and common signals coming out, a form of mode conversion, and the SDC quadrant, in the upper right, describes common signals going in and differential signals coming out, a form of mode conversion.

Everything you ever wanted to know about the electrical properties of a differential pair is contained in these 16 differential S-parameter matrix elements. They are also sometimes called the balanced or mixed-mode S-parameters. These are all different names for exactly the same set of terms.

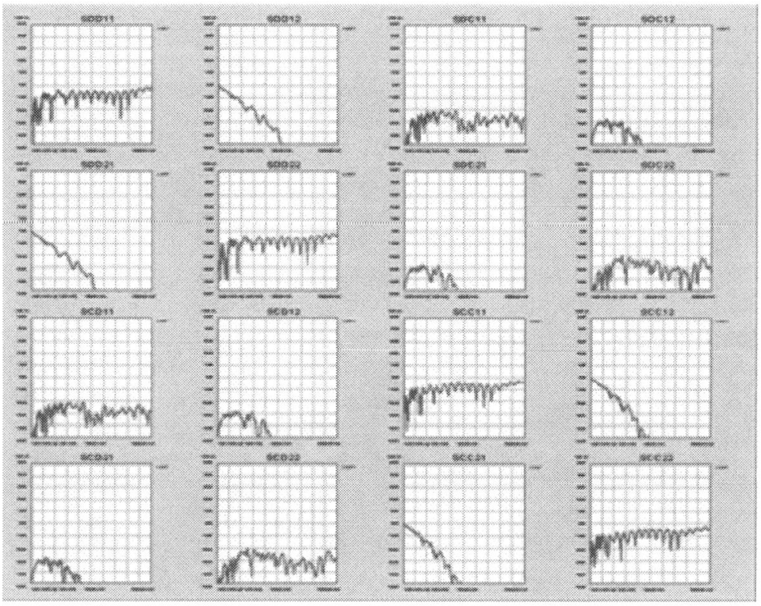

Figure 2.12: *Balanced or Differential S-Parameters*

When all 16 elements are displayed as in *Figure 2.12*, it is conventional to display them in the same orientation as the matrix elements previously described. The four elements that make up the upper left quadrant are the SDD terms, the four on the lower left are the SCC terms, the four in the upper right are the SDC terms, relating mode conversion, and the four in the lower left quadrant are the SCD terms.

Since these are S-parameters, in addition to having a magnitude for each S-parameter plotted as a function of frequency, we also have the phase of each term. For compactness, the phase terms are not displayed in this set of plots but are easily accessible, when needed.

Of course, just as we were able to convert the single-ended S-parameters into the equivalent time domain, T-parameters, each of these differential S-parameter elements can be converted mathematically into their equivalent time domain T-parameters.

The resulting matrix is the balanced or differential T matrix. An example of the differential T matrix for a 16-inch differential pair through a motherboard, and two daughtercards, is shown in *Figure 2.13*. The diagonal elements are all reflection terms and relate to the differential or common impedance profiles.

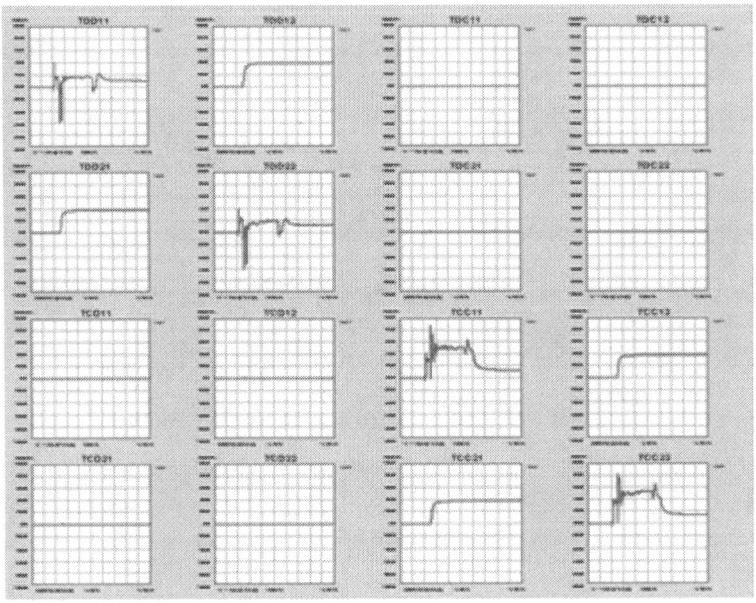

Figure 2.13: *Balanced Time Domain Matrix Elements*

It is important to keep in mind that the various formats to display the S-parameters are all completely interchangeable. As illustrated in *Figure 2.14*, the measurements can be taken in either the time domain with a TDR or in the frequency domain with a VNA and mathematically, using PLTS, converted equivalently into the four formats and two special conditions, and displayed in the time or frequency domains, or as single-ended or differential parameters.

Measurements: TDR, VNA	→	Displayed: PLTS
Time domain		Single ended frequency domain
Frequency domain		Single ended time domain
		Differential frequency domain
		Differential time domain
		Eye Diagram
		W-element RLCG

Figure 2.14: *Summary of Measurement and Display Options*

Depending on the question we are trying to answer, one format or a specific element might be more efficient at getting us to the answer faster. In this application note, we focus on those formats that can provide the most valuable information about differential channels used in high-speed serial links.

High-Speed Serial Links Applications

The information provided by four-port differential S-parameters, either in the time domain or the frequency domain, is ideally suited to analyzing the performance of the differential channels used in high-speed serial links. *Figure 2.15* lists some of the most important problems that can be solved using the techniques outlined in this application note.

1. Characterizing the differential and common impedance profile of a differential channel
2. Characterizing the time delay and group delay of the differential and common signal in a differential channel
3. Measuring the bandwidth of a differential channel
4. Measuring the rise time degradation of a differential channel
5. Direct simulation of eye diagrams
6. Estimating taps for pre-emphasis from transmitted impulse response
7. Estimating possible EMI resulting from mode converted common signal on external cables.
8. Identifying the root cause of mode conversion in a differential pair
9. Extracting first order transmission line models of a differential pair in RLCG format

Figure 2.15: *Some of the Problems That Can Be Solved Using S- and T-Parameter Analysis*

In addition to just the general characterization of the interconnect, by using the information contained in the four-port S- and T-parameters, the resulting performance of the differential channel can be emulated. Direct measurements of the interconnect bandwidth and expected eye diagram can be evaluated.

The rise time of a transmitted signal will be degraded due to loss in the interconnect from the dielectric and conductors, and from impedance discontinuities. By looking at the differential impedance profile, the discontinuities can be isolated and their root cause identified.

If the information displayed by PLTS cannot answer every question directly, the behavioral model of the channel, described by the 16 element differential S-parameter matrix can be used directly in some circuit simulators such as Agilent's ADS or the circuit simulator HSPICE. Using the actual device driver models and the behavioral model of the interconnect, the system performance can be evaluated. Likewise, if a circuit simulator is used that cannot input S-parameter behavioral models, a simplified, uniform differential transmission line model can be exported based on RLGC matrix elements. This can be used by most circuit simulators.

The combination of features in PLTS is a powerful tool to extract the most possible information from any differential channel used for high-speed serial interconnects.

Differential Impedance Profile

In this first application example, we will look at the differential impedance profile of a differential pair, using the differential time domain response. The first DUT is a uniform differential pair four inches long, fabricated in FR4 as a microstrip, and shown in *Figure 2.16*. This particular pair was designed with very tight coupling, having a spacing of about one half the line width. Data is courtesy of GigaTest Labs.

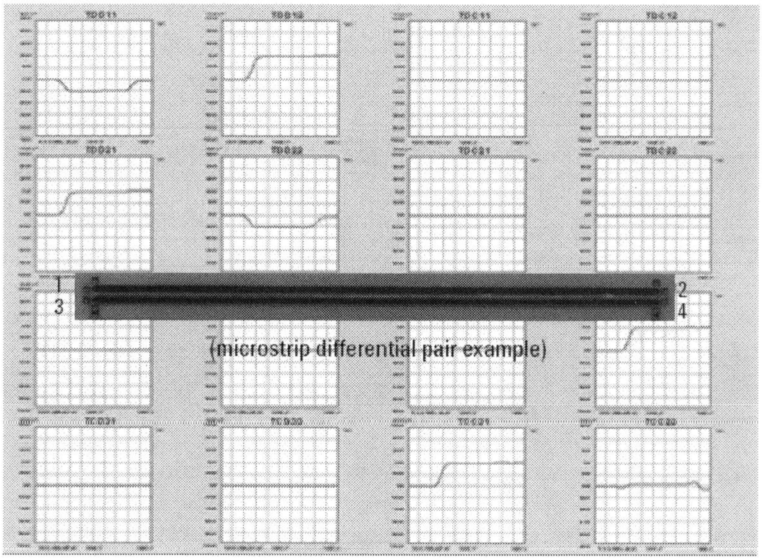

Figure 2.16: *Differential Time Domain Response for Uniform Tightly Coupled Differential Pairs*

All the balanced time domain measured data is displayed in *Figure 2.16*. All the information about how differential and common signals interact with this differential pair in the time domain is contained in these 16 elements. The upper left quadrant has information about how differential signals enter and come out. The lower right quadrant has information about how common signals enter and come out. The two off-diagonal quadrants have information about how differential signals or common signals enter the differential pair and are converted to the opposite type of signal and come out of the differential pair.

To describe the differential impedance profile, we would want to send a differential signal into port 1 and measure the reflected differential signal coming back out of port 1. The reflected signal would be due to encountering changes in the instantaneous differential impedance along the way. This information is found in the TDD11 matrix element, in the upper left quadrant.

The source impedance of each channel of a VNA or a TDR is 50 Ohms. However, when they are sourcing a differential signal, the source impedance is now the series combination of the two impedances, or 100 Ohms. This means that any reflected signal is due to encountering a differential impedance other than 100 Ohms.

The TDD11 term can be displayed with three scales: as the reflected differential voltage, assuming a 400 mV incident differential signal; as the reflection coefficient; or as a first-order calculation of the instantaneous impedance. In *Figure 2.17* is an example of the measured TDD11 element, first as the reflected voltage on a scale of 20 mV/div out of 400 mV incident signal, or 5 percent reflection coefficient per division, and then as the extracted impedance on a scale of 10 Ohms/div.

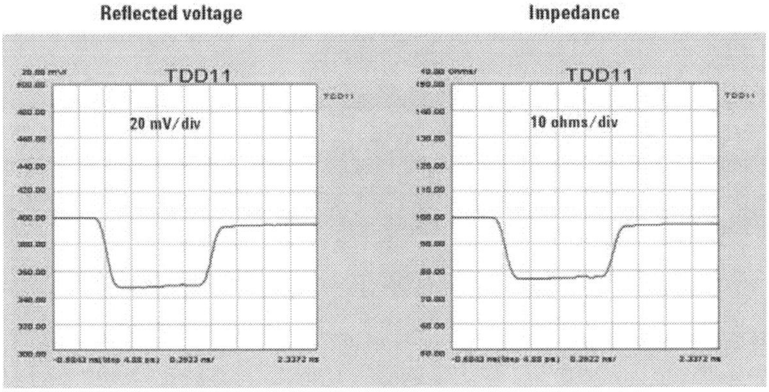

Figure 2.17: *TDD11 of Uniform Pair*

The differential impedance profile of this interconnect can be read off the right impedance screen as about 77 Ohms—very constant from the beginning to the end of the interconnect, as expected for a uniform differential pair.

The beginning of a 22-inch differential channel on a motherboard is shown in *Figure 2.18*. The impedance profile can be read directly off the screen with the aid of the markers.

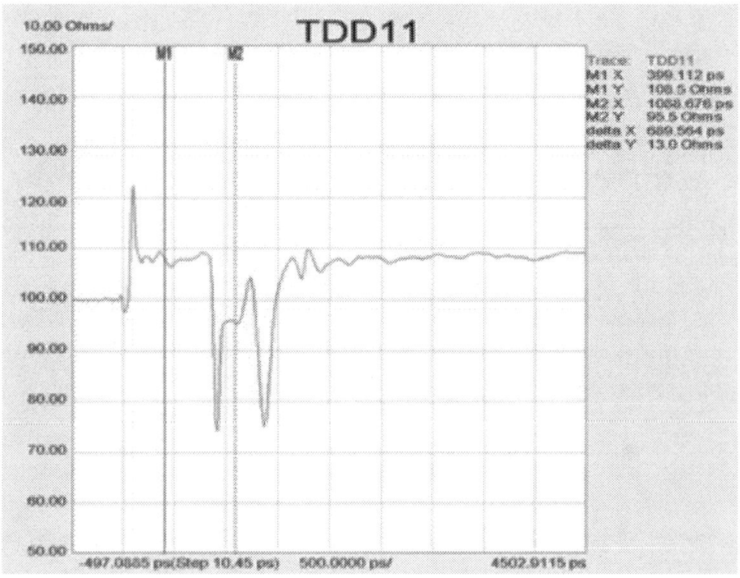

Figure 2.18: *TDD11 of Motherboard Trace*

The very first peak is due to the inductance of the surface-mount assembly (SMA) launch. With too much antipad area around the signal pin, it has a higher impedance. The flat region, which marker M1 intersects, has an impedance read on the scale to the right of 108 Ohms. This is the trace on the daughtercard, which is relatively constant in impedance.

The first large dip, going as low as 75 Ohms on this scale, is the capacitance of the via field where the connector attaches to the daughtercard. The next dip is the capacitance of the via field where the connector attaches to the motherboard. The region between them, where marker M2 spans, shows an impedance of roughly 95 Ohms. This is the region through the connector itself.

The rest of the trace, to the right of the last dip, is the trace on the motherboard, showing an impedance of roughly 108 ohms. This is the typical performance of a motherboard, which shows an impedance within 10 percent of the target impedance of 100 Ohms. The connector itself is a well-matched connector. It is just that the vias the connector is inserted into have an excess capacitance that dramatically degrades the performance of the differential channel. It is not the component causing the problem, but how it has been designed into the board.

Impact from Stimulus Rise Time

The same differential channel is shown in *Figure 2.19* on an expanded scale of 1 ns per division, now showing the connector on the other end of the channel. However, the right end of the differential channel clearly does not match the left end. The impedance mismatch is clearly larger on the left end. Is it possible the connector is different on each end and has a different impedance profile?

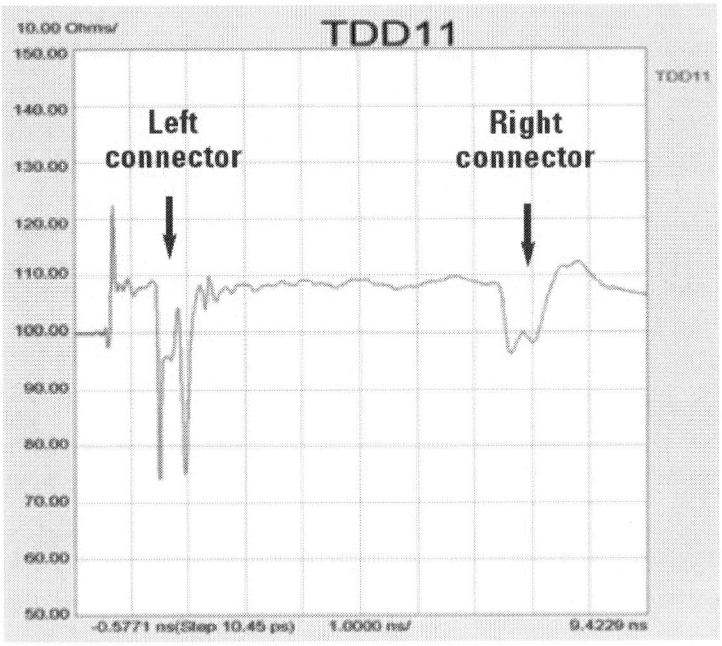

Figure 2.19: *TDD11 of Motherboard Trace on Expanded Scale*

When the time delay of the impedance discontinuity is short compared to the effective rise time of the incident signal, the magnitude of the reflected signal will depend on the rise time. Referring to the peak value of the reflected signal on the impedance scale, and interpreting this as an impedance, is only meaningful if the rise time of the signal at that point is known. To interpret impedance, it is important to know the rise time of the system.

When the data is taken in the time domain with a TDR, the 10–90 rise time of the stimulus entering the DUT can often be read right from the screen of the TDR. When the data is taken in the frequency domain with a VNA and transformed to the time domain, it is not always obvious what the rise time of the signal is that is entering the DUT.

There is a simple way of estimating it, but it will depend on the setting under the time domain window, as shown in *Figure 2.20*. There are three settings that affect the effective rise time of the signal. However, there is a tradeoff between shorter rise time and artificial ripple. This is a natural consequence of the digital filter that is part of the Fourier transform that translates the frequency domain data into the time domain.

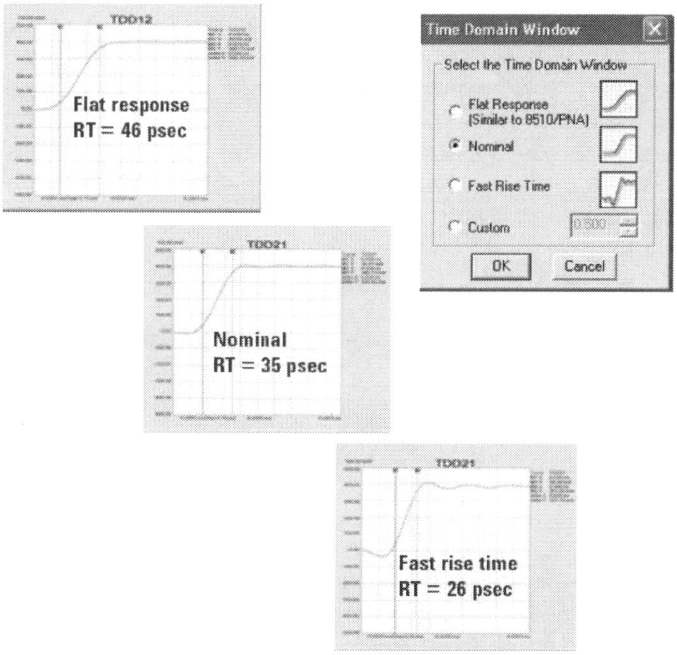

Figure 2.20: *Time Domain Window*

In this example, a thru measurement was measured on a VNA with a frequency range up to 20 GHz. The 10–90 rise time of the received signal, TDD21, was measured using markers, on the screen. As a rough approximation, the 10–90 rise time for the "flat" setting is 46 ps. This corresponds to roughly RT = 0.9/BW. For the nominal setting, which is the default setting for PLTS, the RT = 0.7/BW and for the fast rise-time setting, the RT = 0.54/BW.

While a rough approximation to the rise time of a signal, given the bandwidth of the signal is RT = 0.35/BW, the effective rise time is actually longer in PLTS because of the bandwidth overhead of the digital filter.

This approximation allows us to estimate the rise time entering the DUT if we know the highest frequency in the VNA measurement. Using a nominal setting and a measurement bandwidth of 20 GHz, the rise time entering the DUT is 35 ps. Though this is the rise time entering the DUT, as the incident signal travels down the interconnect, the rise time quickly increases due to the impedance discontinuities and the losses in the channel. Past the first discontinuity, it is not possible to interpret the impedances with any meaning from the screen because the rise time at the location of the discontinuity is unknown. If a faster rise time than 35 ps is desired, then a VNA with higher bandwidth can be used in the measurement (i.e., 50, 67, or 110 GHz).

One way to verify the connectors are identical on the two ends is to compare the TDD11 response with the TDD22 response. This is the differential TDR response, looking from the other end of the differential channel. In *Figure 2.21*, the TDD11 response and the TDD22 response are both displayed superimposed on the same scale of 1 ns per division.

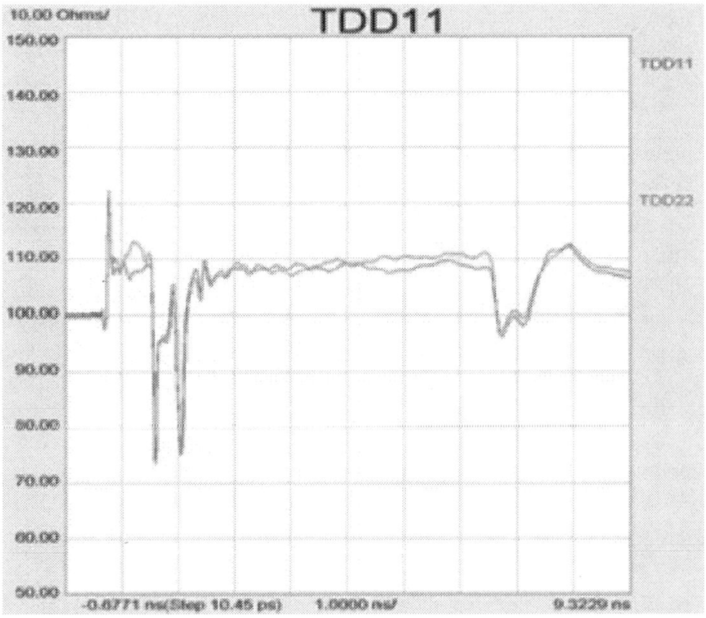

Figure 2.21: *Differential TDR Response of Both Ends of the Interconnect*

We see that the TDD response for each daughtercard and connector are almost identical when viewed from the closest end. The connector on the far end of the interconnect also appears identical. In this particular interconnect example, the interconnect is symmetrical. It can look like the connectors are different on the two ends because of the rise-time degradation of the signal in propagating down the length of the interconnect, smearing out the reflected signal.

Differential or Single-Ended Measurement

It is often believed that a single-ended measurement is good enough and the extra effort of a differential measurement is not necessary. By comparing the TDD11 response with the T11, single-ended response, we can do a direct comparison of the signal-ended and differential response. By definition, the differential impedance is twice the odd-mode impedance. If there were no coupling between the two lines that make up the differential pair, the odd-mode impedance of either line would be identical to the single-ended impedance of either line. How different the two impedances are is a measure of the degree of coupling.

Figure 2.22 shows the same motherboard trace displayed as the differential response, TDD11, and the single-ended response, T11. The time base is identical in the two plots, and the impedance scale for the differential signal is 10 Ohms/div, while the single-ended response is on a 5 Ohms/div scale.

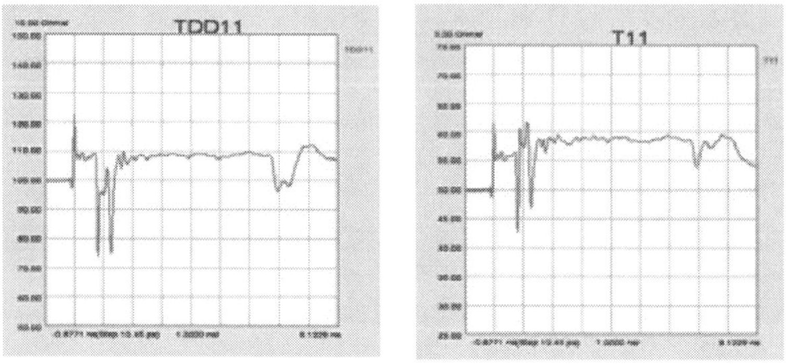

Figure 2.22: *Differential and Single-Ended Impedance Profiles*

The odd-mode impedance of the daughtercard is 108/2 = 54 Ohms. The single-ended impedance of the daughtercard is seen to be 56 Ohms. This suggests that the 2 Ohm lower impedance of the odd mode is due to the coupling on the daughtercard.

The odd-mode impedance of the connector is seen to be about 100/2 = 50 Ohms, a well-matched connector. The single-ended impedance of the same path is roughly 59 Ohms. If we had used this single-ended measurement to evaluate the connector for differential applications and incorrectly called this impedance the odd-mode impedance, we would have estimated the differential impedance was 59 x 2 = 118 Ohms. This large difference is an indication of the strong coupling in the connector.

Connectors used in differential applications are typically designed for tight coupling. To get a realistic measure of the differential impedance of a connector, it would be terribly misleading to use a single-ended measurement. The error in this case would have been about 18 percent.

The odd-mode impedance of the trace on the motherboard is about 110/2 = 55 Ohms. The single-ended impedance of the same trace on the motherboard is about 59 Ohms. The 4 Ohm reduction in the odd-mode impedance suggests tight coupling of the traces on the motherboard interconnect. If the single-ended impedance were used to characterize the motherboard, there would be an error of about 4 Ohms out of 55, or 7 percent. This is why differential impedance measurements are so important in coupled, differential pairs.

Common Impedance Profile

When the stimulus is a common signal, the two ports on each side of the DUT are in parallel and the effective source impedance is 25 Ohms. The TCC11 response can be converted into the common impedance profile. It will be sensitive to impedance changes from the 25 Ohm reference impedance.

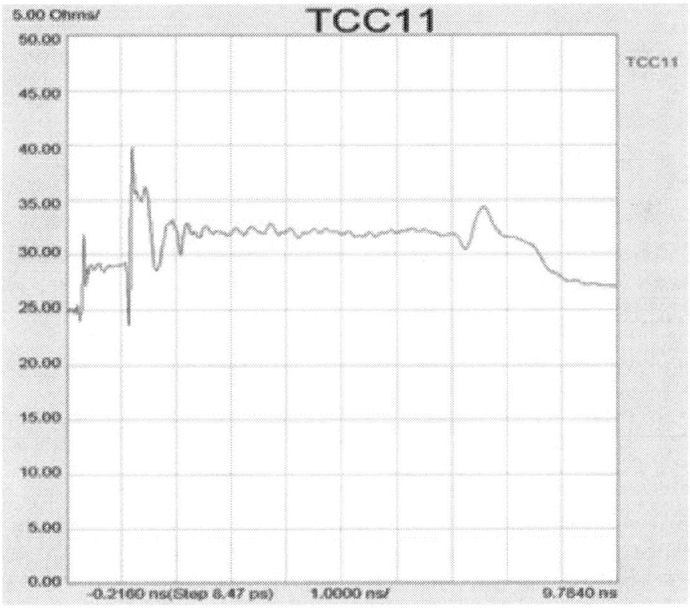

Figure 2.23: *Common Impedance Profile*

Figure 2.23 shows the common impedance profile for the differential channel through the motherboard. The daughtercard is roughly at 28 Ohms, the connector is at 35 Ohms, and the motherboard differential pair is at 32 Ohms. Though it is straightforward and easy to display the common impedance profile of a differential pair, rarely is it of any consequence.

Time Delay and Dispersion

Due to the finite speed of light, there is a time delay between the signal entering a differential channel and that same signal appearing at the far end. This is the second important parameter for a differential channel. The time delay is related to the length of the interconnect and the speed of light in the interconnect medium, and the speed of the signal is due to the dielectric constant of the material. The time delay can be measured in both the frequency domain and the time domain, returning a slightly different piece of information in each domain.

In the frequency domain, the time delay is related to the phase delay of a sine wave entering the interconnect at the reference plane (determined by the calibration setup) and being received at the second differential port. The total phase delay is the number of "unwrapped" wave cycles through the interconnect. The phase delay divided by the frequency is the time it takes for that individual sine wave frequency to travel from one end to the other.

The derivative of this sine wave time delay is called the group delay. It is the time it takes for the shape of a combination of sine waves to travel down the interconnect. Group delay is the term that most closely corresponds to the time delay of a signal through the interconnect and can be displayed directly by a PLTS for any differential channel.

In an interconnect composed of a laminate material that is nondispersive—where the dielectric constant is constant with frequency—the derivative, or slope of the phase delay to the frequency, is exactly the same as the group delay. All frequency components travel at the same speed. Group delay is constant at all frequencies for virtually all differential channels. If the interconnect has a constant impedance and is perfectly matched at the ends to the measuring instrument, the measured group delay is exactly related to the time delay of a signal entering the interconnect and appearing at the far end. It would be constant at all frequencies.

However, if there is any impedance mismatch at particular frequencies, resulting in reflections of the sine waves, the time delay for the reception of the phase at that frequency will be distorted. The measured, transmitted phase is no longer due to just the speed of light in the medium, it is also related to the multiple reflections.

In the frequency domain, the multiple reflections will give rise to variations in the group velocity. In the time domain, the multiple reflections will give rise to a distortion of the leading edge of the signal as it comes out of the interconnect. Some frequency components will arrive at different times compared to others.

In *Figure 2.24* is an example of the measured group delay for a one-meter-long backplane differential channel composed of two daughtercards and 28 inches of backplane trace. The vertical scale is 1 ns delay per division, and the horizontal scale is 2 GHz/div. We see the typical delay is about 6.7 ns for the 40 inches, which is 6 in/ns, exactly the same as the rough rule of thumb that the speed of a signal in an interconnect is 6 in/ns.

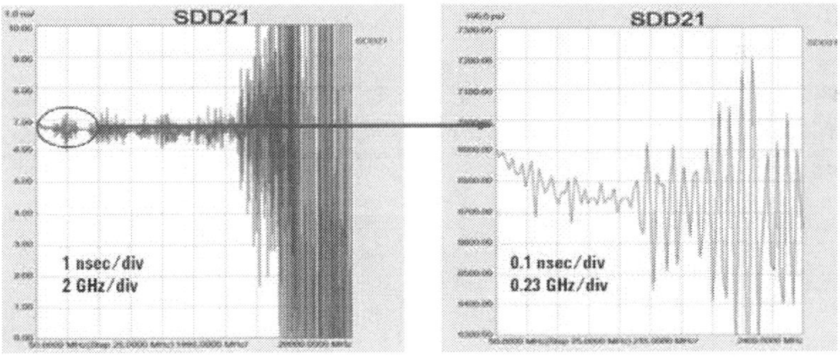

Figure 2.24: *Group Delay of One-Meter-Long Backplane Channel in the Frequency Domain*

However, it is not perfectly constant. The rapid noise and variation after about 13 GHz is due to the large attenuation. After 13 GHz, there is not much signal coming through to be able to measure the phase delay. In addition, there is some noise on the group delay even above 2 GHz. This is due to the multiple reflections from the elements of the differential channel such as the connectors and SMA launches. These multiple reflections will contribute to some distortion of the transmitted signal.

None of these features arise from the natural dispersion of the dielectric material that makes up the multilayer backplane. There is, in fact, dispersion in the laminate, but it occurs only at very low frequencies and has a very small magnitude. By expanding the scale, as on the right side of *Figure 2.24*, and zooming in on the first 2 GHz, we see a small dropoff of the group delay.

At 50 MHz, the group delay is about 6.9 ns. By 500 MHz, the group delay has dropped to 6.75 ns and is relatively constant thereafter. This is a change of about 0.15 ns out of 6.8 ns, or 2 percent. Of course, by 500 MHz, the multiple reflections in the differential channel totally swamp any dispersion in the laminate. This is why worrying about dispersion and frequency dependence to the dielectric constant is often more of a distraction from worrying about the real problems that will cause performance complications.

The time delay through the interconnect can also be measured in the time domain by observing the received differential signal. This is the TDD21 term of the T-parameters. *Figure 2.25* shows an example of the measured TDD21 signal coming out of the same 40-inch backplane trace that was shown previously in the frequency domain.

Figure 2.25: *Time Delay of One-Meter-Long Backplane Channel*

In this example, the bandwidth of the measurement is 20 GHz and the setting for the time domain window was nominal. This means the effective rise time of the incident signal is 0.7/20 GHz, or 35 ps. On this scale of 200 ps per division, the input stimulus would have a 10–90 rise time of less than 0.2 divisions. Clearly the rise time of the leading edge of the output signal has increased considerably. This is due mostly to the attenuation, and to a lesser amount, the group velocity dispersion created by the multiple reflections.

Where do we draw the line to say the time delay of the signal is some value? Which part of the rising edge do we use to measure the time delay? In the frequency domain, we were able to get an average value of 6.8 ns, even though there was as much as ± 0.5 ns of noise. In the time domain, what part of the wave do we use to measure as a reference to arrive at one value for the delay of the signal?

This is why the delay of a signal is only a rough metric of performance. The delay will depend on where we define the reference level of the received signal. This will vary depending on the pattern of bits in the stream and gives rise to a form of jitter classified as deterministic jitter.

As a rough approximation to the delay, we can use the midpoint of the signal as the reference level. This is where the received signal will cross the 50 percent voltage point. With a 400 mV incident signal, this threshold is 200 mV. Using markers on the front screen, we can directly read the time delay to the 50 percent point as 6.84 ns. This is very close to our estimate of 6.8 ns from the frequency domain.

The previous example determined the time delay associated with the differential signal. For high-speed serial interconnects, the differential signal is the only component that the receiver is sensitive to. However, another term that also characterizes the interconnect is the time delay for the common signal component.

While this is typically an unimportant term, it is trivial to measure and might sometimes offer insight into the interconnect with a quick look. In a uniform differential pair with homogenous dielectric, the common signal and differential signal will see the same dielectric distribution, and hence the signals will travel at the same speed and have the same group delay.

Figure 2.26 shows the group delay of the differential and common signal as having exactly the same delay. With a few minor variations due to the different impedance profiles and different multiple reflections, the general features of the group delay are identical for the differential and common signals, as expected for stripline interconnects.

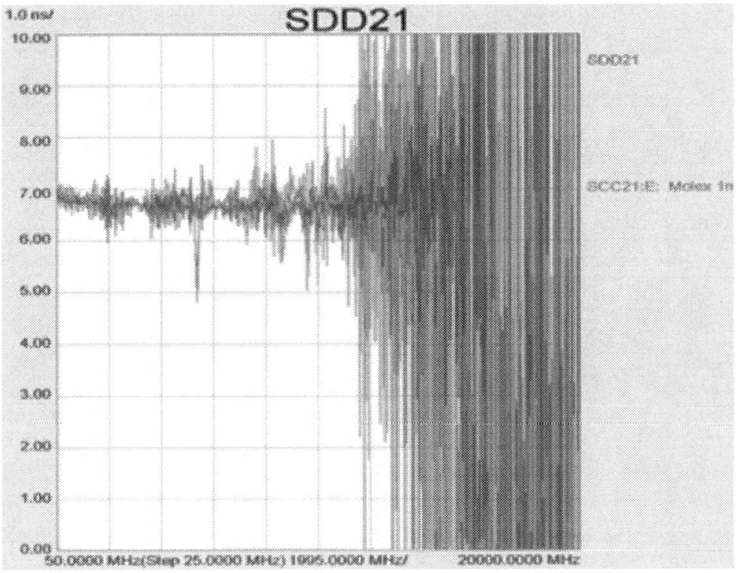

Figure 2.26: *Group Delay of Differential and Common Signal*

In the time domain, the received signals will appear at the receiver at roughly the same time as well. In a homogenous material, the two signals are impacted exactly the same way by the dielectric.

Figure 2.27 shows the measured received signal for the differential and common signal in the time domain. This is the TDD21 and TCC21 terms in the differential T matrix elements. On this scale of 200 ps per division, the two received waveforms are virtually identical.

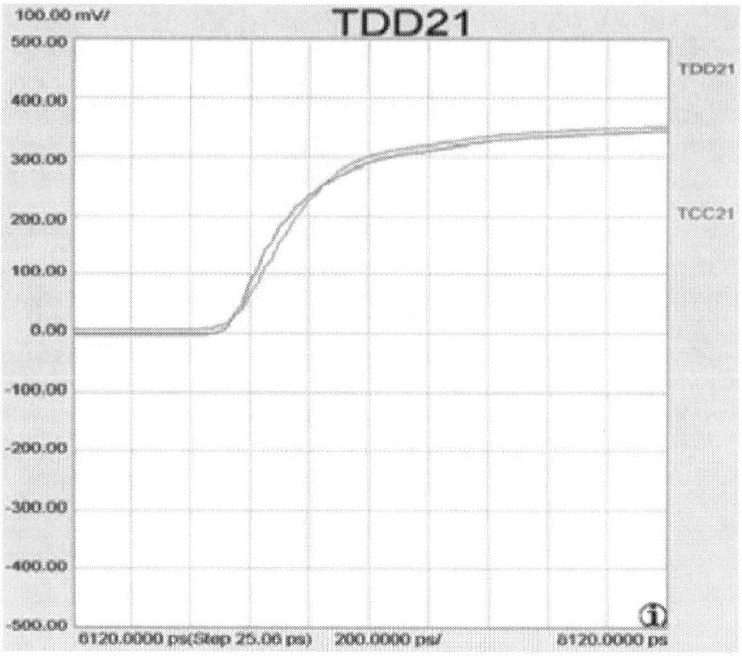

Figure 2.27: *Time Delay of the Received Differential and Common Signals*

However, if there were any asymmetry in the distribution of dielectric materials, so that the electric fields associated with the common signal saw a different effective dielectric constant than for the differential signal, there would be a difference in the group velocity and time delay for the two types of signals.

For a microstrip differential pair, the differential signal will have more field lines in air than the common signal. This will give the differential signal a lower effective dielectric constant, a higher speed, and a lower group delay, compared to the common signal.

Figure 2.28 shows the measured group delay of the common signal transmitted through the four-inch microstrip differential pair, as SCC21, and the group delay of the differential signal, SDD21. We see the general features of the noise from the non-100 Ohm differential impedance and the non–25 Ohm common impedance. On top of this is the clear offset between the average group delay of the common signal at about 680 ps and the group delay of the differential signal of about 600 ps.

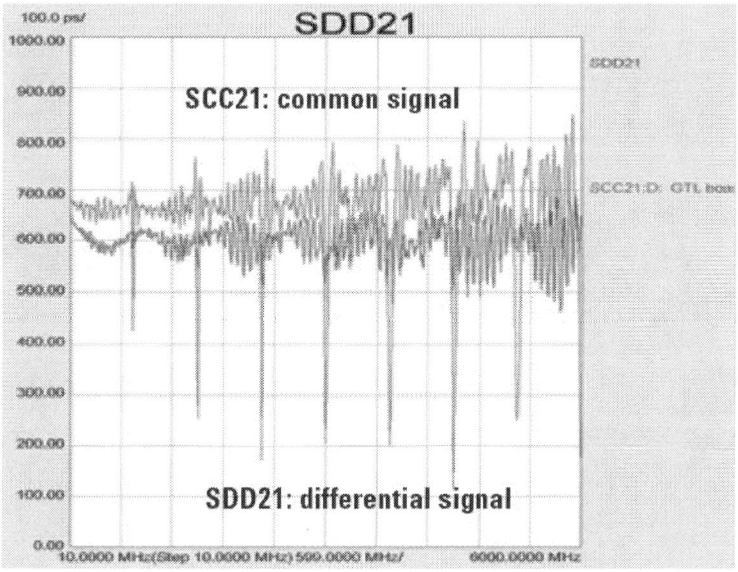

Figure 2.28: *Group Delay in a Microstrip Differential Pair*

Based on these delays and the four-inch-long interconnect, we can estimate the speed of the common signal as 4 in/0.68 ns = 5.9 in/ns, while the speed of the differential signal is 4 in/0.6 ns = 6.7 in/ns. While this difference does not affect the differential signal, it is the effect that ultimately gives rise to far-end noise when this differential pair is considered as two single-ended transmission lines with crosstalk.

Viewed in the time domain, the different arrival times of the common signal and the differential signal in a microstrip is very clear. *Figure 2.29* shows the TDD21 term and the TCC21 term for the four-inch-long microstrip.

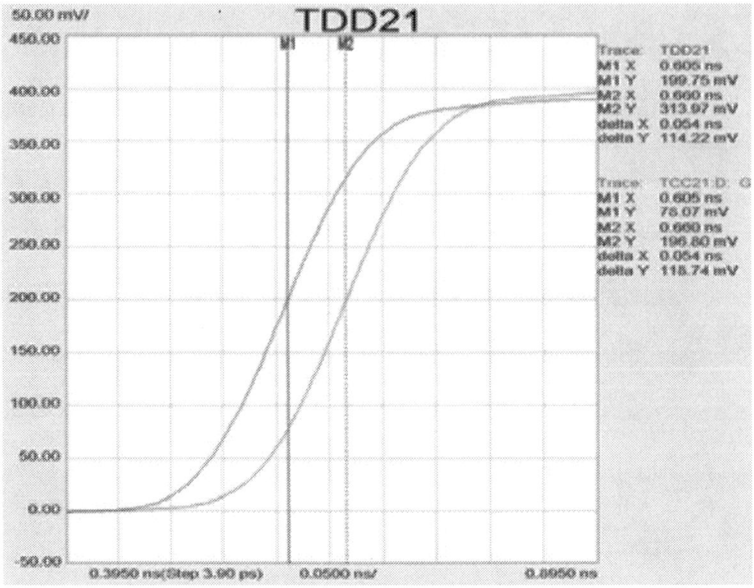

Figure 2.29: *Common and Differential Time Delay in Microstrip*

This time delay difference can be read off the screen using the markers as 54 ps, out of a total delay of 600 ps.

The Bandwidth of an Interconnect

The most important role of an interconnect is to transmit a signal from one point to another with acceptable distortion. It is impossible to transmit a signal with no distortion, so it is often a question of how much is too much.

One metric of the distortion imposed by the interconnect is the attenuation of the signal, and how much amplitude is left coming out of the interconnect. Because the attenuation is different at different frequencies, it is often easier to evaluate the attenuation in the frequency domain. The term that most effectively characterizes the signal degradation in transmission through the interconnect is the SDD21 term. This is also called the differential insertion loss.

Figure 2.30 shows the measured differential insertion loss of a 22-inch channel on a motherboard up to 20 GHz. This defines the behavior of the interconnect. This plot also indicates the noise floor of the measurement as about –70 dB.

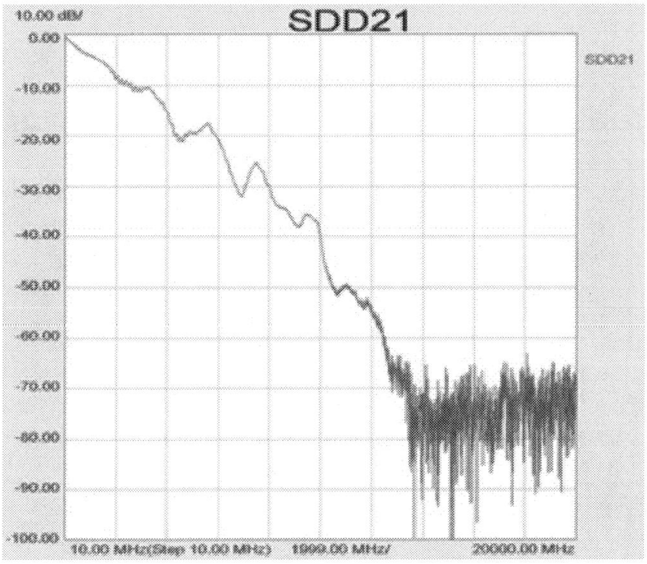

Figure 2.30: *Differential Insertion Loss in 22-Inch Motherboard Differential Channel*

From this insertion loss, we can estimate the highest usable frequency, or the bandwidth of the interconnect. To do this, we need to know how much attenuation is acceptable. This depends on the type of drivers and receivers used in the application. These devices are typically called serializer-deserializer (SerDes) chips.

A low-end driver might work with a differential insertion loss of –10 dB. A mid-range SerDes with some pre-emphasis capability might require at least –20 dB, while a high-end device, with integrated pre-emphasis and equalization, might allow as much as –30 dB. Which device family is used will ultimately determine the usable bandwidth of the interconnect. This is why it is conventional, when referring to the bandwidth of the interconnect, to refer to the –10 dB or –20 dB or the –30 dB bandwidth of the interconnect.

In this example of the 22-inch-long motherboard trace, the –10 dB bandwidth is 2 GHz, while the –20 dB bandwidth is about 4 GHz and the –30 dB bandwidth is 7 GHz.

There are three critical terms that influence the differential insertion loss of a differential channel: the length, the dissipation factor of the laminate, and the presence of impedance discontinuities.

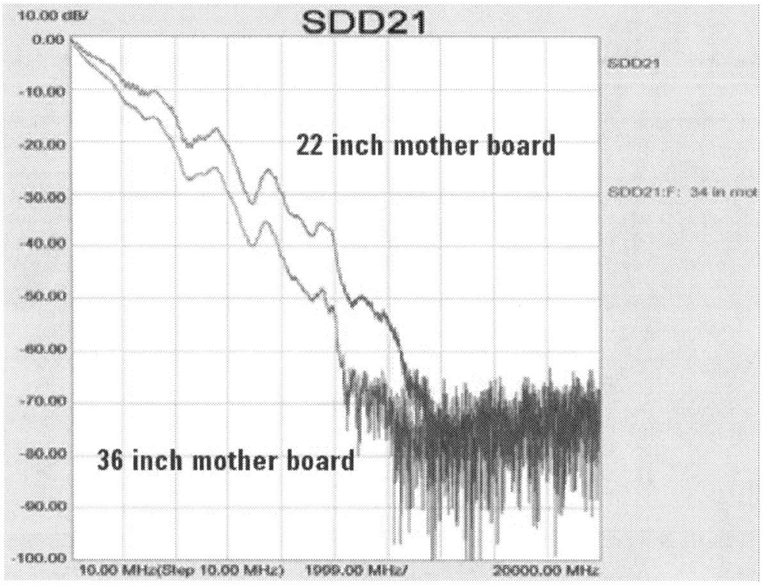

Figure 2.31: *Differential Insertion Loss of Two Length Motherboard Traces*

Usually, there is little that can be done in the design to change the length. This is fixed by the system architecture selected. All things being equal, a longer length interconnect will result in higher insertion loss and lower bandwidth. *Figure 2.31* shows an example of two differential channels on the same motherboard, using the same daughtercards, but with total lengths that are 22 inches and 36 inches. The drop in –20 dB insertion loss is not 60 percent lower in the longer interconnect compared to the shorter one. It is only about 10 percent lower. This is because a large fraction of the insertion loss is due also to the impedance discontinuities.

Rise-Time Degradation

The dropoff in insertion loss with frequency is a direct measure of the higher attenuation seen by the higher-frequency components. The effective rise time of the signal incident to the device is 35 ps for a measurement bandwidth of 20 GHz and time domain window setting of nominal. If this rise time came out, the insertion loss would be above –3 dB all the way through to almost 20 GHz. The losses in the interconnect remove the highest-frequency components of the signal and decrease the bandwidth of the signal.

By the time the signal comes out of the interconnect, the –3 dB frequency has shifted from 20 GHz to closer to 1 or 2 GHz. This means the rise time of the signal will be significantly increased from 35 ps to much higher, into the 200 to 500 ps range.

Figure 2.32 shows an example of the transmitted differential signal in the time domain, the TDD21 term, on a scale of 100 ps per division. When the edge is no longer close to a Gaussian shape, it is difficult to use one number to describe the rise time. The 10–90 rise time has little significance since the tail is so long. The 20–80 rise time or the time to reach the 50 percent point might have more meaningful significance, though both values are only rough approximations to the actual behavior of the edge.

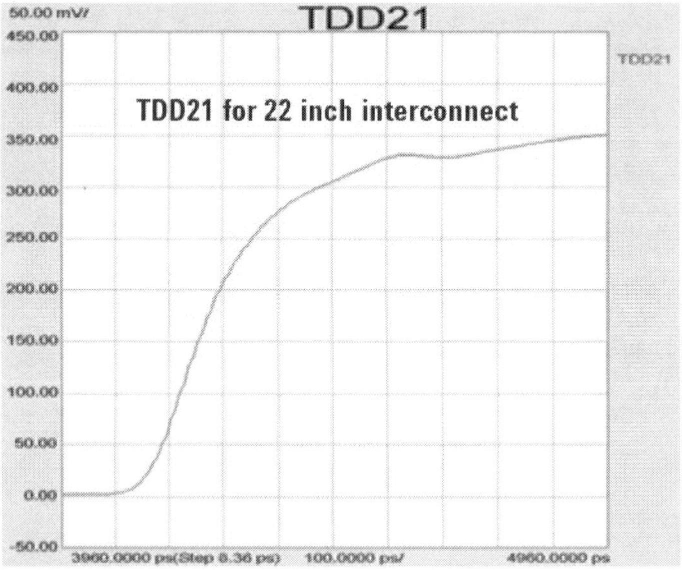

Figure 2.32: *Rise-Time Degradation of Transmitted Signal through 22-Inch-Long Motherboard*

What is more significant is how far the signal rises during a bit period. This will strongly influence the amount of inter-symbol interference (ISI) and collapse of the eye to be expected. For example, a 1 Gbps bit stream will have a bit period of 1 ns. From the measured TDD21 display in *Figure 2.32*, we see that in 1 ns, the received signal will reach more than 85 percent of its final value in one bit period. There will be virtually no ISI and the bit quality should be excellent.

A 5 Gbps signal will have a bit period of 200 ps. In this short a time, the final signal will reach only 250 out of 400 mV, or 62 percent of the final value. This is 50 mV above the midpoint voltage. If the bit pattern had been all highs for a long period, the next low bit would extend only 50 mV below the midpoint. The combination means the maximum eye opening we would expect to see is 100 mV. This is probably below the noise margin of most receivers. This interconnect would have a problem supporting a 5 Gbps bit stream.

As was apparent looking at the SDD21 differential insertion loss term, all things being equal, a longer interconnect means a higher insertion loss. This will also result in a longer rise time of the transmitted signal and offer worse high–bit-rate performance.

Shown in *Figure 2.33* is the measured signal at port 2 for a 22-inch-long interconnect with a 36-inch-long interconnect superimposed to begin at the same location. The much longer rise time, on the order of 300 ps, for the longer interconnect is all due to the dropoff of the insertion loss from dielectric loss and impedance mismatches. The longer rise time will have a bigger impact on high–bit-rate signals than the shorter rise time signal.

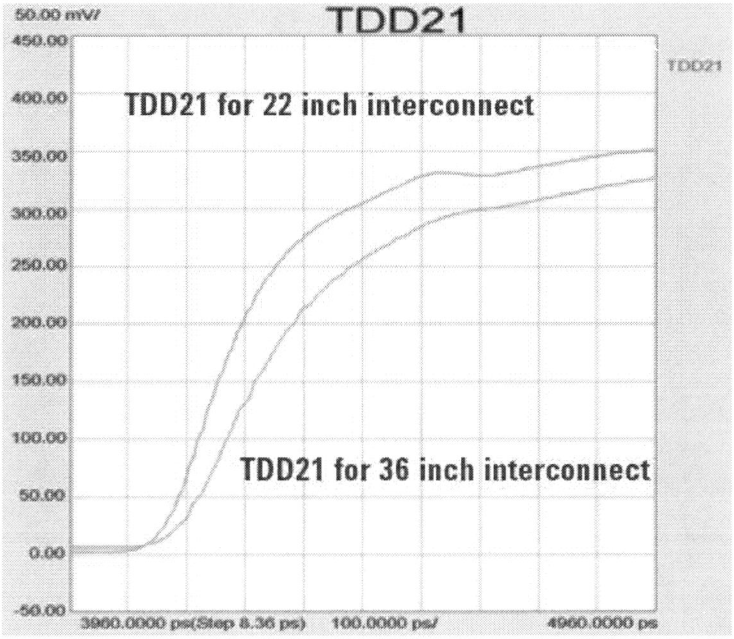

Figure 2.33: *Rise-Time Degradation of Transmitted Signal after 22 and 36 Inches*

A signal with a rise time of 300 ps has an equivalent bandwidth of roughly 0.35/0.3 ns or about 1 GHz. This is a pretty good estimate of the –3 dB insertion loss of the interconnect as seen in *Figure 2.32*. Of course, the concept of bandwidth is inherently an approximation. If knowing the bandwidth to 10 percent accuracy is important, one should not use the concept of bandwidth, but the entire spectrum of the signal.

Eye Diagrams

One of the most important ways of evaluating the performance of a high-speed serial signal is converting the data stream into an eye diagram. A bit stream is a series of high and low signals, synchronous with a clock. Using the clock as the trigger, each bit is extracted from the stream and superimposed. The resulting combination of all possible bit patterns looks a little like a human eye and has been called an eye diagram. An example of an eye diagram is shown in *Figure 2.34*. This is created from the measured TDD21 element of a 34-inch-long motherboard interconnect.

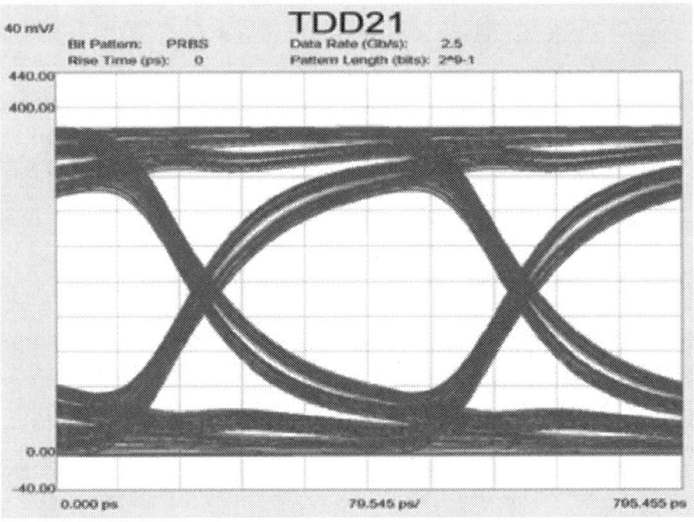

Figure 2.34: *Eye Diagram of 34-Inch-Long Motherboard Interconnect*

The two most important features of an eye diagram are the height of the opening on the vertical scale and the width of the cross-over regions on the horizontal scale. Depending on the noise margin of the receiver, the opening of the eye must be at least 150 to 200 mV, while the cross-over widths should be only a small fraction of the period. This width is often called the deterministic jitter. These two terms fundamentally limit the highest bit rate that can be transmitted down an interconnect.

The eye diagram of a pseudo-random bit stream (PRBS) can be simulated for an interconnect based on the measured S-parameter behavior model. PLTS can synthesize a PRBS signal and simulate the impact on this signal from the interconnect and display the output signal in the form of an eye diagram.

In generating the PRBS signal, one of the parameters is the number of bits that should appear in the signal before it repeats, or the word length. In principle, it could be infinite, but there is always a tradeoff between how accurate the answer needs to be and how long to run the simulation.

Figure 2.35 illustrates the difference in the simulated eye diagram for a 2.5 Gbps PRBS signal, with 27 – 1, 29 – 1 and 211 – 1 as the word length. Based on this analysis, as a good rule of thumb, 29 – 1 bits in the pattern is a good value to start with and take quick looks, while a final simulation might be done with 211 – 1 bits. The computation time is only a few minutes for most situations.

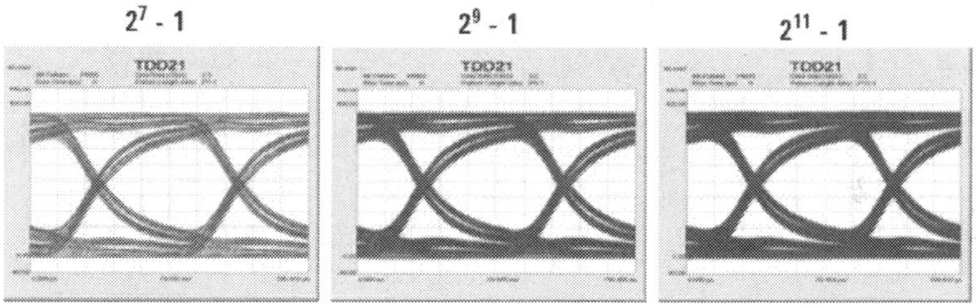

Figure 2.35: *Eye Diagram and PRBS Word Length*

For the same 36-inch-long motherboard interconnect, the eye diagram for different bit rates can be simulated to identify the performance of the interconnect. It is important to note that in this simulation, there is no pre-emphasis or equalization of the signal. It is only a simulation of the eye diagram performance of the interconnect, assuming a PRBS bit stream is incident with a rise time, based on the measurement bandwidth. For 20 GHz, it is roughly 35 ps. The rise time is dramatically increased by the interconnect and contributes to the deterministic jitter and the collapse of the eye diagram.

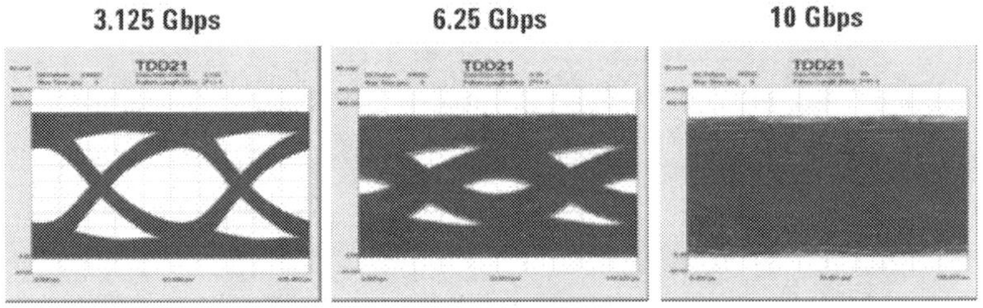

Figure 2.36: *Bit-Rate Collapse of the Eye*

This particular interconnect would be perfectly suitable for XAUI–type signals at 3.125 Gbps, but would probably not work for XAUI 2 interfaces at 6.25 Gbps. As can be seen from *Figure 2.36*, it is completely unusable for 10 Gbps signals unless pre-emphasis or equalization was used in the SerDes chips.

The differential insertion loss is a good indication, but not a total indicator, of the collapse of the eye diagram. The larger the insertion loss, the more the eye will be collapsed. However, it does not directly indicate the impact from the deterministic jitter that will close the eye in from the sides.

Figure 2.37 shows an example of the differential insertion loss of three differential channels in various backplanes. The lowest loss is for an interconnect eight inches long, the next greater loss is from a 25-inch interconnect, and the highest insertion loss is from a 40-inch-long backplane.

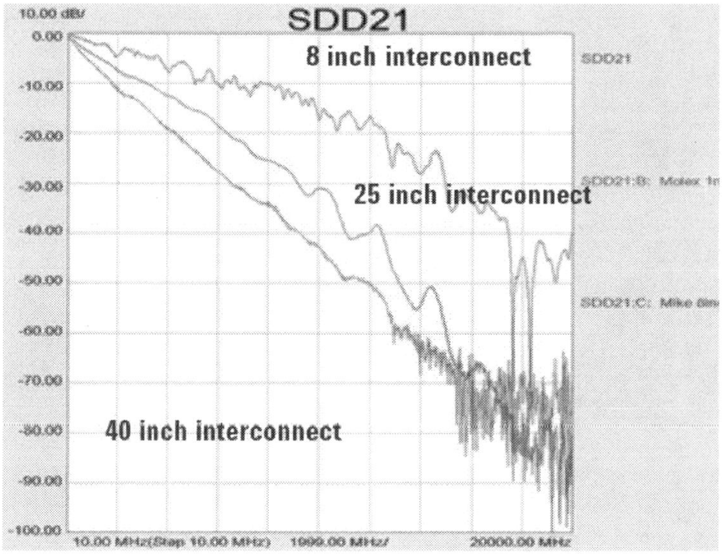

Figure 2.37: Insertion Loss of Three Length Interconnects

By taking the measured S-parameter data from these three interconnects, the PRBS eye diagram can be simulated for a bit rate of 6.25 Gbps. This corresponds to the increasingly popular XAUI 2 interface. It is clear from *Figure 2.38* that at this bit rate, the 40-inch backplane has no hope of supporting XAUI 2 without the use of SerDes features such as pre-emphasis and equalization.

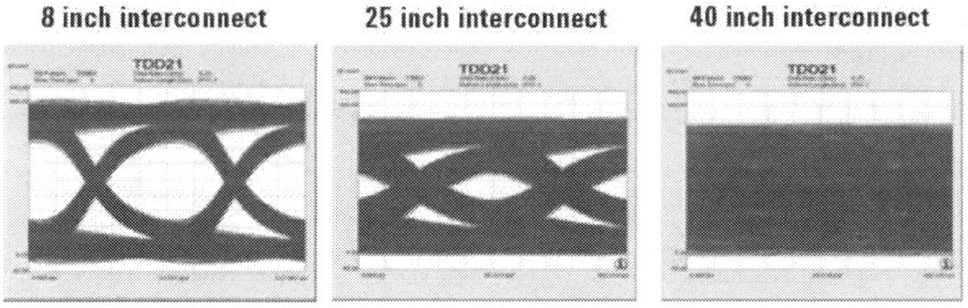

Figure 2.38: Eye Diagrams for PRBS Patterns at 6.25 Gbps

In general, the shorter the interconnect and the lower the insertion loss, the higher the bit rate that can be supported. Even with the insertion loss measurement, there is no substitute to simulating the eye diagram itself.

Impulse Response and Pre-Emphasis Taps

Typical backplanes can exceed 40 inches. This includes six inches on both daughtercards and 36 inches on the backplane itself, for a total of 48 inches. As illustrated earlier, even a 40-inch interconnect can have 100 percent eye closure at 6.25 Gbps. This would seem to limit the use of FR4–based backplanes to applications less than 6.25 Gbps.

One solution is to use lower loss laminates. This will increase the cost of the backplane but may also increase the usable bit rate by 50 percent, depending on the material selected. Another popular method of increasing the bit rate while still using a low-cost laminate is by using pre-emphasis in the SerDes driver. This method adds extra high-frequency components to the signal launched into the interconnect. It is implemented by adding an extra amplitude to the bits based on the specific bit pattern in the signal.

In some versions of SerDes drivers, pre-emphasis is added to not just the first bit in the sequence, but to the next one, two, or three bits, either as positive or negative signals to compensate for multiple reflected signals. Using the TDD21 signal, displayed as an impulse response, we can estimate which consecutive bits should have additional signals added or subtracted to them to compensate for the interconnect.

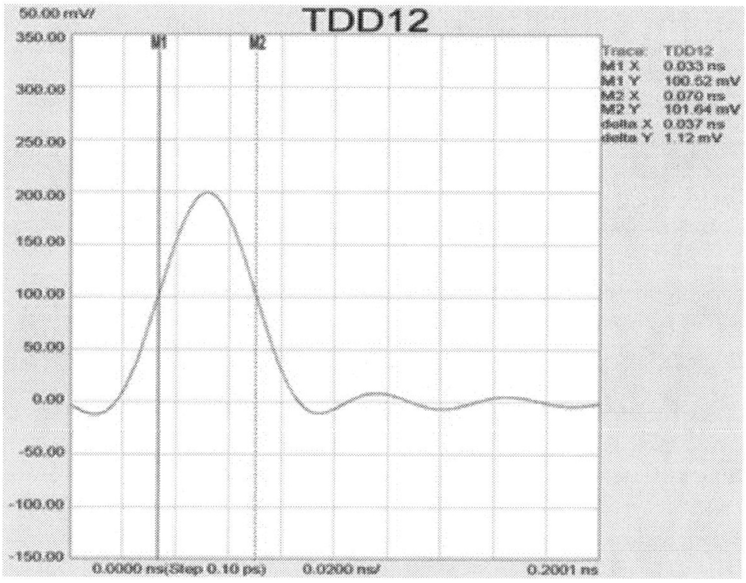

Figure 2.39: *Impulse Response of the Incident Signal, Time Domain Window Set to Nominal*

The time domain response of the interconnect can be simulated as either a step response or an impulse response. The frequency components of a step response drop off like those of a square wave, inversely with frequency. The frequency components of an impulse response are constant with frequency. The full width, half maximum (FWHM) width of the impulse response is the 10–90 rise time of the incident signal. For a 20 GHz measurement bandwidth, the impulse response of the incident signal into the interconnect is about 35 ps. *Figure 2.39* shows an example of the incident signal impulse response having a FWHM of 35 ps.

Of course, due to the losses and impedance discontinuities in the interconnect, this 35 ps wide impulse signal quickly spreads out by the time it exits the interconnect. Shown in *Figure 2.40* is an example of the output impulse response from traveling through 36 inches of a motherboard, on a scale of 200 ps per division. The incident signal would be less than a fifth of a division wide on this scale.

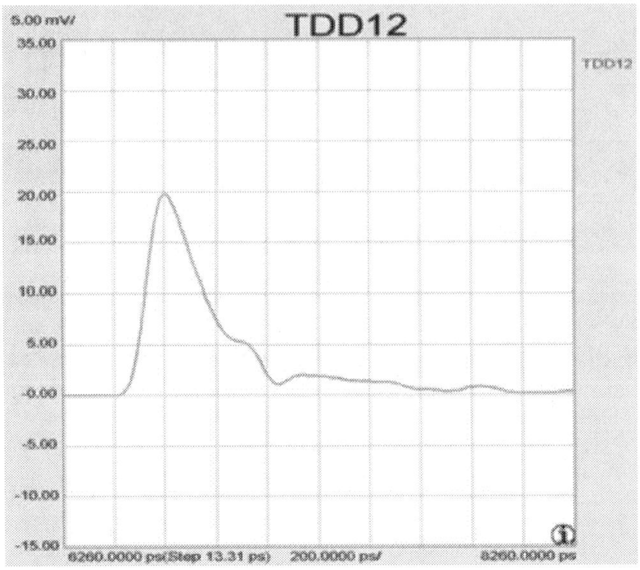

Figure 2.40: *Impulse Response after 36 Inches through a Motherboard*

In addition to a spreading out of the pulse, there are multiple steps appearing after the main peak, due to the multiple bounces of the signal against impedance discontinuities. If the bit rate were 5 Gbps, the bit period would be 200 ps. This profile of the impulse response is a rough measure of what one bit of data would look like at the far end of the line.

If this response is known ahead of time, the data stream can be modified to minimize the amount of crosstalk between successive bits, or ISI, generated due to the multiple reflections. Different SerDes technologies allow adding or subtracting voltage levels to successive bits in the series. Each successive bit is called a tap, and three taps is the typical limit.

For example, if the first bit has pre-emphasis added to it, we would want to subtract about 20 percent of the signal amplitude to the first tap, the second bit, nothing on the second tap, and subtract possibly 10 percent of the signal to the third tap. With this sort of signal, the pulse propagating through the interconnect will have components that help to cancel out the multiple reflections. Using the impulse response of the transmitted differential signal can help guide the design process to quickly reach the optimum pre-emphasis and tap pattern for minimum ISI and maximum opening of the eye.

Mode Conversion and EMI

One of the most difficult problems to fix in high-speed product design is electromagnetic interference (EMI). The largest source of EMI is radiation from common currents that get out on unshielded twisted pair cables, such as cat5 cables.

Normally, the signals launched on twisted-pair cables are supposed to be differential signals. A pure differential signal on an unshielded twisted pair will not radiate very much at all. It poses no problems for EMI. It is when unwanted common signals get into the twisted-pair cable that radiated emissions can happen. As a rough rule of thumb, the radiated field strength, at the three-meter distance in an FCC class B type open field test, from common current on a twisted pair, is about 40 mV/m x V x f, with V the voltage of the common signal and f the frequency in GHz. The typical Federal Communications Commission (FCC) certification failure threshold near 1 GHz is a field in excess of 0.4 mV/m. This suggests that to pass an FCC certification test, the maximum allowable common signal on an external twisted pair should be less than 10 mV at 1 GHz.

In principle, if the drivers produce a perfect differential signal with no common signal and it passes through an ideal differential pair, there should be no common signal generated. In practice, any asymmetry in the interconnect, such as non-equal line widths in the pair, different lengths in the two lines, or different local effective dielectric constant due to the glass weave of the laminate, will convert some of the differential signal into common signal. We call this process mode conversion.

As long as none of this common signal gets out of the box, it will not affect EMI. Of course, if the mode conversion is significant, it may affect the quality of the differential signal's edge, which will have an impact on the eye diagram. But, the common signal will do no harm. It is only if some of this converted common signal gets out on twisted pairs that an EMI problem might arise.

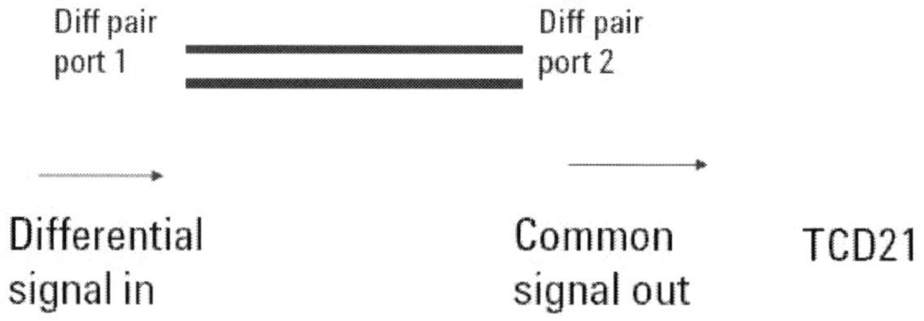

Figure 2.41: *Mode Conversion*

The amount of common signal converted from the differential signal by asymmetries in an interconnect can be measured directly by one of the differential T-parameters. This is illustrated in *Figure 2.41*. Mode conversion is measured by sending a differential signal in at port 1 and looking at how much common signal comes out of port 2. This is the TCD21 term. Measuring the magnitude of TCD21, compared with the 400 mV incident signal, is a measure of the common signal converted.

For the case of a 20-inch backplane, *Figure 2.42* shows the measured TCD21 signal. This is the signature of the common signal that comes out of the interconnect, with a pure 400 mV differential signal step edge going in. This suggests about 1.2 percent of the differential signal is converted into common signal.

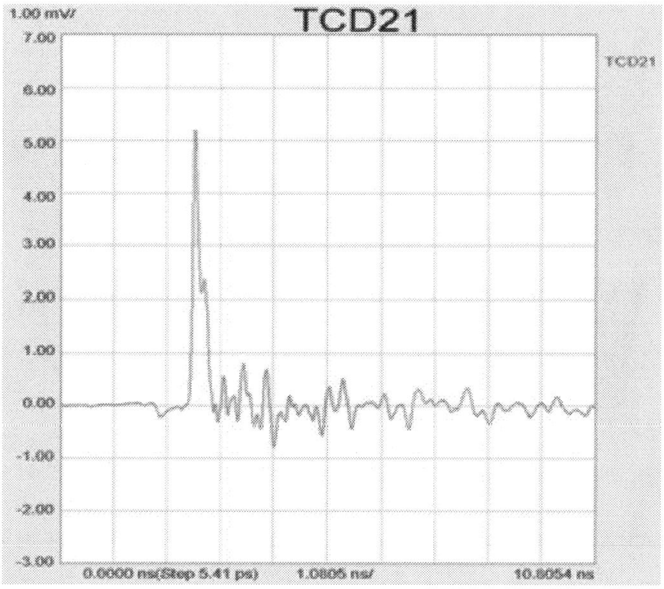

Figure 2.42: *TCD21 from 20-Inch Backplane*

Every bit edge that gets transmitted to external cat5 cable will have 1.2 percent of the differential signal as common signal and contribute to EMI. If the incident voltage is 1 V, approximately 12 mV of common signal may get on the external twisted pair. By itself, this is close to the threshold that would fail FCC certification. In addition, any skew between the drivers will also convert the differential signal into common signal and add to the radiated emissions. This amount of mode-converted common signal might cause a problem.

The first step to solve any design problem is to understand the root cause and optimize the design to fix this problem. To fix this problem, we would like to find out where in the interconnect path is the asymmetry that might be generating this common signal. To find this, we can take advantage of another one of the differential T-parameters.

As illustrated in *Figure 2.43*, if we send a differential signal into port 1, we can look at the converted common signal that comes back out of port 1 as TCD11. The key feature of this signal that enables us to use it to debug the root cause of the mode conversion is its time dependency.

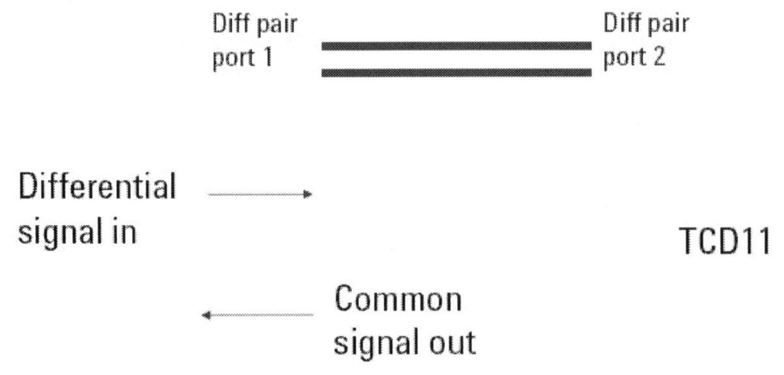

Figure 2.43: *Mode Conversion*

We are effectively launching a step edge into the interconnect. As this edge propagates down the differential pair and encounters any asymmetry, a common signal will be generated. Some of this common signal will propagate in the forward direction, and will be picked up at port 2 as the TCD21 signal, but at the same time, some of this common signal will be sent in the backward direction, back toward port 1. This is the TCD11 signal.

The time between sending the signal into port 1 and picking up the common signal that comes back is the round-trip time of flight for the incident differential signal to reach the asymmetry and then for the common signal to travel back to port 1. By comparing the time response of the TCD11 signal to the TDD11 signal, we can look for what features of the interconnect that we can identify in the TDD11 signal are coincident with the TCD11 signal.

In *Figure 2.44*, the bottom trace is the TCD11 response. The top trace is the TDD11 response. In this signal, we can identify the negative dips from the via fields of the connectors in the daughtercard and the motherboard. The region between them is the connector.

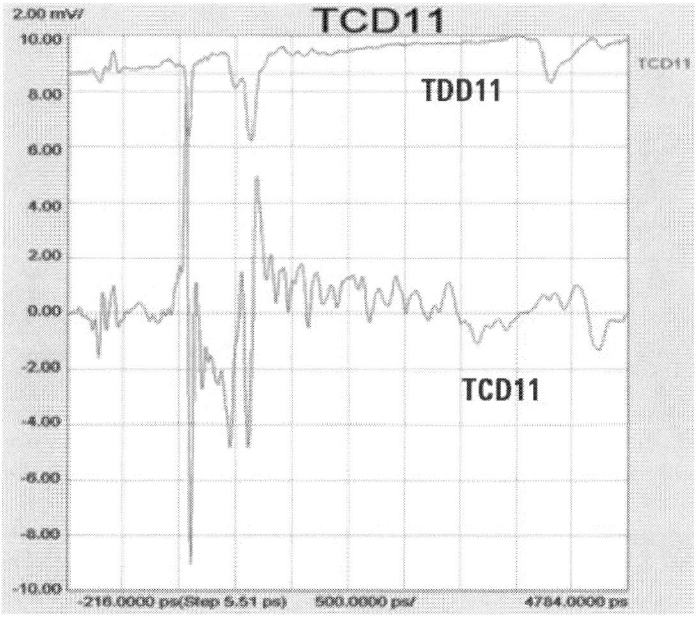

Figure 2.44: *20-Inch Motherboard TDD11 Response*

The TCD11 response shows a large peak in the generated common signal coincident with the via field in the daughtercard and a smaller one in the motherboard side. In addition, a negative common signal is generated and sent back to port 1 by the connector itself. This suggests that an improvement could be made by minimizing the capacitive discontinuities of the connector attach region and adjusting the connector design to be more symmetrical.

Most differential channel interconnects composed of daughter cards and backplanes, will show the dominant source of mode conversion to be in the via field of the connectors. *Figure 2.45* shows another example of a 25-inch backplane interconnect. The common voltage signal as measured by TCD21 is almost 8 out of 400 mV, or 2 percent.

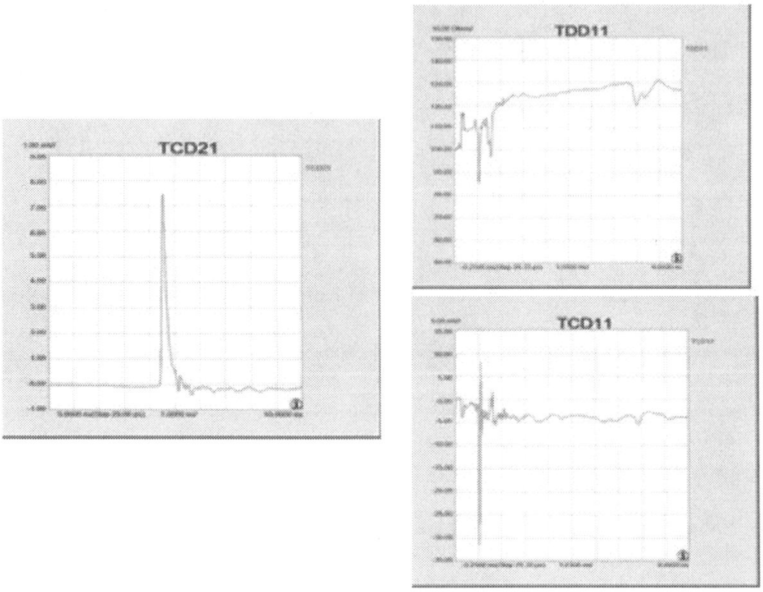

Figure 2.45: *25-Inch Backplane Example*

The comparison of TDD11 and TCD11 shows the source of the common current to be the via field in the connectors. Unless they are optimized to minimize the impedance discontinuity of the excess capacitance of the vias and pads, there will always be asymmetries in the signal-return current flow. These asymmetries will convert the differential signals into common signals. By minimizing the impedance discontinuity by backdrilling the via stub, for example, the bandwidth of the interconnect will be increased and the converted common signal will be decreased.

Modeling Differential Channel Interconnects

The S-parameters of an interconnect, whether measured in the time or the frequency domain, represent a behavioral model of the interconnect. They contain all the information about how a signal entering one port will behave when it exits another port. Depending on the question we are asking about the interconnect, one or more of the S- or T-parameters, either as single-ended or balanced, might get us close to the answer.

In those cases where more detailed information is needed, we can use the exported S-parameters as a behavioral model and integrate them directly into some circuit simulators. This would allow us to perform a system-level simulation of the behavior of the drivers, interconnect, and receivers.

Not all EDA tools allow the use of S-parameter behavioral models. Instead, a commonly used model format to approximate a real differential pair is as two lossy, single-ended, coupled transmission lines. This model, sometimes called a W element model, as it is referred to by HSPICE, describes a pair of coupled transmission lines in terms of their RLCG (resistance, inductance, capacitance, conductance) elements. This model assumes a uniform, coupled, lossy, pair of lines. The distributed elements are defined as their per unit length values. The default units are Ohms/m, H/m, F/m, and S/m.

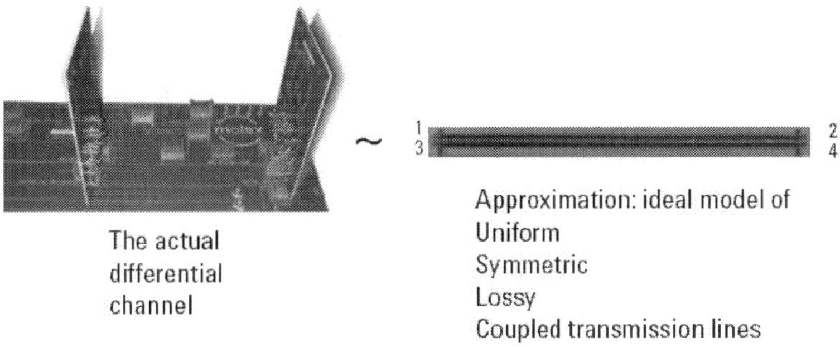

Figure 2.46: *Modeling a Real Interconnect as an Ideal, Uniform Interconnect*

It is important to keep in mind, as illustrated in *Figure 2.46*, that what we are doing is taking the real, measured S-parameter behavior of the differential channel and approximating it as a single, uniform pair of coupled lines. This is not always a good assumption but can sometimes help get us a satisfactory answer quickly.

The W element model can be generated and exported from the S-parameters with one click of the mouse. As illustrated in *Figure*

2.47, there are a few intermediate steps that might sometimes offer useful information.

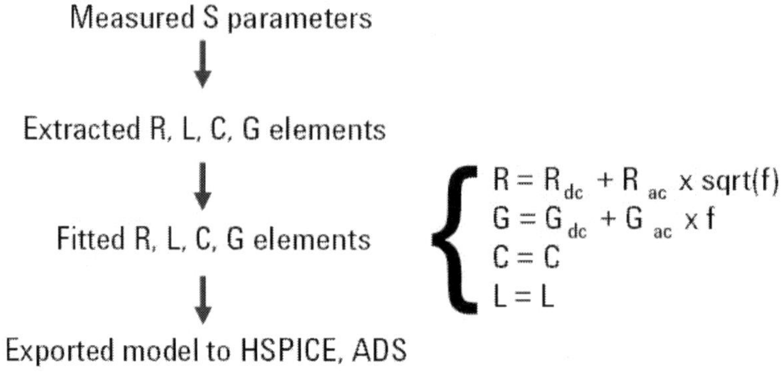

Figure 2.47: *Modeling Process*

From the measured S-parameters, an ideal, uniform, symmetric, pair of coupled transmission lines is used to analytically extract the model parameters of the ideal line. These parameters can be combined transparently back and forth into a number of formats to describe either a pair of single-ended coupled lines or one differential pair. However, in all cases, the model is simplified to assume the lines are perfectly symmetric. This means mode conversion cannot be simulated by this model.

From the extracted circuit element terms, a simple frequency-dependent model is fit for each of the elements based on the assumption of skin depth–limited currents and constant dissipation factor material. This results in the resistance varying like a constant term, plus a term that increases with the square root of frequency, and the conductance being a constant term, and a term that increases proportional to frequency. The capacitance per length and inductance per length are both assumed to be constant with frequency.

It is these terms, as the single-ended and coupled terms, that are exported as the W element. In addition to approximating the real differential channel interconnect as a uniform differential pair with no asymmetry, the additional assumption to generate the W element is the simple frequency dependence, as shown in *Figure 2.47*. In fact, these are often very good assumptions for most real backplane interconnects.

The measured S-parameters are used to extract the line parameters for an ideal, uniform, symmetrical pair of coupled transmission lines. This model can be used to extract a description in terms of the differential or common signal behavior of the ideal transmission line. The assumption made is that the line being measured is uniform.

Figure 2.48 shows the parameters extracted for the case of the differential signal behavior. Of these, the term that has the most value is the real part of the complex impedance. This term is a direct indication of the average differential impedance of the trace. In this example of a uniform differential pair, the extracted differential impedance is seen to be very constant with frequency, up to the full 6 GHz of the measurement. We can read right off the screen that the equivalent differential impedance of this line is 77 Ohms.

It is no coincidence that when this same line is displayed as the TDD11 element, on an impedance scale, that we also measured a uniform differential impedance of 77 Ohms.

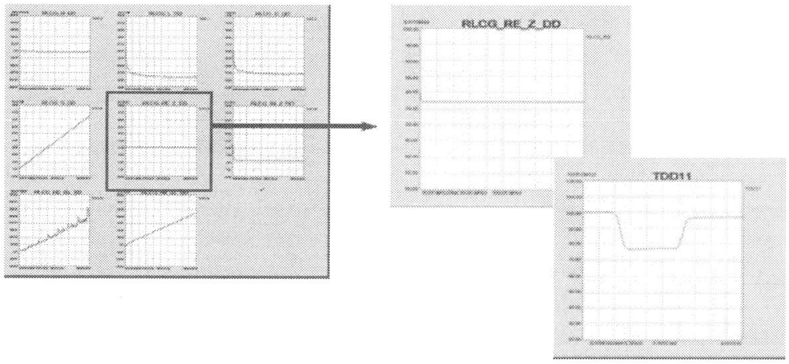

Figure 2.48: *Uniform Four-Inch-Long Differential Pair*

When the interconnect is not a uniform differential channel, the extracted real part of the complex differential impedance term can be used to fit, directly from the screen, the effective, average differential impedance.

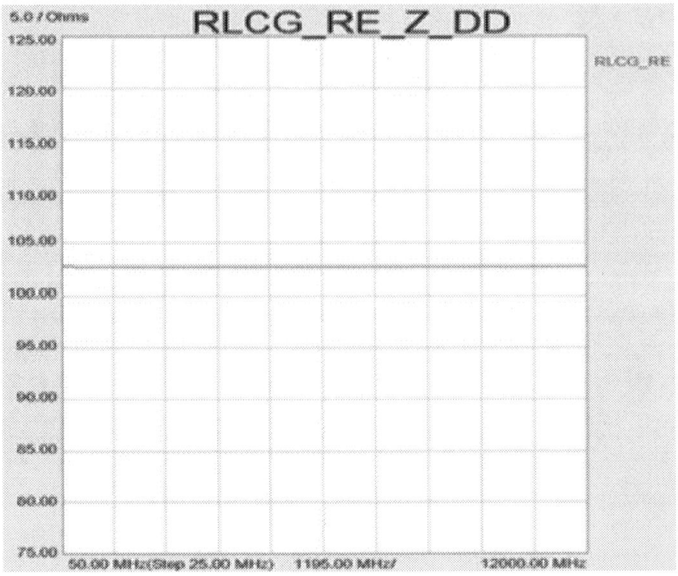

Figure 2.49: *Extracted Differential Impedance of 40-Inch-Long Backplane Interconnects*

Figure 2.49 is an example of the differential impedance extracted from a 40-inch-long backplane. The average differential impedance is about 102 Ohms, very close to the target of 100 Ohms. However, this includes the effects of the daughtercards and connectors in addition to the long trace on the motherboard. From the extracted values of the line parameters, the W element terms are fitted. By comparing the actual extracted parameters based on modeling the real interconnect in terms of a uniform, symmetrical, coupled pair of transmission lines, we can get an estimate of how well the interconnect obeys the frequency dependent description used by the W element.

In the ideal model, the R11 term, often referred to as the self-resistance, is the resistance per length of one of the lines that makes up the differential pair. Both lines are assumed to be identical as part of the approximation. The self-resistance is extracted from the measured S-parameters, based on assuming a uniform pair of coupled transmission lines. It is extracted at each frequency value. If the interconnect were really a uniform line, the self-resistance would increase smoothly with frequency. The non-uniformities such as connectors, vias, and different traces on the daughtercard and motherboard give rise to the jaggedness of the extracted value of the self-resistance.

As shown in *Figure 2.50*, the extracted self-resistance shows some frequency dependence. For this 40-inch backplane, it has a very low resistance, at close to 5 Ohms/m at the lowest frequency, but steadily increases with frequency through 12 GHz, the limit to the display. This increase in resistance can be explained if we assume the current is skin depth–limited. In this case, we would expect the resistance to behave as the W element model being a constant term plus a term that increases with the square root of frequency.

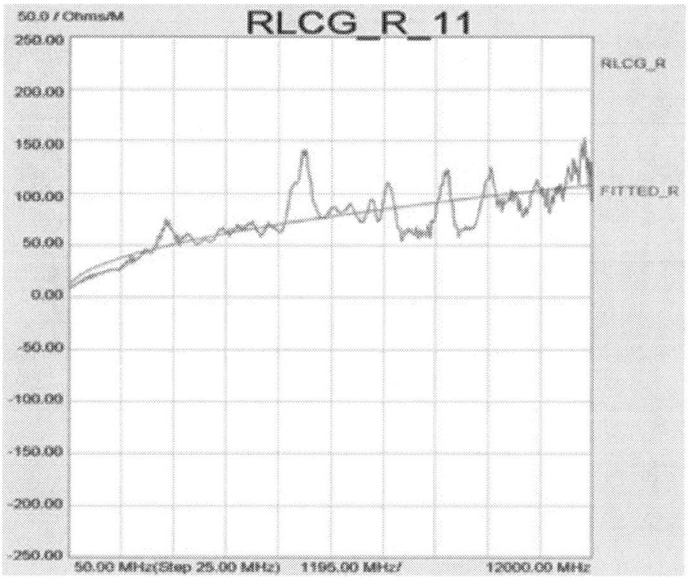

Figure 2.50: *Extracted Series Resistance per Length of either Line in the 40-Inch Backplane Interconnect Compared with the W Element Model for Resistance*

When we use this model and find the best values of the DC term and the AC term that fits the data, we get a resistance behavior, shown in *Figure 2.50* as the smooth curve. This model is a very good approximation to the extracted resistance. This suggests that the series resistance of the actual differential channel really is skin depth–limited.

We see that the W element for the self-resistance is a very good approximation for this real interconnect.

The second loss term in the W element model is the conductance per length. This is the leakage conductivity through the dielectric from the dissipation factor of the material. If the dissipation factor of the laminate is constant with frequency, the conductance will, by definition, increase linearly with frequency.

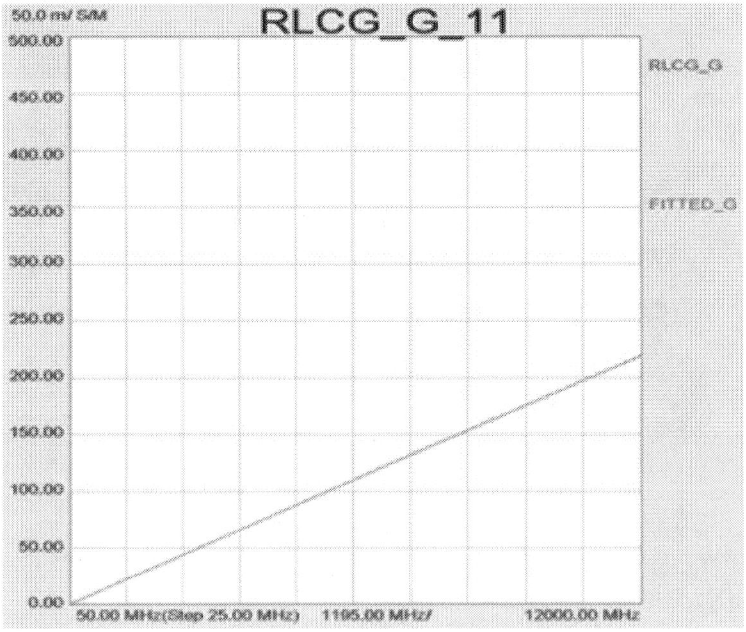

Figure 2.51: *W Element Model for the Conductance per Length*

Figure 2.51 is an example of the extracted and fitted value for the conductance per length for the 40-inch backplane trace. The units are milliSiemen/meter. Two qualities are evident from the extracted terms: it fits the model of an ideal, lossy transmission line with constant dissipation factor really well, and it is a very low value, roughly 50 mS/m at 2.5 GHz.

The W element model of an ideal, lossy, coupled pair of transmission lines assumes the inductance per length and the capacitance per length are both constant with frequency. From the measured performance of a 40-inch-long interconnect in a motherboard, as displayed in *Figure 2.52*, we see that this is a pretty good assumption.

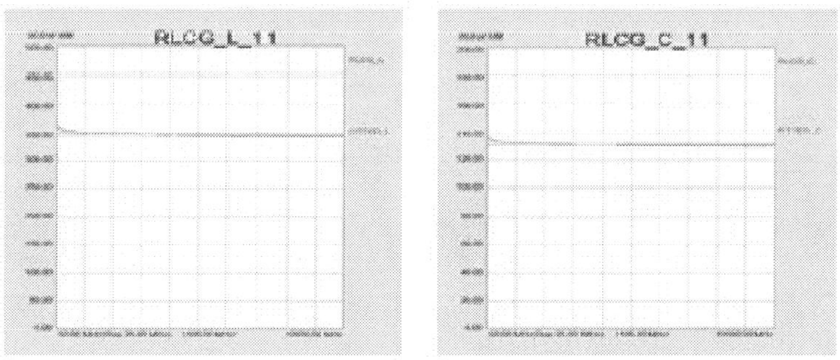

Figure 2.52: *Frequency Dependence of L and C*

The extracted capacitance per length and inductance per length show a small amount of frequency dependence at the low-frequency end, but it quickly reaches a constant value and stays there up to the 12 GHz span of the display.

The loop self-inductance per length of one line in the pair can be read off the screen as 350 nH/m, while the self-capacitance per length of one line in the pair can be read off the screen as 130 pF/m. These are the terms that are exported as the W element coefficients.

The six parameters that define the W element, the two for the resistance, the two for the conductance and the capacitance and inductance terms, can be exported into a text file that can be read directly by HSPICE or other compatible circuit simulators.

```
* RLCG parameters for a 2-conductor lossy
* frequency-dependent line

* N (number of signal conductors)
****************************************
2

* Lo
3.462510e-007
9.231448e-010 3.462510e-007

* Co
1.309606e-010
5.198448e-014 1.309606e-010

* Ro
4.52537
-8.83241e-005 4.52537

* Go
1.249961e-012
-7.500219e-013 1.249961e-012

* Rs
0.000946094
0.000123051 0.000946094

* Gd
1.829429e-011
2.163884e-012 1.829429e-011
```

Figure 2.53: *W Element Model for the 40-Inch Backplane Differential Channel*

Figure 2.53 shows an example of the W element model for the 40-inch backplane trace described previously. Because of the limitation that this model must be symmetric, the diagonal elements of each term are identical. For each of the six elements, there are only two unique terms, the self values and the coupled values. The lower the coupling is, the smaller these off-diagonal terms will be. The units used to describe each term are the default units in SPICE.

2.3 Summary

Everything you ever wanted to know about the electrical properties of a differential channel is contained in the four-port S parameters. These can be transformed into the time or frequency domain, either as single-ended or balanced terms. The variety of options means we can usually find a format that will display the data so that we can extract the most valuable information quickly and effortlessly.

In this application note, we have described nine applications for the analysis of a differential channel to characterize its performance and optimize its design. As the final approach, if we cannot get the required information directly off the screen of the analyzer, we can always use a system simulation tool to predict the precise behavior of a real signal using the measured S-parameters as a behavioral model.

- Characterizing the differential and common impedance profile of a differential channel
- Characterizing the time delay and group delay of the differential and common signal in a differential channel
- Measuring the bandwidth of a differential channel
- Measuring the rise time degradation of a differential channel
- Direct simulation of eye diagrams
- Estimating taps for pre-emphasis from transmitted impulse response
- Estimating possible EMI resulting from mode-converted common signal on external cables
- Identifying the root cause of mode conversion in a differential pair
- Extracting first-order transmission-line models of a differential pair in RLCG format

References

[1] Many of the principles described in this application note are introduced in detail in the book, *Signal Integrity-*

Simplified, by Eric Bogatin, published by Prentice Hall, 2003.

[2] Additional application notes can be found at www.BeTheSignal.com and are available for free download.

[3] Many of the examples of transmission-line structures are available in the circuit boards provided with the Master Class Workshops listed on www.BeTheSignal.com and reviewed in the on-line lectures that can be found on this Web site.

[4] *Signal Integrity Solutions,* Brochure, Literature Number 5988-5405EN, August 29, 2005.

[5] *Limitations and Accuracies of Time and Frequency Domain Analysis of Physical Layer Devices,* Application Note, Literature Number 5989-2421EN, Nov. 1, 2005.

Authors

Eric Bogatin, Signal Integrity Evangelist, Bogatin Enterprises

Mike Resso, Signal Integrity Application Scientist, Component Test Division, Agilent Technologies

Chapter 3

Differential Impedance Design and Verification with Time Domain Reflectometry

3.1 Abstract

Differential impedance circuit boards are becoming more common as low-voltage differential signaling (LVDS) devices proliferate. Yet, there is much confusion in the industry about what differential impedance means, how to characterize its performance, and how to leverage its benefit for noise rejection. This paper reviews the general features of differential pair transmission lines and how they can be characterized with novel time domain reflectometry (TDR) instruments. In particular, some of the real-world effects such as asymmetries in time delays are illustrated.

The behavior of differential transmission lines is analyzed by a high-speed digitizing oscilloscope utilized in both TDR and time domain transmission (TDT) mode. These techniques are used to extract circuit model parameters and to emulate performance in a high-speed digital system. Measurements on a specially designed PCB test vehicle with a gap in the return path are used to illustrate the robustness of differential pairs.

3.2 Overview

- Basic TDR/ DTDR Measurement Processes
- Differential Impedance: a simple perspective
- Coupled Transmission line formalism
- Measuring differential impedance elements
- Emulating received differential signals
- Emulating effects of a split in return path

Figure 3.1: *Overview*

Early Applications for Differential Pairs

MECL I	1962
MECL II	1966
MECL III	1968
MECL 10k	1979
MECL 10kH	1981

ANSI/TIA/EIA-644-1995 is the generic physical layer standard for LVDS. It was approved in November of 1995, and first published in *March of 1996*.

Example: high speed serial transmission

TI 1.8 Gbps LVDS TRX

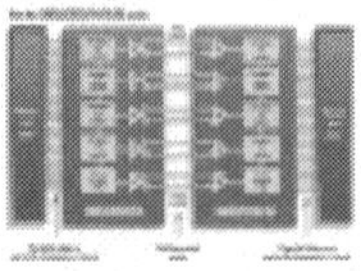

→ IEEE1394
→ IEEE488
→ Gigabit Ethernet

Figure 3.2: *The Growing Importance of Differential Pair Use*

Though differential pairs have been used for high-speed interconnects since the early 1960s, it is only in the last few years that the introduction of LVDS technology has accelerated their use. Differential pairs have proliferated into almost every high-speed application. In addition to their use in many common board-level technologies such as SCSI and Rambus RDRAMs clocks, they are used in virtually all high-speed serial links such as gigabit Ethernet and IEEE 1394. However, even with this widespread use, the properties of differential pairs are often poorly understood by designers.

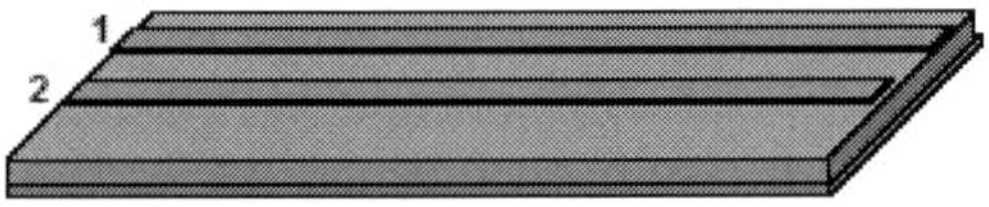

- Physical Design
  - ✓ Balanced or **unbalanced** wrt return path
  - ✓ **Symmetric** or asymmetric geometry
  - ✓ Homogeneous or **inhomogeneous** dielectric
  - ✓ Tightly coupled or **weakly** coupled
- Application use
  - ✓ Single ended vs. differential
  - ✓ Differential drive with virtual ground
  - ✓ Differential signal with DC ground
  - ✓ Common signal with DC ground

Figure 3.3: *What Is a Differential Pair Transmission Line?*
Answer: Any two coupled transmission lines (with their return paths)

A differential pair is simply two transmission lines that are coupled in some way. The formalism for dealing with any arbitrary combination of transmission lines will be developed later in this paper.

There are a number of ways of distinguishing the various types of coupled transmission lines, based on their modal properties. For example, a pair of balanced lines is when the signal paths have the same electrical properties as the return paths. Twisted pairs are balanced lines, while microstrip pairs are unbalanced. The pairs can be strongly coupled or weakly coupled. When each of the two lines has the same cross-sectional geometry, they are often called symmetric and their electrical description is simplified.

Finally, when the dielectric is homogeneous, i.e., all field lines see exactly the same dielectric constant, as will be shown later in this paper, each of the two modes will propagate at the same speed. This is the case for stripline pairs, for example.

In addition to the intrinsic fundamental electrical properties of the interconnects themselves, the nature of the signals on the pair of lines can be described in terms of the voltage patterns applied. When the signal on one line is independent of the signal on an adjacent line, the transmission lines are not really being used as a differential pair. Each line is really being used as a single-ended line.

When the lines are differentially driven but no DC connection to ground exists, the external plane acts as a virtual ground. Its voltage reference is capacitively tied to the midpoint of the two voltages on the signal lines. Alternatively, the plane can be DC connected to ground, as is commonly implemented in high speed differential signaling.

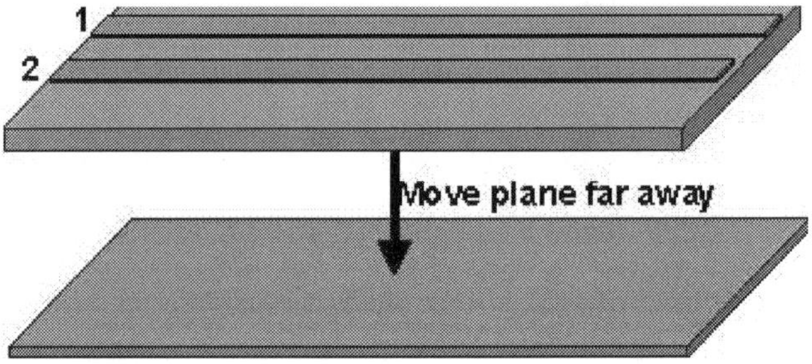

Figure 3.4: *Exploring Single-Ended and Differential Impedance with a TDR*

To begin thinking about what makes a differential pair different from a single-ended transmission line, it is useful to consider the case of a coplanar microstrip with a floating plane below it. In this configuration, the coplanar traces compose a single-ended transmission line.

They are a balanced transmission line with one trace as the signal path and the other as the return path. The impedance of this line will depend on the line parameters of the capacitance per length and the loop inductance per length of the coplanar pair.

Merely by bringing a floating metal plane underneath this coplanar line, the impedance of the line will be changed. How this third conductor influences the single-ended impedance is the basis of understanding what differential impedance really means.

In starting out with this simplified view, a TDR will be used as a tool to measure the impedance of the lines as the plane is moved in proximity.

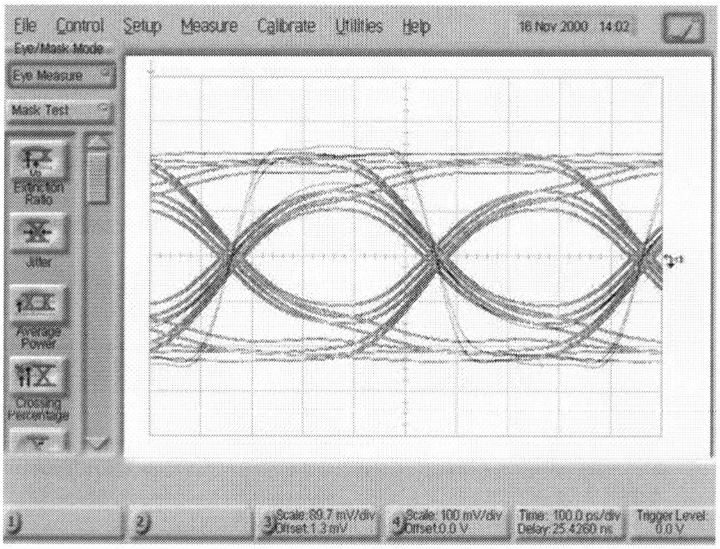

Figure 3.5: *TDR Equipment*

For all the work described in this paper, an HP 83480A mainframe with an HP 54754A differential TDR plug-in was used. This module allows operation as a single-channel TDR as well as a dual-channel TDR with the step waveforms from each channel adjusted for differential drive or common drive. This module allows complete characterization of any two transmission lines, including the odd- and even-mode impedances and the calculation of differential and common impedances.

The addition of an HP 83483A (two-channel 50 GHz module) allows measuring the waveforms that appear at the far end of the transmission line pair. With this plug-in, the actual signal a differential receiver would detect after transmission through the pair can be emulated.

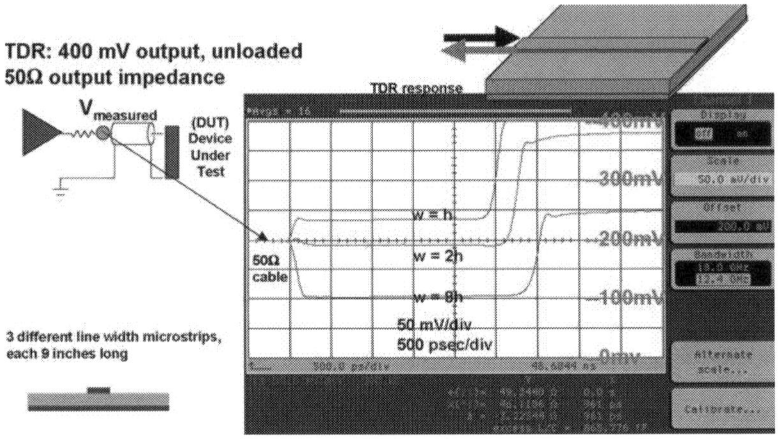

Figure 3.6: Conventional Single-Channel TDR

One channel of the differential TDR plug-in can be used to perform conventional TDR analysis. A 35 ps, fast-edge step signal is generated and launched through a 50 Ohm source impedance to the device under test (DUT). The voltage launched into the internal 50 Ohm transmission line connecting from the internal source to the SMA connector on the front panel is measured with a very fast sampling scope and displayed.

Reflected voltages from impedance discontinuities are displayed as increasing voltages (for higher impedances) or decreasing voltages (for lower impedances). In this way the TDR can act as a very fast time domain impedance analyzer. The reflected voltage is a direct measure of the impedance of the DUT.

For example, the measured TDR response from three microstrip interconnects is displayed in the above TDR plot. In the top trace, the line width is equal to the dielectric thickness. The impedance is about 70 Ohms. In the middle trace, the line width is twice the dielectric thickness. Since there is almost no reflected voltage, the impedance is measured as just slightly less than 50 Ohms. In the bottom trace, the fabricated line width is eight times the dielectric thickness. This is a very wide line and the impedance is measured as very low, less than 20 Ohms.

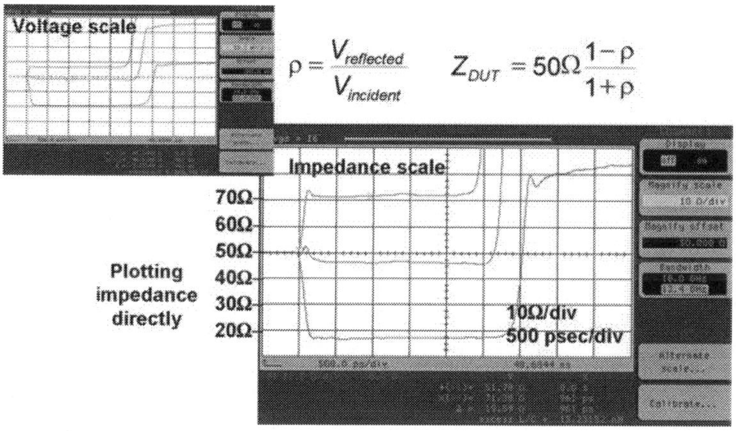

Figure 3.7: *Converting Reflected Voltage into Impedance*

The time-dependent impedance of the DUT can be extracted directly from the measured voltage. The HP 83480A mainframe can perform this analysis automatically, displaying not only the three different microstrip transmission lines measured above, but also their impedance, on the screen. They are seen to be 70 Ohms, 47 Ohms, and 17 Ohms.

What is also of interest to note is that each line has the same physical length of nine inches, yet their electrical lengths are different. The highest impedance line has the shortest electrical length. This is due to the lower effective dielectric constant of the narrow microstrip line. The narrowest line has more fringe fields in air, contributing to a lower effective dielectric constant and hence a shorter round-trip time delay. The widest line has the lowest impedance and least amount of field lines in air, resulting in higher effective dielectric constant and longest round-trip delay.

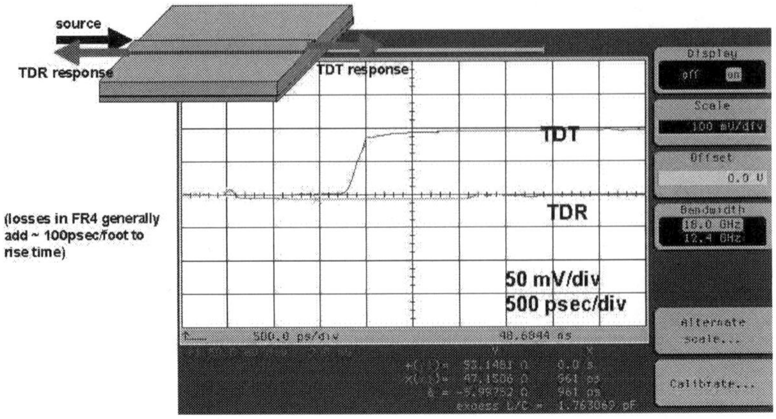

Figure 3.8: *Conventional Dual-Channel TDR/TDT*

The use of a second TDR channel opens up the application possibilities. When the second channel is just used for input, it can measure either the transmitted response of a single line, as in conventional TDT, or the response of an adjacent quiet trace due to crosstalk.

In TDT, the first channel generates the exciting source into one end of the transmission line and the second TDR channel is the receiver at the other end. In this way, the TDR and TDT response of the DUT can be simultaneously measured.

The TDR response gives information about the impedance of the DUT, and the TDT gives information about the signal propagation time, signal quality, and rise-time degradation. In this mode, the TDT is emulating what a receiver will see at the far end.

One limitation of all TDR/TDT instruments is that the source and receiver have impedances of 50 Ohms. This may not match what the actual end-use application is. However, many of the commonly encountered signal integrity effects can be illustrated with this impedance, and these measurements can be used to create or verify interconnect models, which can then be used in simulations with real device models as the sources and loads.

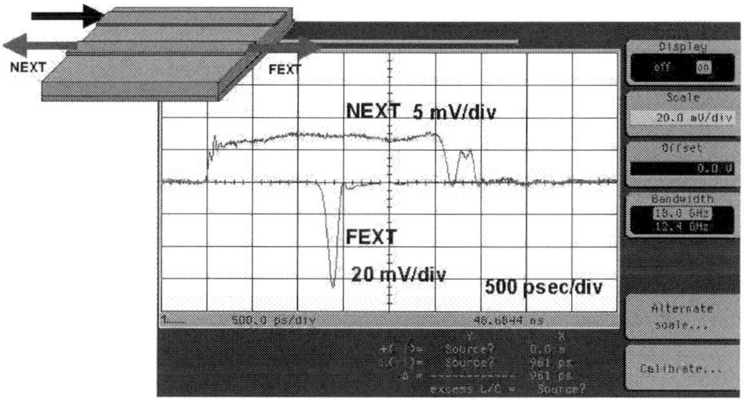

Figure 3.9: *Using Dual-Channel TDR/TDT to Measure Crosstalk*

The configuration of dual-channel TDR can also be used to measure crosstalk between two adjacent traces. The first TDR channel can be used to generate the exciting waveform for the active line. The voltage induced on the quiet line is then measured with the second channel by alternately connecting one end and then the other to this second channel. At the same time, the unattached ends of the two transmission lines should be terminated in 50 Ohms to keep the loads the same as when the cables are attached.

In the example displayed above, the near-end crosstalk (NEXT) and far-end crosstalk (FEXT) of two closely spaced microstrip lines is measured. The line width was 2x the dielectric thickness, h, and the space was equal to the line width. The saturated NEXT is seen to be about 7 mV, which is 3.5 percent. The far-end noise is a peak of 63 mV, strongly dependent on the rise time and coupled length.

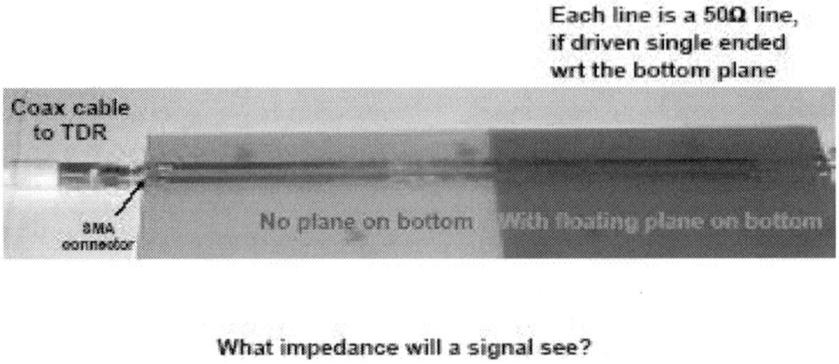

Figure 3.10: *Coplanar Transmission Line*

The simplest way of thinking about differential impedance is to consider first a coplanar transmission line composed of two traces on an FR4 substrate. With no metal plane beneath them, they represent a simple coplanar transmission line. The impedance of this transmission line will depend on the line parameters of the capacitance and loop inductance per length.

What will happen to the impedance a signal sees if the coplanar pair passes over a floating metal plane? To explore this scenario, a simple test board was built up with a coplanar pair of traces mounted to an FR4 substrate. For the first four inches, there is no plane on the backside of the board. For the second four inches, there is a continuous plane.

The front end of the coplanar pair has an SMA connector that is then interfaced to the TDR through a 50 Ohm coax cable. In this way, the TDR can drive a signal into the coplanar pair, with one trace acting as the signal and the other trace acting as the return path. Since this is a balanced pair, it does not matter which line is which.

The TDR allows us to measure directly the impedance the signal sees in propagating down the line.

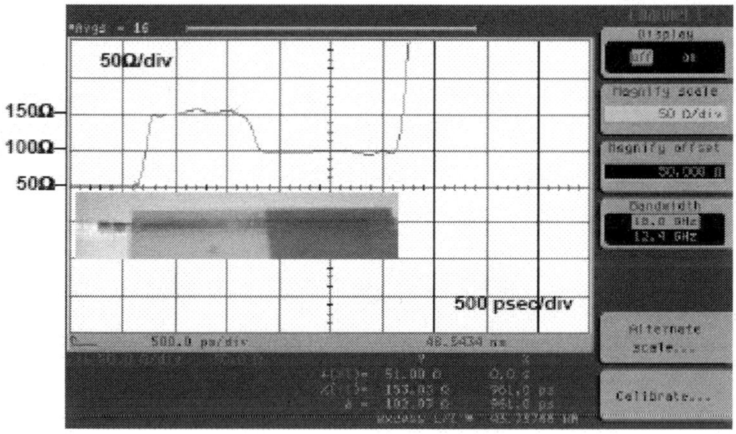

Figure 3.11: TDR of Coplanar Transmission Line

In the first four inches, the impedance is rather high, at about 150 Ohms. This is because of the relatively large separation of the traces, having higher inductance per length and lower capacitance per length than typical of microstrips. This is to be compared with the typical 115–120 Ohm impedance of twisted-pair lines, which has an aspect ratio similar to that of the coplanar lines.

In the second half of the trace, where the plane extends beneath the traces, the impedance the signal sees is dramatically reduced to about 100 Ohms. This drop in impedance is due to the change in the line parameters caused by the proximity of the plane below. The total capacitance between the two lines is dominated by the series combination of the coupling capacitance from one line to the plane and the capacitance of the plane up to the second line. This series capacitance is much larger than the direct line-to-line capacitance.

In addition, the loop inductance is reduced due to the induced eddy currents generated in the plane by the signal edge propagating down the transmission line. The combination results in a reduction of the impedance to only 100 Ohms.

Even though there is no direct electrical connection between the two coplanar lines and the plane below, the electromagnetic coupling has a significant impact on the impedance a signal sees moving down the coplanar lines. It is not a coincidence that the geometry of the second half corresponds to two single-ended lines, each with a single-ended characteristic impedance of 50 Ohms.

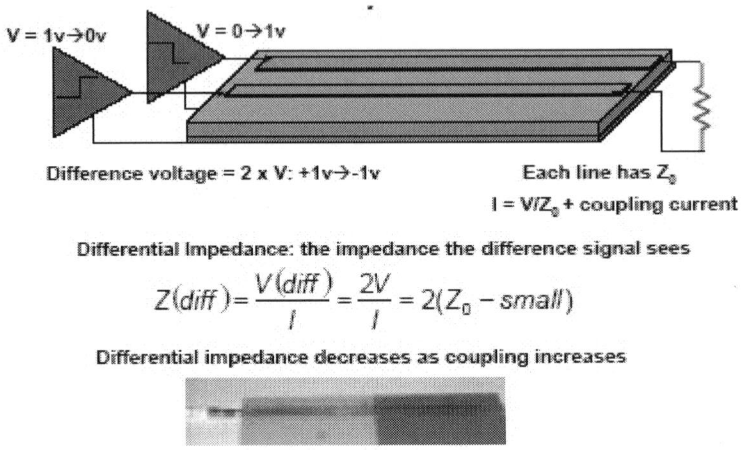

Figure 3.12: *Differential Impedance: The Simple View*

This experiment leads to the simplest possible description of differential impedance. When the two coplanar lines were driven as a single-ended transmission line, the signal was the voltage difference between the two lines. The impedance the signal saw was 150 Ohms where there was no plane and 100 Ohms where there was a plane. In the region where there is a plane below, the transmission line looks like two coupled microstrip lines as part of a differential pair.

When the two transmission lines are driven by single-ended signals that are exactly out of phase, we call this differential driving. As the signals propagate down the differential pair, there is a voltage pattern between each signal line and the reference plane below. In addition, there is a signal between the two signal lines. This is called the difference signal or differential signal. If the differential pair is driven symmetrically, the differential signal voltage is twice the single-ended signal voltage.

The difference signal is the same signal as when the two coplanar traces are driven as a single-ended line in the previous example. In this case, the impedance the signal saw was 100 Ohms in the region where there was a plane. If the two microstrips were driven differentially, the difference signal would see an impedance of 100 Ohms as well. We call the impedance the difference signal sees the difference impedance or differential impedance.

Differential impedance is really the impedance the difference signal sees that is driven between the two signal lines in the differential pair. The impedance the difference signal sees is the ratio of the signal voltage (difference voltage) to the current in the line. The difference voltage is twice the voltage of the edges driven into each line. The current into each line is related to the impedance of each individual line in the pair. There is an additional current between the signal lines that is due to the coupling between the traces themselves. This is in general a small amount, but it cannot be neglected.

In this simple perspective, differential impedance is seen to be the impedance the difference signal sees when opposite polarity edges are launched in a differential pair of transmission lines. And, as we illustrated before, this is also the impedance a signal would see if it were launched between the two signal lines, keeping the external plane as a floating plane.

To quantify the concepts of differential impedance, it is important to introduce the formalism of describing the nature of the coupling between transmission lines. In this way, any arbitrary pair of coupled lines can be analyzed with the same methods.

$$V_1 = Z_{11}I_1 + Z_{12}I_2$$

$$V_2 = Z_{22}I_2 + Z_{21}I_1$$

Example:
Characteristic Impedance Matrix [ohms]:

	1	2
1	49.6	6.4
2	6.4	49.6

Figure 3.13: *Formalism No. 1: The Characteristic Impedance Matrix*

If there were no coupling between transmission lines, the impedance of a line, as defined by the ratio of the voltage across the paths and the current through them, would be dependent on just the line parameters of the one line. However, as soon as coupling is introduced, the voltage on one line may be dependent on the current in an adjacent line. To include these effects, the concept of impedance or characteristic impedance must be expanded to allow for one trace interacting with another. This is handled by expanding the impedance into an impedance matrix.

Any two transmission lines, each with a signal path and a return path, can be modeled using an impedance matrix. The diagonal terms are the impedance of the line when there is no current in the adjacent line. This is sometimes called the self-impedance. The off-diagonal elements represent the amount of voltage noise induced on the adjacent trace when current flows on the active line. If there were little or no coupling, the off-diagonal impedance would be near zero.

As the coupling between the lines increase, the off-diagonal terms will increase. For example, if the microstrip traces, as illustrated previously, were moved closer together, the diagonal impedance would not change very much, but the off-diagonal terms would increase.

Alternatively, if the plane were to be lowered, the diagonal elements would increase and the off- diagonal elements would increase. When the off-diagonal impedance elements are a large fraction of the diagonal elements, the lines would be very strongly coupled.

The definition of the characteristic impedance matrix between any two transmission lines does not in any way depend on assumptions about their size, shape, material composition, or imposed signals. Of course, the values of the matrix elements themselves will strongly depend on the geometry and material properties. For identical signal lines, the matrix is symmetric. This is why this special configuration of transmission lines is often called symmetric lines.

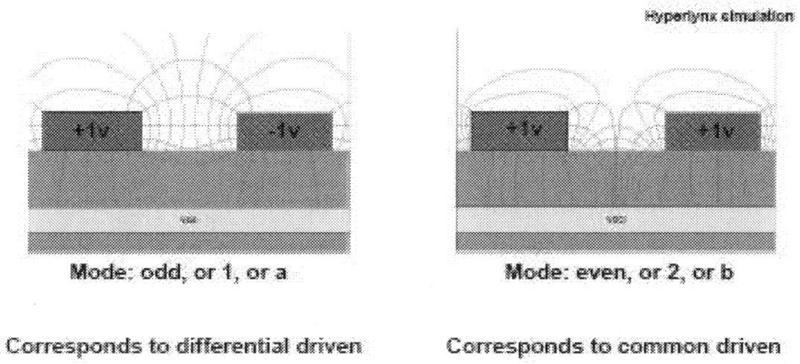

Figure 3.14: *Formalism No. 2: Mode Pattern for Identical Traces*

For a pair of transmission lines, any arbitrary voltage pattern may be imposed. However, certain patterns have special properties in that they will propagate down the line undistorted. These patterns are called modes. When the dielectric is inhomogeneous and the conductors are identical, the mode patterns that propagate undistorted are the same voltage patterns as when driven differentially, with opposite edges or driven in common, with the same voltage-edge polarity. We give these two modes the names of odd and even modes.

When the impedance matrix is symmetric, the odd mode is excited when the pair is driven with a differential signal. The even mode is excited when the pair is driven with a common signal.

It is important to keep in mind that the modes are intrinsic features of the transmission lines. They depend on the precise geometry and material properties. The voltage imposed on the lines is dependent on how the drivers are configured.

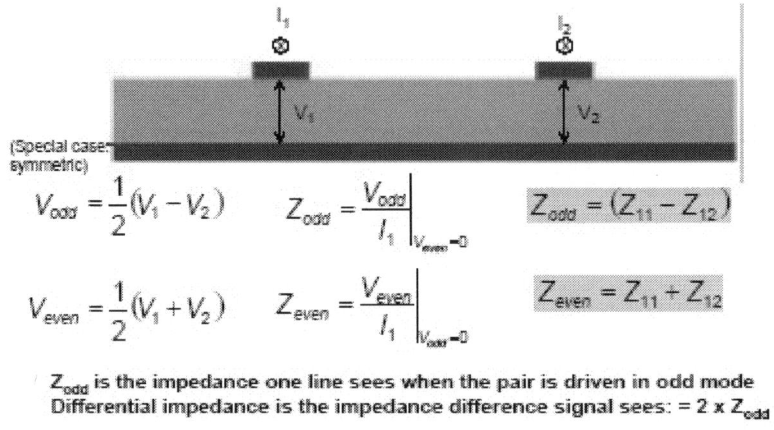

Figure 3.15: *Definition of Odd- and Even-Mode Impedance*

Based on the definition of the impedance matrix and the definition of odd and even modes, the impedance of each mode can be calculated. The odd-mode impedance is the impedance a driver would see, looking into one of the lines, when the pair of lines is driven in the odd mode or with a differential signal. Likewise, the even-mode impedance is the impedance a driver would see, looking into one of the lines, when the pair of lines is driven in the even mode, or by a common signal.

If there were no coupling, both the odd- and even-mode impedances would be equal, and equal to the impedance of just one isolated line, as expected. However, with coupling, there are additional current paths between the signal lines in odd mode, and the odd mode impedance decreases. Some current will flow not only from the first signal line to the return path, but through to the second signal line and then into the return path. This increased current through the coupling path results in a decrease in the odd-mode impedance of one line with increasing coupling.

The even mode is also affected by the coupling. When driven with a common signal, there is no voltage difference between the two signal traces. There is thus no coupled current between the signal lines, and the even-mode impedance is higher than the odd mode.

When the differential pair is driven with a differential signal, the impedance of one line, the odd-mode impedance, defines the current from one trace ultimately into the plane. The voltage difference between the two signal lines, when driven differentially, is twice the voltage of one line to ground. Thus, the differential impedance, as defined by the ratio of the voltage between the two lines to the current between them, is seen to be simply twice the impedance of the odd mode impedance. This quantifies the differential impedance in terms of the characteristic impedance matrix elements.

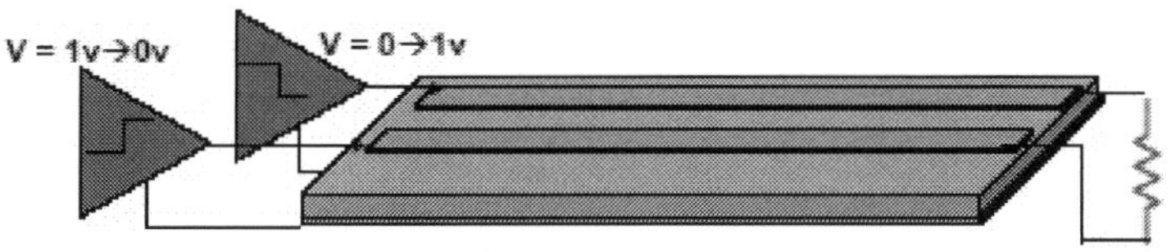

When both lines are driven with a differential signal, line 1 has an impedance of $Z_1 = Z_{odd}$:

$$Z_{odd} = (Z_{11} - Z_{12})$$

When both lines are driven with a common signal, line 1 has an impedance of $Z_1 = Z_{even}$:

$$Z_{even} = (Z_{11} + Z_{12})$$

Figure 3.16: *The Impedance of One Line Depends on How the Other Is Driven*

From this analysis, it is clear that when there is coupling between transmission lines, as in a differential pair, referring to the "impedance" of one line is ambiguous. The impedance will change depending on how the adjacent line is driven.

When both lines are driven in common, the impedance of one line will be the even-mode impedance. When both lines are driven differentially, the impedance of one line will be the odd-mode impedance.

To measure the odd- and even-mode impedances requires applying simultaneous signals to each of the two lines. This requires using a dual-channel TDR that can be configured for differential drive and common drive. With this instrument, the even and odd modes can be measured and the characteristic impedance matrix elements can be extracted.

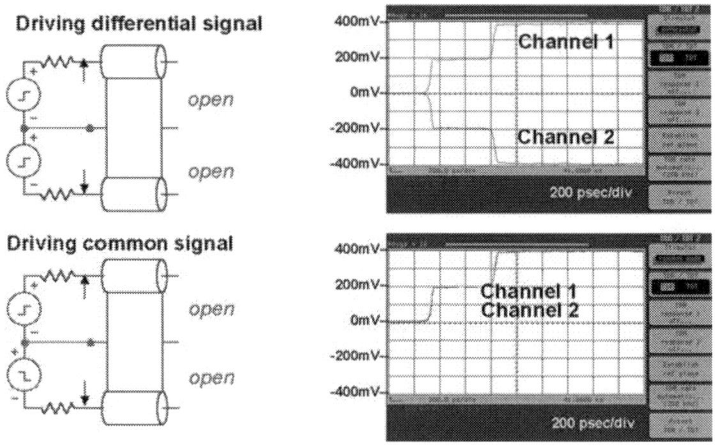

Figure 3.17: *Two-Channel Differential TDR: Differential or Common Driven*

The HP 54754A plug-in module has two independent TDR step generators that can be synchronized. There are four modes of operation: channel 1 on only, channel 2 on only, both channels in phase – common signal, and both channels out of phase with each other – differential signal.

The TDR response from each channel can be measured separately independent of what the other channel is doing. This allows the measurement of the odd-mode impedance separate from the even-mode impedance for each line in the pair.

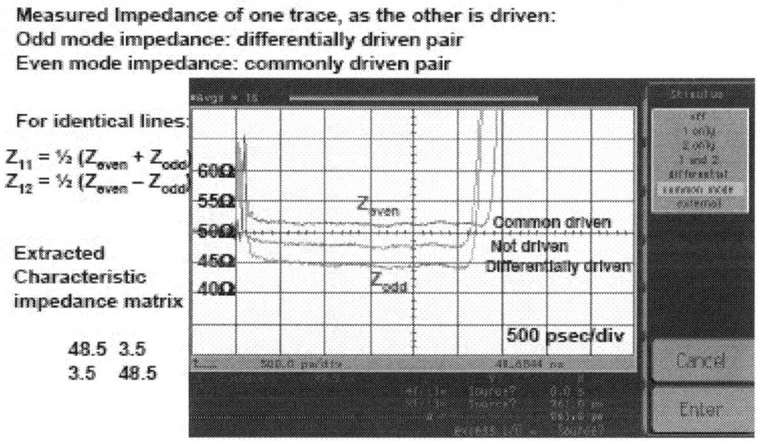

Figure 3.18: *Measuring Odd and Even Impedance of Tightly Coupled Lines*

In *Figure 3.18*, the TDR response of one trace of a closely coupled differential pair of microstrips is measured. The other trace is driven by channel 2 of the HP 54754 module. The TDR response has been converted from a voltage scale directly into an impedance scale to facilitate direct read-out of impedance.

When channel 2 is driven in phase with channel 1, the differential pair is driven with a common signal. The impedance measured by the TDR for one of the microstrip traces is the even-mode impedance of that line and is seen to be about 52 Ohms. By merely changing the signal on the second trace to out of phase or driving the pair with a differential signal, the impedance of the line under test decreases. The odd-mode impedance is seen to be about 45 Ohms.

Finally, when the second channel is turned off and the voltage on line 2 is zero, the impedance of line 1 is measured as the self-impedance, which is the diagonal element of roughly 48.5 Ohms.

From the measurements of the odd- and even-mode impedances, the characteristic impedance matrix elements can be extracted.

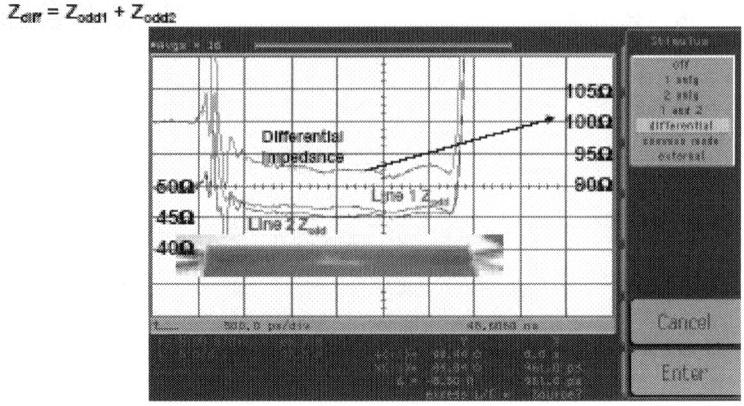

Figure 3.19: *Direct Measurement of Differential Impedance*

When the differential pair is driven differentially by the DTDR module, the impedance measured by each channel is the odd-mode impedance of each line. These can both be displayed directly on the screen. Though for symmetric lines, the odd-mode impedance of each line is nominally the same, in the real world, there are always some asymmetries. These show up as slightly different odd-mode impedances for line 1 and line 2.

In *Figure 3.19*, for two microstrip lines, one line has an odd-mode impedance of 46 Ohms and the other has one of 47 Ohms. There is some variation across the length of the trace, due to line width variations in the tape used to fabricate the trace. When the two traces have different odd-mode impedances, the differential impedance is just the sum of the two odd-mode impedances. After all, the difference signal will see the series combination of the impedances of each line to the plane below.

The differential impedance can be displayed directly on the screen as the sum of the two odd-mode impedances. In this example, it is seen to be about 93 Ohms, with some variation across the length.

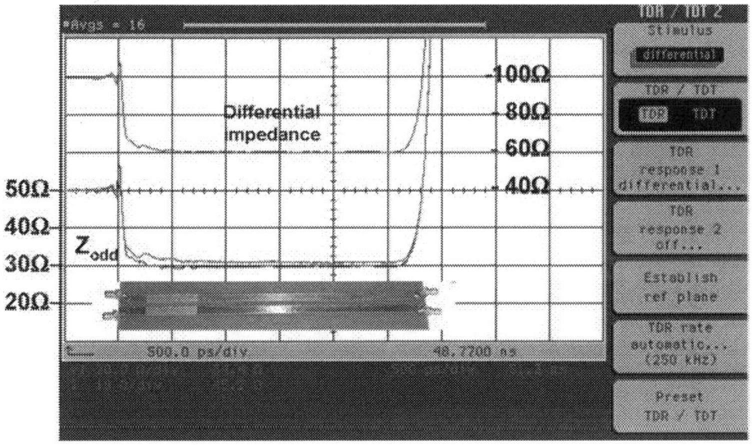

Figure 3.20: *Measuring Differential Impedance of Low-Impedance Traces*

The same measurement technique can be applied to low-impedance traces. In this example, the two traces each have an odd-mode impedance of about 30 Ohms. The differential impedance is calculated and displayed as 60 Ohms. Right near the beginning of the differential line, there was some lifting of the trace from the board. Tape was applied to minimize this problem. The higher impedance due to larger distance between the signal line and the return path is evident in the measured response of the odd-mode impedance. The impedance is elsewhere very uniform, as the natural line width variation is a relatively small amount.

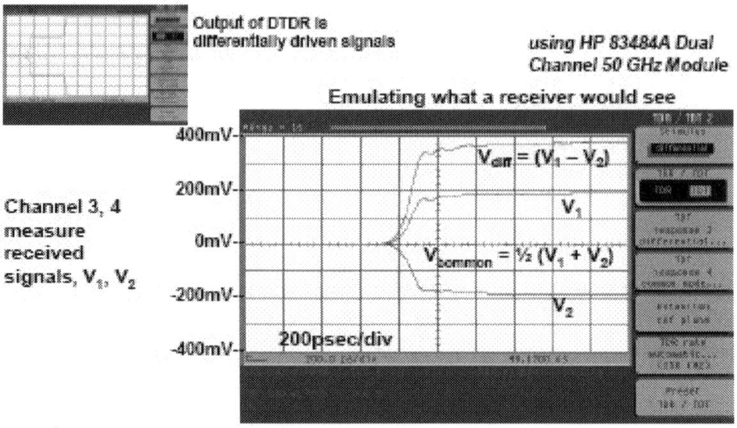

Figure 3.21: *DTDT: Four-Channel Operation Differential Time Domain Transmission*

A dual TDR module allows the measurement of the impedance characteristics of any coupled differential pair. With the addition of an HP 83484A two-channel 50 GHz plug-in module, the signals propagated to the end of the differential pair can be measured. This is an emulation of what the actual far-end receivers might see, given the caveat of 50 Ohm termination.

In this example, the signal at the far end when the pairs are driven differentially is measured. In the upper-left screen shot, the TDR response without the DUT connected is shown. This highlights that one channel is driving a signal of 0 to 400 mV, while the other channel is driving a signal of 0 to -400 mV. What gets launched into a 50 Ohm load is 0 to 200 mV in channel 1 and 0 to -200 mV in channel 2.

At the far end of the roughly 50 Ohm differential pair the two channels of the HP 83484A measure the received voltage, into a 50 Ohm load. This shows the roughly 100 ps rise time from propagating down eight inches of FR4. The individual channels are displayed as directly measured.

In addition, the common signal, being the average of the two, and the differential signal can be automatically displayed. All received signals are displayed on the same scale. When driven differentially, very little common signal is created by the transmission down the pair. When the pair is driven with a well-balanced differential signal, the common signal is virtually non-existent.

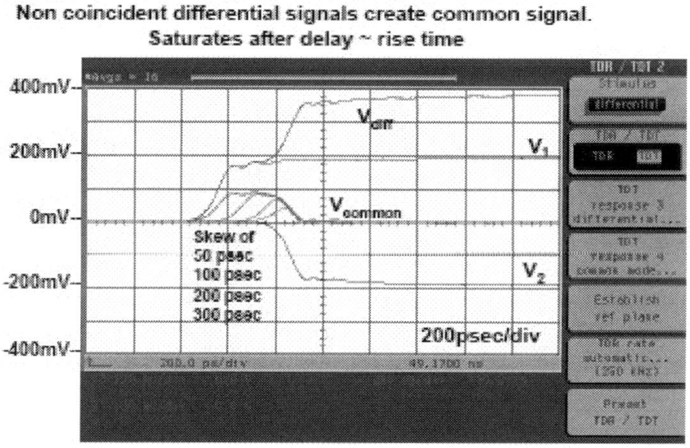

Figure 3.22: *Received Signal with Delay Skew in a Differential Pair*

A common problem with differential drivers for differential pair lines is skew between the two channels. This arises from the mismatching of the drivers, different rise and fall times, or different interconnect delays due to routing differences or different loads on the two lines of the differential pair. Any imbalance in the signals at the receivers will create a common signal.

A variable skew can be introduced between the two driven TDR step generators. This emulates what would happen if there were a skew in the drivers. In this example, the common signal is increased steadily as the skew increases from zero to 100 ps, comparable to the rise time.

Longer than 100 ps, the common signal at the receiver is basically constant. This suggests that, to minimize the common signal, the skew should be kept under just a small fraction of the rise time.

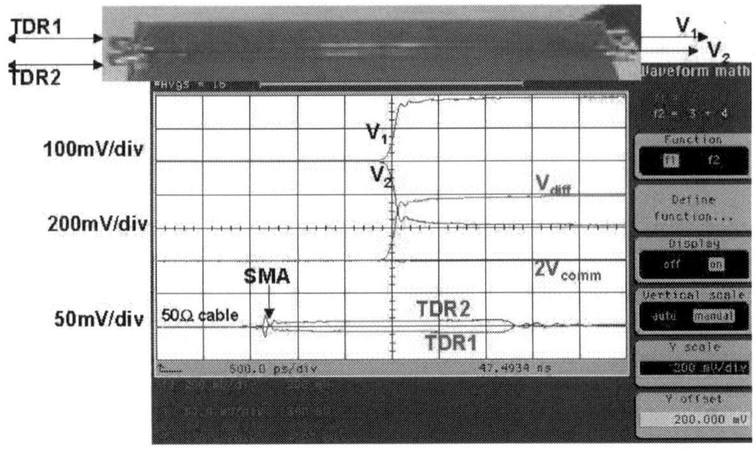

Figure 3.23: *Full Characterization of a Differentially Driven Differential Pair*

The full characterization of the performance of a differentially driven differential pair includes the TDR response of each channel, which relates to the odd-mode impedance of each line, and the received signals of the two channels at the far end, combined as the differential and common signals.

In *Figure 3.23*, the measured response of a uniform differential pair is shown, illustrating all six measurements. In this case, the common signal is very small since the differential drivers are well balanced.

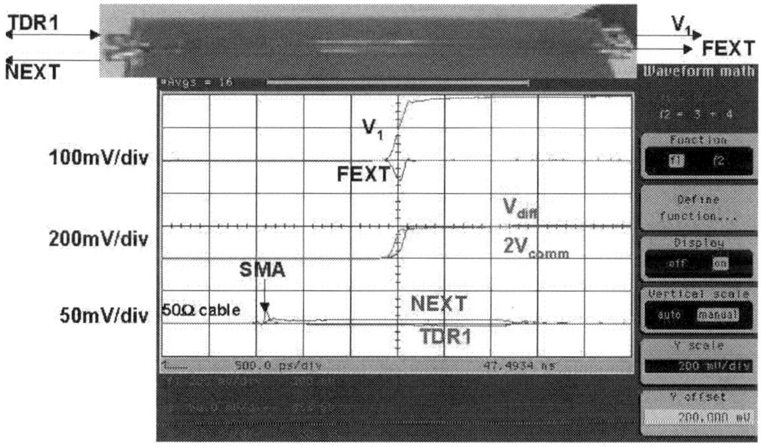

Figure 3.24: *Full Characterization of a Single-End–Driven Differential Pair*

This response should be compared to the behavior of the same pair when one line is driven single-ended, while the other line is held low. In this case, TDR channel 2 is measuring the NEXT and the channel 4 receiver is measuring the FEXT.

When a signal is launched into only one line of a symmetric line, there are equal parts odd-mode and even-mode signal created. These propagate to the end and are received, where they are calculated from the voltages in the two receiver channels and displayed as the differential signal and common signal. As can be seen, the differential signal, corresponding to the odd mode, arrives at the receiver before the common signal, corresponding to the even mode.

This is a direct measure of the difference in velocity of the odd and even modes. The odd mode, having more fringe fields in the air, has a lower effective dielectric constant and hence a higher propagation speed. The even mode has more fields in the dielectric and a higher effective dielectric constant and takes longer to reach the receiver. From this measurement, the connection between crosstalk and modes is also apparent. The common-mode signal is delayed due to the FEXT.

In this example, the differential and common signals are displayed using the Math function of the HP 83480A mainframe, so that the TDR response can be switched between single output and dual output.

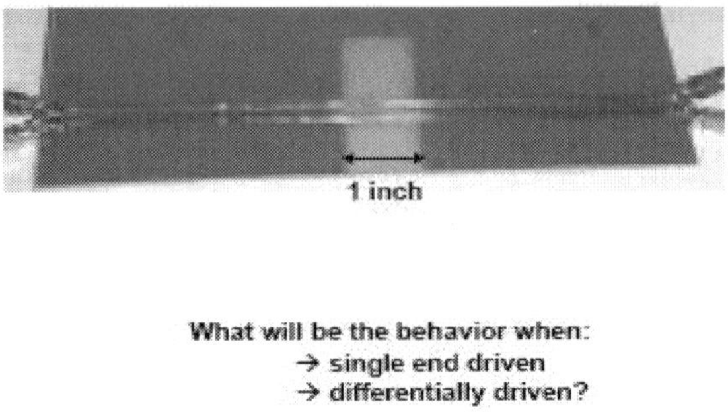

Figure 3.25: *Differential Pair over Split in the Return Path*

With this perspective, we can look at the response of a differential pair that crosses a gap in the return path. The traces are about eight inches long, each about 50 Ohms with weak coupling between them. The gap in the return path is about 1 inch wide. The time delay across this gap is comparable to the rise time of about 100 ps.

In the same way as before, the response at the near end and far end of each line can be measured when one line is driven single-ended and when both lines are driven differentially.

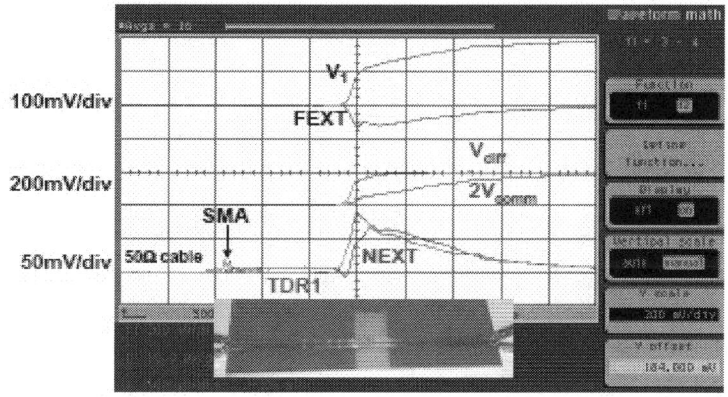

Figure 3.26: *Full Characterization of a Single-End–Driven, Differential Pair over a Split in the Return Path*

As expected, the TDR response of channel 1 shows the uniform transmission line until the gap is reached. Electrically, the gap looks like a large inductive discontinuity and the reflection is a large positive value. The second TDR channel is measuring the reflected NEXT noise. Initially there is the saturated near-end noise until the gap is reached. The mutual inductance between the two traces in the vicinity of the gap is almost as large as the self-inductance that causes the reflection of channel 1. This results in induced noise generated in trace 2 that is almost as large as the reflected signal in trace 1.

The enhanced NEXT and FEXT between the two adjacent traces is due to the high mutual inductance of the return paths around the gap. This is the reason to carefully route signal paths over continuous planes and avoid crossing gaps.

The inductively generated noise in trace 2 propagates down the trace 2 transmission line in both directions and appears at the far end as very enhanced far-end noise. From the time constant of roughly 400 ps, the inductance of the discontinuity can be extracted as roughly 40 nH. This corresponds to the self-inductance of the perimeter of the current path around the gap, which is about 3.5 inches, or roughly 15 nH/inch.

The common and differential signals are also greatly distorted from the case of no gap in the return path. This sort of discontinuity would cause major problems for most single-ended driven transmission lines, and is why design rules recommend routing adjacent traces over continuous return paths.

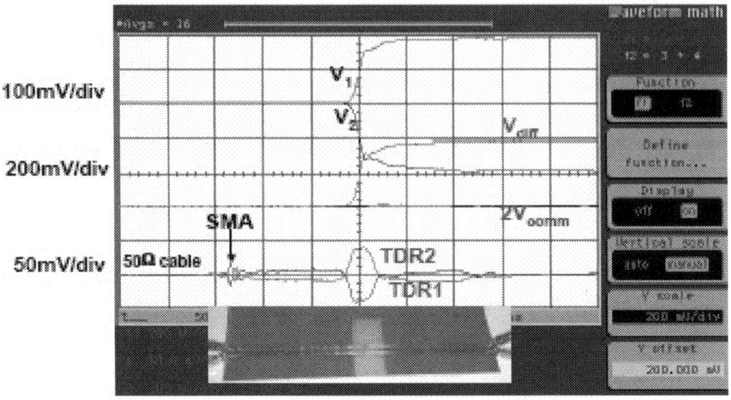

Figure 3.27: *Full Characterization of a Differentially Driven, Differential Pair over a Split in the Return Path*

However, when both lines are driven with a differential signal between the pair of traces, the reflected noise from each line is in the opposite direction and the resulting reflection is reduced considerably. Likewise, since the gap offers a nearly balanced discontinuity to each of the two signal lines, the effect on the common signal is almost negligible. This illustrates a chief advantage of transmitting signals on differential pairs—differential signals are much more robust to imperfections in the propagation paths that are common to both lines. The effects on each line will be better balanced, with less common signal noise generated, as the lines are routed closer to each other and the coupling is larger.

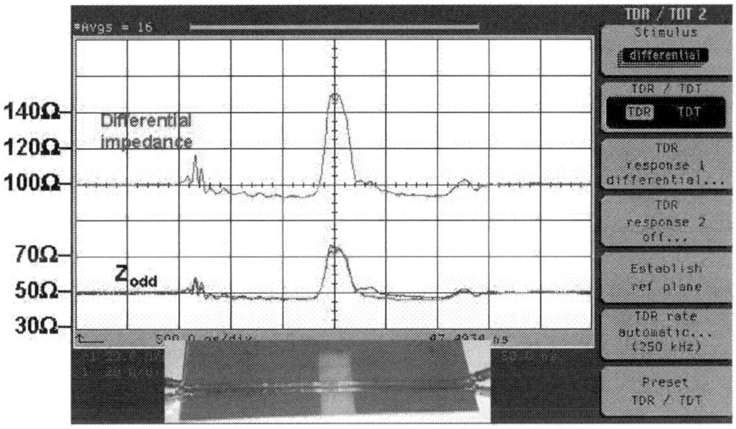

Figure 3.28: Measured Impedances

Another way to look at the gap in the return path is in terms of what impedance the differential signal sees. This can be measured directly with the DTDR module. The differential pair is driven differentially, and the DTDR measures the odd-mode impedance of each line. Their sum is the differential impedance, also displayed.

Before and after the gap, the differential impedance is about 97 Ohms. In the region of the gap, the differential impedance is about 150 Ohms. This corresponds to the impedance that was measured for two coplanar transmission lines with no conducting plane beneath them, which is exactly what the region of the gap appears as. The gap acts as a high-impedance region for the differential signal. This will create a reflection. However, if the lines are terminated at both ends, this reflection may not cause signal integrity problems.

This illustrates that if signals must cross gaps in the return path, routing the signals as differential signals on closely coupled differential pairs is the way to do it.

3.3 Summary

- The impedance of one line in a differential pair depends on how the other is being driven

- Measure odd impedance by driving differentially
    - Measure even impedance by driving in common
    - Requires differential TDR (DTDR)
- Characteristic impedance matrix elements can be extracted from odd and even impedances
- A gap in the return path causes huge increase in crosstalk in single-ended lines due to high mutual inductance
- If you must cross a split plane, better to use a differential pair
    - Some increase in differential impedance
    - Very little distortion of differential signal
    - Very little common voltage created
    - Full characterization of differential pairs is possible with DTDR and dual-channel amplifier module

In this presentation we have presented a simple and intuitive way of thinking about differential impedance. We have expanded this simplified view to include the characteristic impedance matrix formalism for two lines and illustrated how odd- and even-mode impedance can be directly measured using a DTDR module. With coupled transmission lines, the impedance of one line will depend on how the adjacent line is driven.

Finally, by using a dual-channel TDR and dual-channel amplifier, we illustrated how we can emulate some of the unique features of signals in a differential pair of transmission lines.

Resources

[1] www.BogatinEnterprises.com
[2] www.agilent.com/find/si

Authors

Eric Bogatin, Signal Integrity Evangelist, Bogatin Enterprises

Mike Resso, Product Manager, Lightwave Division, Agilent Technologies

Chapter 4

Utilizing TDR and VNA Data to Develop Four-Port Frequency-Dependent Models

4.1 Abstract

Frequency-dependent effects are becoming more prominent with the increasing data rates of digital systems. Differential circuit topology is proliferating throughout design laboratories with the goal of enhancing the data-carrying capabilities of the physical layer (PHY). Simple impedance and delay measurements of copper transmission lines on backplanes are not sufficient to ensure accurate analysis of gigabit interconnects. The challenge to push design rules to the limit now requires the use of concurrent time and frequency domain analysis. This paper will address methods to achieve proper characterization using a time domain reflectometer (TDR) oscilloscope and vector network analyzer (VNA). Measurement accuracy and error-correction techniques will be discussed for both time domain and frequency domain instrumentation. It will be demonstrated that accurate four-port, frequency-dependent models can closely simulate performance of a differential channel.

4.2 Signal Integrity Challenges

With the increase in speed of digital system design into the gigahertz region, frequency-dependent effects become a more prominent challenge than in the past. Yesterday's interconnects could be easily characterized by measuring the self-impedance and propagation delay of the single-ended transmission line. This was true for printed circuit board striplines, microstrips, backplanes, cables, and connectors. However, the proliferation of high-speed serial data formats in today's digital standards demand differential circuit topology. A paradigm shift in measurement technology is required to achieve the design goals of the advanced differential PHY. It is now necessary to consider both time and frequency domain analysis to obtain proper characterization. Tracking the technology adoption curve in *Figure 4.1*, it can be seen that several new implementations of PCI Express and Infiniband reach data rates into the 4 Gbps range. New standards such as XAUI, OC-192, 10G Ethernet, and OC-768 aim even higher—up to and past 40 Gbps. This upward trend creates signal integrity challenges for PHY device designers and the inevitable struggle to keep up with data-processing and storage capabilities.

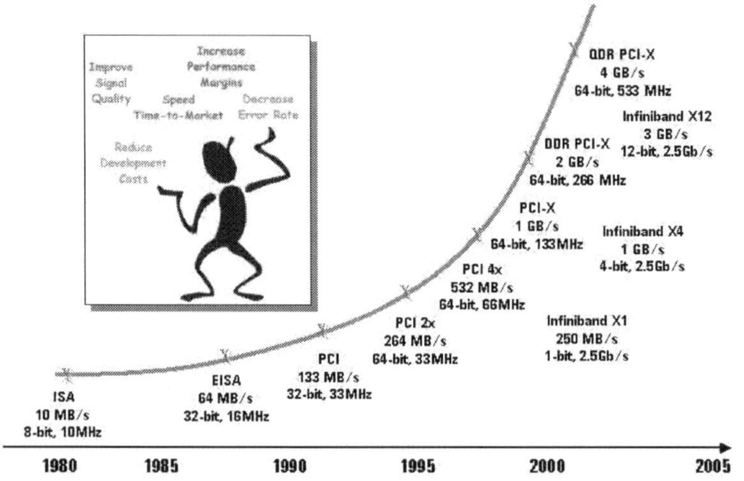

Figure 4.1: *Partial List of Many New High-Speed Serial Link Formats*

4.3 Trend to Differential Topologies

In the discussion of these new signal integrity challenges, it becomes clear why we need to understand the implications of differential topologies and how mode-conversion analysis is an important concept for designing digital interconnects. Ideal differential linear passive interconnects respond to and/or generate only differential signals (two signals of equal amplitude and opposite polarity). These perfectly designed devices exhibit beneficial characteristics noted in *Figure 4.2* and do not generate in-phase signals (e.g., common mode signals). Any radiated external signal incident upon this ideal differential transmission line is considered a common signal and is rejected by the device. This is called common-mode rejection ratio (CMRR) and is the main benefit of differential topology. The radiated common signals are usually generated from adjacent radio frequency (RF) circuitry or from the harmonics of digital clocks. Properly designed differential devices can also reject noise on the electrical ground, since the noise appears common to both input terminals. Non-ideal differential transmission lines, however, do not exhibit these benefits. A differential transmission line with even a small amount of asymmetry will produce a common signal that propagates through the device. This asymmetry can be caused by any physical feature that is on one line of the differential pair and not the other line, including solder pads, jags, bends, and digs. This mode conversion is a source of EMI radiation. Most new product development must include EMI testing near the end of the design cycle. Very often the test results show that the design exhibits EMI radiation or susceptibility. However, there is usually very little insight as to what physical characteristic is causing the EMI problem. Mode-conversion analysis provides the designer with that insight so that EMI problems can be resolved earlier in the design stage.

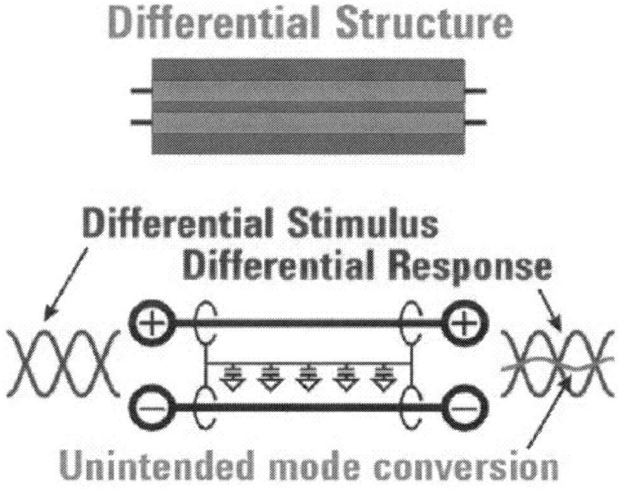

Figure 4.2: *Ideal Differential Structures Exhibit no Mode Conversion, as They Are Perfectly Symmetric*

4.4 Model Extraction Methodologies

In order to describe the test system laboratory configuration used in this design case study, the authors will refer to the flowchart in *Figure 4.3* for clarification purposes. Measurement-based model extraction can be accomplished using a variety of methods. Starting from the top, the goal is to achieve an accurate model that can be simulated in either the time domain or frequency domain. Most digital designers will focus on time domain models, and that will be our focus in this paper, also. Either a topological model or behavioral model may be developed. The topological model is based on the physical structure of the device and can be very complex for a lengthy device exhibiting multiple impedance discontinuities. This requires multiple iterations and is easily done using today's standard PC computational power. The behavioral model is a "black box" approach and describes how the device behaves toward a particular stimulus. One type of behavioral model is scattering parameters, or S-parameters. This flowchart shows that both time domain test equipment (TDR) and frequency domain test equipment (VNA) were used to measure prototype devices. Both test instrument types have strengths and weaknesses, and the specific user application will normally dictate the use of one or the other. In general, the TDR is easier to use and the VNA is more accurate. Most signal integrity laboratories have one of each. In this experiment, measurements were made with a VNA using the Agilent N1930A Physical Layer Test System software to control the VNA via GPIB. This allowed for use of the automated calibration wizard and simplified this typically rigorous and error-prone process. The resultant four-port S-parameter data was exported to the TDA Systems IConnect MeasureXtractor model extraction tool that in turn created an accurate time domain HSpice model.

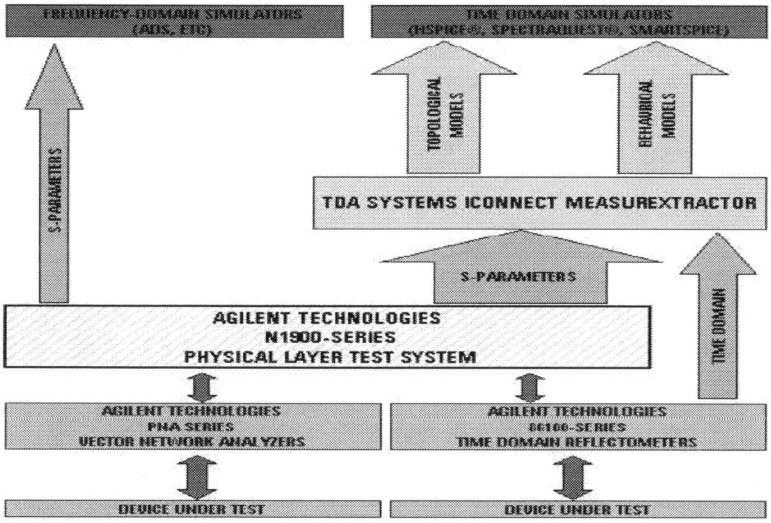

Figure 4.3: *Many Different Methods Exist Today for Model Extraction, but Measurement-Based Model Extraction Is a Relatively New Process Yielding Insight into High-Frequency Effects*

The model extraction tool used in this design case study was the TDA Systems IConnect MeasureXtractor. It was chosen because it was simple and easy to use. This extraction tool imports the impedance profile or four-port S-parameters after the user performs the measurement with either a TDR or VNA. The resultant model can be directly linked to a simulator subsequent to using a laptop to perform multiple iterations of model refinement. The convenience of comparing measured results to simulated results very quickly is an efficient way to check accuracy of models.

4.5 Typical Four-Port Measurement Systems

Measurement-based models for differential devices require a four-port measurement system. A well-calibrated and controlled stimulus will be input into the device under test (DUT) and the response will be measured with receivers co-located within the same measurement system. With a full four-port measurement system, this stimulus/response test is performed on the reflected response and transmitted response in both single-ended mode and differential mode. The TDR instrument accomplishes this task with a fast step with little overshoot in concert with a wideband receiver to measure step response. The VNA uses a precise sine wave and sweeps frequency as a narrowband receiver tracks the swept input response. This narrowband receiver is what enables low noise and high dynamic range of the VNA. Whether the data acquisition hardware is time domain–based or frequency domain–based, mixed-mode data is also compiled in a four-port measurement system. The mixed-mode data refers to two specific test conditions: differential stimulus with common response and common stimulus and differential response. This analysis leads to the discovery of interesting effects due to asymmetry within a differential transmission line.

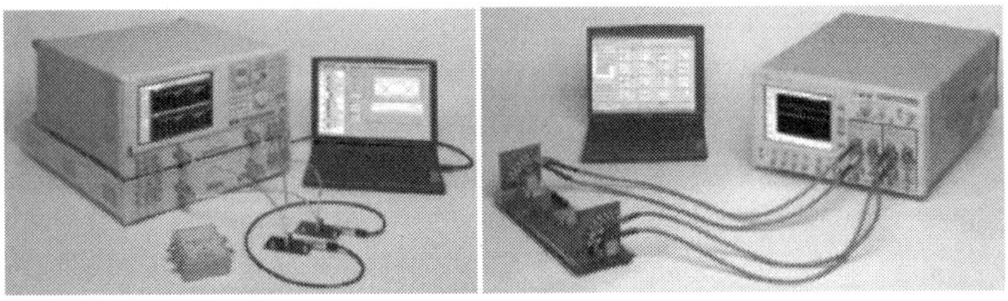

Figure 4.4: *High-Speed Differential Interconnects Need to Be Characterized with a Four-Port Measurement System, Whether It Be a Four-Port VNA (Left) or Four-Channel TDR (Right)*

4.6 Understanding Four-Port Mixed-Mode Analysis

In order to interpret the large amount of data in the differential parameter matrix, it is helpful to analyze one quadrant at a time. The first quadrant is defined as the upper left four parameters describing the differential stimulus and differential response characteristics of the DUT. This is the actual mode of operation for most high-speed differential interconnects, so it is typically the most useful quadrant that is analyzed first. It includes input differential return loss (SDD11), input differential insertion loss (SDD21), output differential return loss (SDD22) and output differential insertion loss (SDD12). Note the format of the parameter notation SXYab, where S stands for S-parameter, X is the response mode (differential or common), Y is the stimulus mode (differential or common), a is the output port, and b is the input port. This is typical nomenclature for frequency domain scattering parameters. All 16 differential S-parameters can be transformed into the time domain by performing an inverse fast Fourier transform (IFFT). The matrix representing the time domain will have similar notation, except the "S" will be replaced by a "T" (i.e., TDD11).

The second and third quadrants are the upper right and lower left four parameters, respectively. These are also referred to as the mixed-mode quadrants. This is because they fully characterize any mode conversion occurring in the DUT, whether it is common-to-differential conversion (electromagnetic interference [EMI] susceptibility) or differential-to-common conversion (EMI radiation). Understanding the magnitude and location of mode conversion is very helpful when trying to optimize the design of interconnects for gigabit data throughput. The fourth quadrant is the lower right four parameters and describes the performance characteristics of the common signal propagating through the DUT. If the device is designed properly, there should be minimal mode conversion and the fourth quadrant data is of little concern. However, if any mode conversion is present due to design flaws, then the fourth quadrant will describe how this common signal behaves.

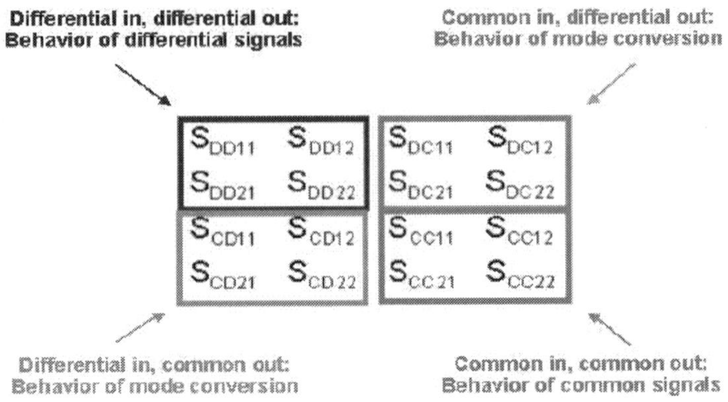

Figure 4.5: *The 16 S-Parameters That Are Obtained by Fully Characterizing a Differential Interconnect Can Be Categorized into Four Stimulus/Response Quadrants: (1) Differential in/Differential out, (2) Common in/Common out, (3) Common in/Differential out, and (4) Differential in/Common out*

4.7 Differential Interconnect Analysis

Figure 4.6 describes the flow-gram of the test and measurement methodology used in the backplane characterization. The first step in the process is to understand the 16 S-parameters and what information can be extracted from the sometimes overwhelming amounts of data. Next, the actual measurement is made to obtain these 16 S-parameters. This can be done with a variety of frequency domain instrumentation (VNAs) or time domain instrumentation (TDRs). Also, finding the amount of losses in transmission lines by observing the input differential insertion loss (SDD21) is important. This will give a very accurate indication of the bandwidth of the DUT. Last, carefully analyzing mode conversion is desirable to high-speed design. For example, pinpointing via field mode conversion in *Figure 4.6*, optimization of the design can be accomplished by determining the magnitude of mode conversion as a percentage of input signal and then locating the physical structure that is causing the mode conversion. One optional analysis that is sometimes interesting is viewing reciprocity. By viewing the forward transmission and reverse transmission data, masking effects of TDR can be removed to clarify directional information.

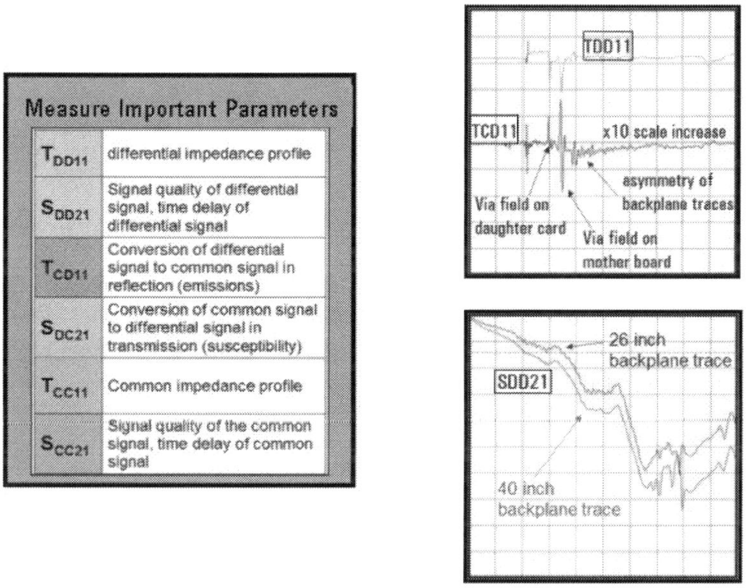

Figure 4.6: *Example of Typical Four-Port Analysis Using Both Time and Frequency Domain Data. Note: This Example Is from a 3Gbps Backplane, not from the Design Case Study Device*

4.8 Measurement Accuracy and Error Correction

Ideally, no test equipment would require any correction. However, in the real world, imperfections exist in even the highest-quality test equipment. Some of the factors that contribute to measurement error are predictable over time and can be removed, while others are random and cannot be removed. The basis of error correction is to measure a known and understood electrical standard and use this device as a reference. All measurement systems can exhibit three types of measurement error:

- Systematic errors
- Random errors
- Drift errors

Systematic errors are caused by imperfections in the test equipment and test setup. If these errors do not vary over time, they can be characterized through calibration and mathematically removed during the measurement process. Random errors vary randomly as a function of time. Since they are not predictable, they cannot be removed by calibration. The main contributors to random errors are instrument noise. Drift errors occur when a test system's performance changes after a calibration has been performed. They are primarily caused by temperature variation and can be removed by additional calibration. The error correction techniques shown in *Figure 4.7* are described as follows: Time-domain gating is easiest to implement. The user defines a start and stop point, and software mathematically replaces the measured data in that section with an ideal transmission line. With the enhanced dynamic range of the network analyzer, multiple gates are possible, but accuracy diminishes in proportion to the number of gates. Port extension will mathematically extend the measurement plane to the input of the DUT. However, it assumes the fixture looks like a perfect transmission line with a flat magnitude response, a linear phase response, and constant impedance. Port extensions are usually done after a coaxial calibration has been performed at the end of the test cables. De-embedding removes a fixture or unwanted structure from the measurement by using the S-parameters or an accurate linear model of the structure. This S-parameter or model representation is mathematically removed from the DUT measurement data in post-processing. Calibration at the DUT reference plane has the advantage that the precise characteristics of the fixture do not need to be known beforehand, as they are measured during the calibration process. While calibration is the technique specific to VNAs, the normalization is the extension of the same procedure to the TDR oscilloscopes.

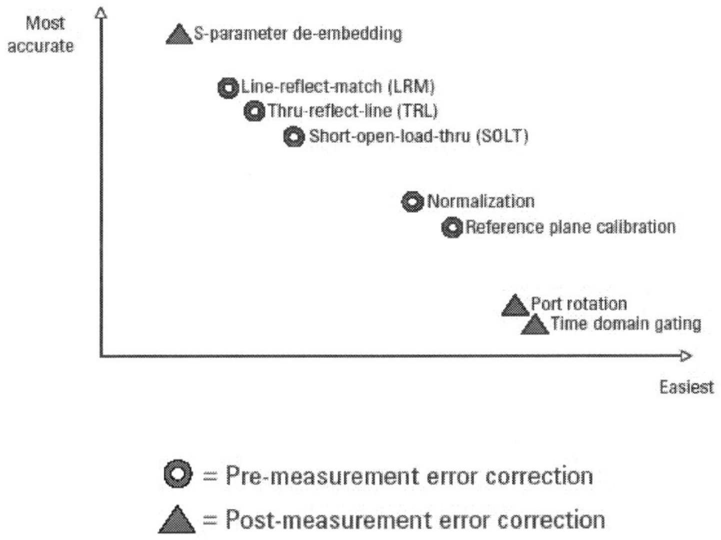

Figure 4.7: *Various Mathematical Computations and Error Correction Techniques Can Be Made to Enhance Measurement Accuracy. Usually, Ease of Use and Degree of Accuracy Are Inversely Proportional to Each Other*

4.9 Design Case Study: Silicon Pipe Channel Plane

The design case study described in the forthcoming analysis employed the use of a prototype provided by Silicon Pipe of San Jose. The ChannelPlane technology developed by Silicon Pipe creates a well-controlled impedance environment in the area surrounding the backplane/connector interface. Using a process conceptually analogous to optical fiber splicing where the two ends of a fiber are highly polished to achieve a closely matched index of refraction to minimize optical reflections, the ChannelPlane cross-sectional copper conductors are polished to minimize electrical reflections. This results in a flush-mount cable assembly compatible with the popular 2 mil Winchester SIP1000 backplane connector. The prototype ChannelPlane cable was constructed from high-bandwidth Gore G4 material that was cut three inches from the SMA end and terminated with a patented coax/twinax flush-mount termination. A coaxial interposer was then used to mate to the Silicon Pipe coax cable. A picture of the ChannelPlane assembly with flush-mount termination is shown in *Figure 4.8*. A functional block diagram in *Figure 4.9* shows the construction details.

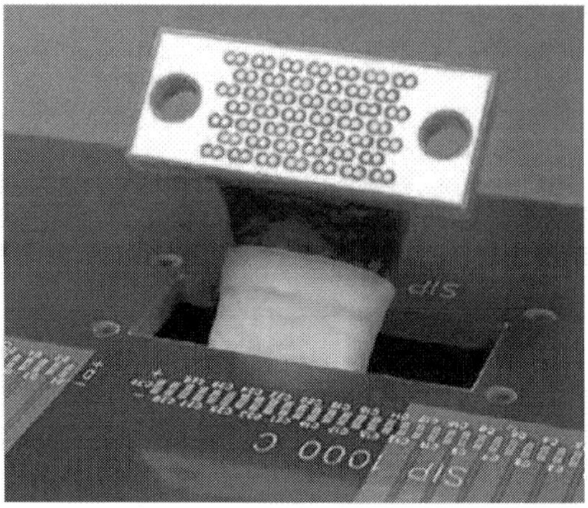

Figure 4.8: *The Silicon Pipe ChannelPlane Shown Above Is a Backplane Cable Assembly That Consists of a Flush-Mount Connector Compatible with the Winchester SIP-1000 Backplane Connector*

Figure 4.9: *The Silicon Pipe ChannelPlane Functional Block Diagram Shows Construction Details*

4.10 Frequency and Time Domain Analysis

There are many ways to develop models for digital interconnects. The first method used in this design case study is shown in the chart in *Figure 4.10*. Measurement-based models constructed in this process combine precision 40 GHz VNA–PLTS testing with IConnect modeling, optimization, and test correlation techniques. This method is capable of resolving femtoFarads over 1/10mm distances even on long cables because of the nature of the VNA instrument. TDR testing, preferred for device model partitioning of paths, can be combined with VNA measurements to resolve small backplane connectors over long distances. In fact, field-solver S-parameter output can also be used to resolve the issue of small devices in large system paths.

Frequency- and time-based correlation was required because of the large bandwidth and dynamic range required for this test. It was important to isolate and test coaxial lines with an 80 dB accuracy in order to optimize impedance, insertion, and return loss models. The model discovery process requires the ability to detect changes in impedance, rise time, and frequency in small connectors embedded in long cables. For short electrical length devices, it is desired to have a test bandwidth of 80 GHz using a 7 ps Gaussian input pulse. Where system geometric dynamic range is large, it becomes easier to detect and optimize measurement-based models within the model extraction tool fitting measurements to HSpice simulations right in the laboratory. This is a more efficient way to discover, study, and optimize the interaction between model attributes.

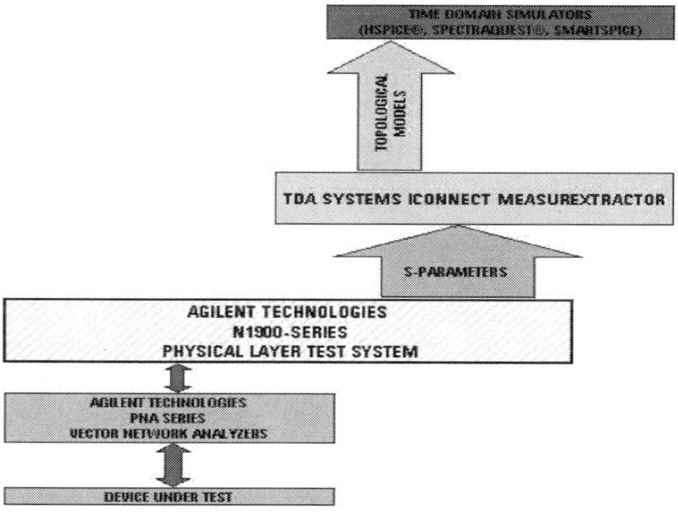

Figure 4.10: *The First Model Extraction Method Used in This Experiment Used Measurements from a VNA, and then Exported the Four-Port S-Parameters from PLTS Directly into MeasureXtractor. A Topological Model Was Then Exported from MeasureXtractor into HSpice*

4.11 Partitioning the Impedance Profile

Optimized HSpice models must be created by compiling measurement-based model schematics in the model extraction tool to make HSpice TDR/T simulations and iterations to refine the match of the measurements as extracted from the S4P files. Both single-ended, differential, even- and odd-mode simulations can be conducted at the discretion of the engineer. Most importantly, the engineer must check the integrity of all DUT components before diving into model extraction and optimization. For this case, two separate coaxial cables were presumed to match physically and electrically. Both differential-leg TDR voltage waveforms were checked for faults before being superimposed as differential or odd/even impedance profiles.

Behavioral S-parameter measurements were exported into the model extraction tool to construct HSpice topological models. Effectively, S-parameters are "time gated" and "re-time gated" in IConnect until HSpice topological simulations match the measurements. Fortunately this iterative optimization sequence requires only PC work and is easily automated. Both impedance and lossy models were constructed using this methodology. HSpice fitted simulations to the TDR, TDT, and S-parameter data, enabling the authors to divide and conquer complex modeling applications.

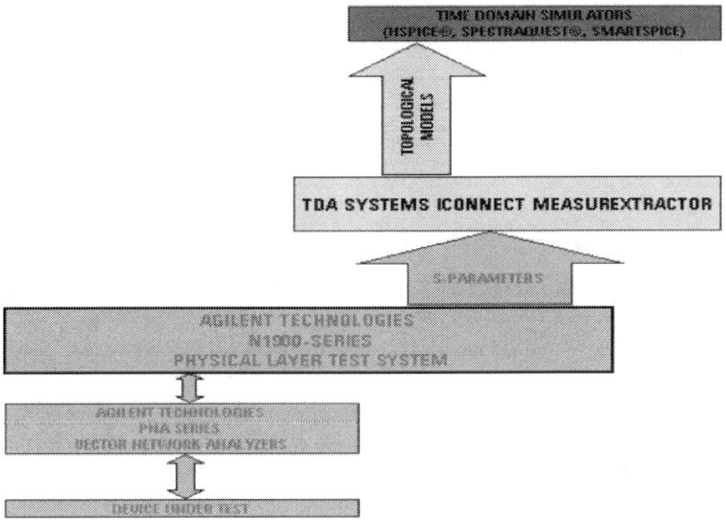

Figure 4.11: *Behavioral S-Parameter Measurements Were Exported into the Model Extraction Tool to Construct HSpice Topological Models.*

4.12 Correlating Measurements and Models

As seen on the upper left-hand side of *Figure 4.12*, the VNA measurement of input differential insertion loss (SDD21) correlates well with the upper right-hand side modeled input differential insertion loss (SDD21) of the model extraction tool that imported the four-port S-parameter file from PLTS. Furthermore, the lower left-hand side eye diagram simulation using the virtual pattern generator of PLTS and the lower right-hand side of the MeasureXtractor eye diagram simulation match quite well with each other. In both eye diagram simulations, a similar algorithm was used in each case. The four-port S-parameter data was used to create an impulse response of the ChannelPlane device. The impulse response was then convolved with an arbitrary binary sequence to achieve a simulated eye diagram. The resultant eye diagram shows the extremely high performance of the ChannelPlane exceeding 40 Gbps. Future experiments will attempt correlation to eye diagrams measured from a digital communications analyzer and 40 Gbps PRBS pattern generator.

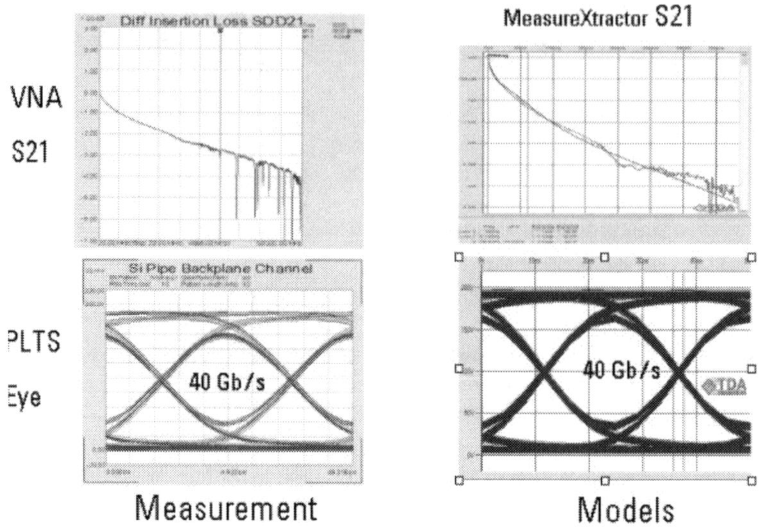

Figure 4.12: *Correlation of Input Differential Insertion Loss (SDD21) and 40Gbps Eye Diagram Was Very Good between Measurement and Simulation*

4.13 HSpice Subcircuit Model for Backplane Only

An HSpice W model was extracted from one leg of a backplane differential cable that had absolutely no physical or electrical defects out to 40 GHz. The HSpice W model is shown in *Figure 4.13* in RLGC format. It includes skin effect resistance (Rac=133 u-Ohms per root Hertz) and dielectric loss (Gac=213 f-Siemens per Hertz) with good agreement between measured and modeled S-parameters, predicting a valid 40 Gbps eye pattern well correlated to VNA/PLTS test data as shown in *Figure 4.12*. However, the second leg of this backplane pair had an internal break, exhibiting a 2 Ohm impedance dip as shown in *Figure 4.14*. The W model for this broken leg was the same as the good leg except for a slight increase in dielectric loss (Gd). It can be concluded that losses associated with the break are shunt losses—linearly dependant with frequency-like dielectric loss.

- .subckt Lossy_Line_34_wire_30in port1 port2 gnd_
- W1 N=1 port1 gnd_ port2 gnd_
  RLGCMODEL=Lossy_Line_34_wire_30in_Model L=1.0
- * RLGC values for W element
- .MODEL Lossy_Line_34_wire_30in_Model W MODELTYPE=RLGC N=1
- + Lo=1.66268e-007
- + Co=6.86542e-011
- + Ro=0.102
- + Go=1.14e-006
- + Rs=0.000133
- + Gd=2.13333e-013
- .ends

Figure 4.13: *HSpice Subcircuit Model of Backplane Cable without Flush-Mount Connector*

When flush-mount connectors were sliced into the backplane cable and re-tested on the VNA, no change in skin-effect resistance was detected in either leg of the differential backplane cable. Only an increase in dielectric loss (Gd) was detected. The flush mount connector appears to behave like a cable break, as shown in *Figure 4.15.*

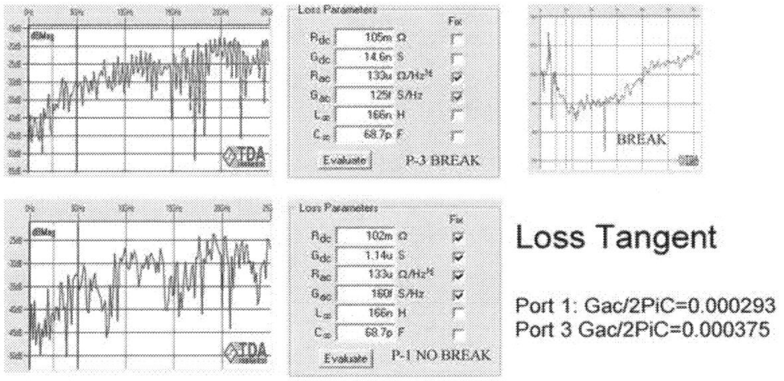

Figure 4.14: *Backplane Cable Assembly Lossy Models Show Loss Tangent with Impedance Discontinuity due to a Cable Break on One Side of the Differential Pair*

4.14 Discontinuities: Flush Connector versus Cable Break

The impedance profile of a cable break shown in *Figure 4.14* is very similar to the impedance profile of a flush-mount connector shown in *Figure 4.15*. Both are modeled as ~2 ohm capacitive discontinuity at a distance from the end of the cable SMA termination. The objective of the flush-mount connection is to maintain a constant impedance through the connection. The observed capacitance discontinuity for both is in the order of ½ pf. The only obvious difference is that there are two flush-mount connectors, or twice the insertion loss, and an elevated return loss due to two breaks rather than a single break.

The impedance profile accuracy of this capacitance discontinuity could be an issue because the connector slice could be only 5 mils wide. The VNA test-step rise time was 10 ps with a propagation velocity in the cable of 120 ps/in. The model extraction tool resolves an impedance profile to ~1/10 of the rise time, or ~1 ps. That translates to 1/120 of an inch of maximum resolution or >10 mils. That means cursor placements used to measure capacitance on the capacitance impedance profile could have resolution issues in the 1–5 mil range. So it is presumed that 400 ff was a maximum and it could be as low as 100 ff. HSpice iterated the capacitance until a return-loss simulation matched the measurement.

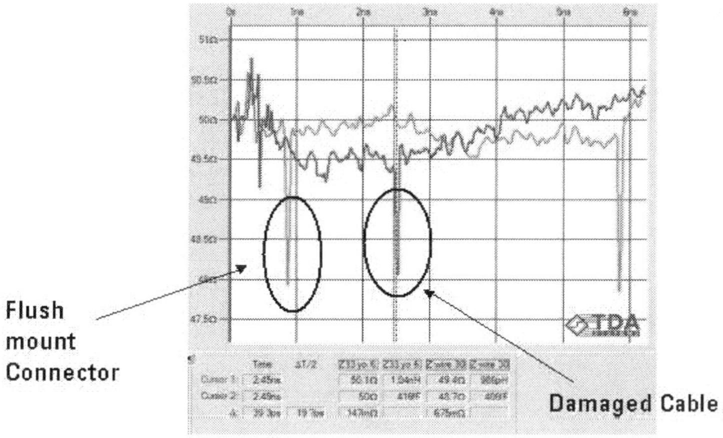

Figure 4.15: *Impedance Discontinuities from Two Different Physical Structures Can Look Very Similar. However, the Intuitive Nature of TDRs Helps Discriminate between a Break in the Cable and a Flush-Mount Connector*

4.15 HSpice Subcircuit Model for Backplane Flush Connector

Connectorized cable modeling yielded a similar HSpice model with the same skin-effect resistance Rs of 111u Ohms per root Hertz. This was extracted on the second VNA 30 GHz test with connector sliced into the previously tested cables. However, Gd increased significantly due to the flush-mount connectors. Again, the model extraction tool S-parameters agreed with VNA measurements. But the eye pattern was degraded over the previous measurements on the cable only. Was the degradation solely related to the flush-mount connectors? Since both the eye height and jitter changed less than 20 percent, it was presumed to be the connectors until proven otherwise. The increase in Gd can be observed with and without connectors by comparing Gd differences in the models shown in *Figures 4.13* and *4.16*.

```
• .subckt Lossy_Line_butt_30in port1 port2 gnd_

• W1 N=1 port1 gnd_ port2 gnd_
  RLGCMODEL=Lossy_Line_butt_30in_Model L=1.0

• * RLGC values for W element
• .MODEL Lossy_Line_butt_30in_Model W MODELTYPE=RLGC N=1
• + Lo=1.62028e-007
• + Co=6.66083e-011
• + Ro=0.106434
• + Go=0
• + Rs=0.000111
• + Gd=2.86667e-013

• .ends
```

Figure 4.16: *HSpice Subcircuit Model of Backplane Cable with Flush-Mount Connectors*

4.16 Model Optimization Schematic: Cable and Connector

The second VNA test simulation schematic is shown in *Figure 4.17*, including both backplane cable and flush-mount connectors. The simulation objective was to build both transmittance and reflectance (TDR/T) waveforms, convert them to S-parameters, and compare the simulations to the VNA measurements. For insertion loss, HSpice was used to overcome the test bandwidth difference between measurements with and without connectors. To do this, simulations first replicated return-loss measurements by adjusting the connector capacitance until a match occurred. It was proven that the return loss was well below -10 dB. Therefore, the simulated TDT waveforms predict valid 40 Gbps eye integrity using flush-mount connectors. HSpice model simulations were fitted out to the 30 GHz, including connector models with elevated return loss and easily extrapolated to more than 40 GHz. A valid 40 Gbps eye was then generated with flush-mount connectors.

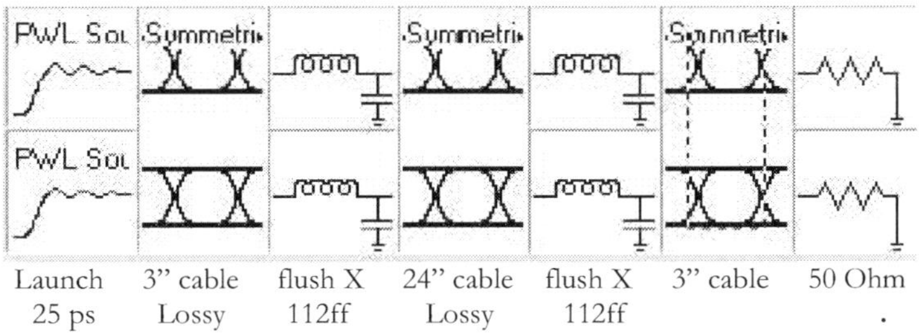

Figure 4.17: *Differential HSpice Schematic Shows Piecewise Linear Source, Lossy Transmission Line Segments, Lumped LC Flush Connector, and Termination*

4.17 Analyzing Connector Shunt Loss

Scaled W-models were interpolated by length from the VNA measurements and used to model three-inch launch and termination cables, as well as the 24-inch backplane cable. The partitioned schematic in *Figure 4.18* shows a TDR source, a three-inch cable launch, a flush connector, a two-foot backplane cable, another flush-mount connector, and a three-inch termination cable to 50 Ohms. Several HSpice runs simulated the TDR voltage waveform with the objective to match a cable insertion and return losses with flush connectors. Only the dielectric loss had to be increased, as shown in the model in *Figure 4.16*, as compared to the model in *Figure 4.13* without flush connectors.

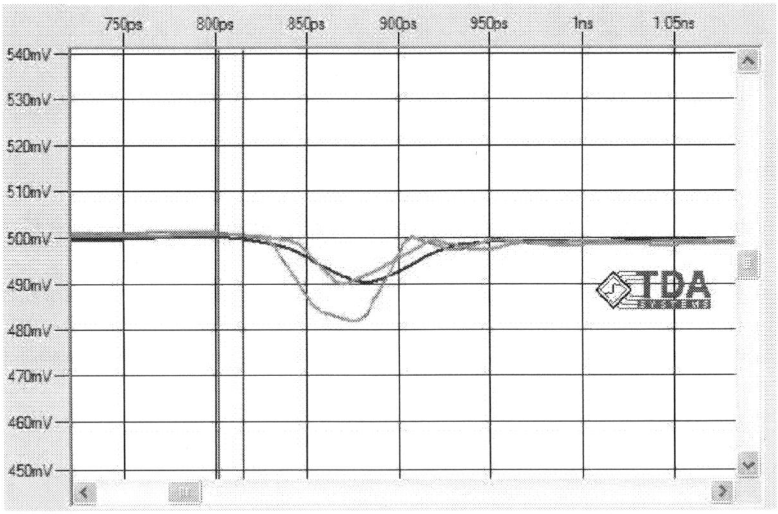

Figure 4.18: *Simulated Cable Differential Schematic for TDR Leg Waveforms Showing Connector Capacitance Discontinuity. Green Simulation Failed at 400 ff but Passed at 112ff*

4.18 Analyzing Fringe Capacitance

Once shunt losses were accurately modeled, the connector impedance profile model was optimized by simulating a TDR waveform and adjusting the connector capacitance to best fit a 10 mv TDR discontinuity as shown in *Figure 4.18*. Once fitted, the modeled TDR waveform was converted back to S-parameters and checked against the measured VNA S-parameters. Different connector capacitance values had to be iterated from a maximum of 400 fF to refine the measurement accuracy. Even if the VNA measurement bandwidth was increased to 100 GHz, HSpice must still be used to overcome errors because the 3D fringe capacitance is physically very small. A -15 dB measured return loss was matched using a 112 fF connector capacitance, which then reached a maximum return loss of -12 dB at 36 GHz as shown in *Figure 4.19*.

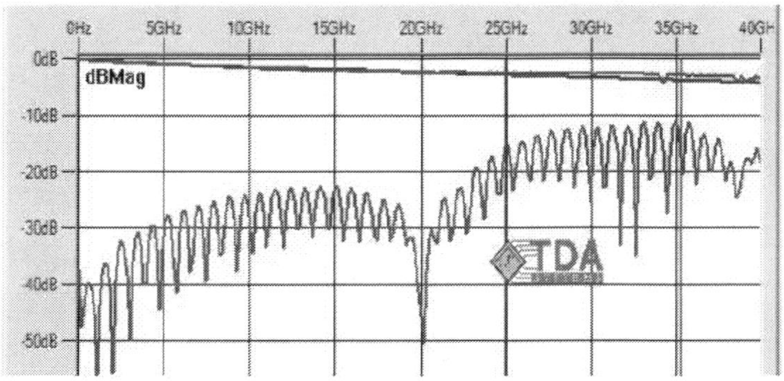

Figure 4.19: *Simulated Insertion Loss (Top) and Return Loss (Bottom) for 30" Backplane Cable Assembly with 112 femtoFarad Connector*

4.19 Modeled S-Parameters from Simulated TDR Waveforms

Simulated TDR waveforms were converted to S-parameters and compared to VNA measurements. The modeled backplane/connector insertion loss (S21) shown as a straight downwards slated line in *Figure 4.19* fits the attenuation measurements at 25 GHz taken by the VNA with flush-mount connectors on the second test. The modeled backplane cable only insertion loss, shown as a continuously curved line, fits the -3 dB at 35 GHz measurement taken by the VNA on the first test. With only a small return loss of -12 dB, the extrapolated insertion loss S-parameters appear valid out to 40 GHz. Given the small (5 ps) rise-time degradation caused by the connectors, this model accurately "predicts" 40 Gbps eye patterns, as shown in *Figure 4.20*. If the backplane, flush-mount connector faces were polished to best match impedance, then return losses could reach -20 dB. Flush-mount connector technology therefore enhances copper interconnect performance out to 40 Gbps, which is a very cost-effective transition point to fiber-optic hardware.

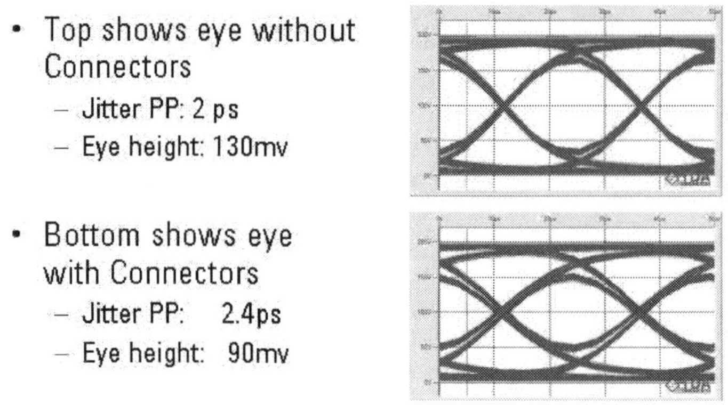

Figure 4.20: *Modeled Eye Diagrams with and without Connectors*

4.20 Modeled Eye Diagram with and without Connectors

The shunt losses in the connectors degraded the input rise time by only 5 ps and lowered the eye height by 10 percent. A 30 ps measured output rise time can be seen in the eye patterns starting at the left side (~10 percent) and reaching 30 ps at the center top (~90 percent). Still, a 20 ps by 75 mv valid eye mask can be constructed inside this eye pattern. A key point is that the connector TDR discontinuity is short enough to minimize reflections sufficiently to not deteriorate the eye opening. For this to occur, the return loss needs to be less than -10 dB. In conclusion, both lossy and capacitance model components are required to accurately simulate backplanes with flush-mount connectors for both insertion and return loss.

4.21 Conclusion

The "polished" cable interface should be considered to minimize return loss in these experimental flush-mount connectors. They were introduced on ends by precision slicing the cable as if it were a fiber-optic connection with a goal to minimize discontinuity across the connection. If a copper interface were "polished" like a fiber-optic interface, then it might achieve a return loss as low as -20 dB.

The model to measurement correlation was excellent for rise time, eye patterns, and S-parameters with no issues identified. An S4P file extracted 40 GHz VNA backplane data set was used to model flush copper cable connectors, comparing S-parameters with and without connectors by slicing them into a test cable. HSpice W model simulations fit VNA S-parameters and TDR/T waveforms well for the backplane cable only, yielding a 150 mv eye height at 40 Gbps with a 200 mv input and 2 ps jitter.

A valid 40 Gbps eye opening was achieved in backplane cables with connectors as modeled in HSpice. The second VNA test evaluated the cable with connectors, transforming 30 GHz S-parameters to TDR/T waveforms via a second S4P file. Results show a 5 ps connector I/O rise-time degradation with return loss below -10 dB, having no effect on the eye opening. But connector insertion loss did diminish eye height by ~10 percent due to shunt loss, which is linearly dependant on frequency-like dielectric loss. A 10 mv capacitance discontinuity was modeled by a 112 ff capacitor placed three inches from the ends of the cable optimized by HSpice simulations. Further study of 3D fringe capacitance should be studied using a 3D time domain field solver to verify dimensions and refine model accuracy.

The advent of high-speed serial channels has driven the circuit topology to differential signaling. While this enables the inherent benefits of coupled pairs of transmission lines, this adds new challenges for signal integrity engineers. Measurement, model extraction, and simulation are critical to an efficient design cycle and meeting time-to-market demands. It sometimes seems as if there are as many design tools as there design engineers, but the message is clear: new techniques that utilize measurement-based modeling are necessary for fully characterizing differential interconnects. It is now possible to use one measurement system for both time and frequency domain information that will quickly identify design flaws that ultimately degrade performance.

Authors

Jim Mayrand, Signal Integrity Consultant

Mike Resso, Product Manager, Signal Integrity Operation, Agilent Technologies

Dima Smolyansky, Product Marketing Manager, TDA Systems

Chapter 5

Accuracies and Limitations of Time and Frequency Domain Analyses of Physical-Layer Devices

5.1 Introduction

The time domain reflectometer (TDR) has long been the standard measurement tool for characterizing and troubleshooting physical layer (PHY) devices and is common in all signal integrity labs. With the push toward higher-speed differential signaling and the need for more accurate characterization and modeling of differential interconnects (e.g., cables, connectors, packages, printed circuit boards [PCBs]) the vector network analyzer (VNA) is becoming more common in signal integrity labs as well. The VNA brings more accuracy, dynamic range, and frequency coverage (faster rise times) to this characterization and modeling. It can cost more than a TDR and is not as familiar to use for the signal integrity engineer.

Depending on the data rates and complexity of the structure, measurements and modeling can be done in either the frequency domain using a VNA or the time domain using a TDR. With commercially available software, it is easy to move between the time and frequency domains and between single-ended measurements and differential measurements, including measurements of mode conversion.

This chapter is a summary of more than a year's work trying to fairly compare and contrast the capabilities of the TDR and VNA. The original work [1] was presented at Design Con 2005 but has been updated and summarized for this chapter. How to perform comprehensive measurements for complete and accurate device or interconnect characterization with either system will be addressed.

To get high-quality measurements, an understanding of the instruments' architecture, calibration, and specifications such as dynamic range, accuracy, noise, and stability will be presented. How each of these affects or limits the quality of the measurement will be addressed in detail. Several calibration techniques are available to remove sources of error in making measurements. These techniques will be compared using results of actual measurements. The differences between the TDR and VNA will be used to show the limitations of specific measurement techniques as well as their impacts on developing models for these structures.

5.2 Equipment Setup

The measurement equipment used in this paper consists of a four-channel TDR with an 18 GHz bandwidth and a four-port 20 GHz VNA. High-quality phase stable cables were used to connect to the devices under test (DUTs), and comparable settings were used on each measurement instrument to achieve as fair a comparison as possible. A typical setup is shown in *Figure 5.1*. The specific description and setup is as follows:

- Agilent N1930B Physical Layer Test System Software (PLTS)
- Agilent Infiniium DCA 86100C with 54754A Differential TDR Modules
- All TDR measurements were taken with a time base of 5 ns, varying rise time, ~2000 pts, and 16 averages.
- All VNA measurements 10 MHz to 20 GHz measured on an E8362B PNA Series Analyzer with a N4419B S-parameter test set.
- VNA measurements are taken over a 10 MHz to 20 GHz frequency range, ~2000 pts, a 300 Hz intermediate frequency (IF) bandwidth, and one average.
- Standard GORE 1 M cables were used for both TDR and VNA measurements of 3.5 mm (SMA) devices.

Note: VNA and PNA will be used interchangeably in this chapter.

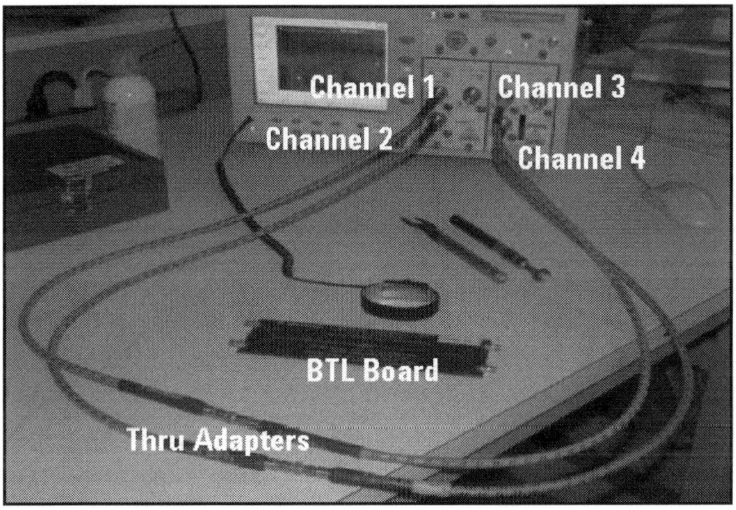

Figure 5.1: *Typical Measurement Setup with Agilent TDR*

5.3 Fundamental Differences between TDR and VNA Instruments

Time and Frequency Domains

There is a duality between the time and frequency domains. The time response can be converted to the frequency domain using a fast Fourier transform (FFT). All the frequencies from the device's characterization in the frequency domain are used to compute the time response using an inverse fast Fourier transform (IFFT).

In the case of the TDR, the measurement is done in the time domain by stimulating the DUT with a voltage step. There is a time delay for the step to travel through the DUT. This delay is related to the length of the DUT. Multiple reflections in the DUT will cause longer delays for the signal to propagate through the device. The size of these reflections can be determined from the magnitude variations.

Measurements with a VNA are done in the frequency domain. The device is characterized at each frequency of interest, one point at a time. The magnitude and phase shift is measured relative to the incident signal. The phase shift is related to the length of the DUT.

The longer the DUT, the larger the phase shifts. Also, the higher the frequency, the larger the phase shifts.

A common measurement in the frequency domain is group delay. Group delay is computed from the phase by taking the derivative of the phase versus frequency. Although the group delay can vary as a function of frequency for most passive interconnects, it is nearly constant and directly related to the time delay measured with the TDR.

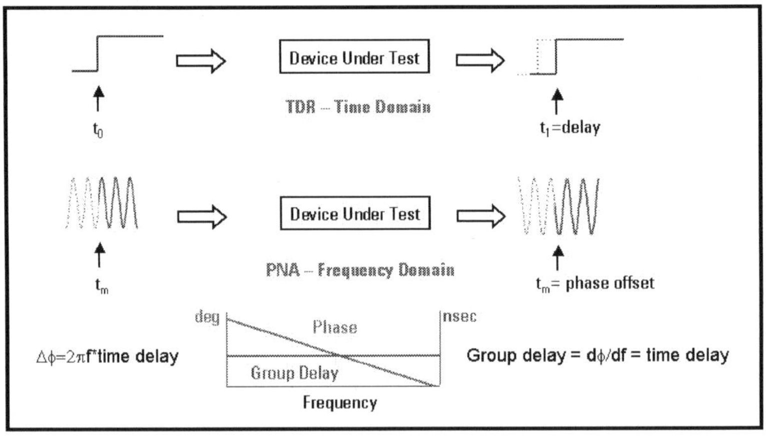

Figure 5.2: *Time and Frequency Measurement Domains*

Measurements in the time and frequency domains are also related. Measurements in the time and frequency domains are also related. A typical measurement in the time domain is a TDR measurement, which is the measure of the signals reflected from the device's input as a function of time. The equivalent in the frequency domain is the S-parameter, S_{11}, which is the input match or the input reflection coefficient. S-parameters are the ratio of the reflected wave (power) to the incident wave. Similarly, a time-domain transmission (TDT) measurement shows the incident pulse after traveling through the device. The corresponding TDT measurement in the frequency domain is S_{21}, the ratio of the transmitted wave to the incident wave. More information on S-parameters is available [2]. Jitter in the time domain is related to phase shift in the frequency domain (*Figure 5.3*). The phase (in radians) is equal to 2 * pi * frequency (in Hz) * delay (in seconds). Microwave engineers talk about a device having a lot of phase, while high-speed digital designers talk about a device having significant length or delay.

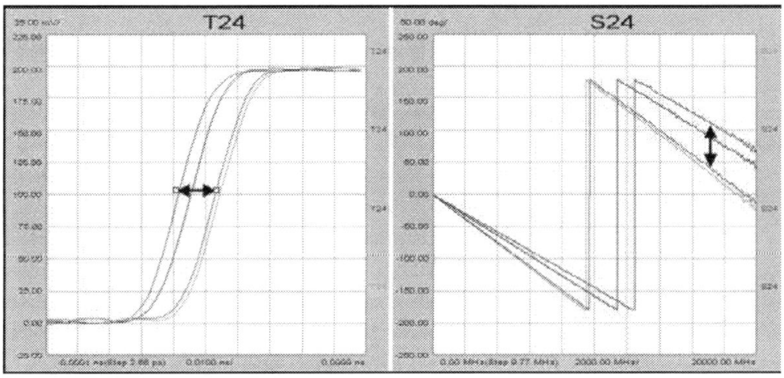

Figure 5.3: *Jitter in the Time Domain Is Phase Shift in the Frequency Domain*

5.4 TDR and VNA Sources

In the case of the TDR, the source is a voltage step generator. The step generator typically puts out a voltage step with a rise time of about 40 ps. The frequency content of the step depends on the rise time of the step, and as the power decreases, the frequencies get higher. This causes loss of dynamic range and accuracy for higher frequencies. The VNA source is a single-tone frequency that is swept across a desired frequency range. The source power is typically leveled in a VNA and is constant over the entire frequency band, which does not cause loss of accuracy for higher frequencies. *Figure 5.4* shows the sources in both domains.

TDR and VNA Receiver Bandwidths

The TDR has a broadband receiver with the choice of 12 or 18 GHz 3 dB bandwidths. The VNA has a selectable IF bandwidth. The bandwidth can be set from 1 Hz to 30 KHz (*Figure 5.5*). This narrow bandwidth significantly reduces the noise floor, to better than –110 dBm. Due to the wideband receiver of the TDR, the noise floor is higher, limiting the TDR's dynamic range to about 40 dB compared to the VNA's dynamic range of greater than 100 dB (*Figure 5.6*). When also considering the source power rolloff at the higher frequencies of the TDR, the TDR signal-to-noise ratio (SNR) above 10 GHz noticeably decreases.

For the TDR, the overall system rise time can be calculated from the following equation:

system rise time = square root of (scope rise time^2 + step rise time^2 + test setup rise time^2)

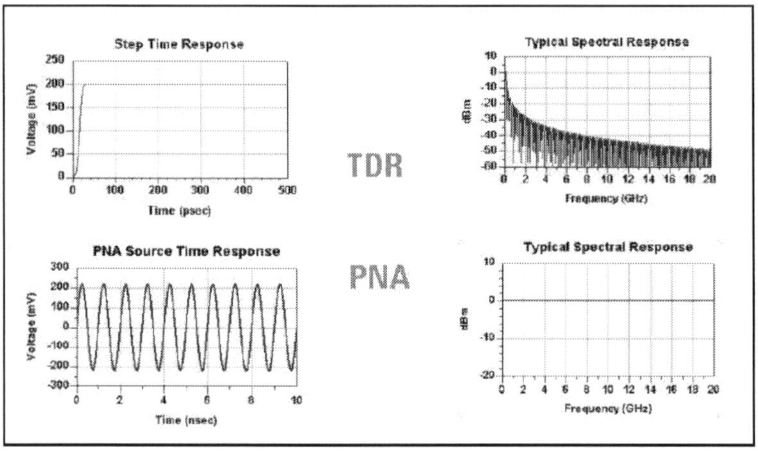

Figure 5.4: *TDR and VNA Sources*

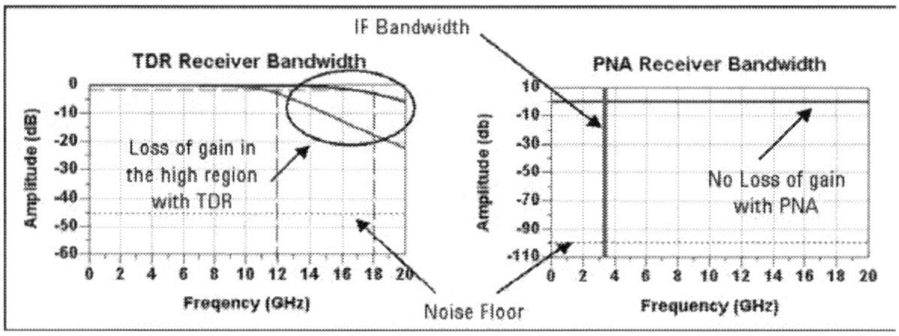

Figure 5.5: *TDR and VNA Receiver Bandwidths*

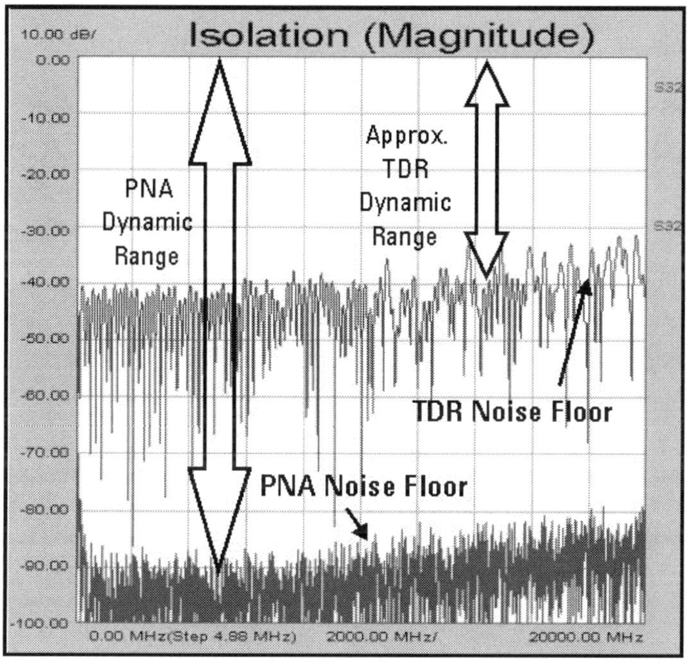

Figure 5.6: TDR and PNA Dynamic Range

5.5 Architectures and Sources of Error

Figure 5.7 shows a simplified block diagram of a two-channel TDR. Each channel has a step generator that generates the stimulus to the DUT, a sampler, and an analog-to-digital converter (ADC) to measure the signal. For TDR measurements the ADC (e.g., channel 1) samples the incident pulse and the reflected signals from the DUT. For TDT measurements, the signal is transmitted through the DUT and sampled by the ADC on channel 2. A common clock triggers each step generator. Jitter, timing, and drift will vary slightly between step generators and will be addressed in detail shortly.

Sources of error for the TDR can be divided into three areas. The first is errors due to the oscilloscope receiver channels. The second area is the step generator itself, and third is the cables and connectors used to connect to the DUT.

- Oscilloscope
  - Finite bandwidth restricts it to a limited measurable rise time
  - Small errors due to trigger coupling into the channels and channel crosstalk
  - Clock stability causes trigger jitter in the measurement
- Step generator
  - Shape of step stimulus (rise time of the edge, aberrations on the step, overshoot, non-flatness)
- Cables and connectors
  - Introduce loss and reflections into the measurement system

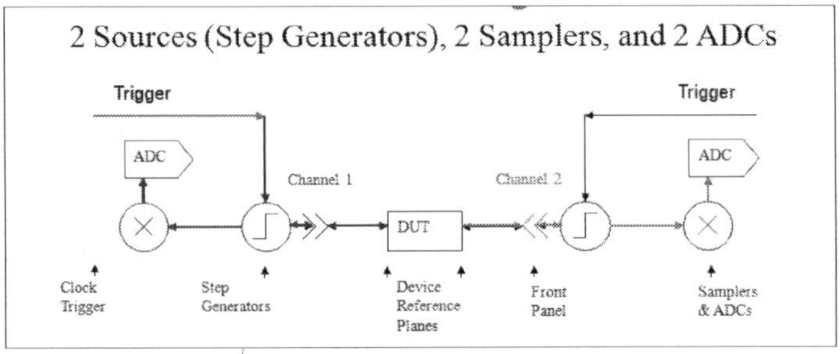

Figure 5.7: *Simplified Block Diagram of a Two-Channel TDR*

Figure 5.8 shows a generic block diagram of a two-channel network analyzer. The single source can be switched to excite port 1 or port 2 of the DUT. The switch also provides a Z_0 termination for the output port in each direction. Directional couplers are used to separate the incident into reflected and transmitted waves in both the forward and reverse direction. Mixers are used to down-convert the RF signals to a fixed low-frequency IF. The local oscillator (LO) source is tuned to the frequency of the RF + IF.

The S parameters of the DUT can be defined as follows:

- $S_{11} = b_0/a_0$, switch in forward direction
- $S_{21} = b_3/a_0$, switch in forward direction
- $S_{12} = b_0/a_3$, switch in reverse direction
- $S_{22} = b_3/a_3$, switch in reverse direction

For a VNA, there are random errors such as noise, switching, and connector repeatability that are not corrected by calibration. There are also systematic errors that are corrected by calibration techniques. There are leakage terms such as directivity errors in each directional coupler and crosstalk between ports. The source and load presented by the VNA are not perfect and result in reflections due the mismatched impedances. Finally there are frequency response errors due to imperfect tracking of the receivers and signal paths. For a two-port measurement, there are 12 error terms and for a four-port measurement there are 48 error terms that need to be corrected in the measurement. For the two-port case the error terms are listed below in *Figure 5.9*. More information of VNA error terms and correction is available [3].

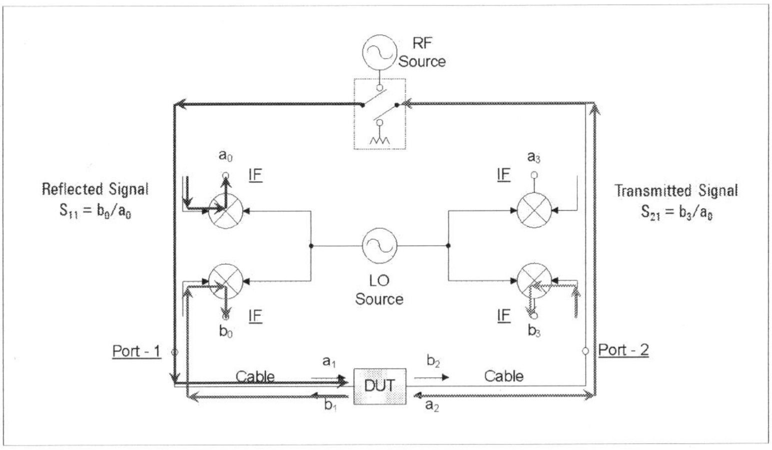

Figure 5.8: *Simplified Block Diagram for a Two-Port (Four-Channel) VNA*

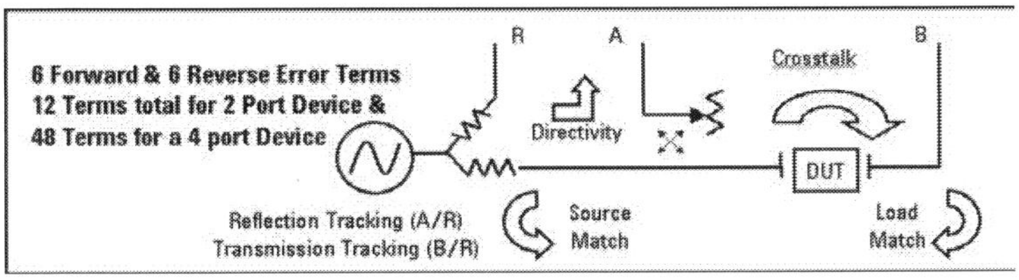

Figure 5.9: *Systematic Errors in a VNA*

5.6 Calibration and Normalization

Overview of TDR and VNA Calibrations

For a VNA one calibration does it all. It removes the systematic errors due to the instrument, test set, and cables used to connect to the DUT. All 48 error terms for a four-port measurement are removed by connecting a short, open, and load to each port and connecting a thru between a minimum of three or all six thru paths. Using extremely accurate calibration kits, this provides the most accurate measurements of S-parameters for linear devices. The S-parameter data taken in the frequency domain can be easily transformed into the time domain by using an IFFT. All of the calibration data (error terms) is stored in a single file on the VNA.

Calibrating a TDR for all the measurements for a four-port device is more complicated. The process requires more than one calibration. First, each of the modules needs to be calibrated. This is referred to as a module or vertical channel calibration. All the test cables are disconnected from both modules, and the calibration requires placing a load on each channel at the directed time in the calibration process. This calibration calibrates the ADCs and timing in the modules. When completed, the modules are calibrated to connectors on the front of the module. When this is completed, the cables are reconnected to the modules and the second calibration begins.

There are two choices for this second-tier calibration when using Agilent TDRs. A reference plane calibration (RPC) is the quickest but least accurate calibration. All that is required is to leave the test cables open and the PLTS will find the end of the cables and set the measurement reference planes to that point. This is done for single-ended, differential, and common-mode reflection measurements for channels. Thrus are then connected to each of the six thru paths. The RPC calibration removes the delay of the test cables by delaying the measurement time appropriately. Note this does not correct for the loss in the test cables, overshoot and ringing of the step generators, or reflections due to mismatch errors. For differential and common-mode measurements, any skew in the test cables and step generators is automatically removed. The reference plane is then set to the end of the "de-skewed" cables.

For Agilent TDRs, a more accurate calibration can be used for the third calibration (part of the second tier). This process is called normalization. After the RPC calibration (leaving an open on the test channel) normalization can begin. For single-ended TDR calibrations, a short and load are placed on each channel. To calibrate channels 1 and 2 and channels 3 and 4 for differential/common-mode calibrations, first, two shorts are placed on channels 1 and 2. Then, two loads are placed on channels 1 and 2. Finally, two shorts are placed on channels 1 and 2. The same is repeated for channels 3 and 4. The normalization process removes the cable loss and reflections due to source and connector mismatches and cleans up the shape of the step generator. To complete all of these normalization steps, 24 normalization and 24 setup files are created and stored on the hard drive in the TDR and two files are stored on the personal computer (PC). These 50 files are recalled and used when measuring the DUT. The management of all of these files is automatically handled when using the PLTS.

Note: For all of these calibrations for both the VNA and TDR, it was assumed that a current factory calibration of the hardware was done.

The most accurate calibration is the VNA calibration, followed by the TDR normalization. The next is the reference plane calibration followed by module calibration only. The least accurate is to do an uncalibrated measurement. An uncalibrated measurement has none of the systematic errors removed and is only useful to get a quick idea of the general response of the DUT.

Comparing TDR Calibration Methods with a VNA Calibration

Figure 5.10 shows the measurement of a thru adapter with the different levels of calibration available with a TDR and a VNA. The TDR results are noticeably less accurate than the measurement obtained with a PNA. The data shows that normalization at faster edge rates get closer to PNA measurement accuracies but will never equal it without additional correction. When looking at phase, the offset seen from the PNA measurement to the TDR measurement with a 30 ps rise time is not due to the slower rise time, but rather the drift of the step generator. This phase correlation (as with the 20 ps edge) will overlay or not overlay depending on when the measurement was taken.

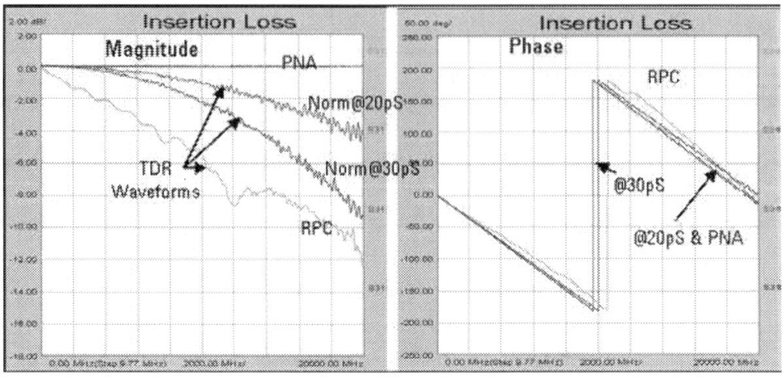

Figure 5.10: *Comparing Calibration Types*

Normalizing at faster rise times gives better accuracy overall and is especially important for accurate measurement data in the higher-frequency region. There are limits to how fast a rise time is acceptable by the TDR instrumentation based on what your time base and number of point settings are, as well as a real minimum. While normalizing at faster rise times results in more accurate S-parameter data, it also increases the noise when looking at the data in the time domain, sometimes making mV or impedance value readings more difficult.

Figure 5.11 illustrates the increase in noise that can be seen when normalizing at faster rise times. This increase in noise can only be partially compensated for by adjusting other parameters such as averaging.

The increase in noise is due to the difference in the bandwidths of the filters used in the normalization process. The basic system response has a predictable cutoff frequency represented by f_c in the left plot of *Figure 5.12*. Through the process of normalization filters are used that accentuate the higher frequencies more than the basic system response. While this provides a faster edge by allowing some of the higher-frequency components of the edge to pass, it also allows some of the high-frequency noise to pass through the filter effectively raising the noise floor of the whole system (right plot).

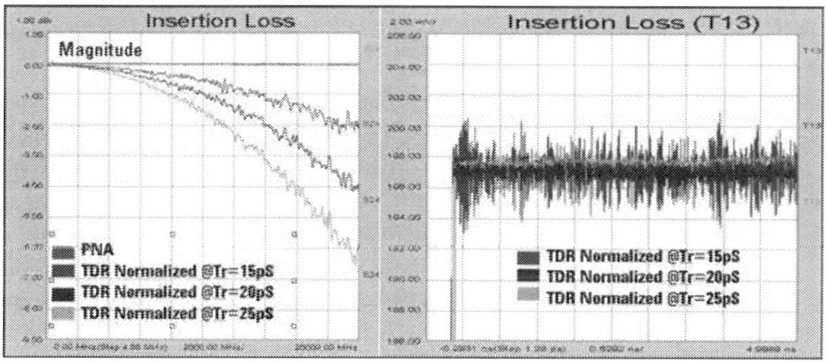

Figure 5.11: *Rise-Time Effects on Frequency Response and Noise*

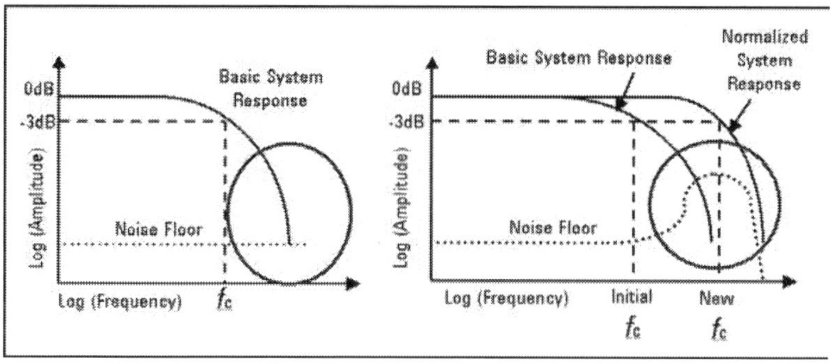

Figure 5.12: *TDR Bandwidth and Noise Floor Changes Due to Normalization*

To correct for the frequency response limitation that is caused by the finite rise time of the step in the TDR, additional correction can be applied. During the calibration process, when a thru (for transmission calibrations) or short (for reflection calibrations) is connected, PLTS saves the time domain data on the PC. The time domain data is then converted to the frequency domain and used to correct the measured data. *Figure 5.13* shows the effect of this correction. The Insertion loss for a short thru is shown in the bottom trace (normalized to 20 ps). The top trace shows the same measurement with the additional correction. It is much closer to the result measured by the PNA (*Figure 5.11*). However, it should be noted that transforming this frequency data back to the time domain will result in a different time response. Therefore, care should be taken when transforming between domains as is done in PLTS.

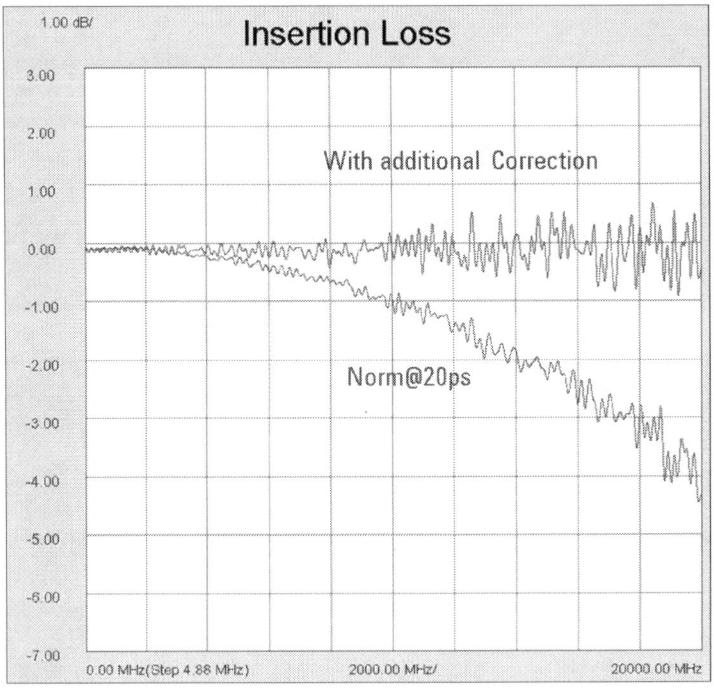

Figure 5.13: *Measurement of a Thru with Additional Correction*

Summary of a Good TDR Calibration

- The real advantage of calibration (and more particularly normalization) is that you can remove unwanted effects of cables and connectors leading up to your device.
- Errors in the magnitude and phase (S-parameters) increase as a function of frequency due to limited bandwidths and increasing noise.

- Faster rise times will result in higher-frequency domain accuracy (in magnitude), but the noise increases in the time domain.
- A good calibration at a reasonable rise time will show acceptable noise in the time domain.

5.7 Measurement Accuracies: Reciprocity, Repeatability, and Drift

In order to help gain insight into the various levels of accuracy available with the TDR and VNA instruments, it is important to understand not only calibration, but also reciprocity, repeatability, and drift. Understanding these attributes of measurements will help determine which instrument should be used based on accuracy needs.

Magnitude and Phase Reciprocity of a TDR and VNA

In its simplest sense, reciprocity maintains that for a passive linear DUT, the insertion loss (magnitude and phase) in the forward direction must equal the magnitude and phase in the reverse direction. This is true for single-ended (SE) devices (*Figure 5.14*) as well as for balanced devices. Instrument architectures play a large part in reciprocity, since in some cases there is a single source and triggers when coming from the different directions (VNA) and in some cases there are different sources (TDR). This is important not only because it is a measure of the quality of the data, but also because some tools when importing S-parameter data require a minimum amount of reciprocity in order for their internal algorithms to operate and converge properly.

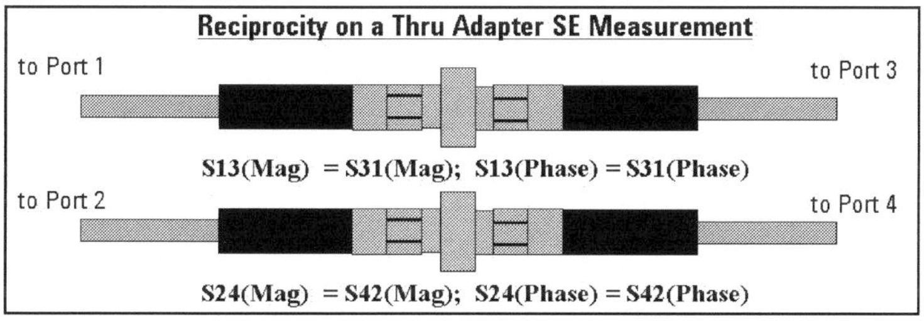

Figure 5.14: *Reciprocity Definitions for Thru Adapter Measurement*

So far our measurements have been on a short thru adapter. The differential device we will be measuring is a small demo board containing a balanced transmission line (BTL) on FR4 material. It is a 100 Ohm line with a 50 Ohm line in the middle. Using the convenient QuickMath function associated with PLTS, it is very easy to get a vector difference between measurements flowing through a device in two directions. *Figure 5.15* shows the TDR measurement of the BTL board. The correlation for both magnitude and phase gets worse as the frequency increases. Below 6 GHz the correlation is very good. From 6–16 GHz it is still good, but gets noticeably worse beyond 16 GHz.

Measuring the same device with a VNA (*Figure 5.16*) after performing an short-open-load-thru (SOLT) calibration, there is excellent agreement in both magnitude and phase over the whole frequency range.

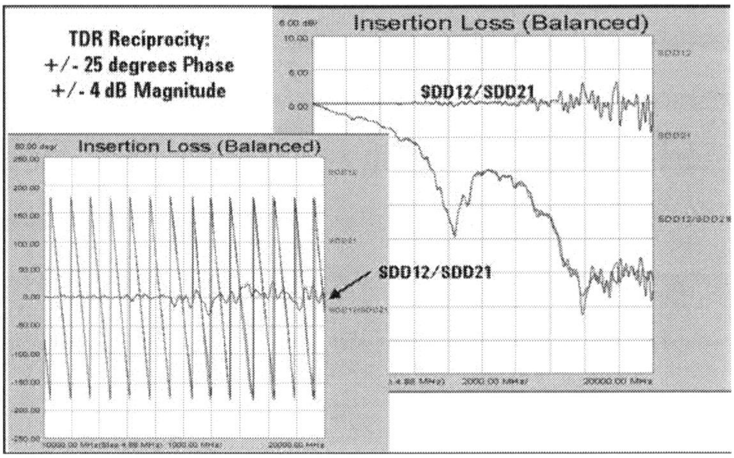

Figure 5.15: *Differential Reciprocity of a BTL Board Measured with a TDR*

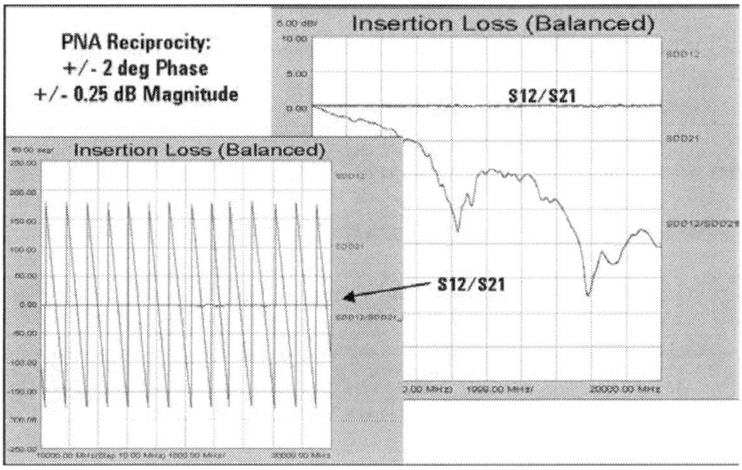

Figure 5.16: *Differential Reciprocity of a BTL Board Measured with a PNA*

Comparing the error in reciprocity for this case study example between TDR measurements and PNA measurements, we can see that the error for the TDR measurements is approximately a factor of 10 higher for TDR.

To understand why reciprocity and repeatability are significantly worse with the TDR relative to the PNA, it is important to understand the architectural differences of the two instruments. The TDR has four sources and the trigger jitter may not be the same for all sources, making near-zero reciprocity (as we get with the PNA) almost impossible. The TDR also has four receivers that will tend to exhibit very small differences in their behavior. The PNA has one source and a few receivers that are switched to the different ports with very stable switches and interconnect. These two measurement approaches and two architectures give two levels of accuracy in our measurements.

Looking at the delay (or phase) of a zero-length thru measured with a TDR, we can see that signals out of the two ports do not arrive at the DUT at the same time (*Figure 5.17*). For this measurement, the normalized reference planes have been set to the end of the cable, taking out any delay and loss associated with it.

Since some time has elapsed since calibration (even a few minutes), trigger jitter and source drift affect the two ports differently. The two plots show how bad this drift can actually get.

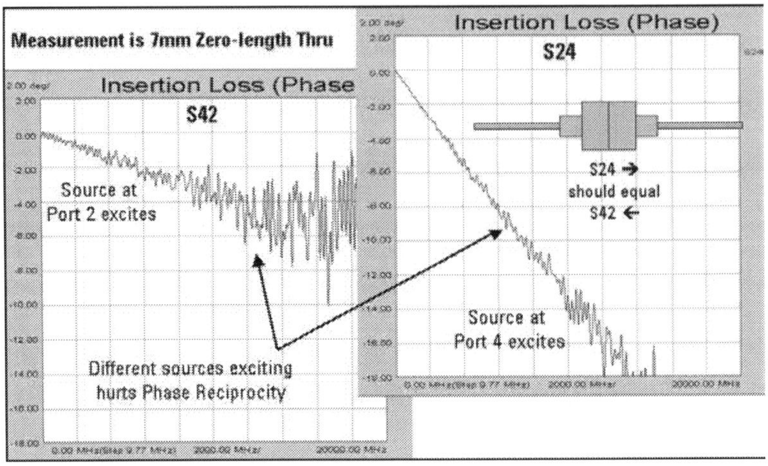

Figure 5.17: *Reciprocity Differences Due to Two Sources in TDR*

Magnitude and Phase Repeatability of a TDR and VNA

Having different sources at the different ports also affects the repeatability of our measurements, even though we perform a calibration immediately before our measurement. The two balanced measurements shown in *Figure 5.18* are an example of the difference that can be seen in the day-to-day drift of TDR channels. The same device (BTL) was measured on two days, with a new calibration performed each time. We can see that the same hardware issues that limit good reciprocity within a measurement also effect repeatability from day to day or week to week. We can easily see ±60 degrees and ±4 dB of repeatability at the higher-frequency ranges. Up to about 10 GHz the repeatability is very good.

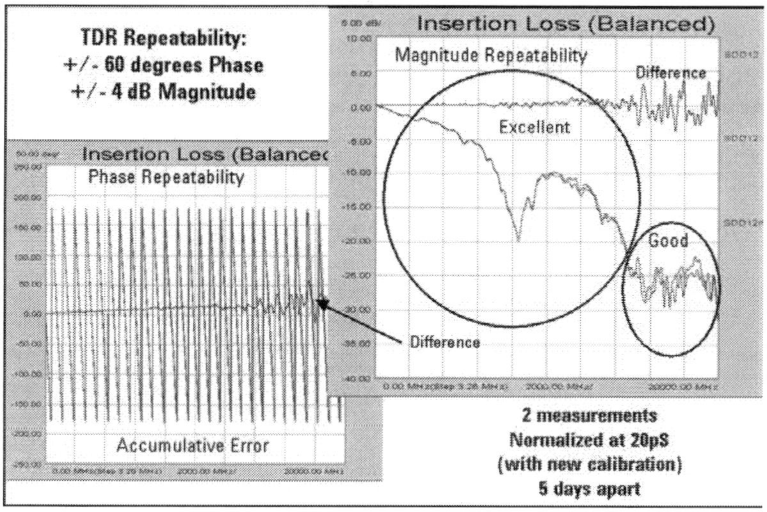

Figure 5.18: *TDR Measurement Repeatability*

By contrast, a PNA measurement strategy exhibits not only very good reciprocity but also excellent repeatability from day to day or week to week mostly because of its superior architecture. With fewer sources and fewer receivers, it follows that there are fewer areas to introduce errors.

With the PNA instrument, phase stability (*Figure 5.19*) is excellent across the entire range and magnitude repeatability is also excellent with <0.5dB of difference between measurements performed on different days.

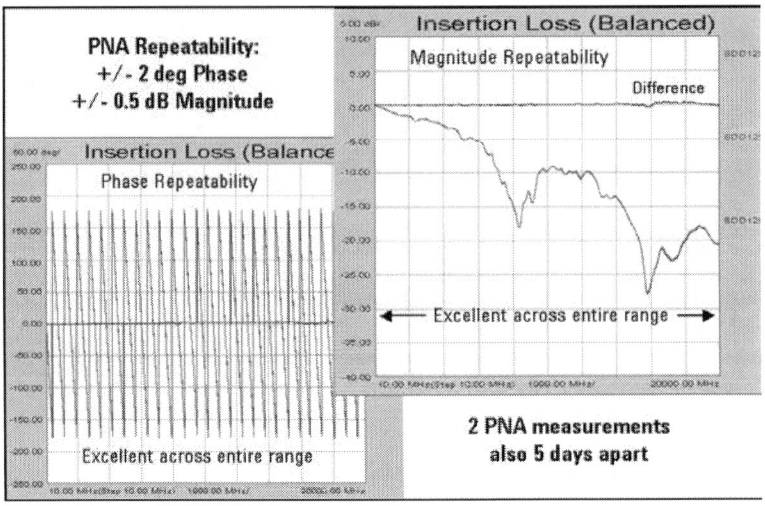

Figure 5.19: *PNA Measurement Repeatability*

Drift of a TDR and VNA

Another issue to examine is drift (differences between measurements taken initially and those taken later in the day without recalibrating). The shifts in the time delay exactly correspond to the shifts in phase, again proving there is a linear relationship between the two. The other important thing that this illustrates is that source and trigger drift are bounded. Instead of drifting unbounded, the TDR channel sources will tend to drift away from an initial result and drift back in a somewhat predictable fashion. In *Figure 5.20*, we see the drift is bounded within a range of about 13 ps, or half of the 25 ps rise time. This drift is fixed and is not dependent on rise time.

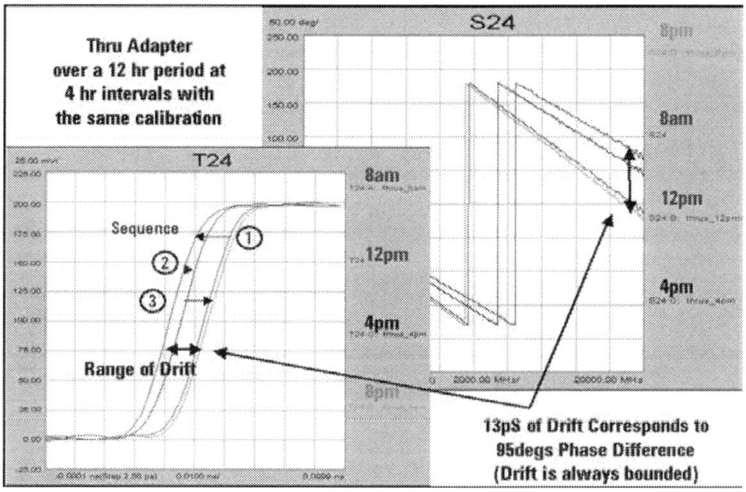

Figure 5.20: *Jitter Drift (Phase Drift) of TDR Source*

Figure 5.21 is a summary comparison of TDR and PNA approximate reciprocity, repeatability, and drift. This data is taken from measuring the same BTL board, so it may or may not be indicative of the overall results that a user would experience with a significantly different device. As we can see, the PNA not only has the capability to collect more accurate measurements as we have seen previously, but it also exhibits more consistent and stable results over time. Most of this, as we have pointed out, is due to the differences in the actual instrument architectures.

	Reciprocity	Repeatability	Drift
TDR*	2-4dB dependent on calibration +/- 25 degrees	3-6dB dependent on calibration +/- 60 degrees	Magnitude within noise of instrument 210deg @ 20GHz
PNA	0.25dB Magnitude +/- 2 degrees	+/- 0.5dB for Magnitude +/- 2 degrees	Magnitude within noise of instrument < 5 degrees

* TDR values may seem larger than expected. It should be noted that these values are at the high end of the frequency range

Note: Values appearing in this table are estimates and do not infer or imply any guarantee by Agilent Technologies as to actual results.

Figure 5.21: *Summary of Reciprocity, Repeatability, and Drift*

5.8 Measurement Comparisons

Devices that will be measured and compared include a SE (two-port) verification standard and a balanced (differential) structure built in FR4. The SE device is taken from the Agilent Technologies 85053B 3.5 mm verification kit. This device is a mismatched airline. It was chosen because its characteristics are very well known and come with measured data that is traceable to National Institute of Standards and Technology (NIST). After understanding the differences found in the SE device, a more complicated and lossy differential device (BTL) will be measured and compared.

The instruments used for these measurements were described earlier and include the Agilent DCA with two 54754A differential TDR modules and an Agilent E8362B PNA series network analyzer with a N4419B test set. The measurements were taken with 2,000 points, covering a frequency range of 20 GHz and a time base of 5 ns. High-quality phase stable 1 Meter Gore cables were used with all the instruments.

Single-Ended Comparisons of TDR and PNA Measurements

A picture of the mismatched line is shown in *Figure 5.22*. It is a 50 Ohm airline with a step in diameter. This step in diameter changes the characteristic impedance to 25 Ohms. This device is basically a 25 Ohm transmission line with a short section of 50 transmission line on each side. The interesting characteristics of this device are that the impedance step is very accurate and causes a well-defined resonance pattern in reflection measurements and a known variation for transmission. Other than the mismatch, it is also a very low-loss device. The distance to the step and the impedance are well controlled.

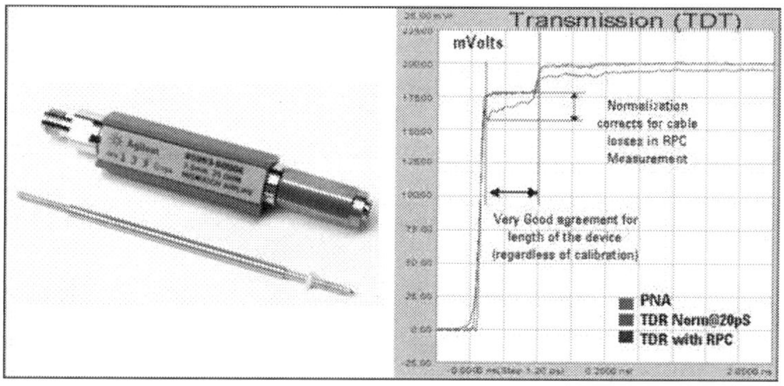

Figure 5.22: *A 25 Ohm Mismatched Airline with TDT Measurement*

Figure 5.22 also shows a plot of the TDT response of the mismatched line. The PNA data (the smooth trace) is the most accurate. The trace that corresponds most with the PNA data is normalized TDR data and has very good agreement with the PNA data. It is just has more "noise" on the trace. The bottom trace is the RPC data from the TDR. The rise time for the RPC data is slower and there are significant differences in amplitude from the normalized data and PNA data. However, all three measurements predict the location of the stepped impedance accurately. The data shows the time references for the instruments are all accurate and calibration techniques accurately remove delays associated with cables to precisely set the measurement reference plane.

The insertion loss (*Figure 5.23*) shows the typical "sinusoidal" variation of the mismatched airline. The variation is from 0 dB to –

2 dB of insertion loss. The normalized (20 ps) and corrected TDR magnitude data correlates well only for the first three divisions. Then the decreasing dynamic range and increased noise due to the normalization filter causes the measurement to become very noisy and the accuracy decreases quickly. Also note that even with just 2 dB of loss, the TDR data has noticeably more noise as frequency increases. The RPC data (without correction) has the general variations, but the loss is significantly pessimistic. At 20 GHz it is showing an additional 12 dB of loss. The phase is reasonable but gets more inaccurate at higher frequencies, with the RPC data being the worst.

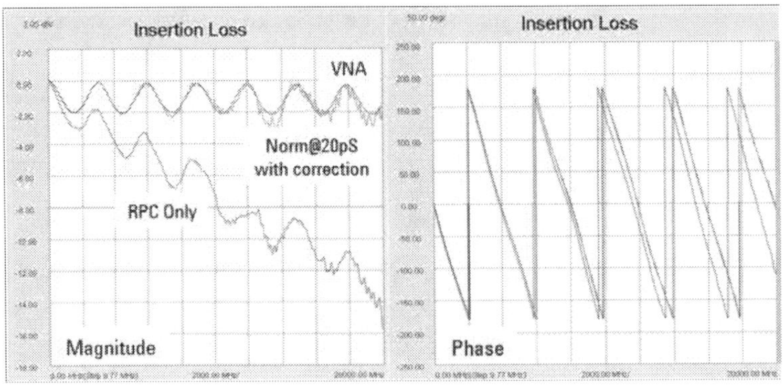

Figure 5.23: *Magnitude and Phase versus Frequency for 25 Ohm Mismatched Line*

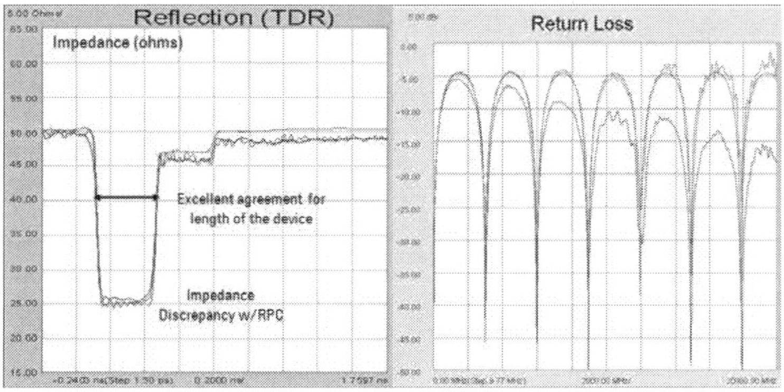

Figure 5.24: *TDR and Return Loss Measurements of the 25 Ohm Mismatched Line*

The TDR response (*Figure 5.24*) of the mismatched line again shows good agreement for the positions in time for the steps in impedance. The normalized TDR data agrees with the PNA data through the mismatch area of the line (again more noise). However the agreement for the reflected step and rest of the line has an offset. The RPC data missed the 25 Ohm impedance of the step by a couple of Ohms. The return loss shows the PNA correctly measured the resonances of the mismatched line and the 4–5 dB peaks. The normalized data (and corrected) again does well for the first four or five divisions and then starts to show more noise and variation due to loss of power and dynamic range. The RPC data (without correction) catches the resonances but shows 10 dB too much loss at 20 GHz.

Balanced (Differential) Comparisons of TDR and PNA Measurements

The balanced device to be measured is the balanced transmission line demo board that is included with the physical layer test system. It is a coupled microstrip transmission line on FR4. The line has a step width (impedance) change in the middle of the line and then returns to 100 Ohms balanced. It is in the picture (*Figure 5.1*).

Studying the results of the SE measurements, will give insight into what is happening in the more complicated balance devices. The same trends seen in the SE measurements will be seen here.

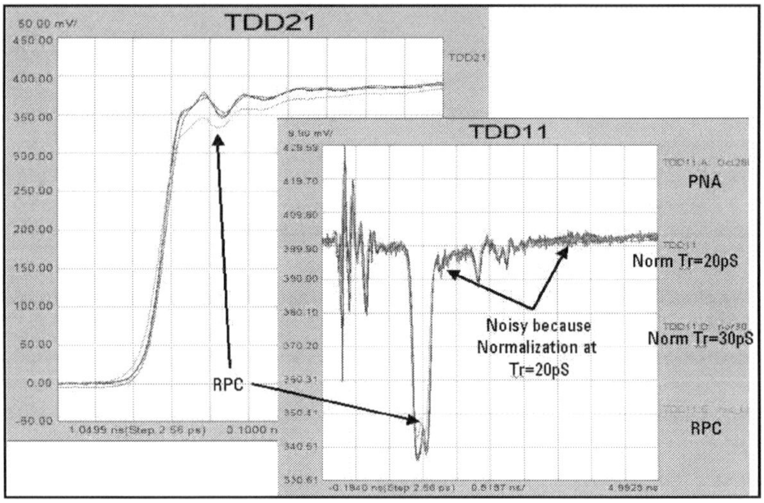

Figure 5.25: *Differential TDT and TDR Responses of the BTL Board*

In both the differential TDR and TDT measurements, there is good agreement between the PNA (most accurate) and the normalized TDR measurements. Again the TDR measurements has noticeably more noise than the PNA measurement but still goes a good job of measuring the time domain response of the device. The RPC data again misses the value of the step in voltage (impedance) for the TDR measurement and predicts too much loss for the TDT measurement.

Figure 5.26 shows the return loss in the frequency domain. The 20 ps data (without correction) comes closest to matching the PNA data. There is very good agreement in the lower third of the frequency range, good agreement in the midband, and OK agreement in the high band. The 30 ps data is less accurate, and the RPC data is only good for the lower band and then predicts too much loss.

The phase agrees well for the lower band but starts deviating in midband and continues to deviate and get noisier at higher frequencies. The RPC data has a problem around 12 GHz, where the resonance is, and there is a larger phase error.

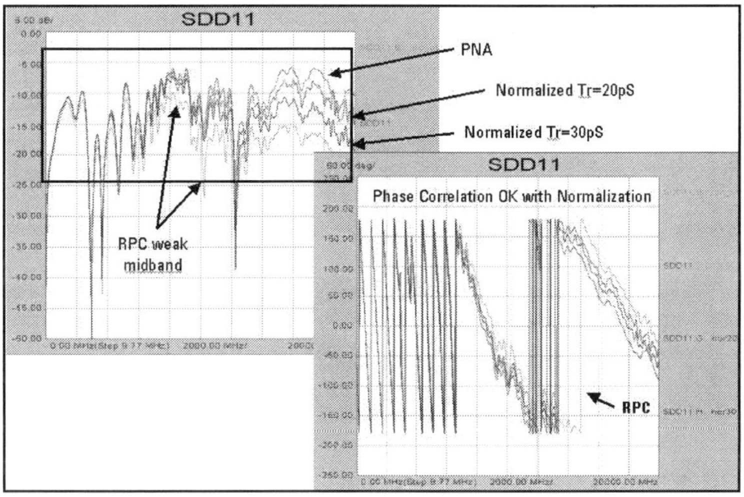

Figure 5.26: *Differential Return Loss for BTL Board*

Just looking at the PNA data and the 20 ps normalized data (this time with the additional correction), the difference can be seen in the top trace in *Figure 5.27*. The top trace is the vector difference (division) of the two traces. The top trace can be broken into three bands. Below 6 GHz the error is less than 1 dB. From 6 GHz to 14 GHz there is about 2 dB of error, and above 14 GHz the error increases to 6 dB. Note: we are ignoring the spikes in the top trace that are caused by the slight differences in the large resonances in the data.

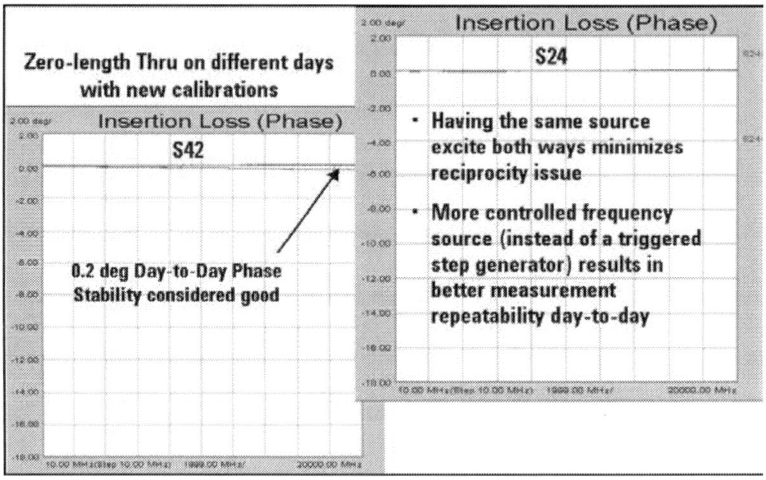

Figure 5.27: *Reciprocity Drift over a 24-Hour Period for VNA*

Figure 5.28 shows the insertion loss for the BTL board. Again for this measurement the normalized TDR data at 20 ps is the closest to the PNA data. The RPC data shows the loss to be about 12 dB too much. The phase is good at the low-frequency ranges and deviates more as frequency gets higher.

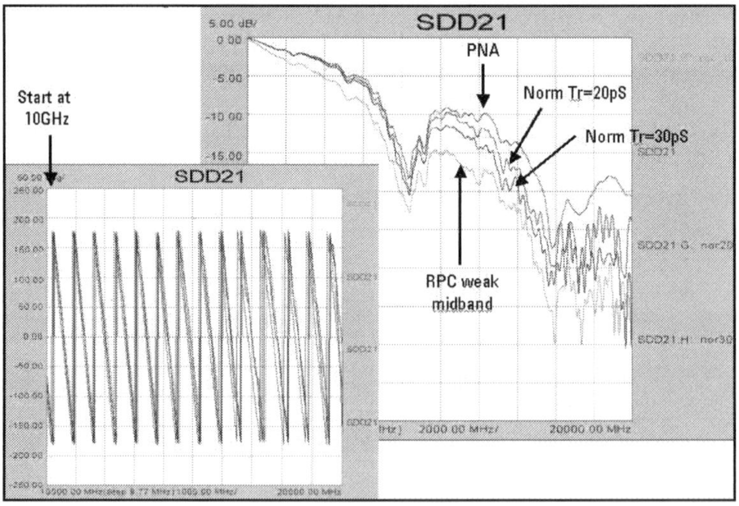

Figure 5.28: *Differential Insertion Loss for BTL Board*

In *Figure 5.29*, the top trace is the error of the jagged bottom trace, assuming the PNA data is the most accurate. The jagged bottom trace is the TDR data normalized to 20 ps with additional correction. Again the first three divisions (up to 6 GHz) are very good with about 1 dB of error. From 6 GHz to 14 GHz the errors increase to 2 dB. Above 14 GHz they continue to increase to 6 dB.

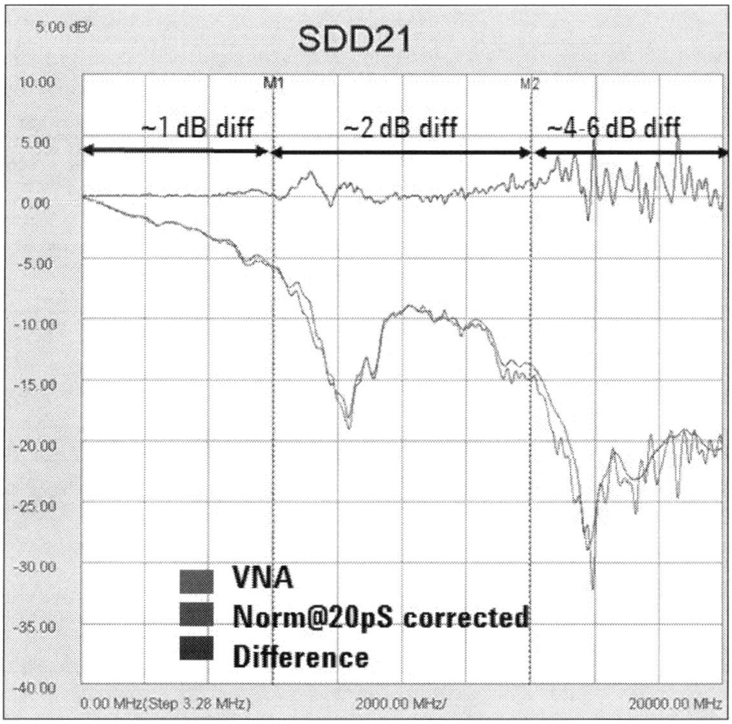

Figure 5.29: *Errors in Differential Insertion Loss TDR Measurements*

5.9 Summary

The TDR has long been used in signal integrity labs for characterizing passive structures. The VNA is becoming more popular in labs as data rates increase and digital standards require frequency domain characterization. Models can be developed from either TDR or VNA data. The VNA clearly provides the most accurate data in both time and frequency domains. Models using S-parameters directly will be the most accurate when measured by a VNA. The Agilent TDR 86100 with normalization gives time domain data very close to that derived from a 20 GHz VNA. To get this close correlation, a fast rise time needs to be selected after normalization. This leads to noisier data in the time domain than data from the VNA. Frequency domain data derived from TDR data rolls off at higher frequencies. The rolloff is dependent on the rise time selected. Without additional correction this rolloff leads to error that can be interpreted as pessimistic insertion loss data and optimistic return loss data for frequencies greater than 10–12 GHz. With the additional correction the data looks good to about 14 GHz with the accuracy decreasing and the noise increasing out to 20 GHz. Without TDR normalization, the data rolls off much quicker and is much less useful, except at very low frequencies (data rates). As data rates cross the 6.25 Gb rate and continue to increase, the accuracy provided by VNA data will be required for accurate designs and validation.

I would like to acknowledge Greg Fitzgerald, an independent signal integrity engineer, who helped extensively with this project and contributed to the case study. He spent numerous hours making measurements, plotting data, and contributing to the content of this paper.

Acknowledgment

I would like to acknowledge Greg Fitzgerald, an independent signal integrity engineer, who helped extensively with this project.

Author

Robert Schaefer, Technical Leader and R&D Project Manager, Signal Integrity Group, Agilent Technologies

Chapter 6

Data Mining 12-Port S-Parameters

6.1 Abstract

Twelve-port differential S-parameters contain the complete behavior of up to three independent differential channels in a high-speed serial interconnect. There are 78 unique elements, but when including magnitude and phase information, single-ended and differential forms, and frequency and time domain descriptions, there are more than 400 elements. This paper puts in perspective the most important terms and the valuable information that can be extracted about interconnect performance from these measurements. In particular, we show how the information about coupling regions can be mined from 12-port measurements.

6.2.1 High-Speed Serial Links and the Bandwidth of Interconnects

High-speed serial links proliferate in data centers between servers, in backplanes between plug-in cards, and between devices on a board. Each protocol, such as InfiniBand, PCI Express, and SATA, undergoes generational advances, with typically a factor of two increase in bit rate per channel. For example, InfiniBand was introduced at 2.5 Gbps, with a second generation announced as double data rate (DDR) at 5 Gbps and a quad data rate (QDR) at 10 Gbps.

The bandwidth of the signal components that make up the bit stream is difficult to quantify because it changes as it propagates down the channel. The signal with the highest bit-transition density looks like a clock with a clock frequency of half the data rate. If the rise time of the signal were about 7 percent of the clock period, the bandwidth of this bit pattern would be the fifth harmonic, or 5 times 0.5 times the bit rate, or 2.5 times the bit rate.

While this might be the bandwidth of the signal at the transmitter, as it propagates down the interconnect, high frequencies are attenuated and the bandwidth reduces. In the typical case of a lossy line, only the first harmonic is left and the bandwidth is close to the clock frequency, or 0.5 times the bit rate at the receiver.

This is why the bandwidth of a serial data stream is typically reported as anywhere between 0.5 and 2.5 times the bit rate. This is a factor of five difference between values. As a safe estimate, with bit rates in volume production of 5 Gbps, the signal bandwidth is in excess of 10 GHz and interconnects should be designed to support bandwidths in excess of 10 GHz.

In this Gigahertz frequency regime, interconnects contribute four important signal integrity problems, above and beyond the lower bandwidth problems such as terminations, switching noise, ground bounce, and power distribution noise. These are losses, reflection noise from vias, mode conversion, and crosstalk. Any one of these problems can cause failures in the channel if they are not specifically identified and designed out of the system right at the beginning.

Once built, the next step is evaluating the performance of the interconnect to a specification or compliance standard. If it does not pass, it is critical to identify the root cause of the performance limitation so it can be redesigned. Measurements based on S-parameters can be a powerful tool to describe the measured electrical properties of the interconnect, and by manipulating the information into various formats, can almost at a glance provide a first-order estimate of the source of the design limitation.

6.3 Four-Port S-Parameters

S-parameters are defined in terms of how sine waves interact with a device. A sine wave with an amplitude, phase, and frequency is incident on a port of the device, coming from a 50 Ohm environment. The change in the amplitude and phase of the scattered wave has information about the device. Each port of the device under test (DUT) is labeled with an index number and the ratio of the sine wave scattered to the sine wave incident is tracked by the index numbers.

To interpret the various S-parameters the same way, everyone has to agree on the same port assignments. Unfortunately, there is no standardization and this is a source of confusion. When multiple channels are described, the port assignment that provides the greatest flexibility and scaling is shown in *Figure 6.1*.

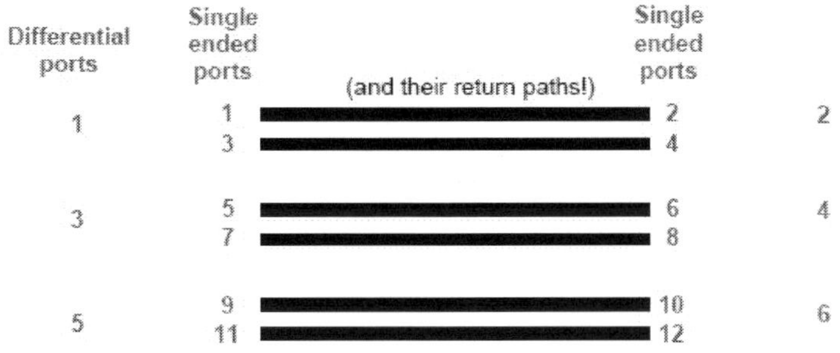

Figure 6.1: *Port Assignments for Single-Ended and Differential Channels*

The first port is labeled as port 1, with its far end labeled as port 2. A second, single-ended channel would have its ends labeled as port 3 and port 4. In this way, additional interconnect channels can be added while maintaining a consistent labeling scheme. An odd-numbered port always has an even-numbered port connected to it. This approach can be scaled to label an unlimited number of ports.

With this approach, the return loss of the first channel is S11 or S22 and the transmitted signal would be S21. The near-end crosstalk (NEXT), from a sine wave going into port 1 and coming out port 3, would be S31, while far-end noise from one line to the adjacent would be labeled with S41.

The NEXT between adjacent lines that make up a parallel bus, for example, would be labeled as S31, S53, S75, etc. The NEXT from the first line to all other lines in the bus would be S31, S51, S71, etc. It would be expected that the magnitudes of these terms drop off with spacing, if the coupling were due to short-range effects.

This labeling scheme can be applied to the same interconnects if adjacent traces are grouped as one differential pair. Two separate traces, with four single-ended ports, would have just two differential ports. In the same labeling scheme, an odd-numbered differential port would connect to an even-numbered differential port.

While the complete description of a collection of interconnects is contained in its single-ended S-parameters, when describing the behavior of differential signals and common signals on a differential pair, it is convenient to convert the single-ended S-parameters into balanced S-parameters. This balanced S-parameter format is also called the mixed-mode or differential S-parameter format. These are three names used interchangeably in the industry for the same S-parameters.

The balanced S-parameters describe the behavior of differential and common signals on the differential pairs. In addition to the standard responses of a differential signal reflected and transmitted through the channel, or a common signal reflected or transmitted through the channel, the balanced S-parameters can describe how a differential signal is converted into a common signal and vice versa.

When describing the interactions of differential and common sine waves with each differential port, a D or C suffix is used in addition to the port index to describe the nature of the signal going in and coming out. In the normal S-parameter notation, the first letter or index is the coming-out signal, while the second letter or index is the going-in signal.

In this format, SCD21 refers to the ratio of the common sine-wave signal coming out of port 2 to the differential sine wave going into port 1. In a single differential-pair channel, there is port 1 on the left and port 2 on the right. Differential and common signals can interact with this channel in four combinations. The S-parameters associated with each combination of differential or common signal going in or out are grouped into four quadrants. These quadrants, shown in *Figure 6.2*, are differential in differential out (DD), common in and common out (CC), and the mode conversion terms differential in and common out (CD) and common in and differential out (DC).

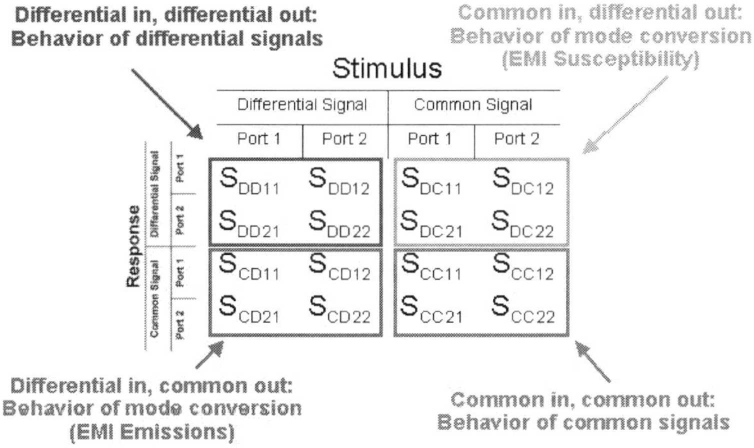

Figure 6.2: *Four Quadrants of the Differential S-Parameters*

Within each quadrant are the return and insertion loss elements. This is a total of 16 elements. All the electrical properties of a differential channel are contained in these 16 matrix elements. Also, all the electrical properties of the two interconnects as single-ended channels are contained in their 16 single-ended S-parameter elements.

Both of these matrices are equivalent ways of describing the same interconnects. The information is the same between them, they are just in different forms. They can be converted back and forth between each form using linear combinations of one matrix element to describe the other.

Though the S-parameter formalism is defined in terms of the frequency domain and the behavior of sine waves, the information about the frequency domain behavior of interconnects can be transformed into the time domain using Fourier transforms. When viewed in the time domain, the S-Parameters define the way time domain waveforms reflect from and transmit through the interconnect.

When the time domain waveform is a step-edge wave, the response is identical to the time domain reflectometer (TDR) response. The transmitted response is the time domain transmission (TDT) response of the interconnect. These two-time domain responses can be either single-ended or differential responses.

The simple single-ended S-parameter matrix has information that can be converted into a variety of other forms. By redefining the interconnect as a single differential pair, the single-ended matrix and be converted into the differential matrix. By converting either of the waveforms into the time domain response, they can display the TDR and TDT response of the interconnect. This transparency of the same information in each format, just displayed differently, is illustrated in *Figure 6.3*.

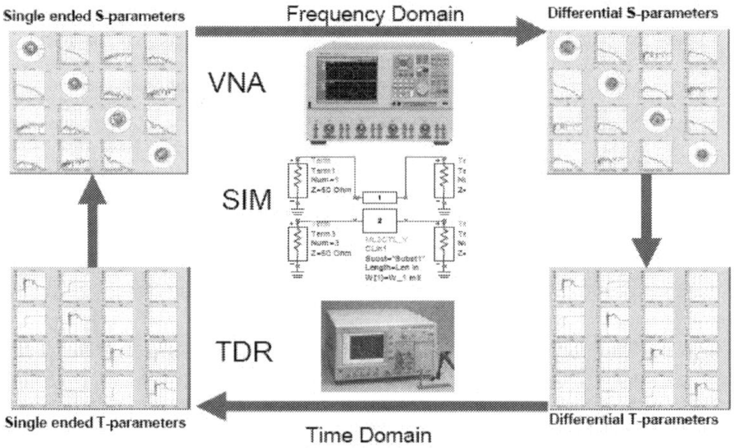

Figure 6.3: *Originating from a Time or Frequency Measurement or a Simulation, the S-Parameters Can Be Transformed between Single-Ended, Differential, and Frequency and Time Domains*

The information is the same in each format, just displayed differently. Each element in each format reveals a different behavior more clearly than another element. All the important electrical properties of a differential channel are contained in these four-port single-ended or two-port differential S-parameter elements.

6.4 Twelve-Port S-Parameters and Information Overload

Crosstalk between differential channels is an important interconnect property that is not included in the two-port differential S-parameters of a single differential channel. However, the interactions between two or more differential channels can easily be described with the same S-parameter formalism, just expanded to incorporate the additional channels.

Each differential channel has two ports, an odd number on the left side and an even number on the right side. In the case of three differential channels, there are six differential ports.

As differential S-parameters, there are still four quadrants describing the interactions of each combination of differential and common signal. Within each quadrant, there is a 6 x 6 element matrix, corresponding to every combination of going-in and coming-out ports. This is a total of 36 x 4 = 144 matrix elements in the differential S-parameter matrix.

If the three differential channels were described by their single-ended S-parameters, there would be six interconnects each with two single-ended ports, or 12 ports. The S-parameter matrix would be 12 x 12, to describe each combination of a going-in port and a coming-out port. This is also 144 elements. However, of these 144 elements, only (13 x 12)/2 = 78 of them are unique. Between these two matrices, there are 156 unique terms.

Each matrix element in the single-ended and differential form has two sets of data: a magnitude versus frequency and a phase versus frequency. This means that there are really 156 x 2 = 312 sets of data in a 12-port S-parameter matrix.

The 156 unique S-parameter matrix elements describe the behavior of sine waves interacting with the interconnect. Each of these elements can be transformed into the time domain step response. In addition, another useful time domain response is the impulse response, also referred to as the Green's function response.

The impulse response of an interconnect describes how a unit pulse of input voltage is scattered by the interconnect over a period of time. The reflected or transmitted behavior of any arbitrary incident waveform, such as a pseudo-random bit sequence (PRBS), can be simulated by taking the convolution integral of the impulse response and the incident waveform.

The 156 frequency domain elements, displayed in the time domain as either a step-edge response or an impulse response, result in 156 x 2 = 312 additional elements. Add to this the 156 phase terms, and there are really 624 elements contained in the single-ended S-parameter matrix. Each element displays its information in a slightly different way.

Of these 624 elements, nine of them are especially useful in answering high-speed serial-link performance questions almost by inspection. Focusing on these nine most useful elements, and not being distracted by the other 615 elements, will dramatically improve productivity.

6.5 Serial-Link Performance Analysis

To measure the 12-port S-parameters of three channels, an instrument capable of at least two-port measurements is required. Each matrix element would be measured one at a time. While the two ports of the instrument are connected to two of the 12 ports, the other 10 ports would have to be terminated in 50 Ohms. For a 12-port single-ended system, a total of 72 pairs of connections and reconnections would have to be done to cover all 78 unique elements in the single-ended S-parameter matrix. Measuring the 12 diagonal elements would only require six pairs of connections.

A much simpler process is possible if a 12-port instrument is used, as shown in *Figure 6.4*.

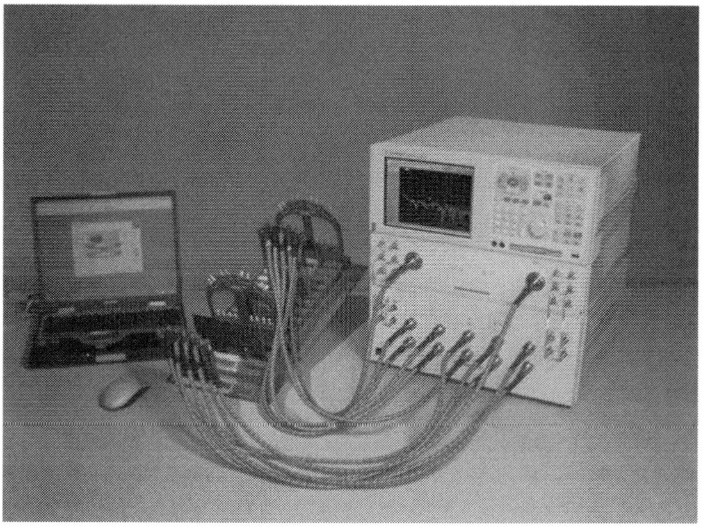

Figure 6.4: *Example of Typical 12-Port VNA Measurement System*

Regardless of how the initial measurements of the 12-port S-parameters are performed, they can be transformed into any of the formats with straightforward mathematical operations.

The most important issue to address is the performance of each measured channel. For high-speed serial data channels, the shape and features of the eye diagram can directly assess performance.

A measurement of the SDD21 time domain impulse response contains information about how any arbitrary waveform will propagate through the channel. To turn this into an eye diagram, a PRBS signal at the test bit rate is synthesized and the convolution integral between the waveform and the impulse response is calculated. The resulting time domain waveform is well correlated to what appears at the receiver on an oscilloscope when a signal generator is input into the DUT. This is sliced synchronous with the clock, and each consecutive pair of bits is superimposed to create an eye diagram.

Figure 6.5 is an example of the measured impulse response of a backplane channel and the resulting eye diagram at 2.5 Gbps and 5 Gbps.

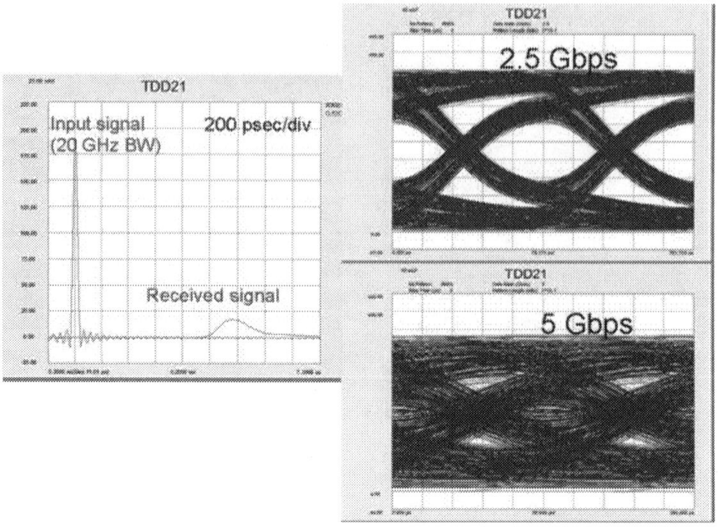

Figure 6.5: *Impulse Response of the SDD21 Element and Resulting Simulated Eye Diagrams for Synthesized 2.5 Gbps and 5 Gbps PRBS Signals*

The impulse response provides an immediate view of the inter-symbol interference (ISI), which will arise in the interconnect. With a 20 GHz measurement bandwidth, the impulse response is about 20 ps. As this impulse propagates through the interconnect, the losses and impedance discontinuities remove the higher-frequency components of the signal, causing it to spread out.

The time base in this example is 200 ps/div. This is one unit interval for a 5 Gbps signal. It is apparent how a single bit would spread out over at least two bit intervals. This is for a 20 ps wide impulse response. If the input signal were a 200 ps wide bit, the transmitted bit would have spread out even more. With this much ISI, we would expect the eye diagram at 5 Gbps to show considerable collapse and deterministic jitter. This is apparent in the synthesized eye diagrams.

While the eye diagram describes the performance of the interconnect, there is no information about why the interconnect has such poor 5 Gbps performance. The first step in optimizing performance is identifying the root cause of the limitation.

6.6 Losses

In the gigabit regime of high-speed serial links, interconnects are not transparent due to four families of problems: losses, impedance discontinuities from vias, mode conversion, and channel-to-channel crosstalk. The impact each problem has on interconnect performance can be mined from specific S-parameters.

The differential insertion loss, SDD21 in the frequency domain, has information about the nature of the losses. *Figure 6.6* shows an example of SDD21 for three channels measured in the same backplane.

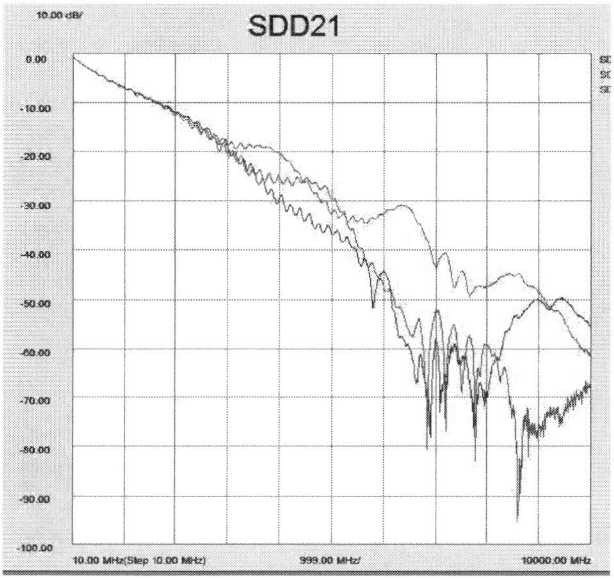

Figure 6.6: *Measured SDD21 of Three Channels in the Same Backplane*

Below 3 GHz, the insertion loss is monotonic, suggesting the drop is dominated by conductor and dielectric losses. The slope is about 20 dB/3 GHz, or 6.7 dB/GHz. In this example, the total interconnect length was about 40 inches, so the normalized loss is about 0.17 dB/inch/GHz. This is to be compared with the rough rule of thumb for a 5 mil wide line in FR4 of about 0.15 dB/inch/GHz.

Below a bandwidth of 3 GHz, the behavior suggests losses dominate performance. This would apply to bit rates as high as 6 Gbps. This suggests that the dominant root cause of the collapse of the eye at 5 Gbps is probably due to losses.

Above 3 GHz, the variations in SDD21 suggest the presence of impedance discontinuities. Even though these three channels are adjacent in the same backplane, they have very different insertion losses above 3 GHz, suggesting specific structural differences in the channels.

6.7 Impedance Continuities

The details of the impedance discontinuities, which might give rise to the insertion loss behavior above 3 GHz, can be explored from the return loss measurements, displayed in the time domain for a step response. This is sometimes referred to as the SDD11 time domain or TDD11 differential response. *Figure 6.7* shows the measured TDD11 response for these same three channels.

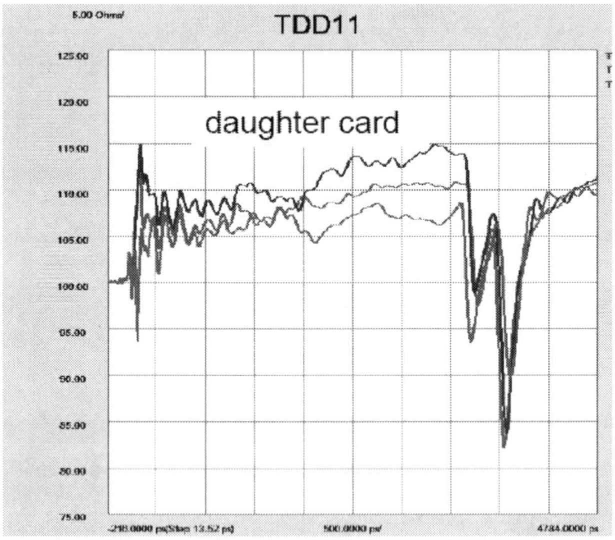

Figure 6.7: *Measured Differential TDR Response of Three Differential Channels*

The differential TDR response immediately identifies the impedance discontinuities from the 100 Ohm differential source. Though there are small variations in the differential impedance of the daughtercard, including the launch, the dominant discontinuities are from the vias on the daughtercard side and then on the backplane side.

While much focus is placed on the connectors between the daughtercard and motherboard, the connectors themselves are often very well matched to the 100 Ohm environment. Rather, the biggest source of discontinuity is the vias. In this example, the variation in the magnitude of the via reflections probably are the source of the insertion-loss variation.

Improving performance for bandwidths greater than 3 GHz would require improving the vias. They appear as low-impedance discontinuities because of the stubs on the top and bottom of the signal transition layers. By reducing the capacitance of the vias and minimizing their lengths, their impedance can be greatly reduced to the point where they do not influence performance well into the 10 GHz region.

6.8 Mode Conversion

Another source of differential insertion loss is mode conversion of some differential signal into common signal. Any asymmetry between the two lines that make up a differential pair will contribute to mode conversion. Mode conversion will cause two problems: a reduction in differential signal quality, which will increase the differential insertion loss, and creation of the common signal.

Most differential receivers have a high threshold for common signal rejection, so the common signal by itself may not cause a problem. However, if the common signal were to reflect from common impedance variations and encounter the asymmetry again, it could reconvert back into a differential signal, but with an added skew. This will further distort the differential signal, increase ISI, and ultimately cause a higher bit-error rate.

If any of the common signal were to get out of the system, especially on twisted pairs, it can contribute to radiated emissions and possibly cause an electromagnetic interference (EMI) failure. It only takes about 3 microAmps of common current on an external cable to fail a Federal Communications Commission (FCC) class B test. Even if the common impedance were as high as 300 Ohms, it only takes a common signal of about 1 mV to fail an FCC test. When the typical high-speed serial link signal is at least 100 mV, only 1 percent conversion is required to fail an FCC test.

Mode conversion by asymmetries in the interconnect is characterized by the SCD21 term. This is a measure of how much common signal emerges on port 2 from a differential signal incident on port 1. *Figure 6.8* is an example of the measured SCD21 signal in the time domain, using a 400 mV incident differential signal.

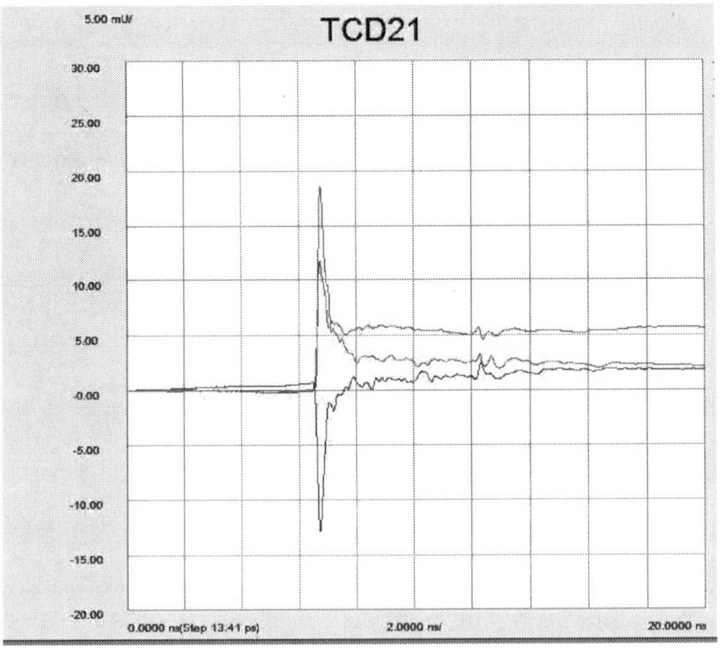

Figure 6.8: *Measured SCD21 in the Time Domain for Three Different Channels on a Scale of 5 mV/div*

In this example, the three channels have about the same mode conversion, of about 15 mV out of 400 mV, or 4 percent. The difference in the sign between the three channels is an indication that the slow line in the pair varies between the three channels.

While this amount of common signal might cause a problem if it were to escape onto external twisted pairs, it would not cause a problem if it were confined to the backplane interconnect.

If a common signal poses a problem, the first step to reduce it is to determine where the asymmetry is that causes the problem. This can be deduced by observing the SCD11 term, also in the time domain. This term is the common signal coming out of port 1 when a differential signal goes into port 1. When viewed in the time domain from a step response, the instant in time when the common signal comes back out is a measure of where it might have been produced. *Figure 6.9* is an example of the SCD11 time domain response compared to the SDD11 response of the same backplane.

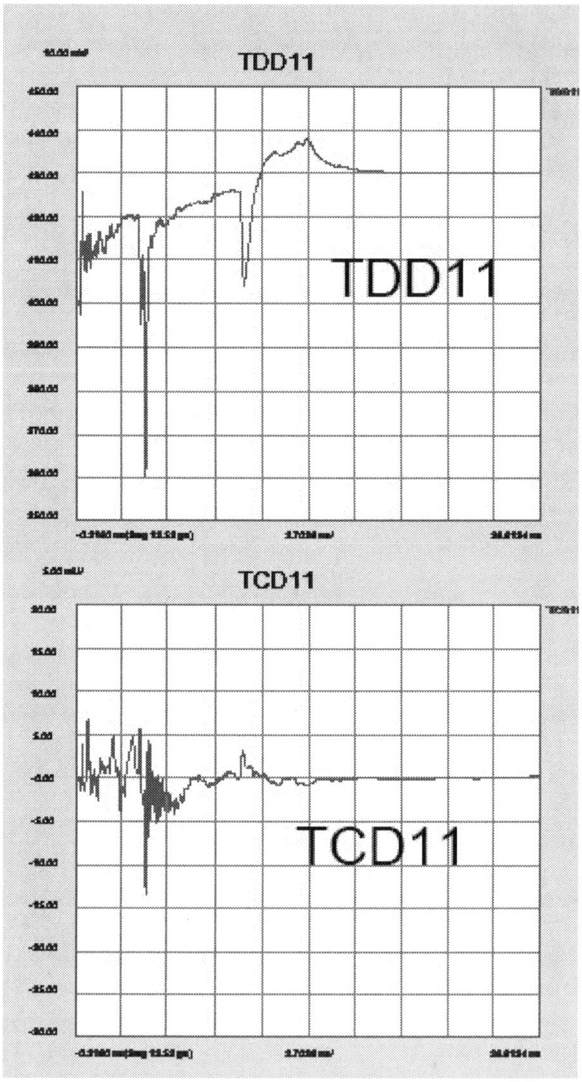

Figure 6.9: *Measured SCD11 Displayed in the Time Domain from a Step Response Showing the Possible Location of the Mode Conversion as the Connector and Via Field*

In this example, most of the common signal reflected back seems to occur coincident with the reflection from the connector and the via field. A possible fix might be to select different paths in the connector, match the vias structures, or better match the escapes from the via field. The root cause of mode conversion is only suggested by the SCD11 response, as not all conversion processes result in a reflection of the converted common signal back to the source.

It is not possible to further refine the location directly from the front screen. The only way to refine the source of the mode conversion would be to model various possible mechanisms and compare the simulated responses with the measured behavior, comparing the SCD11, SCD21, SDD11, and SDD21 responses.

Of course, it is almost impossible to eliminate mode conversion in any real interconnect. In addition to reducing it, all external connections must use common choke filters to attenuate the common signal before it can launch on an external cable. However the more it can be reduced at the source, the further reduction there is through the filter and the more robust the product is to EMI problems.

6.9 Channel-to-Channel Crosstalk

The final high-speed serial-link problem from the interconnect is channel-to-channel crosstalk between any two channels. The noise on one pair from a signal on the other pair will be picked up at either the near end or far end of the quiet pair. The magnitude and shape of the noise signature at the near end and far end of the quiet pair will be different. The differential near-end noise between adjacent channels is described by SDD31, while the far-end noise is described by SDD41. Likewise, in the channel two away from the active line, the near-end noise is SDD51 and the far-end noise is SDD61. An example of the measured noise is shown in *Figure 6.10.*

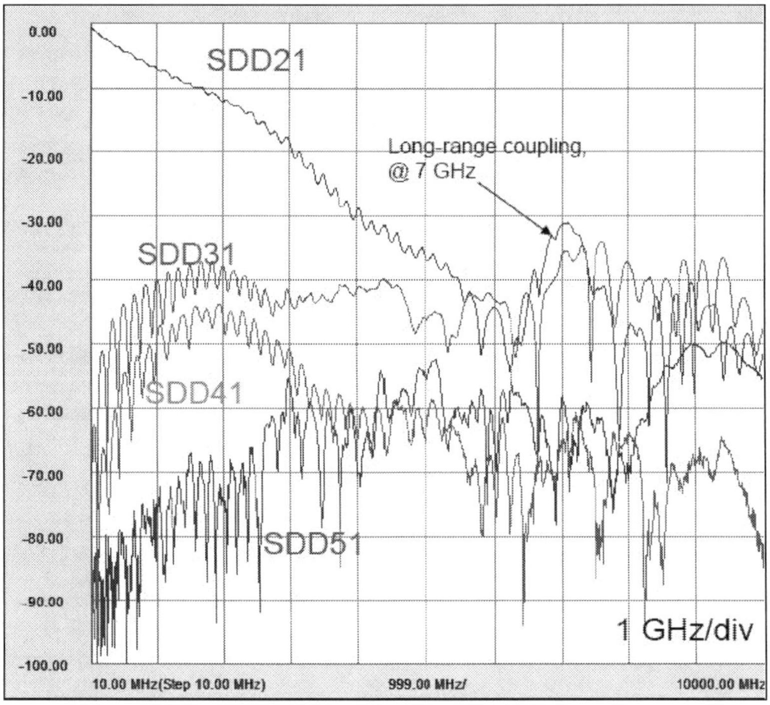

Figure 6.10: *Measured Differential NEXT and FEXT in a Backplane System*

For comparison, the SDD21 response is also shown on the same scale. As expected, in this stripline-based interconnect system, the far-end noise, SDD41, is much less than the near-end noise, SDD31. In fact, the presence of any far-end noise in a stripline system is usually due not to noise generated propagating in the forward direction, but to the backward-propagating noise reflected into the forward direction by impedance discontinuities. In general, the far-end noise is typically about 10 dB lower than the near-end noise.

When the crosstalk is dominated by the distributed coupling between the transmission lines that make up each differential pair, the near-end noise to a channel two lanes distant is expected to be lower and is shown in this example to be more than 30 dB lower than to an adjacent channel. However, at about 7 GHz, the near-end noise between the active channel and the adjacent and two-away channel are almost the same. This is usually an indication of coupling in the connector or via field and can be a longer-range coupling.

One way of identifying the dominant source of the coupling is by observing the SDD31 response in the time domain, as illustrated in *Figure 6.11*.

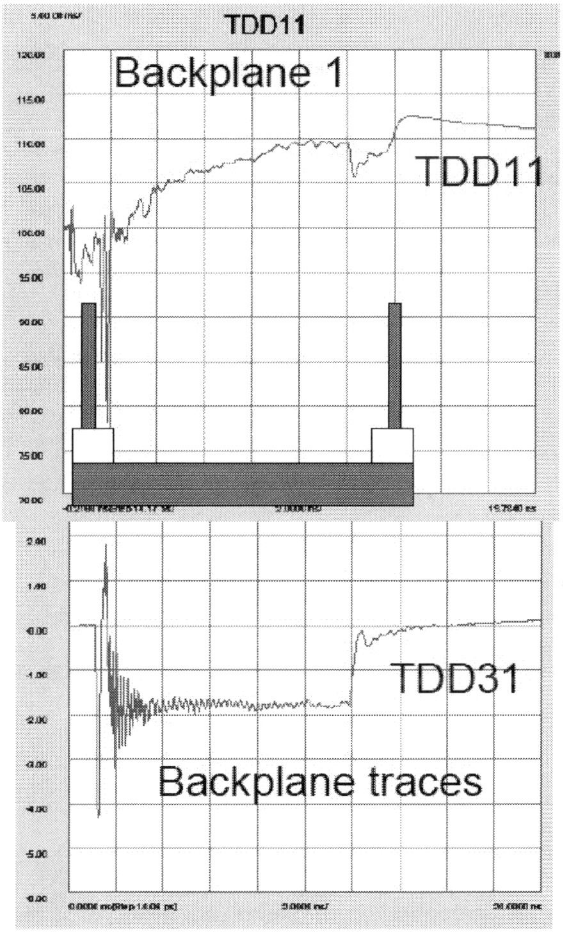

Figure 6.11: *Measured SDD31, NEXT Response, Displayed in the Time Domain Compared with the SDD11 Response*

The SDD11 response, displayed in the time domain, shows the impedance profile of the backplane system. It identifies the time at which the reflection from each physical feature is received at port 1. This timing can be used to interpret the near-end noise, SDD31 on the second channel.

There is a large NEXT noise pulse picked up in the quiet channel right when the incident signal reflects from the connector and via field. This is probably the origin of the 7 GHz noise.

After the connector, a large amount of near-end noise can be seen as the incident signal is propagating down the backplane trace. This suggests that most of the noise, especially at lower frequency, is due to pair-to-pair coupling in the backplane interconnect. To reduce this noise would require increasing the spacing between the differential pairs in the backplane.

6.10 Conclusion

All the electrical properties of three differential channels are contained in the 12-port S-parameters matrix elements. These can be seamlessly transformed to the differential format, and between the frequency and time domains. Each different element in each different form tells a slightly different story about the behavior of the interconnects.

The first step in any interconnect characterization is to use the information as presented on the front screen to quickly and routinely obtain a first-order analysis of the interconnect. This does not require any model building. Nine important elements carry more valuable information than the others, from which can be obtained the intrinsic performance limitations of the interconnect and indications of the root causes of these limitations.

It is important to note that additional information about the interconnect performance is also buried in the 12-port S-parameters. However, this information cannot be as easily mined by simply observing the measured response. Rather, to dig deeper, specific models would have to be constructed and then fitted to the measured responses. In this way, all the secrets of the interconnect can be revealed. By identifying the details of the root causes of performance limitations, the design knobs that influence performance can be adjusted to find the optimized cost-performance balance to interconnect performance.

Authors

Eric Bogatin, Signal Integrity Evangelist, Bogatin Enterprises

Mike Resso, Signal Integrity Application Scientist, Component Test Division, Agilent Technologies

Part II

Backplane Measurements and Analysis

Chapter 7

A Design of Experiments for Gigabit Serial Backplane Channels

7.1 Abstract

Today's backplane environment presents significant challenges for high-speed digital designers. Tradeoffs between signal integrity performance, cost, and reliability must be made to achieve the proper architecture for a robust physical layer (PHY) channel. The right combination of connectors, dielectric materials, and topology must be used to accomplish this engineering task. This paper will address an in-depth design of experiments using combinations of three high-speed connectors, three dielectric materials, and three channel lengths. Data will be gathered with a 12-port vector network analyzer (VNA), and the results will be presented in time, frequency, and eye diagram domains.

7.2 Introduction

Tomorrow's generation of consumer products will exploit the triple play of telecommunication—voice, video, and data. This new development will merge two broadband services, high-speed Internet access and television, with one narrowband service over a narrowband service such as telephone. In order to support the extraordinary amount of bandwidth required by this broadband service, the Internet infrastructure is transforming into a superhighway of information. High-speed Internet switch and router equipment performance is a critical component that will dictate the outer limits of this network. The high-speed backplane in the aforementioned network equipment is the fundamental backbone of this PHY that will sustain future technologies of advanced line cards containing ultra-fast serializer-deserializer (SerDes) chipsets.

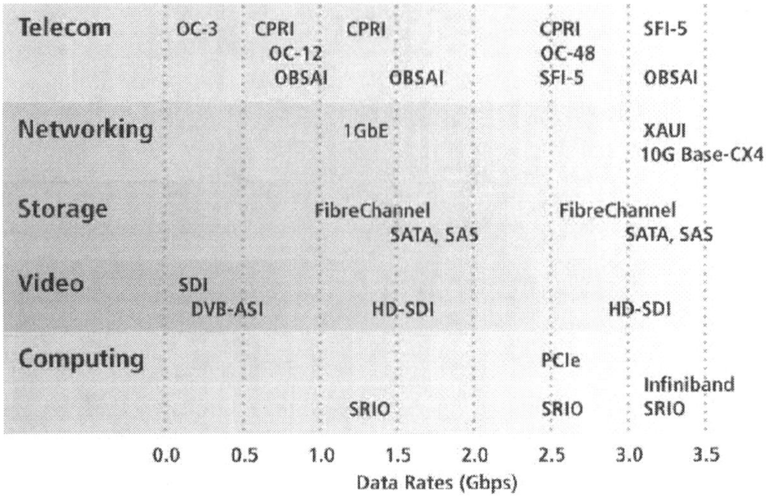

Figure 7.1: *Serial Standards*

This market demands that backplanes provide more bandwidth than ever. The answer for this demand is backplanes with many multi-gigabit serial channels. Designing, building, and characterizing these backplanes are becoming more challenging with every increase in the serial bit rate. A network equipment manufacturing company's whole product line depends on the longevity of the backplane. Upgrade and innovation are implemented with daughtercards, but the backplane is the anchor that holds the customer base.

Because of this, backplanes have to be designed and built to last, often through several line-card product generations. Although the backplane usually has no active components, a significant effort must be expended to characterize and verify its performance. There are three commonly used tools for characterizing multi-gigabit interconnects: the time domain reflectometer (TDR), the VNA, and the eye diagram.

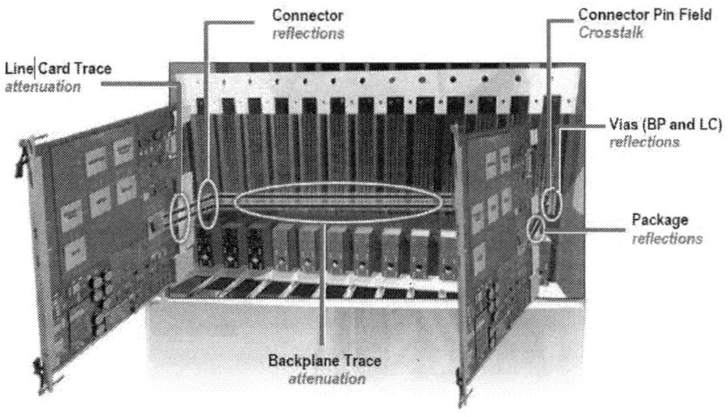

Figure 7.2: *The Serial I/O Channel*

7.3 Serial Backplane Channels

Any backplane-based system comprises the same basic components: the line cards/daughtercards and the backplane. The signal path for communication between two daughtercards over a backplane channel consists of a number of transitions. Starting with the transmitter die, the signal path includes transitions from the transmitter die to the package, to the printed circuit board (PCB) transmission line, to the backplane connector, to the backplane transmission line, to the far-end daughtercard backplane connector, to the daughtercard transmission line, to the package, and finally to the die at the receiver. Each of these transitions is an opportunity for the signal to be degraded.

The transitions result in degradation as a result of reflections, attenuation, and interference. At the connector pin field, impedance mismatch results in reflections that add to the jitter in the signal at the receiver. Likewise, attenuation in the signal from the transmission line further increases jitter by generating inter-symbol interference (ISI). In addition, signals on other channels can induce crosstalk noise in the signal path and thereby increase even further the eye closer at the receiver. Since all of these can work together to reduce the quality of the signal at the receiver, we must characterize the backplane channels to ensure acceptable performance for the application.

7.4 Backplane Platform Description

At Xilinx, we have built a backplane that would be a platform for demonstrating the usage and operation of the Xilinx Virtex-5 GTP multi-gigabit transceivers in a backplane application. The demonstration backplane provides a variety of channel behaviors. These behaviors are the result of a number of variations in materials, components, and routing structures. The demonstration backplane also has structures that allow for the decomposition of the channel path. With these structures, the root cause of some channel responses can be isolated.

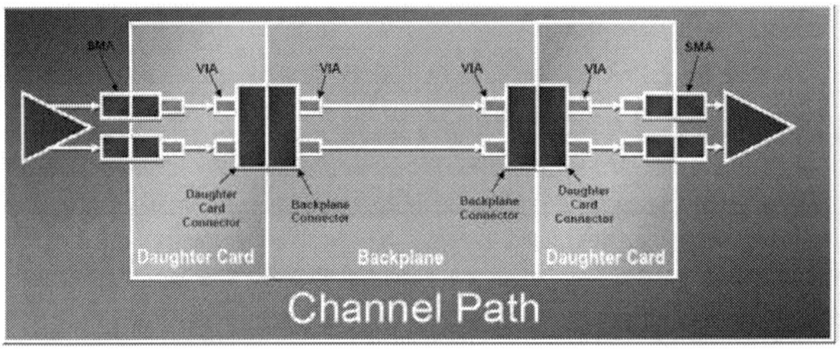

Figure 7.3: *Platform Channel Path*

This demonstration platform consists of a backplane board and daughtercard boards. On the daughtercard, the channel path, as shown in *Figure 7.3*, starts with a surface-mounted assembly (SMA) launch into the daughtercard, then propagates through a transmission line from the SMA to the backplane mating connector. On the backplane, the channel path runs from one backplane connector through a transmission line to another backplane connector that mates with the destination daughtercard. By varying the connector type, the PCB dielectric material, the routing length, and the routing layer, the backplane and daughtercards can provide a variety of channel behaviors.

In addition to the standard channel path, the demonstration backplane has reference channels. These channels provide a means to decompose the channel path and give visibility to some intermediate channel-path structures. The demonstration backplane actually consists of a number of experiments that involve variations in PCB dielectric materials, backplane connectors, channel length, channel physical structure, and channel physical path routing.

There are three connector types on the backplane. The connectors were selected to cover a range of backplane channels that an engineer may encounter when designing for a multi-gigabit transceiver. To represent a legacy channel, the HM 2mm connector was selected. With the advent of advance TCA backplanes, the HM-Zd connector was selected to represent a popular contemporary serial backplane channel. The Amphenol eHSD connector was selected to represent a higher-performance channel.

Material	Dielectric Constant	Loss Tangent
Nelco 4000-13	3.7	0.009
Nelco 4000-13si	3.4	0.008
ISOLA FR408	3.7	0.012

Notes
1. 50% resin content
2. Test frequency is 1 GHz

Table 7.1: *Dielectric Material Properties*

The backplane was built using three types of dielectric material: ISOLA FR408, Nelco 400013, and Nelco 4000-13si. Although all of the materials are upgrades to standard FR-4, they do provide variation in performance, as shown in *Table 7.1*. The values in this table were obtained from the manufacturer's published product brochures.

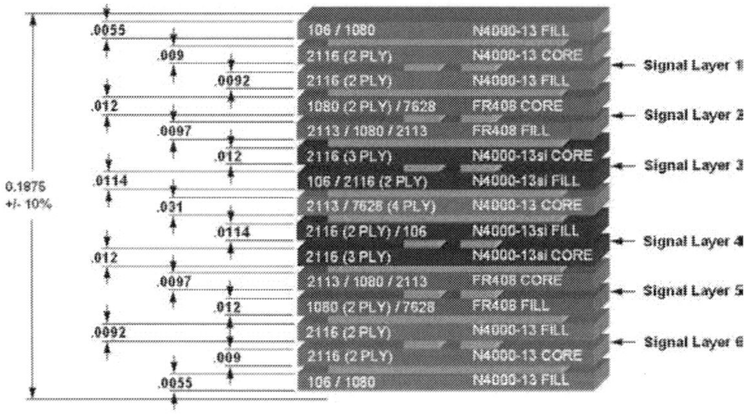

Figure 7.4: *Backplane Stack-Up*

The backplane stack-up consists of six signal layers and eight internal plane layers for a total of 16 layers. With the cooperation of Sanmina-SCI, a single stack-up was designed that included all of the dielectric material types. As shown in *Figure 7.4*, the stack-up was designed so that each material type is represented by two routing layers, one in the upper half of the stack-up and another place symmetrically in the lower portion of the stack-up. As we will see, besides allowing for manufacturing issues, having the opposing pairs of routing layers for each of the board material types adds variation to channel performance. As also shown, the overall thickness of the backplane is 187 mils.

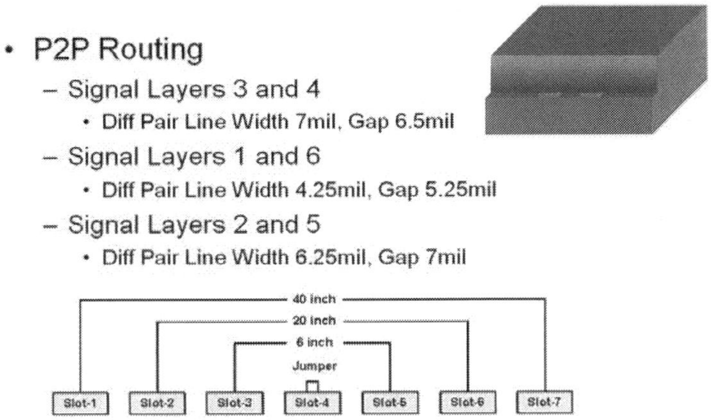

Figure 7.5: *Backplane Trace Geometry*

The backplane has four lengths of point-to-point channels on each of the connectors. The shortest length is actually a jumper from the one set of pins on a backplane connector to another set of pins. This path virtually eliminates signal degradation due to transmission line attenuation in the backplane. The other lengths are 6, 20, and 40 inches. The trace geometries for each of the layers are also shown in *Figure 7.5*.

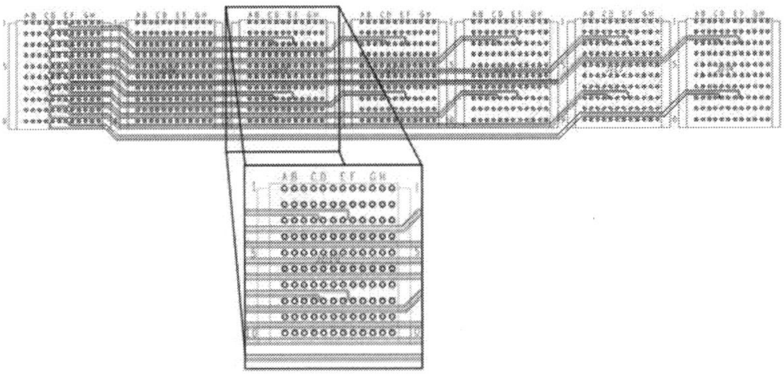

Figure 7.6: *HM-Zd Routing*

As there are many combinations of trace geometry and layer thickness, it should be noted that for layers 1 and 6, the trace geometry was driven by the desire to implement the quad-route routing as recommended by Tyco for HM-Zd connectors. This routing technique offers a greater signal density in the backplane because it doubles the number of traces that can be routed on each layer between the HM-Zd pin rows.

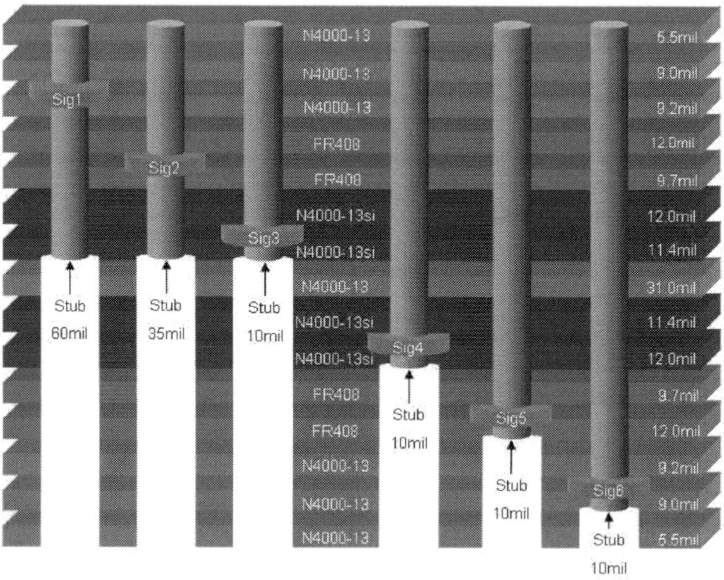

Figure 7.7: *Backplane Via Stubs*

Under normal circumstances a backplane must be built with enough mechanical strength to support the stresses of insertion and removal of daughtercards. Because of this, and because of the need to have a significant number of signal routing layers, a backplane is usually thicker than most other types of circuit boards. For this reason, we built the demonstration backplane with an overall thickness of 187 mils.

Because of the backplane board thickness, channels routed on the upper layers exhibit a significant via stub length. These via stubs can be a source of channel signal degradation because of the reflections generated by the stub. A widely accepted practice to minimize the effects of via stubs is to backdrill the vias so as to eliminate the stub. The board fabricator, in this case Sanmina-SCI, is able to backdrill a via to within 10 mils of the signal layer. Therefore, the specified backdrill depth for each of the layers was calculated based on this requirement. That is except for signal layers 1 and 2. Because press-fit–compliant pin connectors were used on the backplane, the upper two layers could not be backdrilled to within 10 mils of the signal layer. The compliant pins require a minimum via barrel depth of 62 mils. Therefore, the vias for signals routed on these layers could only be backdrilled to within 72 mils of the top of the board. As shown in the figure, this means that channels routed on the upper layers still had some via stubs. The effects of these stubs on channel performance are included in the characterization.

In addition to the normal channel paths from slot to slot, the backplane also has a series of reference channels. There are six reference channels on the backplane, one for each of the signal layers. Each reference channel uses SMA connectors for launching and retrieving signals. The reference channels are routed as balanced signal paths using the same trace geometry as that of the other traces on the layer and have a trace length of 20 inches. The SMA connectors are compliant pin press-fit connectors. By removing the daughtercard and backplane connectors from the channel path, the reference channel provides for a more simplified signal path on the backplane, and allows visibility into the behavior of the backplane transmission lines on each layer.

7.5 Daughtercard Description

In addition to the backplane, the demonstration platform includes daughtercards. These cards use SMA connectors to launch signals into and retrieve signals from the backplane. For each of the connector types, the daughtercard has a number of channels. For the eHSD connector, the daughtercard has 24 channels. For the HM-Zd connector, there are 32 channels and for the HM 2mm connector, there are 16 channels. The trace length for all connections between the SMA connectors and the backplane connector pins is eight inches. All of the channels are routed as balance pairs. There are also four reference channels that are routed as single-ended connections between two SMA connectors on the daughtercard.

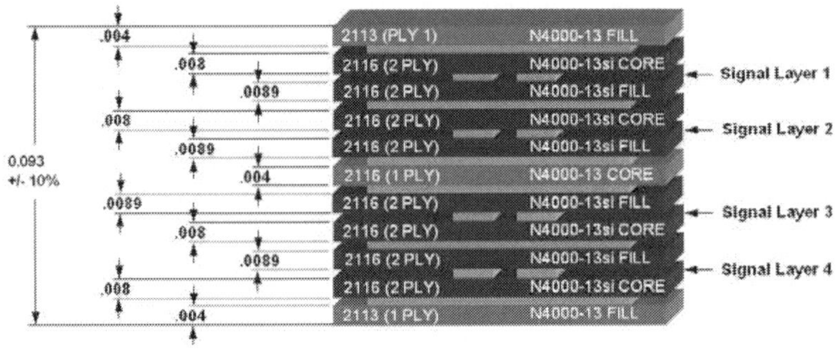

Figure 7.8: *Daughtercard Stack-Up*

The stack-up for the daughtercard has four signal layers and six internal plane layers for a total of 12 layers. Signals are not routed on the top and bottom layers. The daughtercard uses Nelco 4000-13si as the dielectric material around all of the signal layers. Since this is the highest performing material that we used on the backplane, we chose it for the daughtercard in order to limit the attenuation in the daughtercard signal path. The overall thickness of the daughtercard is 93 mils, a common daughtercard thickness.

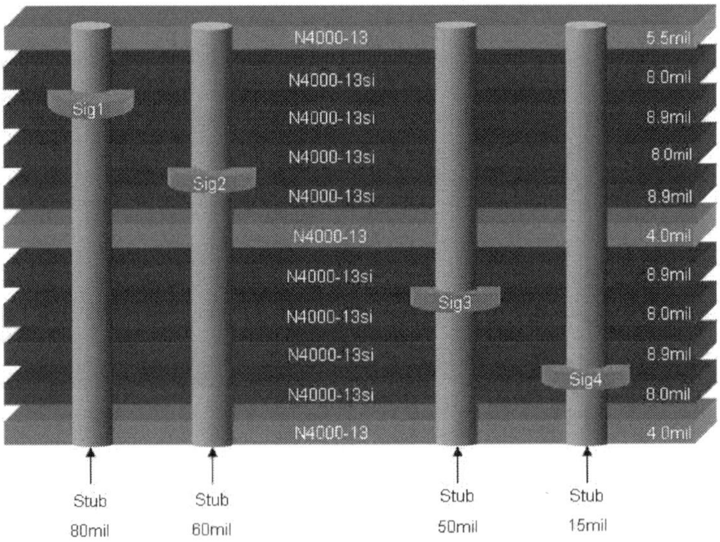

Figure 7.9: *Daughtercard Via Stubs*

As in the case of the backplane, the routing layers that are closer to the top of the board have longer via stubs. With the effects associated with varying stub lengths on the daughtercard, the demonstration platform is able to provide a variety of channel behaviors.

7.6 Backplane Characterization

The performance of the demonstration platform was characterized and verified using a combination of time domain and frequency domain test and analysis. The insertion loss for the channels was evaluated for performance over the band of interest.

As with any product, the development effort must be bounded by a set of performance requirements. Because backplanes are usually required to be usable over several product performance upgrades, it must be designed to meet the immediate and future performance demands of the product. To emulate such a product development, the demonstration platform was characterized for an immediate requirement of 3.125 Gbps and a future requirement of 6.25 Gbps.

For the demonstration platform, each of the channel types was tested and analyzed over the required ranges of performance. The analysis includes insertion loss, TDR, and eye diagram analysis. The insertion loss data provides a view of the overall frequency response of the channel. Combining insertion loss with TDR data gives a more complete picture of the performance of the channel by providing information on the effects of each transition in the channel path on the overall response of the channel. Eye diagram analysis using data collected from the VNA was performed to acquire an understanding of the effects of attenuation and reflection on the performance of each channel.

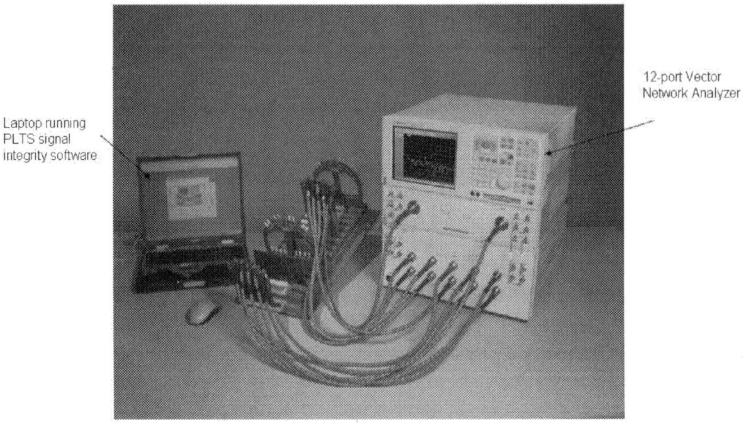

Figure 7.10: *Test Setup*

The test setup for this design of experiments was a 12-port VNA controlled by a laptop running physical layer test system (PLTS) signal integrity software. The resultant data files was a Touchstone format S-parameter file with an *.s12p suffix. This is a standardized file format used frequently in the modeling and simulation industry that allows import and export into many design tools. Advanced design system (ADS) is one popular tool that is starting to migrate from the microwave industry to the high-speed digital industry. In any case, PLTS was used to gather differential data in all domains of analysis, including frequency, time, eye diagram, and RLCG modeling. The most pertinent data obtained from PLTS was used to quickly optimize the design of this backplane and will be shown in this paper.

The channel performance was evaluated by connector type and by channel path. For each of the connector types, there are a number of channel paths. These channel paths vary by layer on the daughtercard and on the backplane. All of the tests were performed on the channels with 20 inches of backplane trace length. To aid in managing the testing, a matrix was developed for each connector type. This matrix shows the path of each channel on the daughtercard and on the backplane.

					Backplane					
					Sig1	Sig2	Sig3	Sig4	Sig5	Sig6
					4000-13	FR408	4000-13si	4000-13si	FR408	4000-13
		Diff Pair	Row		A-B	A-B	A-B	C-D	C-D	C-D
				Pins	9-12	5-8	13-16	13-16	5-8	9-12
Line Card	Sig1	105-108	A-B	13-16			4000-13si			
		121-124	A-B	5-8		FR408				
	Sig2	113-116	A-B	9-12	4000-13					
	Sig3	109-112	D-C	9-12						4000-13
	Sig4	101-104	D-C	13-16				4000-13si		
		117-120	D-C	5-8					FR408	

Table 7.2a: *eHSD Channel Path Matrix*

7.7 eHSD Connector Channels

Table 7.2a is the path matrix for the Amphenol eHSD connector channels. Each channel on the daughtercard is identified by a differential pair number. The channels are routed in groups of four channels for each channel path type. On the daughtercard, there are four signal layers. Each of the signal paths on the daughtercard are routed to a signal layer on the backplane.

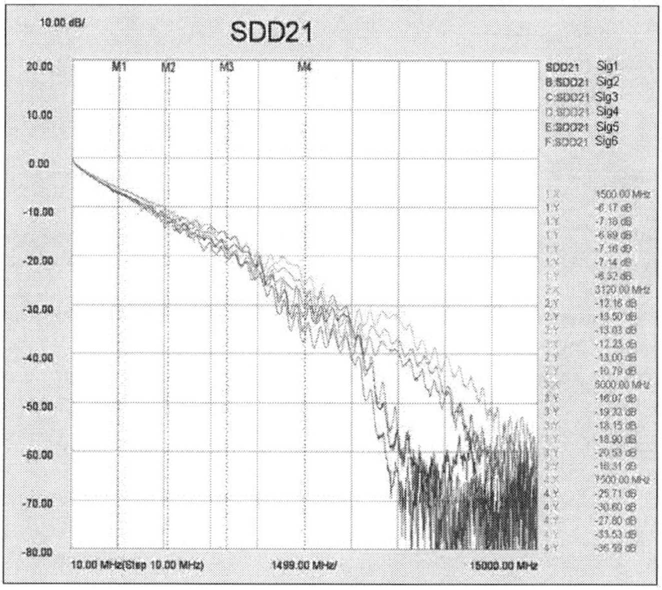

Table 7.2b: *eHSD Channel Path Matrix*

The insertion loss data for the Amphenol eHSD channels is in *Figure 7.11*. The channel behavior tends to fall into one of two groups. As would be expected, the channels that are routed on the upper layers of the daughtercard and backplane have lower operating bandwidth than do the channels that are routed on the lower layers. We attribute this behavior to the effects of the via stubs on signal integrity.

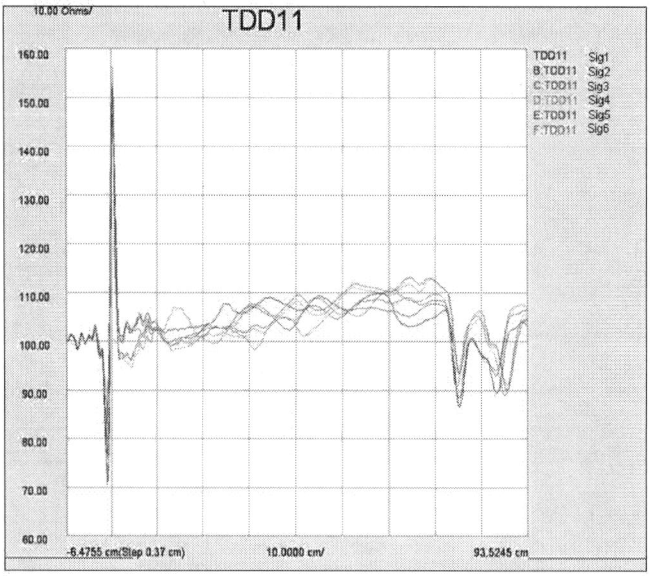

Figure 7.11: *eHSD Channel Insertion Loss*

With the TDR data, the impedance discontinuities are greater for the daughtercard SMA launch than for the backplane connectors. Also, the magnitude of the impedance discontinuities is relative to the amount of stub length.

					Backplane							
					Sig1		Sig2	Sig3	Sig4	Sig5	Sig6	
					4000-13		FR408	4000-13si	4000-13si	FR408	4000-13	
		Diff Pair	Row		C-D	E-F	A-B	A-B	G-H	G-H	C-D	E-F
				Pins	2-5	2-5	6-9	2-5	2-5	6-9	6-9	6-9
Line Card	Sig1	205-208	A-B	2-5				4000-13si				
		221-224	A-B	6-9			FR408					
	Sig2	209-212	C-D	2-5	4000-13							
		225-228	C-D	6-9							4000-13	
	Sig3	213-216	E-F	2-5		4000-13						
		229-232	E-F	6-9								4000-13
	Sig4	201-204	G-H	2-5					4000-13si			
		217-220	G-H	6-9						FR408		

Table 7.3: *HM-Zd Channel Path Matrix*

7.8 HM-Zd Channels

Table 7.3 shows the path matrix for the HM-Zd connector channels. It should be noted that the channels routed on backplane signal layers 1 and 6 are routed using the Tyco-recommended quad-route method.

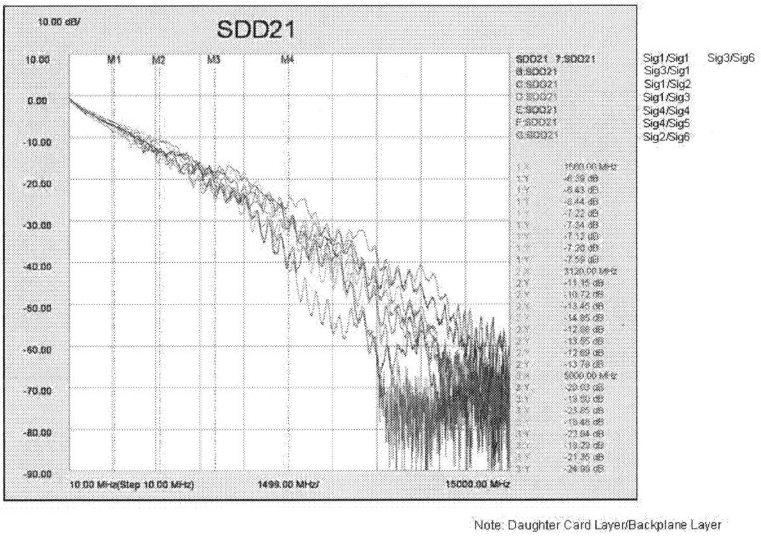

Figure 7.12: *HM-Zd Channel Insertion Loss*

Figure 7.12 shows the insertion loss for the HM-Zd channels. The traces on the plot are labeled to show the daughtercard and backplane signal layers that were used to route the signal. So as in the case of second label, Sig3/Sig1, the path is from signal layer 3 of the source daughtercard to signal layer 1 of the backplane to signal layer 3 of the destination daughtercard.

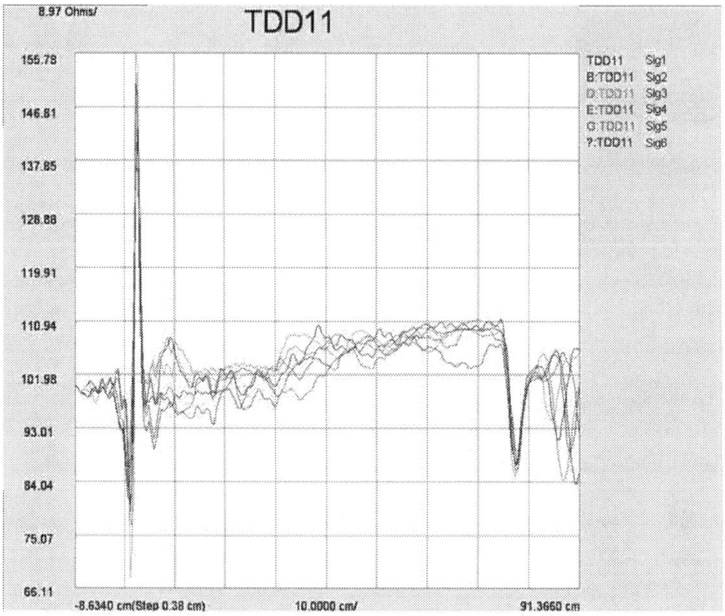

Figure 7.13: *HM-Zd Channel TDR*

The TDR data for the HM-Zd channels shows that the SMA launch on the daughtercard has a greater impedance discontinuity than does the backplane connector interface.

					Backplane															
					Sig2				Sig3				Sig4				Sig5			
					FR408				4000-13si				4000-13si				FR408			
		Diff Pair	Row		A-B				A-B				D-E				D-E			
				Pins	11	12	14	15	2	3	5	6	2	3	5	6	11	12	14	15
Line Card	Sig1	307	A-B	3						13si										
		305	A-B	6								13si								
		314	A-B	12		408														
		316	A-B	15				408												
	Sig2	308	A-B	2					13si											
		306	A-B	5							13si									
		313	A-B	11	408															
		315	A-B	14			408													
	Sig3	304	D-E	2									13si							
		302	D-E	5											13si					
		309	D-E	11													408			
		311	D-E	14															408	
	Sig4	303	D-E	3										13si						
		301	D-E	6												13si				
		310	D-E	12														408		
		312	D-E	15																408

Table 7.4: *HM-2mm Channel Path Matrix*

7.9 HM-2mm Channels

Table 7.4 shows the path matrix for the HM-2mm connector channels. As previously mentioned, the HM-2mm connector channels are designed to represent a legacy backplane. For this reason, backdrilling was not specified for any of the channels on the HM-2mm connector. The signal pin assignment includes a liberal use of ground pins. Channel signal differential signal pairs are grouped by twos with a ground connections assigned to all of the connector pins that are adjacent to them.

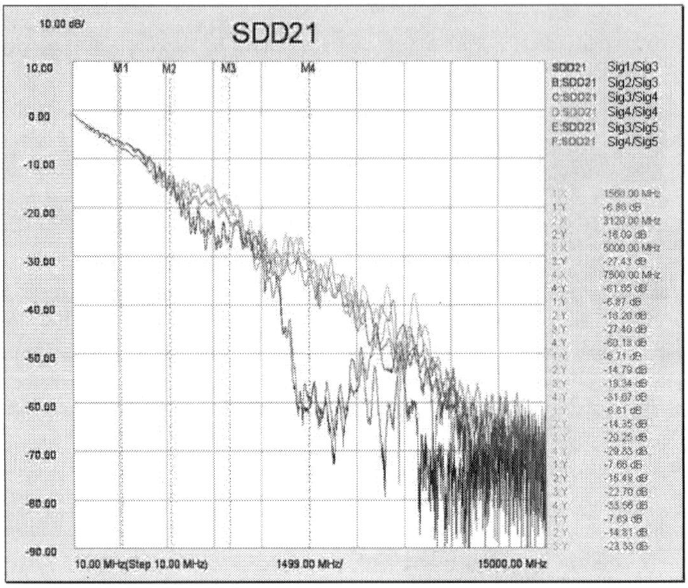

Figure 7.14: *HM-2mm Channel Insertion Loss*

The insertion loss data for the HM-2mm channels is shown in *Figure 7.14*. Once again, we see from the data that the channels that are routed on the upper layers of the daughtercard and the backplane have a lower channel bandwidth than do the channels that are routed on the lower layers. Because of the lack of backdrilling on these vias, the channel bandwidth is even lower than that of the other two connector types.

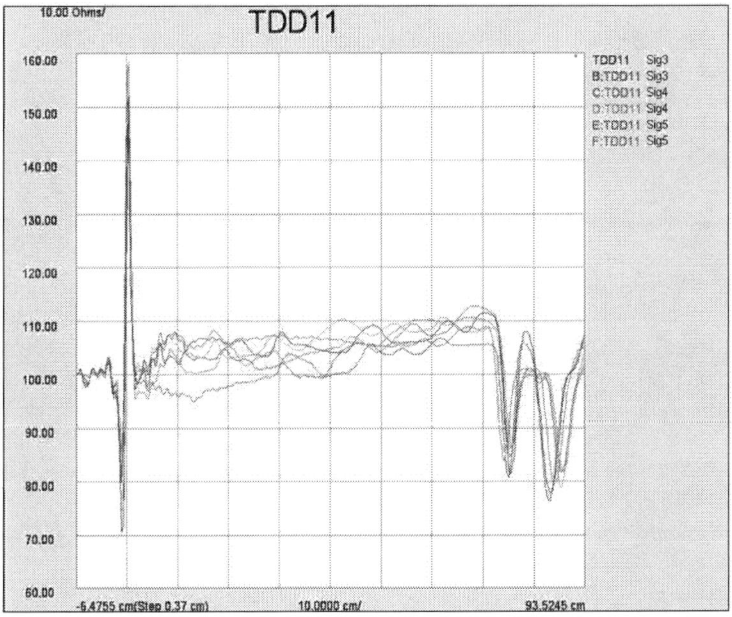

Figure 7.15: *HM-2mm Channel Insertion Loss*

The TDR data for the HM-2mm connector reveals the same large discontinuity in the SMA launch on the daughtercard. It also shows a larger impedance discontinuity at the backplane connector interface.

7.10 Crosstalk Measurements

The channel-to-channel crosstalk will now be investigated for each of the connector types using a 12-port VNA. We set up each test to evaluate the crosstalk from two aggressor channels on a single victim channel. We tested the crosstalk effects based on the physical location in the connector of the victim channel pins relative to the location of the aggressor channel pin.

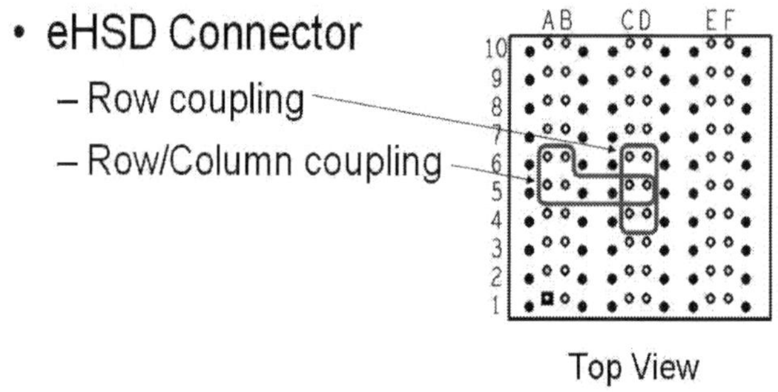

Figure 7.16: *eHSD Crosstalk Configurations*

For the eHSD connector, we tested the row coupling and row/column coupling as shown in *Figure 7.16*.

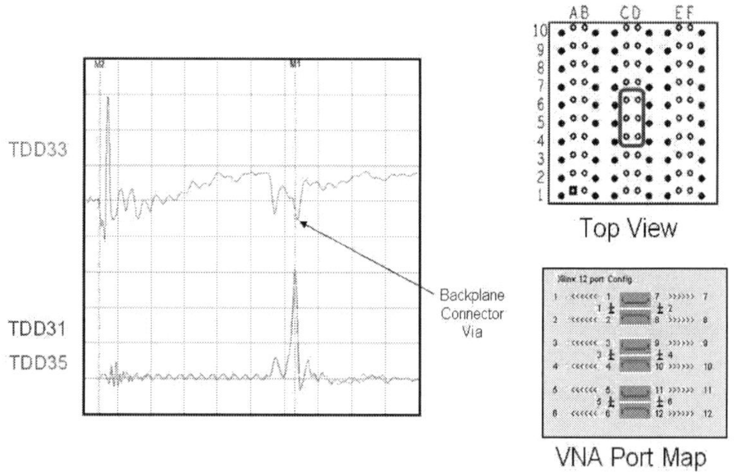

Figure 7.17: *eHSD Channel Row Crosstalk*

Observing the time domain differential (TDD) data for the NEXT between the aggressor channels and the victim channel reveals an area of significant crosstalk relative the other portions of the channel. By temporally marking the crosstalk region on the time domain NEXT TDD, the location of the vertical marker in the differential time domain reflection of the victim channel, TDD33, shows the channel structures that are contributing to the crosstalk (because the waveforms in *Figures 7.17* and *7.18* are "time-aligned" sharing the same horizontal time base). In this case, the pin field via for the backplane connector is the major contributor. It should be noted that the temporal plot shows that the contribution to crosstalk is primarily from the pin field vias and not the connector itself.

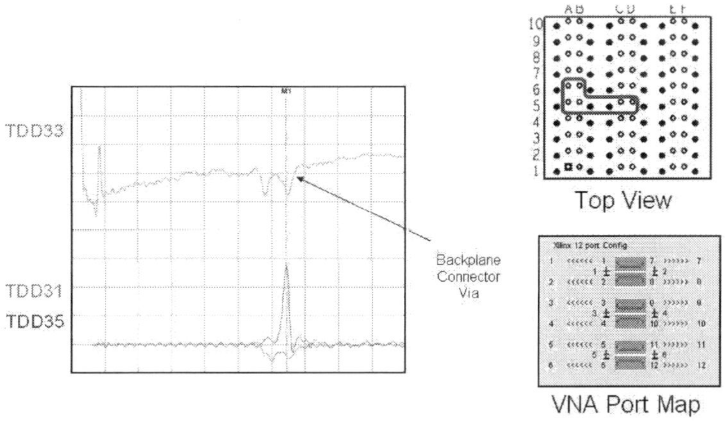

Figure 7.18: *eHSD Channel Row/Column Crosstalk*

The same measurements were performed for NEXT for row/column channel pin pair patterns, as shown in *Figure 7.18*. As expected, the row coupling in this test was the same as was measured for the previous row coupling test. Intuitively, the column coupling magnitude should be less than that of the row coupling due to the addition of ground pins between the two differential channels. The data reveals that this is the case.

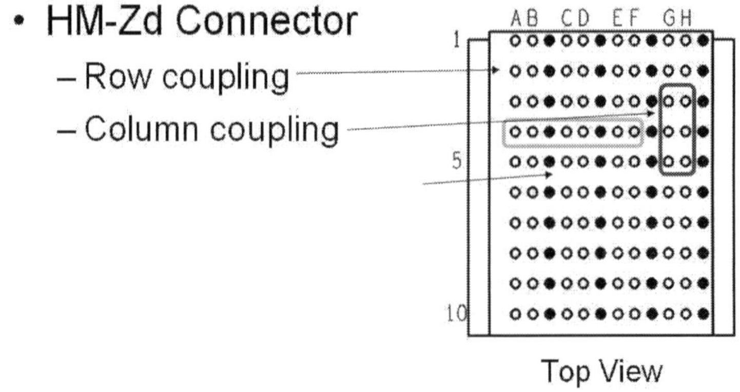

Figure 7.19: HM-Zd Crosstalk Configurations

The HM-Zd connector was tested for NEXT in both row and column configurations.

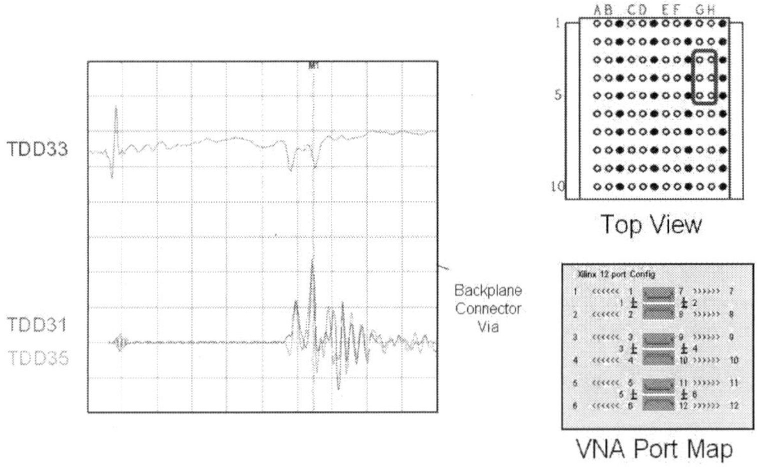

Figure 7.20: HM-Zd Channel Row Crosstalk

The data from the row coupling tests shows that the major contributor is once again the connector pin via field on the backplane.

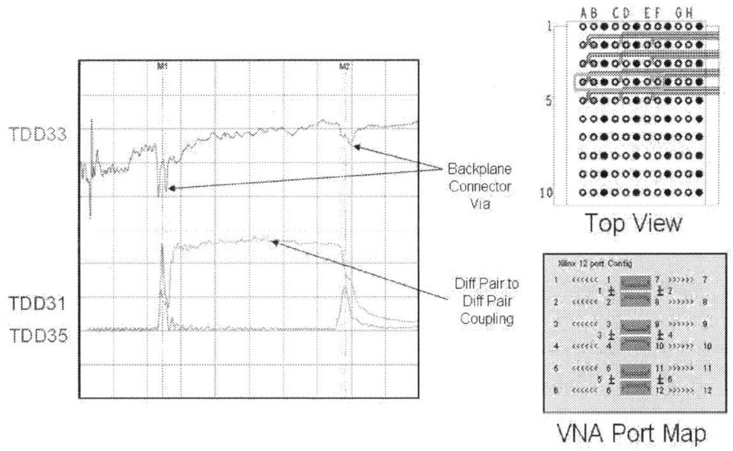

Figure 7.21: HM-Zd Channel Column Crosstalk

The column coupling tests gave some interesting results. For the row coupling case, the crosstalk results were similar to those of the previous test that tested only row coupling. This was not a surprise. The column channel arrangement gave a unique NEXT result. As can be seen in the figure, a significant amount of the crosstalk between the two channels occurred in the backplane traces.

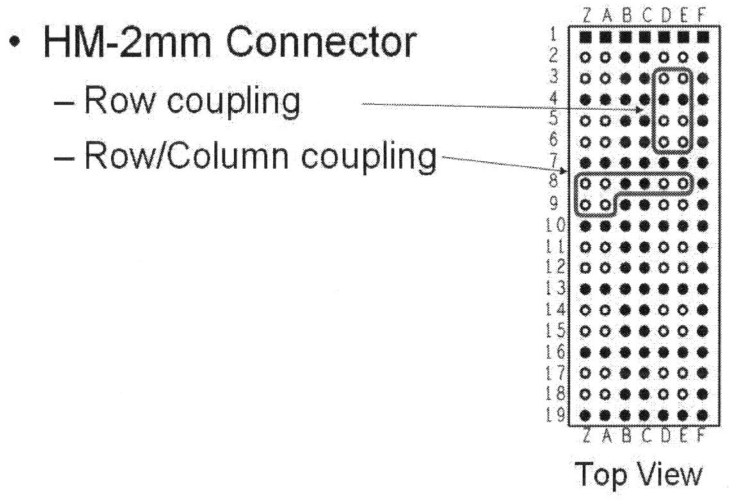

Figure 7.22: HM-2mm Crosstalk Configurations

Pins were assigned on the HM-2mm connectors so that rows of pins are separated by rows of ground pins. The usage of ground pins in this manner is not an uncommon practice for these types of connectors.

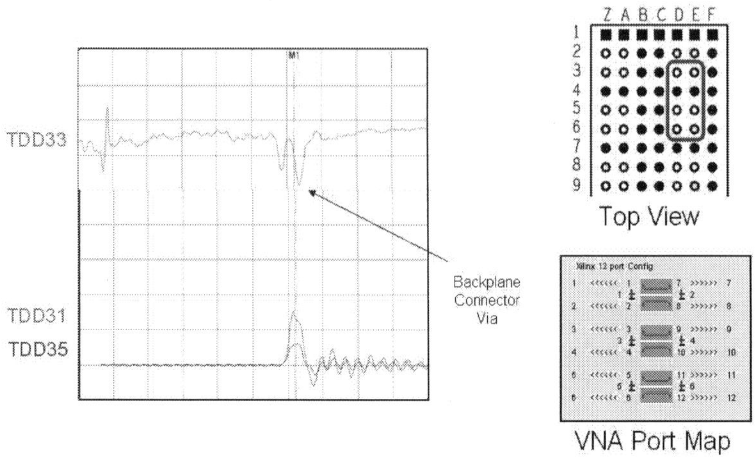

Figure 7.23: *HM-2mm Channel Row Crosstalk*

The NEXT test data shows that unlike the other two types of connectors, the major portion of the coupling is in the HM 2mm connector. As can be seen in the TDD33 plot in the figure, the crosstalk peak appears between the backplane connector vias on the daughtercard and on the backplane. Also notice there is less crosstalk from the aggressor channel that is separated from the victim by a set of ground pins.

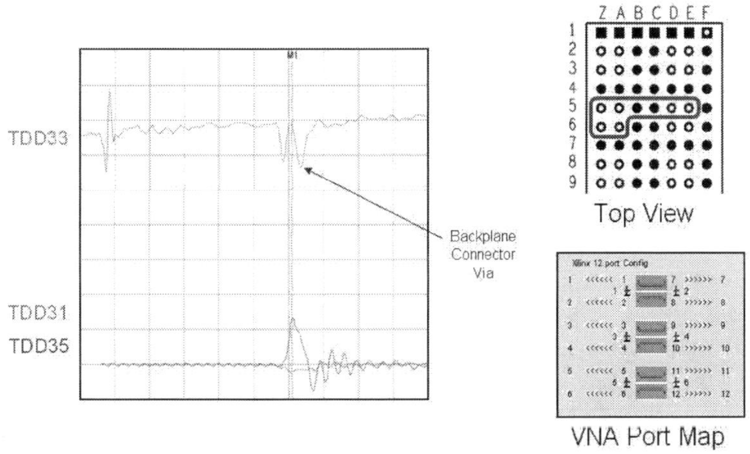

Figure 7.24: *HM-2mm Channel Row/Column Crosstalk*

For the row/column tests the HM-2mm shows crosstalk coupling in the connect for the row coupling portion. For the column coupling the amount of crosstalk coupling is substantially less than the row crosstalk coupling.

7.11 Eye Diagram Analysis

After evaluating the channels for specific parametric responses, they were analyzed using the collected channel measurements to perform an eye diagram analysis. This analysis provides a method to qualitatively evaluate the overall performance of the channel. As in the previous tests, the channels were analyzed by connector type.

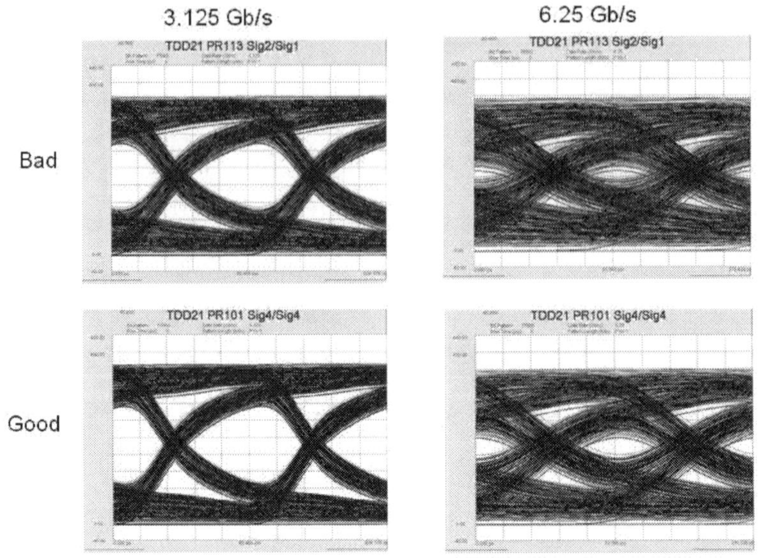

Figure 7.25: *eHSD Eye Diagrams*

The best eHSD channel uses N4000-13si with signal layer 4 on the line card and signal layer 4 on the backplane. The worst channel uses daughtercard signal layer 2 and backplane signal layer 1. As previously mentioned, signal layer 1 on the backplane has a via stub of over 60 mils. Although the effects of the signal degradation are barely discernable at 3.125 Gbps, the eye is substantially affected at 6.25 Gbps.

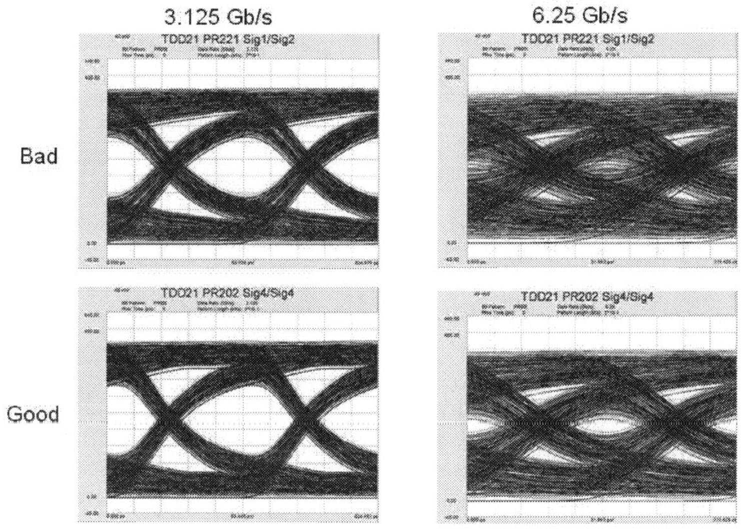

Figure 7.26: *HM-Zd Eye Diagrams*

For the HM-Zd connector channels, the best channel uses the bottom layer on the daughtercard and a lower backdrilled layer on the backplane. Given that the backplane layer is N4000-13si, the attenuation due to dielectric material loss is minimized on this channel also. The worst channel uses the top layer on the daughtercard and an upper signal layer on the backplane. This channel uses signal layer 2 on the backplane. This layer is one of the upper layers that are above the region that can be backdrilled. Therefore it has a significant via stub that impacts the overall performance of the channel.

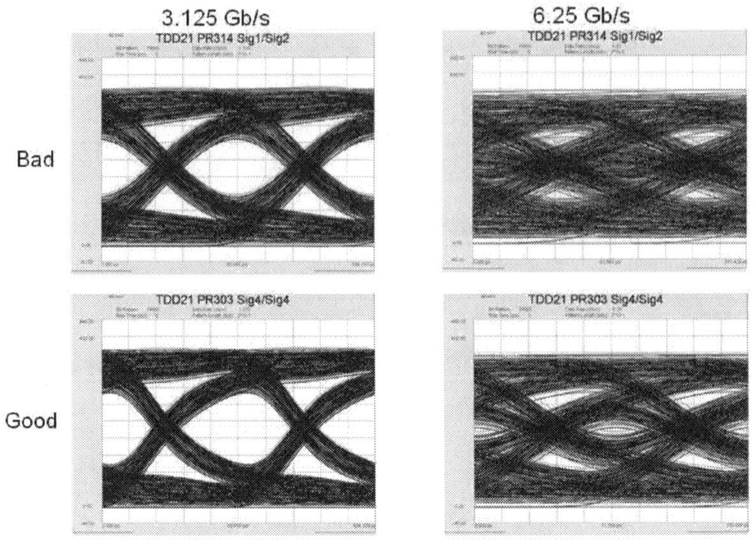

Figure 7.27: *HM-2mm Eye Diagrams*

For the HM-2mm connector the best channel is routed on the bottom signal layer on the daughtercard and on the lower Nelco4000-13si layer on the backplane. This backplane layer is backdrilled to a minimal stub length. Even with these advantages, the eye opening of the best channel at 6.25 Gbps is almost closed. Even with the lack of performance of these channels, at 3.125 Gbps the channel appears serviceable.

Figure 7.28: Backplane Reference Channels

7.12 Reference Channels

As mentioned previously, there is a reference channel on the backplane for each of the routing layers. These reference channels use an SMA connector for signal launch. Each reference channel is routed as a balanced differential signal pair and use the same trace geometry as that of the traces on that layer that run between the backplane/daughtercard connectors. That is with the exception that for a short distance, the traces are routed from the SMA connectors as single-ended traces before they are transitioned to a balanced, differentially coupled trace pair. Since there are six signal layers on the backplane, there are six reference channels.

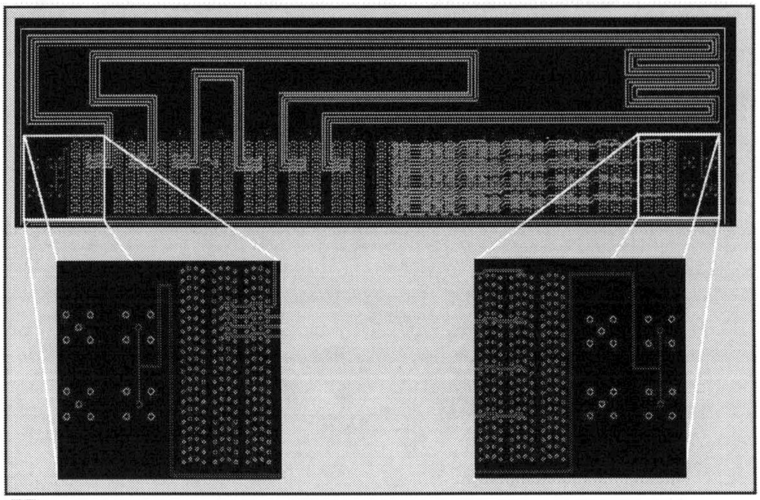

Figure 7.29: *Reference Channel Routing*

Figure 7.29 shows the routing for one of the reference channels. As you can see, there is a short distance where the traces are routed as striplines before they transition to differential striplines. As differential striplines, the reference channels use the same trace geometry, width, and separation as is used for the other traces on that signal layer. The total length of each reference channel trace is 20 inches.

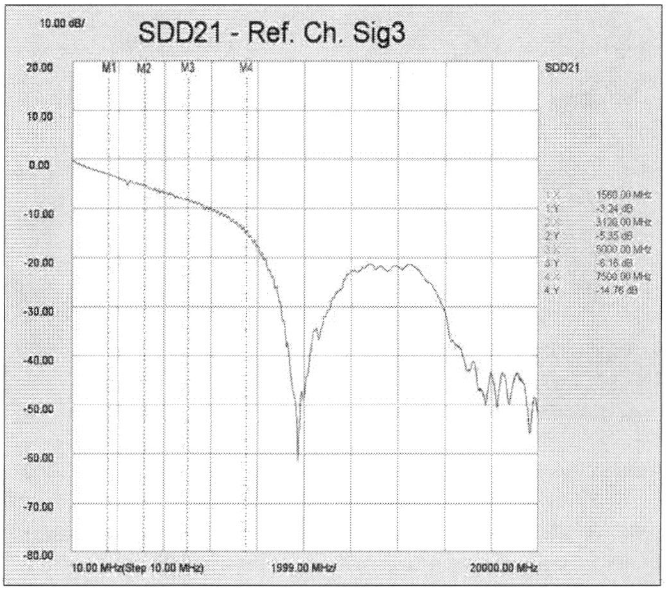

Figure 7.30: *Reference-Channel Insertion Loss*

The insertion loss on one of the reference channels is shown in *Figure 7.30.* There is a resonance at approximately 9.7 GHz. This resonance does not appear in the other channels routed on this layer. As mentioned previously, the trace geometries for the reference channel are the same as those of the other signals on this layer. Therefore the investigation focused on the SMA connector launch. The via barrel geometry has a different diameter from that of the compliant pin vias for the backplane connectors. More significant is the fact that the pin depth for the SMA connector is much greater than that of the backplane connectors. Whereas the backplane connector pin depth is on the order of 50 mils, the reference channel SMA pin depth is more than 190 mils. So even though the via barrels for the reference channel SMA connectors were backdrilled to the same depth at each layer as those of the compliant pin backplane connectors, the SMA signal pin was acting as a stub.

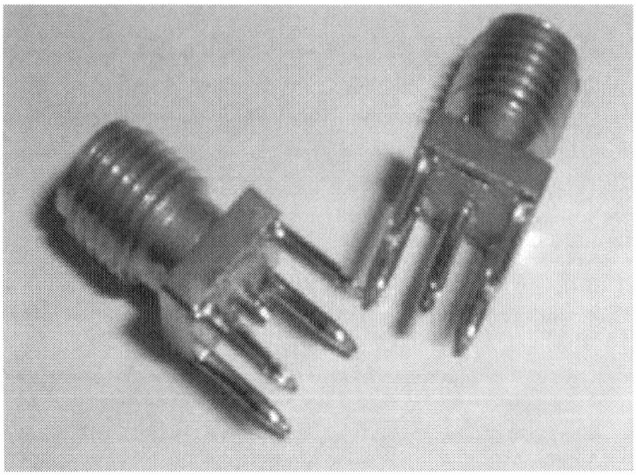

Figure 7.31: *Trimmed and Untrimmed SMA Connectors*

To address the issue of the SMA–compliant pin generating a stub, the center pin on the SMA connector was trimmed to approximately 50 mils. *Figure 7.31* is a photograph of an SMA connector with the center pin trimmed and an SMA connector with the center pin untrimmed. *Figure 7.32* is a diagram of the compliant-pin SMA connector mounted in the backplane with a trimmed and untrimmed center pin. It shows that the untrimmed connector extends beyond the backdrilled via barrel, while the trimmed connector does not extend beyond the backdrilled via barrel.

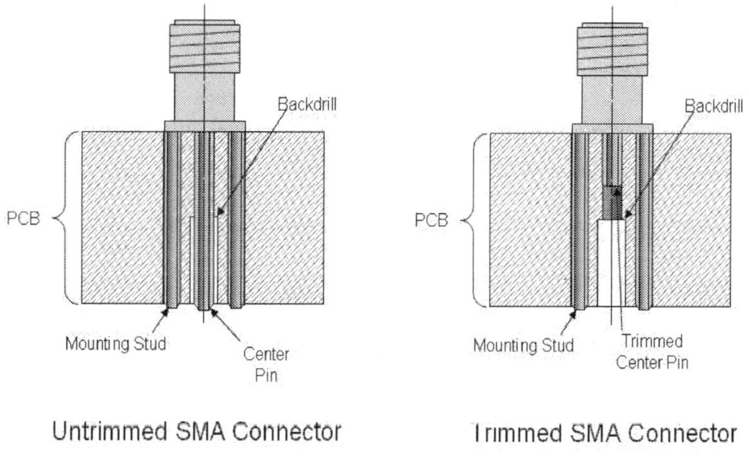

Figure 7.32: *SMA Connector Pin Stub Diagram*

With trimmed and untrimmed connectors, the insertion-loss measurement was repeated. The tests were performed with the reference channel connectors on one side of the reference channel trimmed and with both of the reference channel connectors trimmed. As can be seen in the figure, the resonance is eliminated by trimming the center pin of all of the SMA connectors.

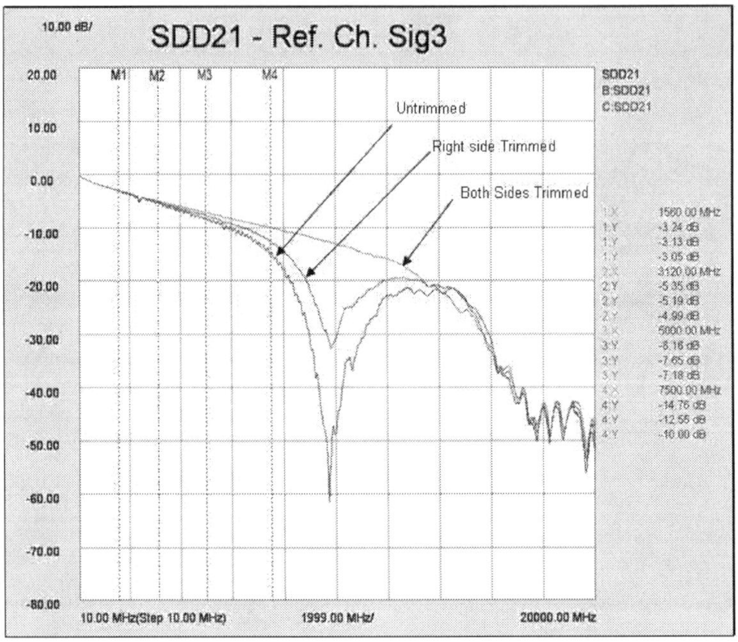

Figure 7.33: *Reference Channel Insertion Loss with Trimmed SMA*

Figure 7.33 shows the effects of the SMA connector pin stub on the eye diagram at 3.125 Gbps and at 6.25 Gbps. Although the effect is noticeable at 6.25 Gbps, the difference in the performance of the channel at 3.125 Gbps between the trimmed and untrimmed connector pin is barely perceivable.

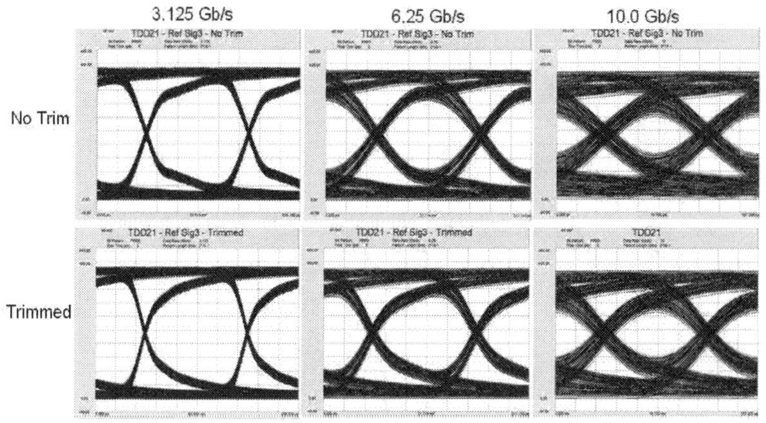

Figure 7.34: *Eye Diagrams for Reference Channel*

As shown in *Figure 7.34*, the eye diagram is barely discernable at 3.125 Gbps. It is noticeable at 6.25 Gbps. At 10 Gbps, it has much more effect. The SMA connector stub caused a resonance at 9.7 GHz. Even at 10 Gbps, the fundamental bandwidth is only 5 GHz, and because the resonance primarily affected the region around 9 to 11 GHz, the eye for the 10 Gbps signal still had a healthy fundamental and third harmonic response for passing the data.

7.13 Summary

The backplane demonstration platform was built to provide a range of channel behaviors. By using a variety of connectors, PCB materials, PCB routing structures, and routing paths, a range of channel behaviors was achieved. These channel behaviors were characterized and analyzed using a 12-port VNA. The stub length on the connector vias had a significant effect on channel behavior. The crosstalk test and analysis showed that most of the crosstalk appeared in the backplane connector via field. Although some of the channel responses appeared to be suspect, the eye diagram analysis showed that a serviceable eye opening can be achieved with over 36 inches of daughtercard and backplane trace length for most of the channels at bit rates up to 6.25 Gbps.

References

[1] E. Sayre, J. Chen, M. Baxter, G. Patel, J. Goldie, and M. Resso, "Minimizing Crosstalk in High-Speed Interconnects Using Measurement-Based Modeling," DesignCon Proceedings.

[2] H. Johnson and M. Graham, "High-Speed Signal Propagation: Advanced Black Magic," Prentice Hall PTR, 2003.

Authors

Jack Carrel, System IO Specialist, Xilinx

Bill Dempsey, Owner and President, Redwire Enterprises

Mike Resso, Signal Integrity Application Scientist, Component Test Division, Agilent Technologies

Chapter 8

Gigabit Backplane Design, Simulation, and Measurement: The Unabridged Story

8.1 Introduction

This paper focuses on high-speed point-to-point links, using low-voltage differential signaling (LVDS) technology across a GETEK–based backplane and plug-in cards using the VHDM–HSD connector system. Board design, SPICE simulations, and channel performance measurements are discussed in detail. The paper concludes with recommendations to achieve maximum throughput for tomorrow's high-performance backplanes operating with 1–3 Gbps channel speeds.

8.2 Gigabit Backplane Design Case Study

This section addresses the design of the point-to-point gigabit backplane used as the test bed.

The following were used in the backplane design:

- Point-to-point bus configuration
- GETEK backplane with 10-inch and 20-inch traces
- GETEK plug-in cards
- VHDM–HSD connectors
- Gigabit LVDS driver test silicon

The bus configuration is an uncomplicated point-to-point link. Due to the desired high throughput and the required signal-edge rate, a multi-drop/multipoint bus configuration was eliminated. The plug-in card is connected to the load via a direct connection in the backplane as with a simple point-to-point link or in a cross-bar application.

Material for both the backplane and the plug-in cards was selected to be GETEK over FR4, since the cost differential has lessened and the GETEK material has become more common in the industry. GETEK offers slightly better high-frequency performance and stable performance over temperature. This paper does not compare materials and their respective performance, as that subject has been covered adequately by many other papers to date.

Noting that this is a gigabit link, the Teradyne VHDM–HSD differential connector was selected.

The LVDS driver and receiver used was test silicon designed by National Semiconductor. The edge rate of this device was targeted for 1.5–2.0 Gbps operation.

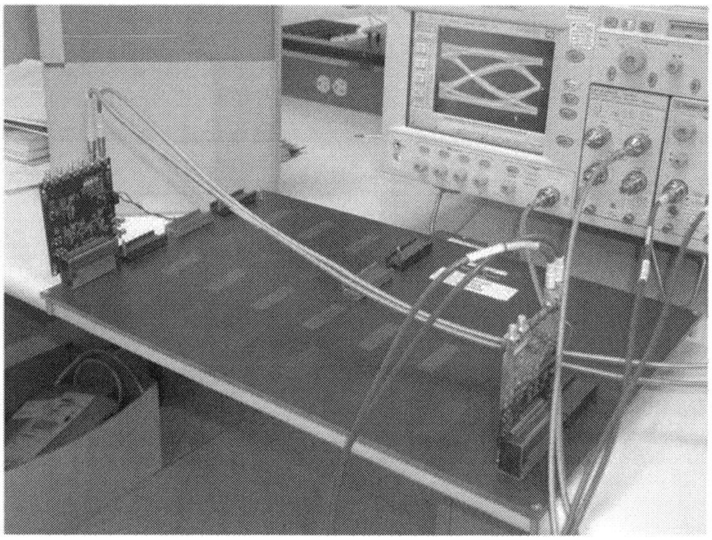

Figure 8.1: *Design Test System Picture*

Figure 8.1 shows the system under test (SUT) that was the subject of this paper. The backplane was designed by Teradyne and the plug-in cards were designed by National Semiconductor. The backplane provided both 10-inch and 20-inch interconnects.

Figure 8.2: *Close-Up View of Card and Connector*

In *Figure 8.2*, the test pair can be seen in the lower right next to the HSD connector. This allowed for test access for the time domain reflectometer (TDR) and generator measurements shown later in the presentation. The surface-mounted assembly (SMA) connectors on the top of the card provided the differential input to the test silicon, which was configured as a LVDS line driver and standard LVDS receiver (without clock-data recovery [CDR]).

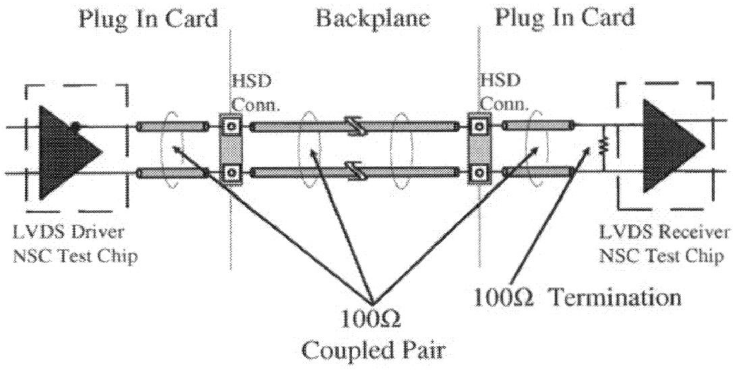

Figure 8.3: *LVDS Signal Path Topology*

The signal path is shown in *Figure 8.3*. This is known as an uncomplicated point-to-point link and is optimal for high–data-rate applications due to the pure and clean signal path.

The logic card featured a two-inch 100 Ohm coupled trace from the LVDS output pads to the HSD connector.

The backplane also used a 100 Ohm coupled pair between connectors. Trace lengths of 10 inches and 20 inches were available for test.

The plug-in card for the load had a two-inch interconnect to the termination location. A 100 Ohm differential termination resistor was used across the pair, and a quarter-inch stub connected the LVDS receiver inputs to the line.

Probing of the LVDS signals was done at the load end. The NS Test Silicon was packaged in a system-on-package (SOP) 14-lead package.

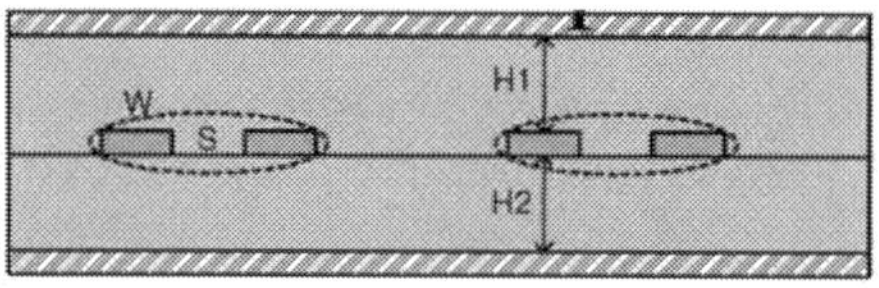

Stackup for the backplane:
H1 = H2 = 10 mils
W = 8 mils
S = 8 mils
Getek material
target differential impedance: 100 Ω
1/2 oz copper

Stackup for the plug in card:
H1 = H2 = 12 mils
W = 8 mils
S = 10 mils
Getek material
target differential impedance: 100 Ω
1 oz copper

Figure 8.4: *Backplane and Plug-In Card Stackup*

Figure 8.4 shows the cross-sections of the backplane and the plug-in cards. For this project edge-coupled, differential striplines were chosen. It is not the intention of this paper to compare broadside lines to edge-coupled lines. Edge-coupled lines were chosen due to ease of manufacturing and routing reasons.

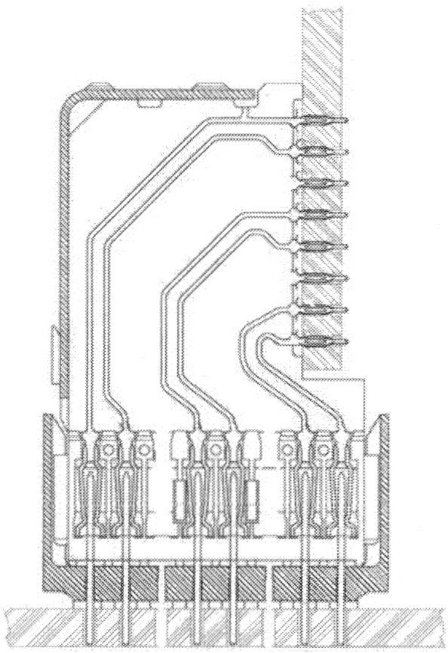

Figure 8.5: *VHDM–HSD – Side View*

The VHDM–HSD connector is uniquely constructed for differential signal transmission both in the daughtercard and backplane halves. As is shown in *Figure 8.5*, the signal lead frame in the connector is tightly coupled. This was done to minimize the skew within the differential pair. The measured skew within the pairs range from 6 to 10 ps. Also, by effectively moving the pairs further apart, the crosstalk is greatly reduced. For 200 ps edge rates, the crosstalk ranges from 1.56 to 0.85 percent. In order to achieve these electrical results, density had to be sacrificed: the VHDM eight-row connector has 50 pairs per linear inch, whereas HSD eight-row gives 38 pairs per linear inch.

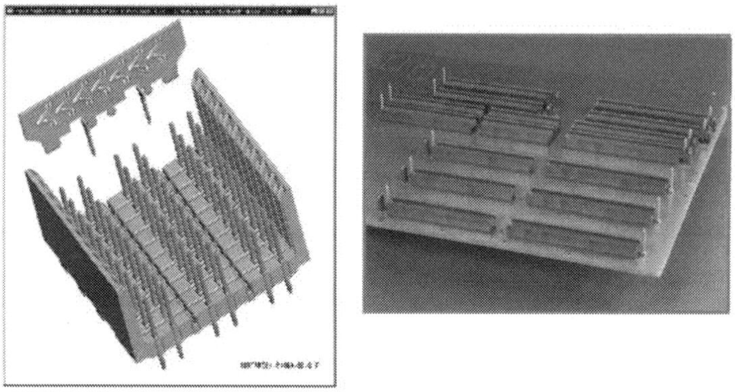

Figure 8.6: *VHDM–HSD – Backplane Shroud*

As shown in *Figure 8.6*, the daughtercard portion of the connector was optimized for differential signaling. In order to make the entire connector perform well at high-speed differential data rates, the backplane module had to be modified as well. This was done by removing two signal pins within a column and sliding two ground legs in their place (see *Figure 8.6*). The VHDM eight-row connector uses eight signal pins with seven ground pins, and in the HSD connector, six signal pins are used with two ground pins. The reason that seven ground pins can be reduced to two ground pins is due to the nature of differential signaling.

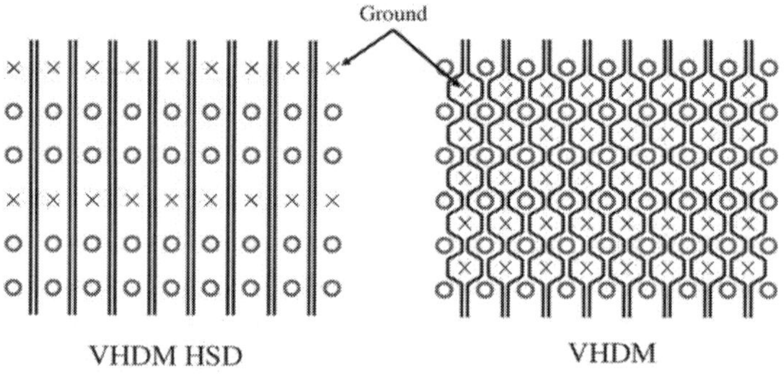

Figure 8.7: *VHDM–HSD – Board Routing*

Another benefit of HSD is in its routing. As shown in *Figure 8.7*, the VHDM connector, though it offers density, is not ideal for differential signal routing. The HSD connector eliminates the routing bottleneck by moving the interstitial grounds seen in VHDM in line with the signal pins in HSD. This could only be done by sacrificing ground pins, as described earlier. The jogged routing of VHDM effectively adds 23 percent to the overall trace length. The additional unnecessary trace length can have a severely negative impact at high data rates (more than 2.5 Gbps). This is because the backplane material becomes very lossy at high frequencies and long lengths.

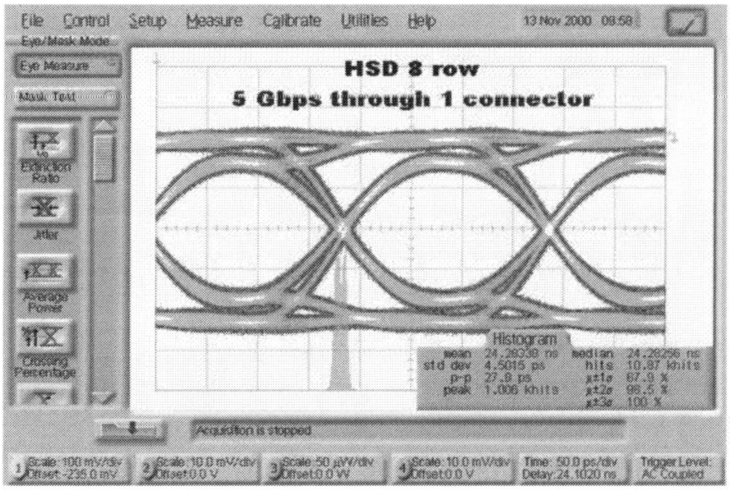

Figure 8.8: *Eye Pattern – VHDM HSD*

Figure 8.8 shows an eye pattern running at 5 Gbps through a single HSD eight-row connector. The total trace length was six inches in FR4 plus two feet of cable. *Figure 8.8* demonstrates that the connector in a stand-alone environment performs very well at 5 Gbps. The problem arises when the same data rate is passed through a more realistic system that includes two connectors plus some trace length. In this environment, the effects of the dielectric become the dominant factor.

8.3 Simulations

The next phase of the project was to look at the simulations of the system. Simulations were completed by NESA for both impedance and wave shape using Avanti Corp.'s Star-HSpice analog circuit simulator.

The simulated interconnect included the two test cards and the backplane, which were connected with two Teradyne HSD backplane connectors.

The cards and backplane were fabricated with GETEK dielectric material. HSpice lossy W-element models (with NESA–supplied parameters) were used for the transmission-line models.

Both differential TDR profiles and eye patterns are presented in the following slides. The backplane length was set to 10 inches and 20 inches, and data rates of 1.5 Gbps, 2 Gbps, and 2.5 Gbps were simulated using the K28.5 data pattern.

Note: additional via capacitances were included in the simulations as needed—card via capacitance (cvia) = 1 pF; backplane via capacitance (bvia) = 2 pF. The TDR for a short path shows the effects of the discontinuities suffered by a waveform traversing the semiconductor package, plug-in card paths, and backplane connectors to the matched 100 Ohm termination. Note that the card via generally has a lesser effect than the backplane via due to the relative differences in thickness between the two. The slight rise in the TDR impedance on the backplane is due to the series resistance of the etch. As the signal waveform travels through the connector, it suffers some reflections shown as ripples in the TDR. The discontinuity of the second connector is substantially less due to the loss in rise time suffered by the waveform due to dielectric losses. The via capacitances that were included in the simulations are: cvia = 1 pF; bvia = 2 pF.

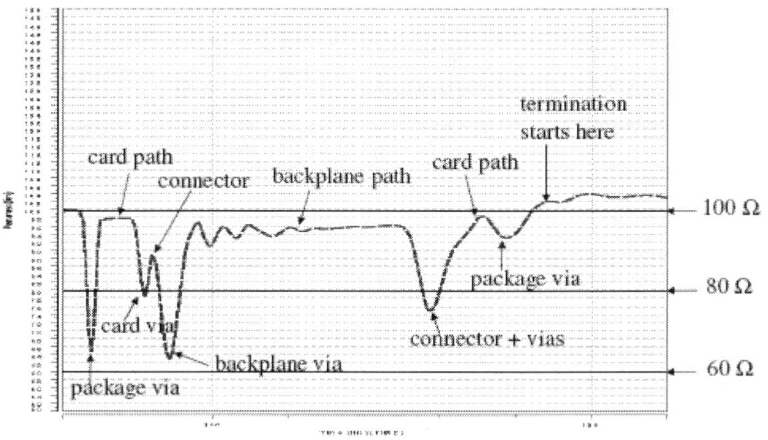

Figure 8.9: *Differential TDR Simulation with 100 ps TDR Rise Time (with Two Two-Inch Cards, a 10-Inch Backplane, and Two HSD Connectors)*

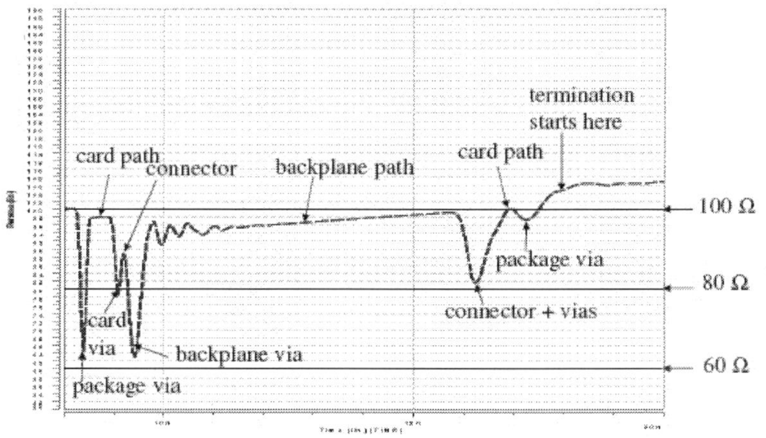

Figure 8.10: *Differential TDR Simulation with 100 ps TDR Rise Time (with Two Two-Inch Cards, a 20-Inch Backplane, and Two HSD Connectors)*

The TDR for the longer path in *Figure 8.10* shows similar effects of the discontinuities suffered by a waveform traversing the semiconductor package, the plug-in card paths, and the backplane connectors to the matched 100 Ohm termination. Note that the card via generally has a lesser effect than the backplane via due to the relative differences in thickness between the two. The more pronounced rise in the TDR impedance on the backplane is due to the longer path series resistance of the etch. Similar reflections, shown as ripples in the TDR, occur at the near-end connector but are largely missing after the second. The discontinuity of the second connector is even less than that exhibited over the shorter path due to the greater loss in rise time suffered by the waveform. The via capacitances that were included in the simulations are: cvia = 1 pF; bvia = 2 pF.

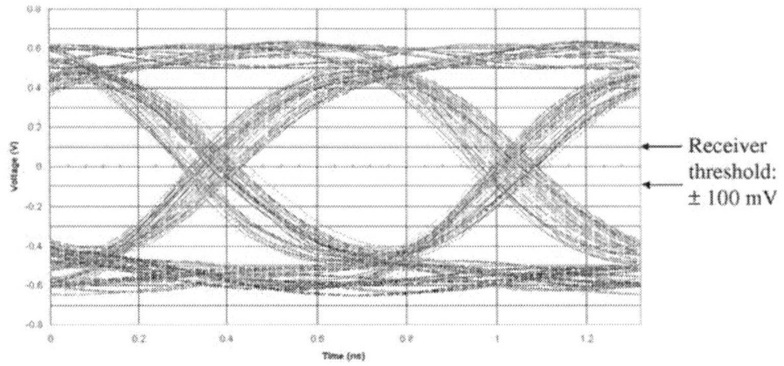

Figure 8.11: *Differential Eye Pattern, 1.5 Gbps Data Rate (Total 14-Inch PCB, Including 10-Inch Backplane)*

The eye diagram of a 1.5 Gbps data rate over a 10-inch backplane, shown in *Figure 8.11*, shows that the voltage margin for this path is more than satisfactory and is approximately 320 mV above the specified differential LVDS thresholds. The time jitter through the short backplane path is on the order of 160 ps. The attenuation of single bits is only slightly greater than bit patterns where the peak voltage excursion has been reached, indicating that the principal loss mechanism is high-frequency in nature. The via capacitances that were included in the simulations are: cvia = 1 pF; bvia = 2 pF.

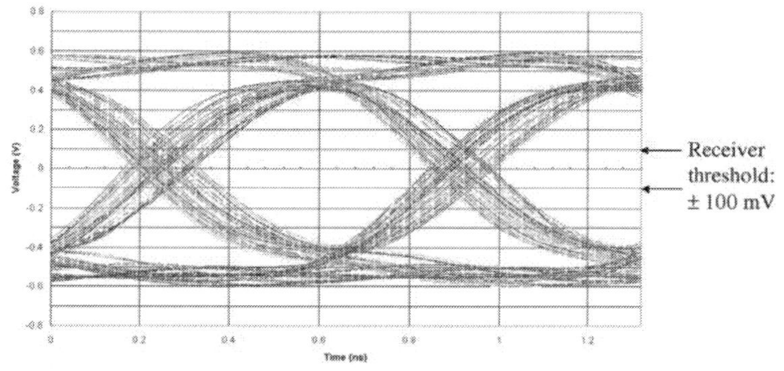

Figure 8.12: *Differential Eye Pattern, 1.5 Gbps Data Rate (Total 24-Inch PCB, Including 20-Inch Backplane)*

The eye diagram of a 1.5 Gbps data rate over a 20-inch backplane, shown in *Figure 8.12*, shows that the voltage margin for this path is more than satisfactory and is approximately 300 mV above the specified differential LVDS thresholds. The time jitter through the short backplane path is on the order of 180 ps. The attenuation of single bits is somewhat greater than bit patterns across a 10-inch backplane. This indicates that the principal loss mechanism is high-frequency in nature and not DC or skin-effect etch loss. The via capacitances that were included in the simulations are: cvia = 1 pF; bvia = 2 pF.

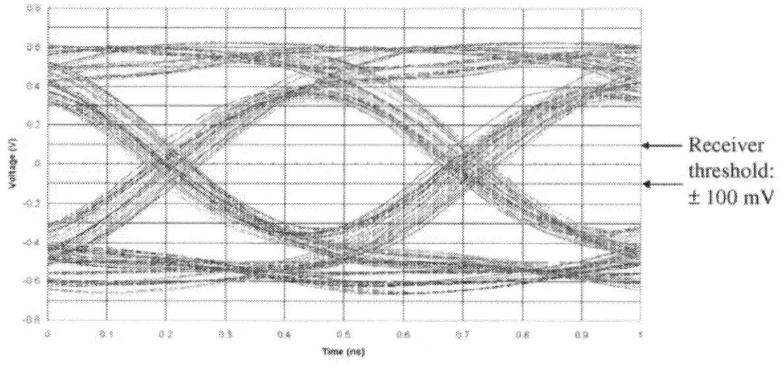

Figure 8.13: *Differential Eye Pattern, 2 Gbps Data Rate (Total 14-Inch PCB, Including 10-Inch Backplane)*

The eye diagram of a 2.0 Gbps data rate over a 10-inch backplane, shown in *Figure 8.13*, shows that the voltage margin for this path is less satisfactory than at 1.5 Gbps and is approximately 210 mV above the specified differential LVDS thresholds. The path should work satisfactorily. The time jitter through the short backplane path is on the order of 120 ps. The attenuation of single bits is only somewhat greater than bit patterns at 1.5 Gbps, indicating that the principal loss mechanism is high-frequency in nature. The via capacitances that were included in the simulations are: cvia = 1 pF; bvia = 2 pF.

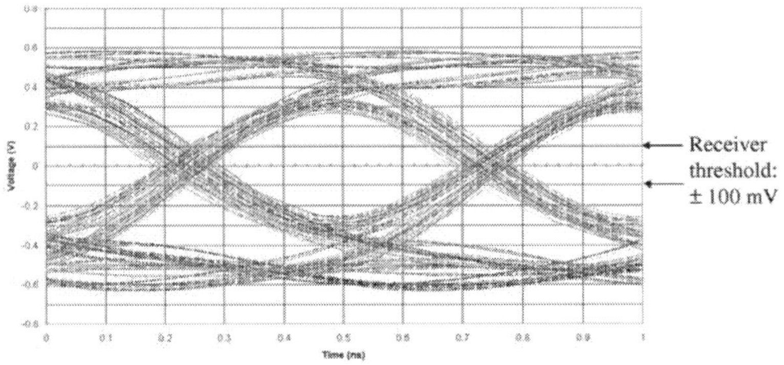

Figure 8.14: *Differential Eye Pattern, 2 Gbps Data Rate (Total 24-Inch PCB, Including 20-Inch Backplane)*

The eye diagram of a 2.0 Gbps data rate over a 20-inch backplane, shown in *Figure 8.14*, shows that the voltage margin for this path is less satisfactory than at 1.5 Gbps and is approximately 180 mV above the specified differential LVDS thresholds. The path should work satisfactorily. The time jitter through the short backplane path is still on the order of 120 ps. The attenuation of single bits is somewhat greater than bit patterns at 1.5 Gbps, indicating that the principal loss mechanism is high-frequency in nature. The via capacitances that were included in the simulations are: cvia = 1 pF; bvia = 2 pF.

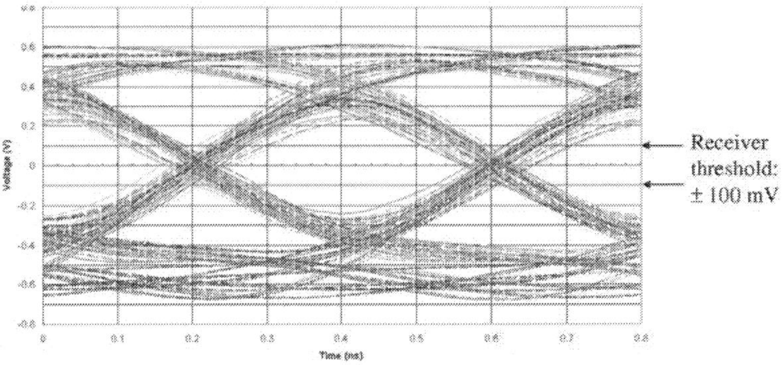

Figure 8.15: *Differential Eye Pattern, 2.5 Gbps Data Rate (Total 14-Inch PCB, Including 10-Inch Backplane)*

The eye diagram of a 2.5 Gbps data rate over a 10-inch backplane, shown in *Figure 8.15*, shows that the voltage margin for this path is less satisfactory than at 2.0 Gbps and is approximately 120 mV above the specified differential LVDS thresholds. The path should work satisfactorily. The time jitter through the short backplane path is still on the order of 110 ps. The attenuation of single bits is greater than bit patterns at 2.0 Gbps, indicating that the principal loss mechanism is high-frequency in nature. The via capacitances that were included in the simulations are: cvia = 1 pF; bvia = 2 pF.

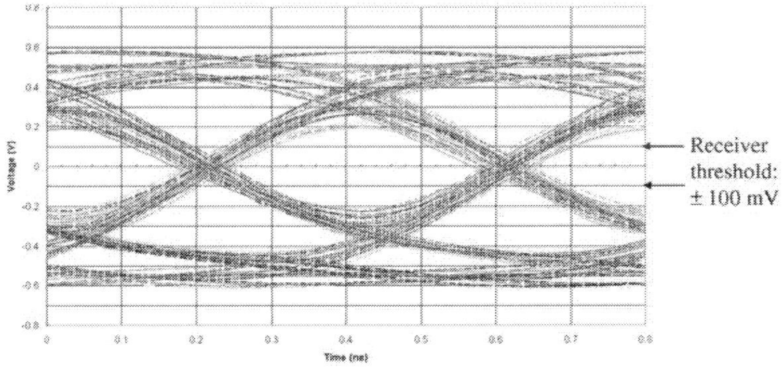

Figure 8.16: *Differential Eye Pattern, 2.5 Gbps Data Rate (Total 24-Inch PCB, Including 20-Inch Backplane)*

The eye diagram of a 2.5 Gbps data rate over a 20-inch backplane, shown in *Figure 8.16*, shows that the voltage margin for this path is less satisfactory than for the 10-inch backplane at 2.5 Gbps and is approximately 100 mV above the specified differential LVDS thresholds. The path should work satisfactorily, especially if the real thresholds are less than 100 mV. The time jitter through the short backplane path is still on the order of 100 ps. The attenuation of single bits is greater than bit patterns at 2.0 Gbps. Double bit effects are also apparent, indicating that the principal loss mechanism is high-frequency in nature. The via capacitances that were included in the simulations are: cvia = 1 pF; bvia = 2 pF.

8.4 Measurements

The final phase of the project was to check predictions and simulations against actual bench measurements. For this, a variety of Agilent equipment was used to make TDR and wave-shape measurements.

Connection to the test equipment was done with 50 Ohm coax cables and edge-launch SMA connectors.

Probing of the LVDS signals was done with a passive divider and biasing circuit to allow for a connection to high-bandwidth 50 Ohm scope channels.

Baseline measurements of the equipment were taken along with channel measurements. These are addressed in the following section.

$$V_1 = Z_{11}I_1 + Z_{12}I_2$$

$$V_2 = Z_{22}I_2 + Z_{21}I_1$$

Example:
Characteristic Impedance Matrix [ohms]:

	1	2
1	49.6	6.4
2	6.4	49.6

Coupling Factor (6.4)
Self Impedance (49.6)

Figure 8.17: *Differential Impedance: The Characteristic Impedance Matrix*

If there were no coupling between transmission lines, the impedance of a line, as defined by the ratio of the voltage across the paths and the current through them, would be dependent on just the line parameters of the one line. However, as soon as coupling is introduced, the voltage on one line may be dependent on the current in an adjacent line. To include these effects, the concept of impedance or characteristic impedance must be expanded to allow for one trace interacting with another. This is handled by expanding the impedance into an impedance matrix.

Any two transmission lines, each with a signal path and a return path, can be modeled using an impedance matrix. The diagonal terms are the impedance of the line when there is no current in the adjacent line. This is sometimes called the self-impedance. The off-diagonal elements represent the amount of voltage noise induced on the adjacent trace when current flows on the active line. If there were little or no coupling, the off-diagonal impedance would be near zero.

As the coupling between the lines increase, the off-diagonal terms will increase. For example, if the microstrip traces, as illustrated in *Figure 8.17*, were moved closer together, the diagonal impedance would not change very much, but the off-diagonal terms would increase.

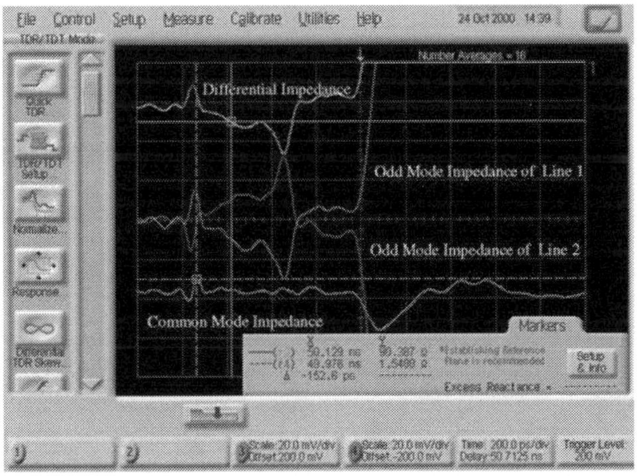

Figure 8.18: *Differential and Common-Mode Impedance of the Backplane*

The TDR instrument setup state for the top three measurements on the TDR display is as follows: TDR step generators are in differential stimulus state. This means the two TDR steps being launched into the backplane are of equal and opposite polarity. The steps are 40 ps rise time with 200 (and -200 mV) amplitude. The top waveform is the differential impedance, defined as channel 1 – channel 2. Since the stimulus is differential, channel 1 – (-channel 2) is actually channel 1 + channel 2. Thus, the differential impedance measurement is made by placing the marker on this waveform near the middle of the backplane path and noted as 90.39 Ohms.

The two middle waveforms are the odd-mode impedance of each of the differential lines. TDR stimulus is still differential. This measurement is made by selecting channel 1 or channel 2 as the marker reference channel and reading directly from the marker tab in the lower right portion of the screen.

The bottom waveform is the common-mode impedance. The TDR stimulus for this measurement has been changed to common-mode drive (in-phase and driven on each line of the pair). This TDR configuration yields common-mode stimulus and differential response (mixed-mode analysis). The result is channel 1 – (+ channel 2). This measurement is made by placing the marker on this waveform near the connector and can be read as 1.55 Ohms.

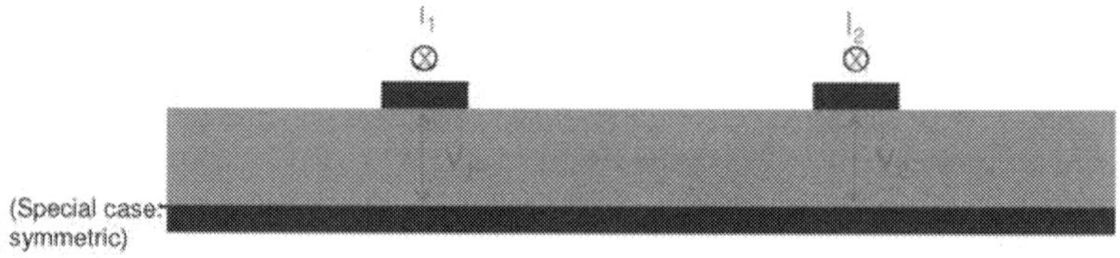

$$V_{odd} = \frac{1}{2}(V_1 - V_2) \qquad Z_{odd} = \left.\frac{V_{odd}}{I_1}\right|_{V_{even}=0}$$

$$V_{even} = \frac{1}{2}(V_1 + V_2) \qquad Z_{even} = \left.\frac{V_{even}}{I_1}\right|_{V_{odd}=0}$$

Figure 8.19: *Definition of Odd- and Even-Mode Impedance*

Based on the definition of the impedance matrix and the definition of odd and even mode, the impedance of each mode can be calculated. The odd-mode impedance is the impedance a driver would see, looking into one of the lines, when the pair of lines is driven in the odd mode or with a differential signal. Likewise, the even-mode impedance is the impedance a driver would see, looking into one of the lines, when the pair of lines is driven in the even mode or by a common signal.

If there were no coupling, both the odd- and even-mode impedances would be equal, and equal to the impedance of just one isolated line, as expected. However, with coupling, there are additional current paths between the signal lines in odd mode, and the odd-mode impedance decreases. Some current will flow not only from the first signal line to the return path, but through to the second signal line and then into the return path. This increased current through the coupling path results in a decrease in the odd-mode impedance of one line with increasing coupling.

The even mode is also affected by the coupling. When driven with a common signal, there is no voltage difference between the two signal traces. There is thus no coupled current between the signal lines, and the even-mode impedance is higher than the odd mode.

A universal equation for a directional coupler contains a coefficient of coupling, k, defined as the ratio of the difference, Zoe-Zoo, to the sum, Zoe+Zoo, of the even- and odd-mode characteristic impedances. The overall characteristic impedance is equal to the square root of the product of the even- and odd-mode characteristic impedances, Zo^2=Zoe x Zoo. These two equations are thus used to calculate the even- and odd-mode impedances for the desired coupling and overall Zo.

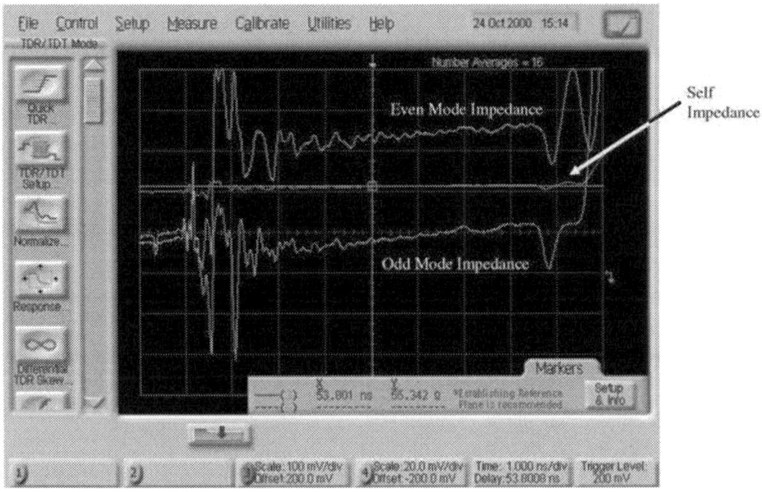

Figure 8.20: *Even- and Odd-Mode Impedance on LVDS Backplane*

The even- and odd-mode impedance measurements can be made by selecting only one of the differential lines and changing the TDR step stimulus from differential-mode to common-mode drive. This is simply changing from equal and opposite polarity steps to equal and same polarity steps, respectively. Waveform memory was implemented to first store the odd-mode impedance, then stimulus was changed to common mode and even-mode impedance was obtained. The vertical separation of the even- and odd-mode impedance waveforms on the display of the TDR is exhibiting the phenomena of good differential coupling.

A more subtle waveform is shown in between the even- and odd-mode waveforms. This is the self-impedance of the one differential line. This measurement is obtained by selecting a single-ended TDR stimulus and not driving the second differential line at all.

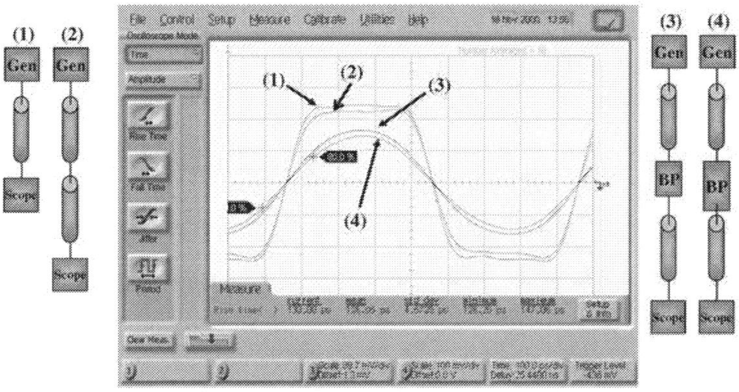

Figure 8.21: *Clock Wave Shapes*

Four waveforms are shown in *Figure 8.21*. From the signal comes the following, with the fastest rise time to the slowest:

- The signal with the fastest rise time is the generator connected directly to the scope via a 50 Ohm coax cable.
- The signal with the next fastest rise time is the generator connected directly to the scope via two 50 Ohm coax cables connected in series.
- The next fastest signal is the clock signal passing through the 10-inch backplane interconnect.
- The slowest signal is the clock signal passing through the 20-inch backplane interconnect.

The bandwidth of the backplane filters the signal and causes rise-time degradation and attenuation. The 10+ inch interconnect increased the rise time by 80–100 ps, and the 20+ inch interconnect increased the rise time by about 120 ps.

Note that the time base is 100 ps/div.

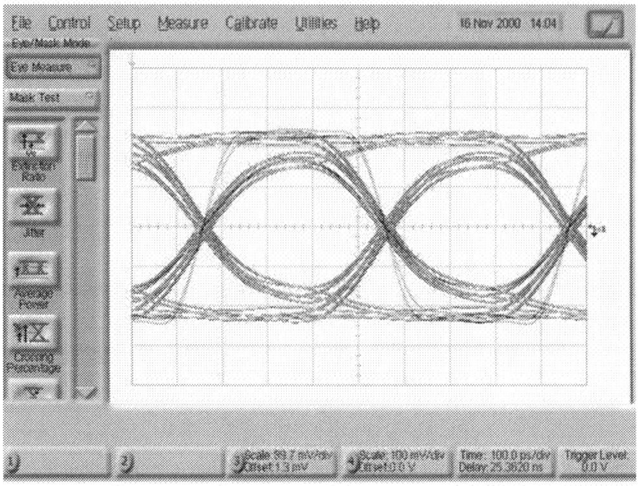

Figure 8.22: *K28.5 Eye Pattern (Ten Inch)*

The K28.5 pattern is driven across the backplane from the signal generator to the scope in the same configuration as in test case 3 in *Figure 8.22.* Also shown are the three clock wave shapes for comparison. The K28.5 pattern has five rising edges and five falling edges, which can be seen in the figure. The data rate is 2.5 Gbps, and a differential waveform is shown. The bandwidth inter-symbol interference (ISI) can be seen in the form of increased jitter at the zero crossing. If the prior data bit was in the same state, the line charges to a higher value, thus when the transition occurs there is a different starting point compared to that of a bit that had just switched to that state. The result is increased deterministic jitter, as shown in *Figure 8.22.* This plot should also be compared to that of a pseudorandom binary sequence (PRBS) pattern, which is worst-case. The PRBS pattern does not force transitions to occur; in fact it includes long strings of 1s and 0s, which fully charge the line. This is the benefit of encoding data. An example of encoding is the popular 8b/10b code that guarantees transitions and DC balancing of the data on the line, which improves the eye opening and thus reduces jitter. The K28.5 pattern is commonly used to represent the worst-case pattern, as it includes the highest- and lowest-frequency patterns of 8b/10b.

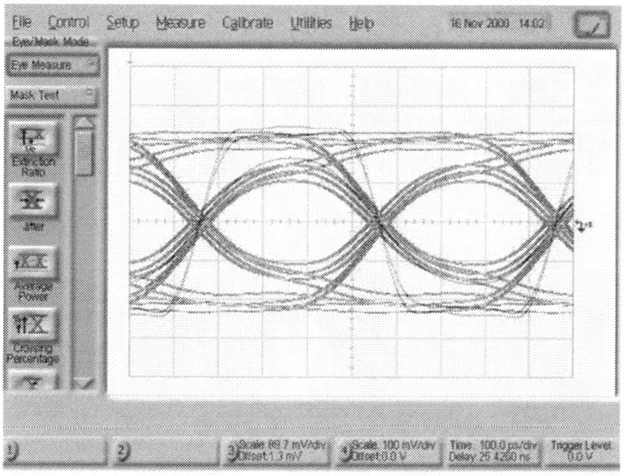

Figure 8.23: *K28.5 Eye Pattern (20-Inch)*

The loading effect of the backplane can be seen even clearer in *Figure 8.23*. The interconnect was changed from the 10-inch backplane path to the 20-inch path. With the longer length, the loading effects are greater and easier to see. Note that on the longer path, the rise time is slowed further, thus a drop in amplitude occurs and the eye closes more.

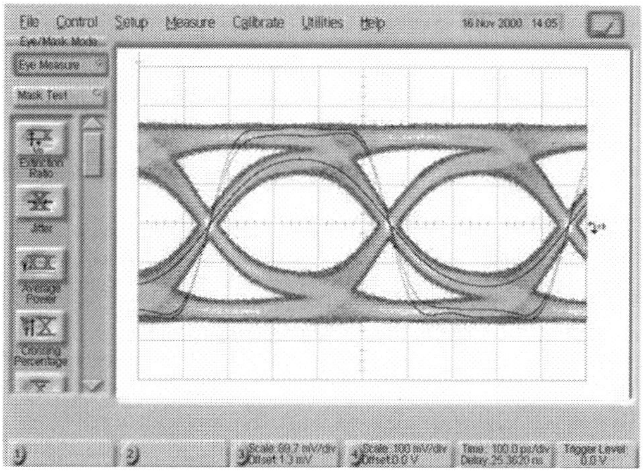

Figure 8.24: *PRBS–31 Eye Pattern (10 Inch)*

The scope in *Figure 8.24* is the same as the K.28 pattern shown in *Figure 8.22*, except the pattern has been changed to PRBS–31. The impact is more jitter at the zero crossing point and also a wider distribution at the top and bottom base lines.

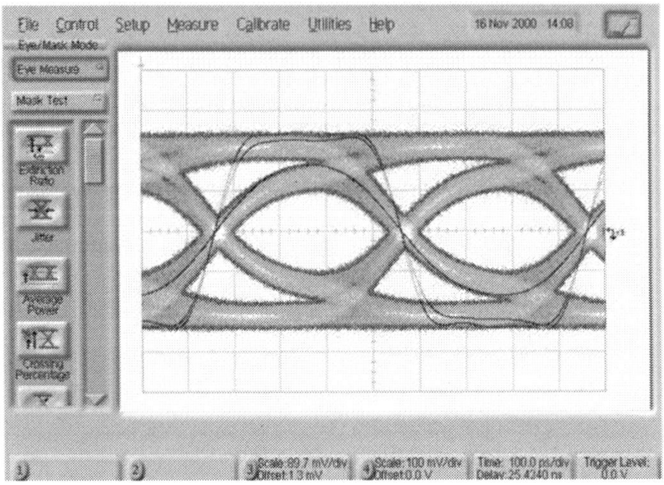

Figure 8.25: *PRBS–31 Eye Pattern (20 Inch)*

The scope in *Figure 8.25* is the same as the K.28 patter shown in *Figure 8.23*, except the pattern has been changed to PRBS–31. The impact is again more jitter at the zero crossing point and further closing of the amplitude of the signal at the center of the eye pattern due to the slower edge.

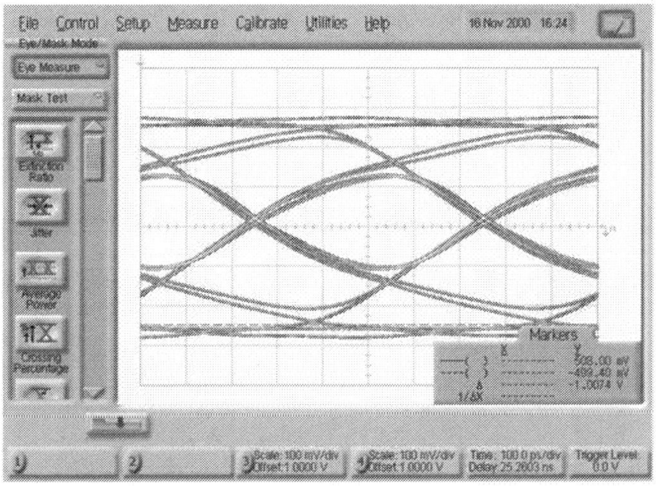

Figure 8.26: *LVDS Driver – 10 Inch/2 Gbps/K28*

The scope shot in *Figure 8.26* uses the National test silicon low-voltage differential signaling (LVDS) driver. A complicated passive load has been used to allow direct measurement of the LVDS driver into the 50 Ohm scope on the receiver card in place of the LVDS receiver. This allows the signal quality to be checked at the receiver input pads. This divider provides an equivalent 100 Ohm load to the driver and also a 2:1 divider to the scope. Some additional rise-time degradation is induced by this probing method, thus reducing the amplitude further. This can be seen when comparing this figure to the simulation eye pattern.

Even though the eye is closing down, the design of the receiver and CDR circuitry will recover the data. The LVDS receivers tend to have very tight thresholds that can switch with as small as 10 mV signal amplitudes. CDR circuitry, depending upon implementation, tends to be able to recover data from a signal with jitter of 50 to 70 percent of the unit interval.

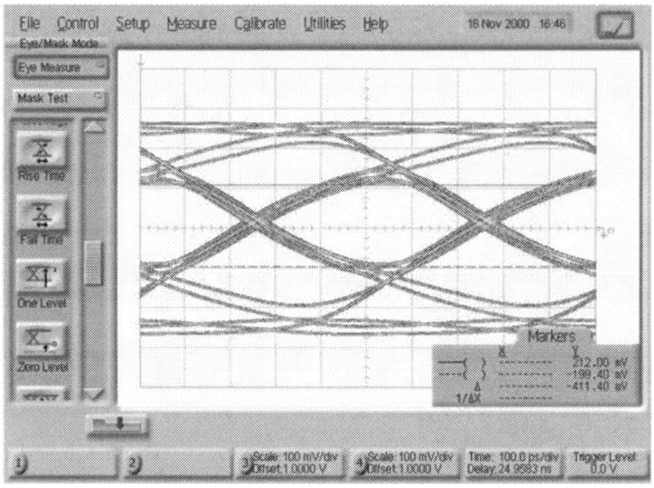

Figure 8.27: *LVDS Driver – 20 Inch/2 Gbps/K28*

Figure 8.27 shows the additional loading effects of the longer backplane interconnect. Once again, this is illustrated by the reduced amplitude and also the increased jitter.

8.5 Recommendations

- Optimize interconnect for best differential signal transmission:
  - Limit the number of vias on the line
  - Match impedance and trace length, maintain balance of pair
  - Require proper termination
  - Keep stubs as short as possible
- Predict and verify signal quality—evaluate eye patterns at the load for signal quality
- Signal edge rates quicker than 300 ps should be used in point-to-point links only
- For 200 ps signal edge rates, equipment should have a rise time of 100 ps for less than 10 percent error and 29 ps for less than 1 percent error.

8.6 Summary

This case study has shown that it is feasible to design a 1–3 Gbps backplane link using standard materials, VHDM–HSD connectors, and LVDS signaling.

Additional enhancements to the LVDS driver to speed up the test silicon's driver-edge rate will allow for operation at 2.5/3.125 Gbps. Above these rates, additional tuning of the backplane would be required to address the interconnect's bandwidth.

The TDR plots provide great insight into the interconnect to determine which structures impact the signal path. Analyzing the signal quality at the load gives a good indication of the bandwidth of the interconnect and also the amount of jitter.

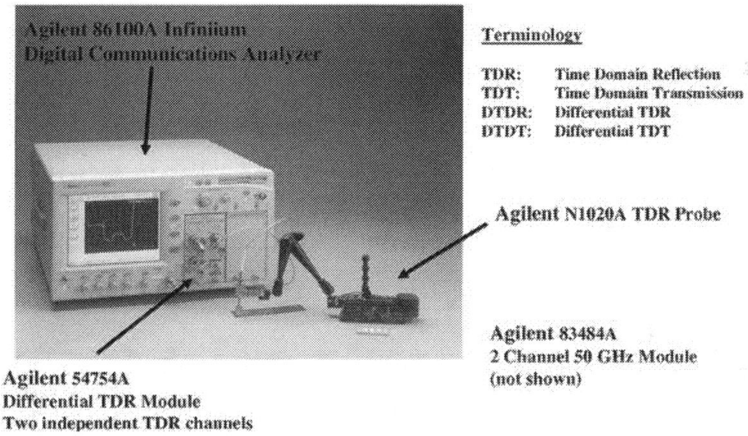

Figure 8.28: *The Test Equipment Used for the Bench Measurements*

Resources

- www.nesa.com
- www.national.com/appinfo/lvds
- www.teradyne.com/prods/bps/home.html
- www.agilent.com/findsi

Authors

Edward Sayre, Owner and Director, NESA

Jinhua Chen, Signal Integrity and EMI Engineer, NESA

Michael Baxter, Signal Integrity Engineer, NESA

Gautam Patel, Signal Integrity Engineer, New Product Development, Teradyne

John Goldie, Member of the Technical Staff, National Semiconductor

Mike Resso, Product Manager, Lightwave Division, Agilent Technologies

Part III

Assuring Quality Measurements: Probing and De-Embedding

Chapter 9

The ABCs of De-Embedding

9.1 Introduction

No test equipment is perfect, including vector network analyzers (VNAs). However, using a powerful toolbox of special operations enables the measurement errors to be dramatically minimized. These operations improve the laboratory measurement data, turning it into an excellent representation of the device under test (DUT). While calibration is a pre-measurement step that minimizes errors, the most important operation to reveal information about just the DUT is a post-measurement process called de-embedding. Traditionally, de-embedding has been used only by experienced users. By understanding the principles and how to use the new generation of built-in de-embedding features of VNAs, this powerful technique can be leveraged by all users and the quality of information extracted from all measurements dramatically improved. This practical guide to de-embedding will enable all users to take advantage of this important feature.

9.2 Why De-Embedding?

Although a VNA is a powerful tool for component characterization, it can only measure between well-calibrated reference planes. Often there is some type of fixture that makes the physical connection between the reference plane of the VNA to the ends of the DUT.

Figure 9.1 shows an example from Altera Corporation. The DUT is a ball grid array (BGA) component on the back side of the board that is accessed from the edge where there are surface-mounted assembly (SMA) launches and traces on the circuit board. The SMA and traces on the board contribute a larger measured impact than the DUT itself. How do you isolate the DUT performance when all you have are the DUT and the fixtures?

This is the value of de-embedding techniques. When you have a composite measurement of a DUT/fixture combination, you can isolate the performance of the fixture and use de-embedding to extract or de-embed the fixture from the measurements.

Figure 9.1: *De-Embedding: Removing Fixture Effects from a Measurement*

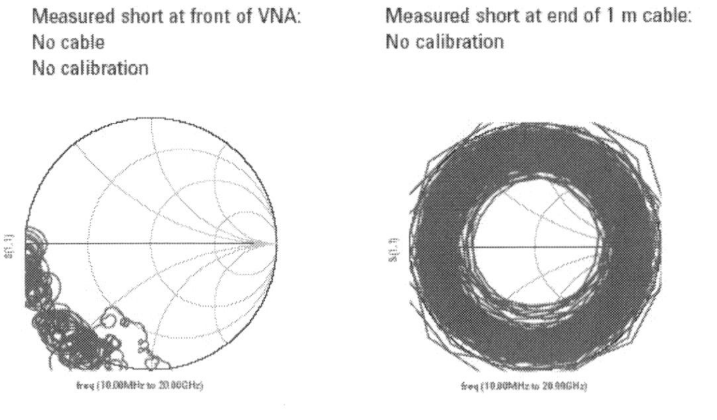

Figure 9.2: *Why Calibration and De-Embedding Are Important*

Between the sources and receivers at the core of the VNA are directional couplers, switches, and connectors, all designed to make the measurement of the S-parameters of the DUT effortless and transparent to the user. Each of these internal components contributes to measurement artifacts that hide and obscure the intrinsic measurement of the DUT.

For example, an ideal short on a Smith chart should be a dot on the left-hand side of the chart. The actual measured Smith chart of a short connected directly to the front connector of the VNA is shown on the left of *Figure 9.2.* This measurement is basically on the left side of the Smith chart but, other than that, is nothing like an ideal short. Clearly, the internal interconnects of the VNA are hiding the true nature of the DUT.

It is virtually impossible to connect a DUT directly to the front panel of test instrumentation. To further complicate measurements, cables and other mounting fixtures are almost always used to interface the DUT to the test instrumentation. Even the most precise interconnects will dramatically distort the measured response of the DUT.

The Smith chart on the right of *Figure 9.2* shows the measured S11 response from a short located at the end of a meter-long, precision 50 Ohm, low-loss cable. Deciphering any information about the DUT is virtually impossible from this measured response.

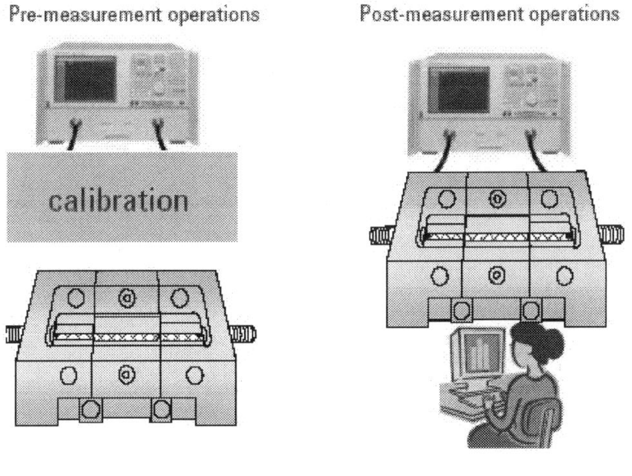

Figure 9.3: *Error Correction Techniques*

Over the years, many approaches have been developed for removing the effects of the internal VNA features and the test fixtures from a measurement to reveal the behavior of just the DUT. They fall into two fundamental categories: pre-measurement and post-measurement operations.

Pre-measurement operations require specialized calibration standards that are inserted at the ends of the test fixture and measured. This process moves the calibration plane to the end of the fixture. All the effects of the fixture are calibrated out up to this plane. The accuracy of the subsequent device measurement relies on the quality of these physical standards. This is why most VNA calibration kits include very precise air dielectric coaxial standards with calibration coefficients of inductance and capacitance that are read into the VNA firmware. Using any 50 Ohm loads out of lab stock is not recommended. Post-measurement operations involve taking a measurement of the DUT and all the fixturing leading up to it, then mathematically removing the fixturing, leaving only the DUT behavior. Of course, the essential ingredient for a post-measurement operation is accurate information about the fixture. This process is called de-embedding. The intrinsic DUT behavior is embedded in the total measurement and de-embedding removes the fixture effects, leaving just the DUT behavior.

This powerful technique can be used when the DUT is remote from the calibration plane or when there are non-coaxial connections from the VNA cables to the DUT. De-embedding is commonly used with circuit board traces, backplane channels, semiconductor packages, connectors, and discrete components. In signal integrity applications, de-embedding is the most important technique besides calibration for obtaining artifact-free device measurements. It is noteworthy to mention that full de-embedding requires all S-parameters for the fixture. With a differential fixture, this means the .s4p Touchstone file with all 16 elements in the 4x4 matrix.

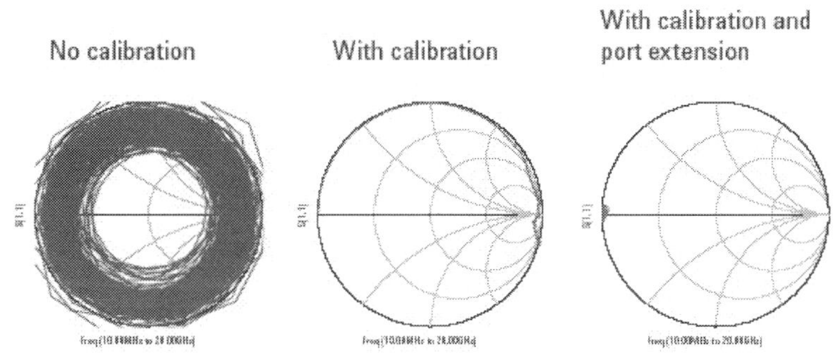

Figure 9.4: *Impact of Calibration on a Short at the End of a 1 m Precision Cable*

With the use of pre- and post-measurement compensation techniques, you can turn a network analyzer with all the artifacts from the cabling and fixturing into a nearly ideal network analyzer that can display the intrinsic S-parameters of the DUT.

With no calibration, a short at the end of a cable connected to a VNA looks anything but ideal. See the Smith chart measurement on the left of *Figure 9.4*. By applying the pre-measurement calibration process to the end of the cable, the resulting measured response of the short looks closer to what is expected for a short. See the center measurement. However, the real short is not an ideal short. As a calibration standard, it does not have to be. It just has to be a known standard. In this case, the real short is a very good coaxial short connected to the end of a coaxial transmission line that is approximately 100 mil long.

For measurements up to 20 GHz, the real short actually looks like an ideal, lossless transmission line shorted at the far end with a time delay of 13 ps. This is why it loops around the Smith chart in the clockwise direction. The coaxial short can be de-embedded from the measurement.

If you assume the connection to the short is an ideal, lossless, 50 Ohm transmission line segment with a time delay of 13 ps, then the S-parameter performance of this interconnect can be calculated analytically. The fixture S-parameters are used to "de-embed" the short from the composite measurement. This technique is called port extension.

The last Smith chart on the right of *Figure 9.4* shows an almost ideal short. This has the internal VNA circuitry, the cable effects, and the transmission line of the DUT all de-embedded. This is why de-embedding is such a powerful technique; it removes the unwanted artifacts of the system and fixtures to reveal the true characteristics of the DUT.

Pre- and post-measurement compensation techniques enable you to routinely make accurate S-parameter measurements of interconnect structures such as backplane traces, semiconductor package leads, surface mount components, and circuit board traces.

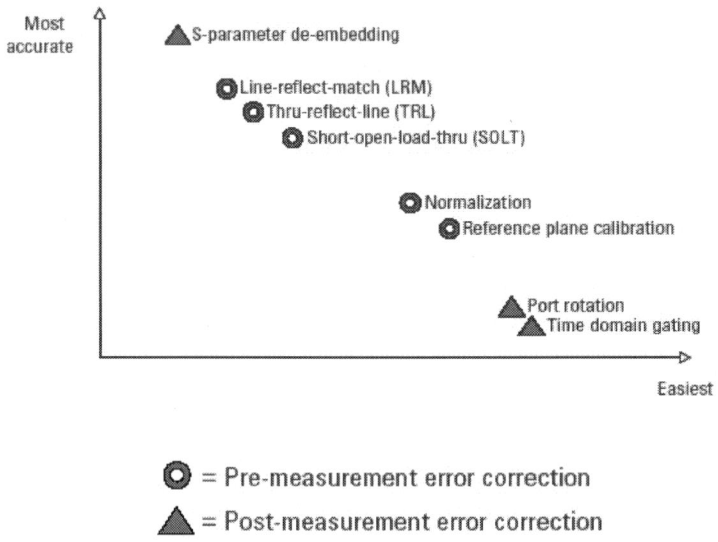

Figure 9.5: *Fixture Error Correction Techniques*

The various methods of turning real laboratory measurements into a close approximation of ideal measurements of the DUT are summarized in *Figure 9.5*. The simplest to understand and implement is time domain gating. It involves converting the frequency domain return loss measurements into a time domain response (TDR). A narrow region of the TDR response is selected, and only this TDR response is converted back into the frequency domain to be interpreted as the return loss from a specific region of the DUT. Though it is easy to implement in virtually all network analyzers, it is limited in application to just the return loss measurement and decreases in accuracy with increasing loss in the interconnect. It is most useful when the DUT dominates the total measurement such as when a connector adaptor is used with the DUT.

Port rotation is a lesser form of de-embedding in which a phase shift, increasing linearly with frequency, is subtracted from a measurement. The phase shift corresponds to a short length of ideal, 50 Ohm interconnect between the calibration plane and the DUT. Port rotation is a built-in feature in all network analyzers and can compensate for extra lengths in both the return loss and the insertion loss measurements. In modern network analyzers, the skin depth and dielectric losses of a uniform transmission line can also be simply and routinely removed from a measurement by de-embedding. In typical applications, the accuracy of port rotation drops off significantly as the length of the transmission line length increases. When a port rotation of more than 360 degrees is needed, non-uniformities in the interconnect often limit the accuracy of the resulting measurement.

While many of the calibration techniques listed here are straight forward, some of them such as thru-reflect-line (TRL) require careful design in the fixturing. This can often be a challenge in both fabricating the calibration standards and implementing their measurements in the calibration procedure.

Historically, the limitation of de-embedding techniques has not been in the technology used to implement it, but in the challenges faced by the user due to its complex nature. This application note will lower this user barrier.

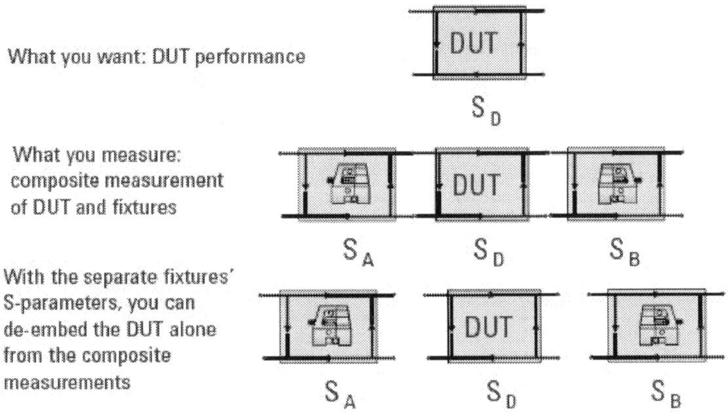

Figure 9.6: *What Is De-Embedding?*

10.3 Principles of De-Embedding

The goal in any measurement is to extract an accurate value of the S-parameter of the DUT with minimum effort and artifacts. The DUT can be measured directly when it is connected directly to the calibration plane. Unfortunately, this is rarely possible.

There is usually some type of connection between the calibration plane and the DUT. This can be a coaxial to microstrip transition, stripline traces, and even some intervening vias. If you can obtain an accurate description of the S-parameters of the fixture, you can mathematically take the measured composite S-parameters of the DUT and fixtures and extract just the DUT performance.

- Definition of a port on a DUT
- Definition of the S-parameter
- S-parameter in power flow and matrix notation
- Cascading S-parameter networks with T-parameter matrices
- De-embedding T and S-parameter networks

Figure 9.7: *The Five Principles of De-Embedding*

The principals in *Figure 9.7* are also useful for general S-parameter measurements. The sections that follow introduce five simple principles that will explain the process of de-embedding the DUT from the measurements of a composite structure. With a brief glimpse into what goes into de-embedding, the requirements on setting up a measurement system to de-embed will be clearer.

Definition of a Port on a DUT

The first principle is the definition of a port. The definition of each S-parameter comes from the idea of a port. The popular way of describing the signal flows into a network based on the S-parameter is the basis of concatenated networks. The interactions of the S-parameter in a cascaded collection of multiple networks are the starting place for the actual de-embedding process.

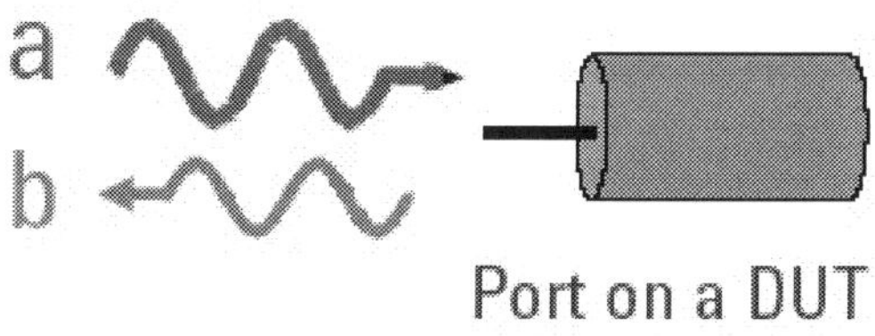

Figure 9.8: *Principle No. 1: The Port*

Any device to be measured has some number of signal-return path connections, referred to as ports. Though the connections can be any transmission-line geometry, such as microstrip, stripline, coplanar, or even twisted wire, it is easiest to conceptually think about a port as coaxial. This emphasizes that every port always has a signal and return connection, and just one signal and return connection.

At each port, there may be two signals present simultaneously. One will be moving toward the DUT while the other signal, superimposed on the same conductors, will be moving away from the DUT toward the VNA. Part of the formalism of S-parameters is using the letter "a" to designate signal sine waves moving into the DUT and the letter "b" to designate signal sine waves moving away from the DUT. A subscript with the port index number identifies at which port the signal sine wave is present. Consider the following scenario:

- Ports define the interface to a DUT.
- A port is a signal-return path connection.
- Into each port are simultaneously ingoing and outgoing sine waves superimposed on the same conductors. They do not interact on the port.
- A voltage waves enter the DUT.
- B voltage waves leave the DUT.

$$S_{jk} = \frac{b_j}{a_k}$$

Figure 9.9: *Principle No. 2: S-Parameters*

Definition of the S-Parameter

The second important S-parameter principle is the definition of each S-parameter. The "S" stands for scattering. The S-parameter values describe how the DUT scatters incoming a waves into outgoing b waves.

Every combination of a wave going in at one port to a wave coming out at another port has an S-parameter value. Each S-parameter is defined as the ratio of the outgoing b wave at one port to the incoming a wave at another port, provided there are no incoming waves at any other ports.

To keep track of all the combinations of waves going out at each port and waves coming in at each port, subscripts corresponding to the port where the action is are used and are carried over to the S-parameter index. An a wave coming in on port 1 would be designated as "a1." An outgoing b wave on port 2 would be designated "b2."

Each S-parameter is the ratio of a b wave to an a wave and the indices of each S-parameter identifies the two ports involved. But, there is one subtle and confusing twist to the order of the indices for the S-parameter.

It would be convenient if in the definition of each S-parameter, the index was read in the order the signal moved. If the signal came into port 1 and went out of port 2, it would be reasonable to expect the S-parameter element that described this to be labeled as S12. This is not the case. In order to take advantage of the power of matrix algebra, the order is reversed. The S-parameter that describes how the network transforms a1 into b2 is designated S21. The S-parameter that describes how a network transforms the a3 wave into the b1 wave is S13.

S-parameter in power flow and matrix notation

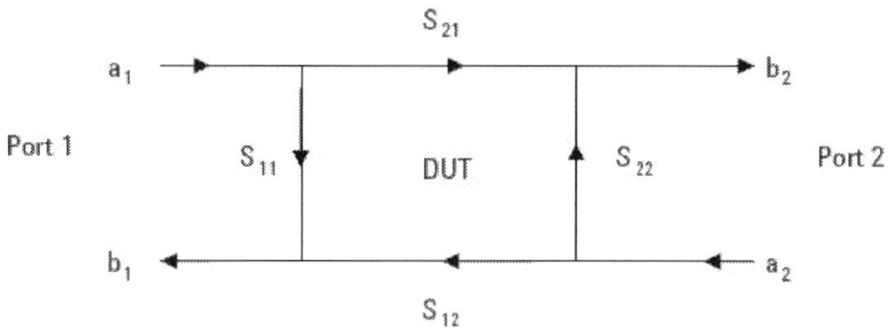

$$\begin{pmatrix} b_1 \\ b_2 \end{pmatrix} = \begin{pmatrix} S_{11} & S_{12} \\ S_{21} & S_{22} \end{pmatrix} \begin{pmatrix} a_1 \\ a_2 \end{pmatrix} \qquad b_j = S_{jk} a_k$$

Figure 9.10: *Principle No. 3: Signal Flow Diagrams*

S-Parameter in Power Flow and Matrix Notation
The flow of the signals from an input to an output of any DUT can be illustrated using a signal flow diagram.

This is a schematic way of illustrating the flow of sine waves into and out of ports. Of course, since each port is really like a coaxial connection, the input and output waves flow over the same conductors in each port. If the signal paths were shown with arrows in the same location as their actual position, the diagram would be confusing with arrows on top of each other.

For this reason, the convention is to space apart the incoming and outgoing signal arrows. All arrows along the top flow in the same direction, from left to right. All arrows on the bottom flow from right to left.

At a glance, you can see that S11 transforms the a1 wave into the b1 wave, and S21 transforms the a1 wave into the b2 wave. There is no additional information content in a signal flow diagram, it is just a convenient visual aid to display the function of each of the S-parameter elements.

When an S-parameter is used to describe a device, the device is represented as a network that converts incoming sine waves into outgoing sine waves. It does not matter how physically complex the device is, the network description is simply related to how it transforms the various incoming a waves into outgoing b waves at each of the ports.

Though the values of the S-parameter are affected by the nature of the network, the formalism to describe what happens at each port is completely independent of what the interconnect network between the ports looks like. It is all described by its S-parameters.

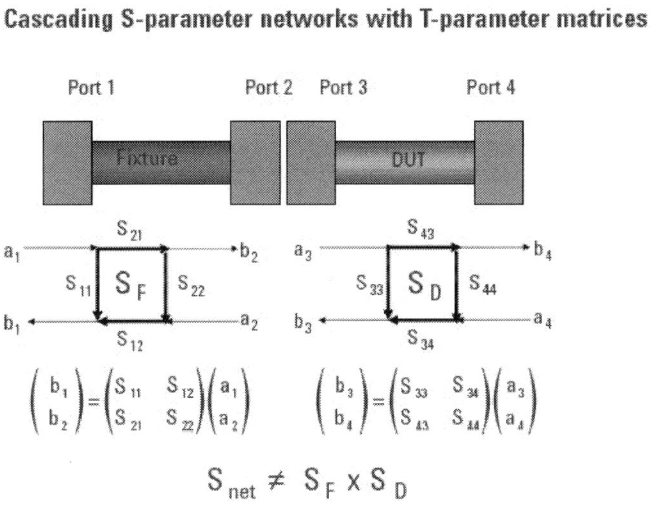

Figure 9.11: *Principle No. 4: Cascading Signal Flow Diagrams*

Cascading S-Parameter Networks with T-Parameter Matrices
An important property of the signal flow diagram is its ability to show the signal flow for devices that are connected in series or cascaded. When two two-port networks are connected in series—for example, a small fixture and a DUT or two DUTs—the signal flows from one into the other.

In this example, the connection is between ports 2 and 3. Port 1 is the entrance to the first structure and port 4 is at the far end. Using this formalism for the signal flow diagram, multiple networks can be cascaded in series. Two two-port networks in series are equivalent to one two-port network with the two outermost ports defining its ends. This is shown in the cascaded signal flow diagram in *Figure 9.11*. There is a new S-parameter matrix that describes the two networks in series. It would be defined based on ports 1 and 4 in this example.

Though the formalism of signal flow diagrams makes it easy to cascade multiple networks in series, unfortunately, the definition of the S-parameter matrix does not provide a simple process to calculate the equivalent series combination of the S-parameter of two networks.

The S-parameter matrix is defined by how incoming waves are transformed into outgoing waves. At the interface between the networks, an outgoing b wave from the first network is transformed into an incoming a wave of the second network.

The formalism for signal flow diagrams is designed to illustrate this series connection. As the b2 wave exits port 2, it becomes the a3 wave entering port 3. Likewise, the b3 wave leaving port 3 becomes the a2 wave entering port 2. As it is defined, you cannot take the two S-parameter matrices and simply multiply them together. This makes the calculation of cascaded networks difficult if you are limited to the S-parameter matrix.

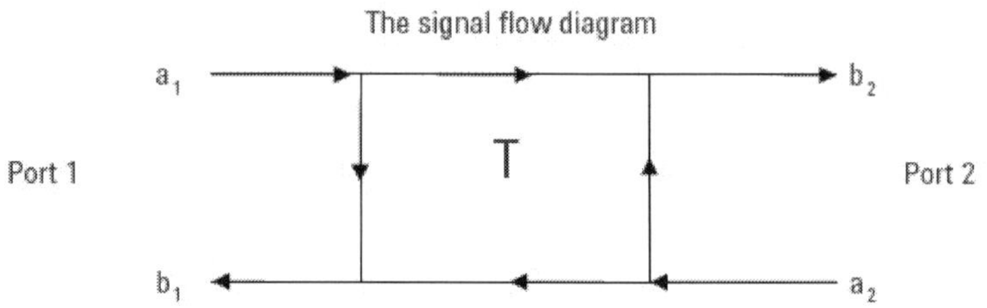

$$\begin{pmatrix} b_1 \\ a_1 \end{pmatrix} = \begin{pmatrix} T_{11} & T_{12} \\ T_{21} & T_{22} \end{pmatrix} \begin{pmatrix} a_2 \\ b_2 \end{pmatrix}$$

$$b_1 = T_{11}\, a_2 + T_{12}\, b_2$$
$$a_1 = T_{21}\, a_2 + T_{22}\, b_2$$

Figure 9.12: *The Transfer-Parameter, or T-Parameter, Matrix*

To be able to describe the equivalent network performance of two networks in series as the product of their two matrices, you need a matrix that relates the inputs and outputs at port 1 to the inputs and outputs at port 2. This way, the output of network 1 can become the input to network 2 in the matrix representation. The "transfer" scattering matrix, or T-parameter matrix, is designed to translate the input and output waves at port 1 to the input and output waves at port 2. The T matrix elements are defined so that the inputs and outputs on port 2 are the inputs and outputs to the next network at port 3.

With this definition of the T-parameter matrix, each of its element can be defined in terms of the a and b sine-wave signals. Unfortunately, it is sometimes difficult to understand what each element means.

For example, T22 is the ratio of a1 to b2, when a2 = 0. The T12 term is the ratio of b1 to b2 when a2 = 0. Likewise, the T11 term is b1 divided by a2 when b2 = 0 and T21 is a1 divided by a2 when b2 = 0.

When the interconnect is transparent, the T-parameter matrix has a special form. To be transparent to the interconnect, it must have 50 Ohm input and output impedance, no loss, and no phase shift.

In this case, T11 is 1. Everything goes right through. T12 is 0 since b1 has to be 0 as nothing reflects. T21 = 0 as a2 is 0 and T22 = 1, since everything in the reverse direction goes through. The T-parameter matrix for a transparent interconnect is the identity matrix.

The purpose of the T-parameter matrix description of a network is to facilitate the calculation of multiple networks in a series. It is important to keep in mind that the S-parameter matrix and the T-parameter matrix description of the same network use the same content. They are just different combinations of the same features.

$$\begin{bmatrix} S_{11} & S_{12} \\ S_{21} & S_{22} \end{bmatrix} = \begin{bmatrix} \frac{T_{12}}{T_{22}} & \frac{T_{11}T_{22}-T_{12}T_{21}}{T_{22}} \\ \frac{1}{T_{22}} & -\frac{T_{21}}{T_{22}} \end{bmatrix}$$

$$\begin{bmatrix} T_{11} & T_{12} \\ T_{21} & T_{22} \end{bmatrix} = \begin{bmatrix} -\frac{S_{11}S_{22}-S_{12}S_{21}}{S_{21}} & \frac{S_{11}}{S_{21}} \\ -\frac{S_{22}}{S_{21}} & \frac{1}{S_{21}} \end{bmatrix}$$

Figure 9.13: *Converting between S-Parameter and T-Parameter Matrices*

With a little algebra, each T element can be translated from the S-parameter matrix and each S-parameter element can be converted into the T-parameter matrix.

Given the translations shown here, it is possible to perform this conversion with a spreadsheet. However, the conversion is also built in to most network analyzers and is a key feature in Agilent's advanced design system (ADS). In practice, there should never be a need to manually transform a T-matrix into an S-matrix. Each matrix has its use. When you want to describe the electrical properties of a DUT, it is most convenient to use the S-parameter representation. When you want to cascade multiple networks in series and find a resulting network, use the T-parameter matrix representation of the network. Depending on the application, you can transparently go back and forth without gaining or losing any information.

This transformation is similar to the transformation performed in a time domain or frequency domain description of interconnect behavior. They both have exactly the same information content, it is just that depending on the question asked, one format will get you to the answer faster.

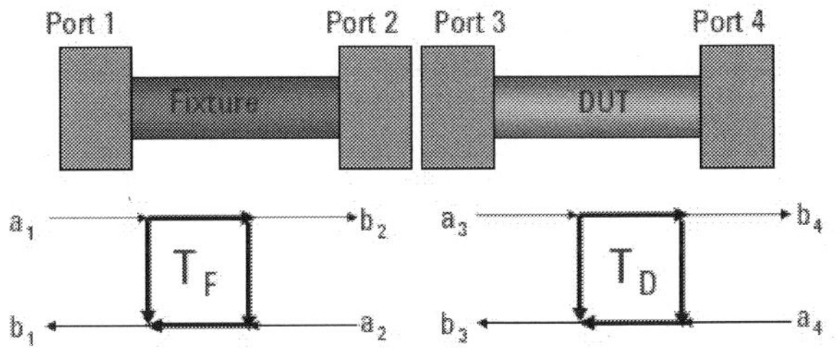

$$\begin{pmatrix} b_1 \\ a_1 \end{pmatrix} = \begin{pmatrix} T_{11} & T_{12} \\ T_{21} & T_{22} \end{pmatrix} \begin{pmatrix} a_2 \\ b_2 \end{pmatrix} \qquad \begin{pmatrix} b_3 \\ a_3 \end{pmatrix} = \begin{pmatrix} T_{33} & T_{34} \\ T_{43} & T_{44} \end{pmatrix} \begin{pmatrix} a_4 \\ b_4 \end{pmatrix}$$

$$\begin{pmatrix} b_1 \\ a_1 \end{pmatrix} = \begin{pmatrix} T_{11} & T_{12} \\ T_{21} & T_{22} \end{pmatrix} \begin{pmatrix} T_{33} & T_{34} \\ T_{43} & T_{44} \end{pmatrix} \begin{pmatrix} a_4 \\ b_4 \end{pmatrix} \qquad T_{net} = T_F \, T_D$$

Figure 9.14: *Cascading T-Parameters*

The special format of the T-parameter matrix is designed so that the equivalent T-parameter matrix of two networks in series is just the product of the T-parameter matrices of each of the two networks. The output of one T-parameter matrix is the input to another T-parameter matrix.

In *Figure 9.14*, you can see that the a2 wave is the same as the b3 wave and the b2 wave is the same as the a3 wave. This lets you substitute the a2, b2 vector for the b3, a3 vector, which results in the product of the two T-parameter matrices.

With this formalism, the net T-parameter matrix of two separate networks is the product of the two T-parameter matrices of each individual network. This is an incredibly powerful principle and is the basis of all calibration and de-embedding methods.

$$T_M = T_F\, T_D$$

$$T_F^{-1}\, T_M = T_F^{-1}\, T_F\, T_D = T_D$$

Figure 9.15: *Matrix Math for One Port*

This definition of the T-parameter matrix description of a two-port network makes describing the de-embedding process simple and concise. When you perform a one-port measurement of a DUT that has a fixture in series with it, you will have two S-parameter matrices.

First you will see the two-port S-parameter of the fixture and the one-port S-parameter of the DUT. Each of these S-parameter matrices can be converted into T-parameter matrices using algebra.

The resulting measured T matrix (Tm) is the series combination of the fixture matrix (TF) in series with the DUT matrix (TD). The series combination is calculated by the product of the two T-matrices, as TM = TF x TD.
The de-embedding process uses matrix algebra to de-embed the DUT matrix from the measured matrix and the fixture matrix.

As shown in *Figure 9.15*, applying matrix algebra, it is a simple matter of multiplying each side of the equation by the inverse matrix of the TF matrix. This results in the T matrix for the DUT alone as the inverse of the fixture T matrix times the measured T matrix.

$$T_M = T_{F1}\, T_D\, T_{F2}$$

$$T_{F1}^{-1}\, T_M\, T_{F2}^{-1} = T_{F1}^{-1}\, T_{F1}\, T_D\, T_{F2}\, T_{F2}^{-1} = T_D$$

Figure 9.16: *Matrix Math for Two Ports*

With two ports, the DUT is embedded between two series two-port fixture matrices. To de-embed the TD matrix, multiply each side of the measured T matrix by the inverse of each fixture T matrix.

Using algebra, the T matrix of just the DUT can be extracted. Of course, from the T matrix, it is a simple step to convert this into the S-parameter matrix. This is the basic de-embedding process.

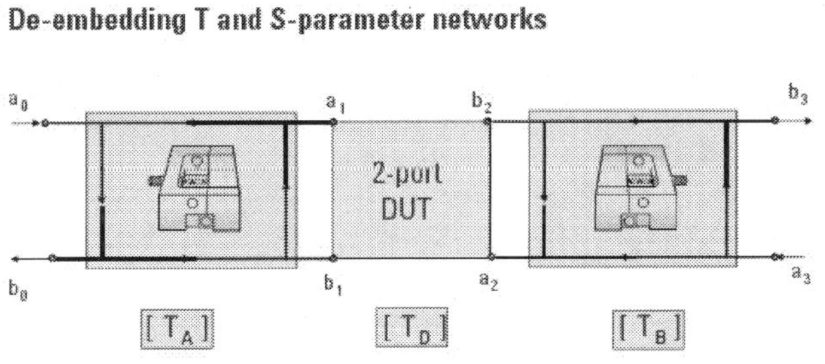

Measured $\qquad\qquad\qquad\qquad$ DUT

$$[T_m] = [T_A][T_D][T_B] \qquad [T_D] = [T_A]^{-1}[T_m][T_B]^{-1}$$

Figure 9.17: *Principle No. 5: De-Embed Fixtures from Measurements Using T-Parameter Matrices*

De-Embedding T- and S-Parameter Networks

Principle no. 5 of the de-embedding process tells you how to de-embed the DUT. If you have the description of the performance of each fixture in terms of their T-parameter matrix and the performance of the composite of the DUT and the fixtures on either side as a T matrix, then the T matrix of the DUT can be extracted mathematically.

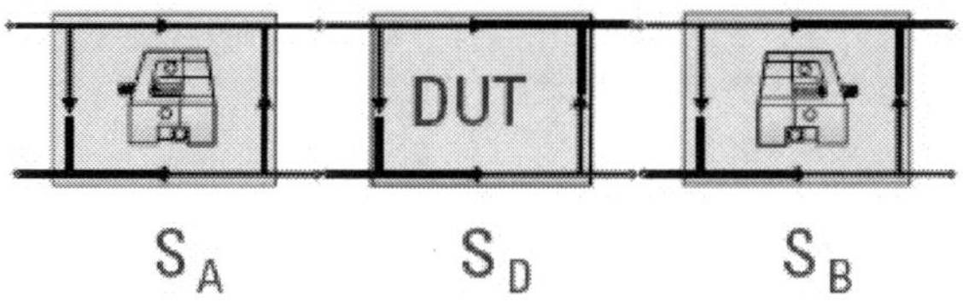

$$T_M = T_A T_D T_B$$

$$T_A^{-1} T_M T_B^{-1} = T_D$$

S_D

Figure 9.18: *The De-Embedding Process*

The following are six basic steps to perform a de-embedding operation:

- Measure the composite system of the DUT and the fixtures.
- Obtain the two-port S-parameter of each fixture; this is usually the most challenging part of de-embedding.
- Convert each of the S-parameter matrices into T-parameter matrices, which is just a mathematical process.
- Perform the matrix math to convert each of the fixture T matrices into their inverse matrices
- Multiply the measured T matrix by the inverse of the fixture T matrices using matrix math.
- Convert the de-embedded DUT's T-matrix into a S-parameter matrix using matrix math.

In most network analyzer and processing software, such as Agilent's physical layer test system (PLTS) and ADS, the matrix operations are completely hidden from the user. You simply supply the combined series S-parameter and fixture S-parameter measurements, and the conversions and matrix math are performed automatically. In practice, there is no need to understand the details of how the matrices are combined together.

9.4 Obtaining the S-Parameter of the Future

The secret to successful de-embedding is to obtain a good value of the S-parameters of the fixture. The following four basic methods used for obtaining S-parameters work on any size network—two-port, four-port, or more:

- Direct measurement
- Extracting equivalent model from a measurement
- Calculate S-parameters from a scalable, analytical approximation
- Calculate S-parameters from a 2D or 3D field solver model

Direct Measurement

The first method is direct measurement. This requires being able to connect to both ends of the fixture at the calibration plane of the VNA. While not the most common configuration, it is the simplest and most direct way of getting the S-parameter measurement of the fixture.

Extracting Equivalent Model from a Measurement

When a direct measurement of the fixture is not possible, the next best solution is to perform a measurement of the fixture in such a way that a circuit model can be fit to the measurement and the S-parameters of just the fixture created.

Calculate S-Parameters from a Scalable, Analytical Approximation

Sometimes, the fixture can be as simple as a uniform, lossy transmission line. Or a part of the fixture could be a uniform, lossy transmission line. In such cases, an accurate analytical model can be used to generate the S-parameter of the line segment. The advantage of this sort of model is that it can be scaled to any appropriate length. This is the basis of the port extension calibration procedure.

Calculate S-Parameters from a 2D or 3D Field Solver Model

Finally, the S-parameters of the fixture can be calculated using a 3D planar or full-wave field solver. Most full-wave tools are accurate enough to extract S-parameters for practical use, if their geometry can be accurately described and the material properties of the fixtures are well known.

When designing the fixturing for the DUT, it is important to think about how the S-parameter of the fixture can be generated.

One of the most important concerns is the accuracy of the fixture S-parameters. The fixture connected to the DUT must be identical to the structure that is measured and the S-parameter extracted to be used for de-embedding. This is always a concern for all cases of calibration or de-embedding.

The following case studies illustrate the principles of de-embedding and show that a routine process can be used to generate S-parameter files that can then be used to provide direct measurements of the DUT.

9.5 Direct Measurement of Fixture S-Parameters

In *Figure 9.19*, the fixture is a microprobe used to make contact with circuit boards and semiconductor packages. On one end is a standard 3.5 mm coaxial connector and the other is a coplanar S-G tip. The path between these two ends is a short length of rigid coaxial cable.

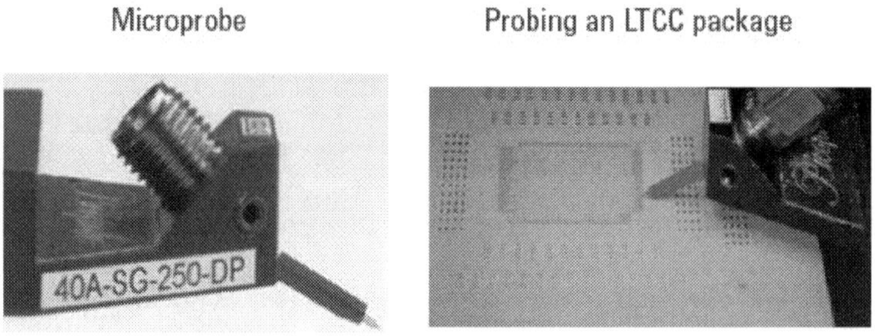

Figure 9.19: *Example No. 1: De-Embedding Microprobes*

In most typical applications, the microprobe is calibrated at the probe tip using reference calibration structures. This is perfectly acceptable and is the common practice. In general, there are three calibration measurements—open, short, and load—performed on each port. Then for multiple ports, there are n–1 thru measurements plus one isolation measurement. For n ports (n = number of ports), this is a total of 4 x n individual measurements in the calibration process. In a two-port measurement, this equals eight measurements. There are eight chances to break a probe tip. In a four-port measurement, there will be 16 measurements in the calibration procedure. In a 12-port measurement, there are 48 chances to break a tip, just in the calibration process.

By leveraging de-embedding, the calibration procedure using microprobes can be dramatically simplified.

Using De-Embedding in the Calibration Procedure

The calibration procedure with de-embedding is simple. The ends of the 3.5 mm connectors on the cables from the VNA are connected to an e-cal module, in this case an Agilent N4433A. With one mouse click, the calibration plane of the VNA is moved to the ends of the 3.5 mm cables.

Next, the microprobes are attached to the ends of the cable and the measurements of the semiconductor package performed. These measured S-parameters have the DUT with fixtures on all sides. However, if you have a good set of S-parameters for the fixtures, their influence can be removed from the measurements.

The de-embed operation can be started from the calibration menu tool bar of the VNA. The S-parameter files for the fixture are selected and the VNA now displays the S-parameter for just the semiconductor package leads.

For a four-port measurement, the calibration process has changed from 16 measurements—which might take three or four hours and introduce opportunities for mistakes and damage to the delicate probes—to literally a three-minute automated operation with robust connectors.

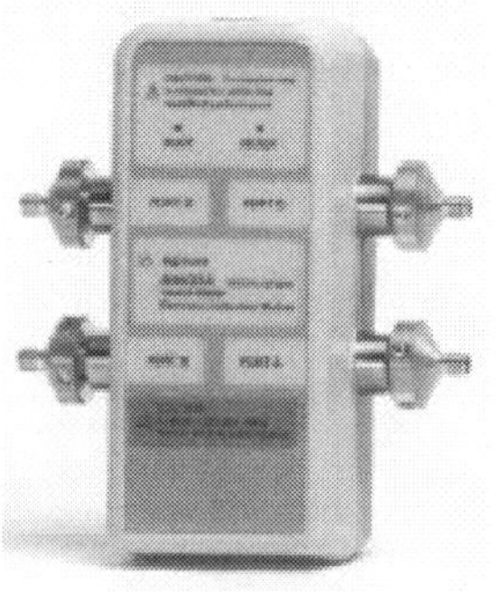

Figure 9.20: *Steps for Using De-Embedding in Calibration.*
Step 1: Calibrate All Ports at the End of the Coax Cable Using the Electronic Calibration Module

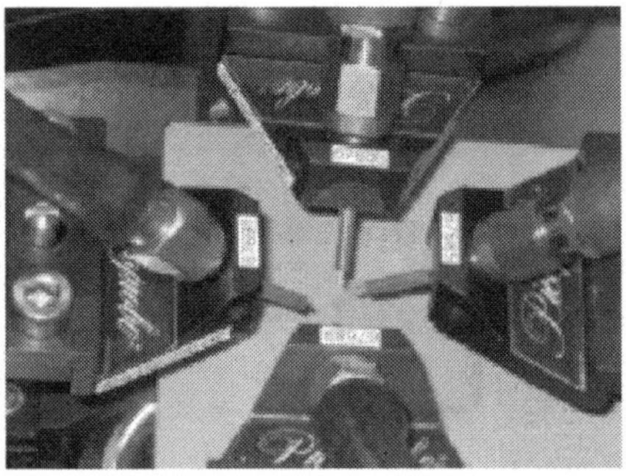

Step 2: Attach Microprobes to the Ends of the VNA Cables.
Step 3: Perform All the Measurements with Microprobes

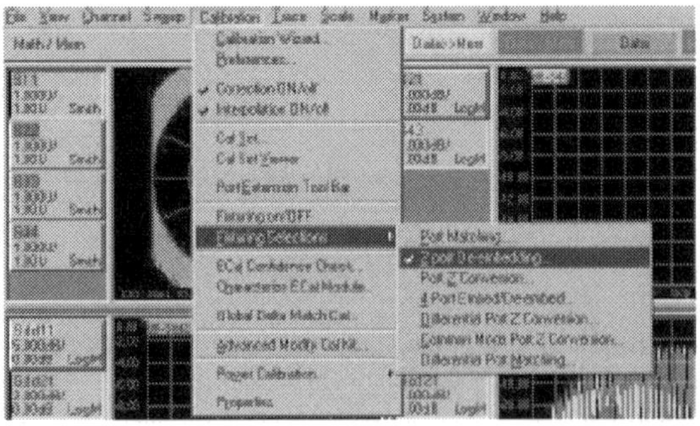

Step 3: De-Embed the Microprobe S-Parameters

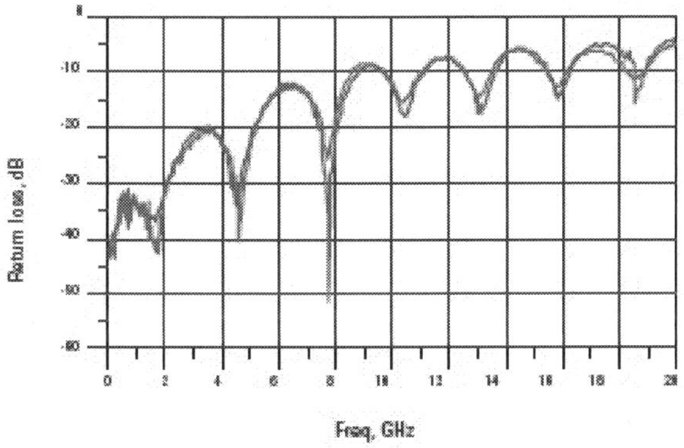

Step 4: Analyze the Measurements of the DUT

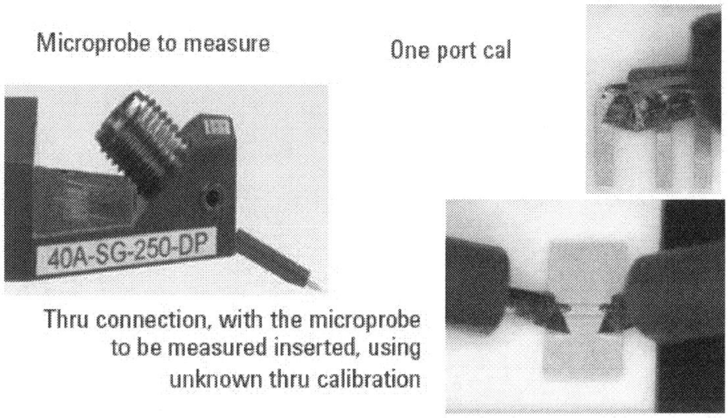

Figure 9.21: *Measuring Microprobe S-Parameters*

De-Embedding the Microprobe in Series

A two-port model of the microprobe is required to de-embed the microprobe in series with the DUT from a composite measurement. This is accomplished using a very straightforward, three-step process. A two-port VNA is used to perform the fixture measurement.

First, one port of the VNA is set up with a precision cable ending with a 3.5 mm connector. The other port has a precision microprobe on the end of a cable that has the same pitch as the microprobe to be measured.

Next, the VNA is calibrated with this configuration. To calibrate a VNA with different adaptors on the ends, a standard "two port with unknown thru" technique is used. This requires first an open, short, and load calibration on each port. On the coaxial port, coaxial standards are used. On the microprobe end, a precision calibration substrate is used as the reference. This is the standard process for calibrating with a microprobe.

Next the thru path is calibrated. All the latest VNA models, and PLTS, have a new integrated calibration routine called "unknown thru" calibration. This is a breakthrough technology that enables the use of "non-insertable" connectors on the ends of the cables connected to the VNA ports. This calibration procedure allows the use of different connector sizes, types, or geometries on the ends. This could be a 2.4 mm connector on one end and 3.5 mm connector on another, for example, or in this case, a coaxial connector and a microprobe.

All that is required is a thru connection with only a few dB of insertion loss between the two connectors and a small phase delay at the lowest frequency. In *Figure 9.21*, the unknown thru connection is made with a second microprobe on the cable end, and the tips of the two probes are in contact with a very small thru connection pad. This is the same configuration as used in the thru calibration of the microprobes.

Using this calibration process, the reference planes of the VNA are moved to the end of the coaxial cable for one port and the end of the microprobe on the other. Finally, the two-port S-parameters of a microprobe are measured by inserting the microprobe to be measured between the coax on one port and the other, calibrated microprobe tip on the other. This is just the unknown thru calibration configuration. After the calibration, a measurement of the two-port S-parameter in this configuration is a direct measurement of the return and insertion loss of the second microprobe itself.

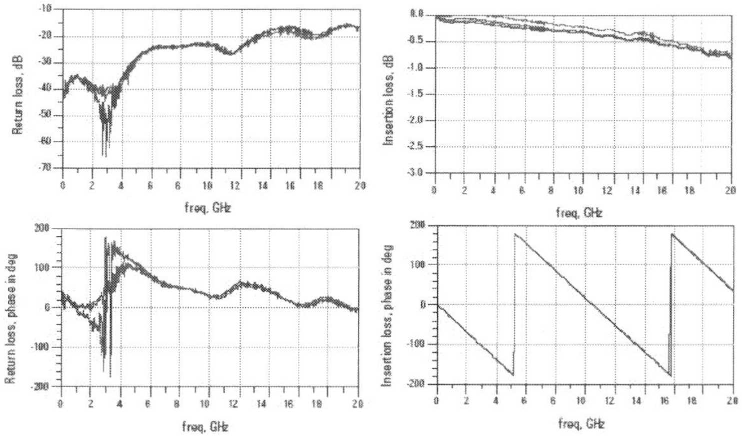

Figure 9.22: *Measured Performance of Four Microprobes*

Using this routine procedure, the two-port S-parameter of any microprobe can be measured. *Figure 9.22* shows an example of the measured performance of four individual microprobes with the same geometry. The return loss, sensitive to impedance discontinuities, is very similar in all four units. This indicates that they are physically indistinguishable.

The insertion loss shows identical phase up to 20 GHz, indicating very similar lengths. There is a slight variation in insertion loss between the microprobes. This could be due to slight variations in how the probes are touching down on the substrate or the quality of the thru connection between the two tips.

The variation apparent among these four probes is small enough that one average value set of S-parameters can be used for all the nominally identical probes. In fact, it is routine now for probe suppliers to provide an .s2p file for each of the probes they sell. These 2-port S-parameter data sets are created using this procedure.

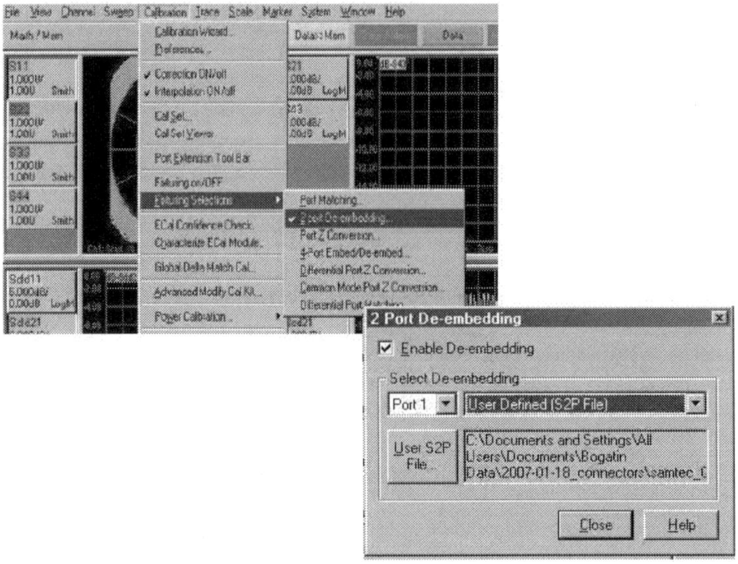

Figure 9.23: *De-Embedding on the VNA*

Since the actual de-embedding operation that extracts just the DUT behavior is performed after the measurement, there is flexibility in how it is performed. Since the actual de-embedding operation which extracts just the DUT behavior is performed after the measurement, there is some flexibility in performing this type of error correction. There are three ways to accomplish de-embedding: VNA firmware internal to the hardware, specialist signal integrity software (PLTS), or modeling software (ADS).

De-Embedding Using a Vector Network Analyzer

De-embedding a DUT directly on the Agilent VNA is as simple as selecting Calibration > Fixturing Selections > 2-port De-embedding. Once this is selected, a new window opens up allowing the user to select the .s2p file of the fixture on each of the two ports.

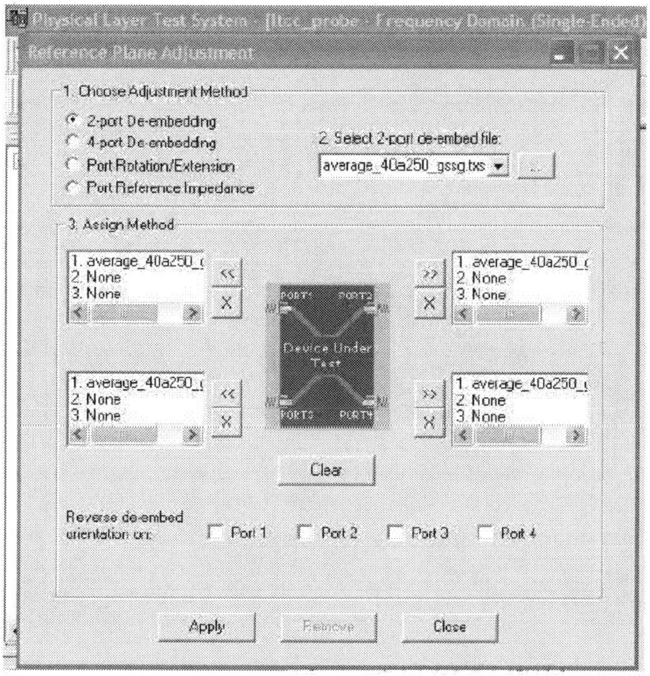

Figure 9.24: De-Embedding in PLTS

De-Embedding Using PLTS Software with the VNA

This same process can be implemented using Agilent's PLTS to control the VNA and collect the measurement results, or in post-processing of the measurements. In PLTS, de-embedding can be accessed by selecting the Utilities > Reference Plane Adjustment. The .s2p de-embed file is selected for each port required, and then the de-embedded measurement of the DUT is displayed in the user window.

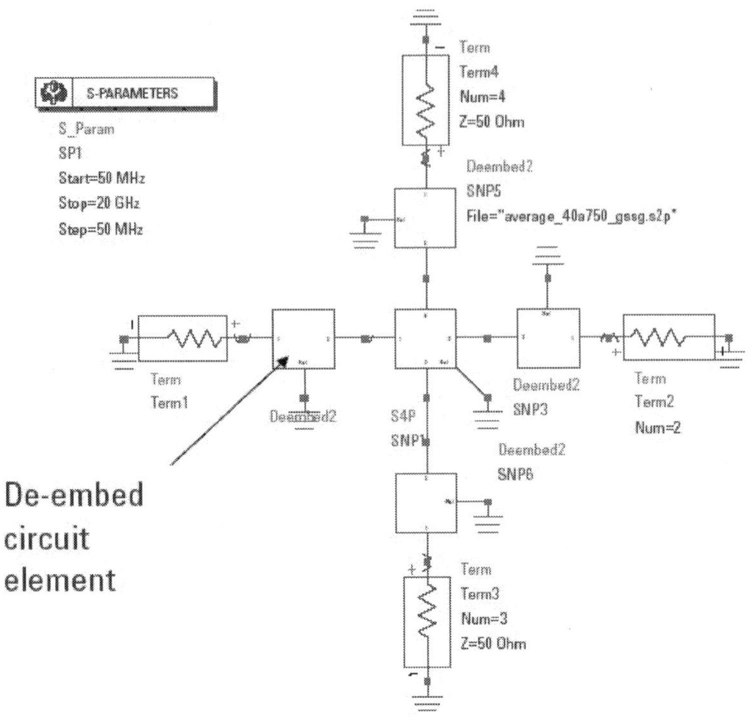

Figure 9.25: *De-Embedding in ADS*

De-Embedding Using ADS Modeling Software

In addition, de-embedding can be performed using Agilent's ADS. The measured, four-port data file of the DUT with the microprobes in series can be brought into the ADS simulation environment. The two-port de-embed circuit element, native to ADS, is loaded with the measured .s2p file of the microprobe and the resulting simulated S-parameters are of just the DUT with the microprobes removed.

These new four-port S-parameter data sets can be used in all the typical operations such as transient simulation, eye diagram simulation, and model building through optimization.

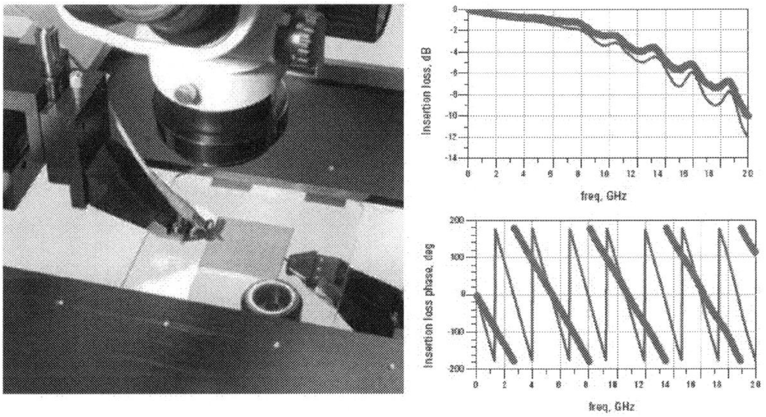

Figure 9.26: *Impact of the Microprobe*

The microprobes on each end of the cable add a significant amount of phase to a thru measurement of a package interconnect. As seen in *Figure 9.26*, the measured insertion loss of a package interconnect is obtained by probing opposite sides of the package, from the C4 pads on one side to the land grid array pads on the other side.

The calibration plane of the VNA was moved to the end of the cables. Then microprobes were attached to the cables and the DUT measurement performed. The thin-line trace is the measured insertion loss with the microprobe fixtures in the measurement. The thick-line trace is the same measurement with the microprobe fixtures removed from the measurement and the package interconnect de-embedded. The thick-line trace is a better approximation to the package interconnects alone.

The large impact on the phase is the result from the fixture adding length to the measurement. The magnitude of the insertion loss is only slightly affected by the probe, since the probes are designed to be as transparent as possible. However, the impact on the 3 dB bandwidth is important. With the fixture in the measurement, the 3 dB bandwidth of the package interconnect is measured as about 8 GHz. With the fixture effects removed, the 3 dB bandwidth is 11 GHz. This is a 20 percent increase in bandwidth.

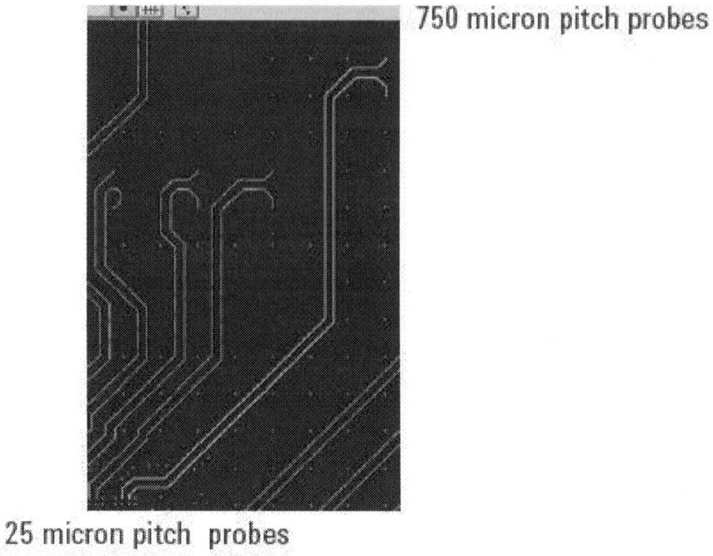

Figure 9.27: *Measuring the Differential Pair of an LTCC*

As a final verification of this technique, two methods were used to measure the same differential pair in a low-temperature co-fired ceramic (LTCC) package. First, the differential channel was measured using microprobes that were calibrated directly at their tips. Then the same differential channel was measured using the new calibration technique of first calibrating to the cable, and then using the .s2p file for the microprobe to de-embed the package interconnect from the composite measurement.

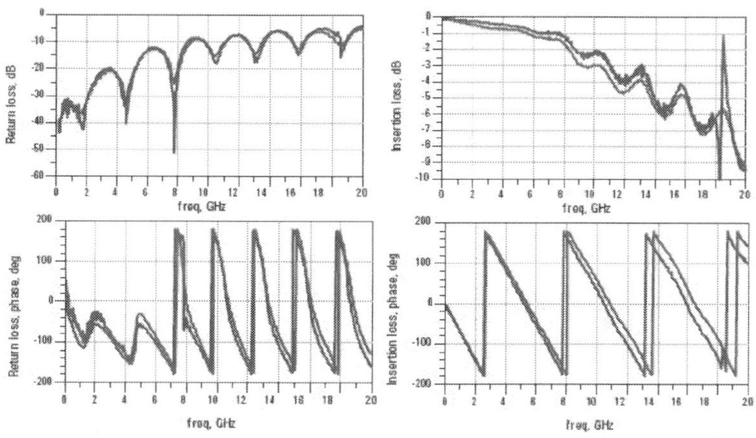

Figure 9.28: *Final Verification*

In these measurements, up to 20 GHz, the agreement is seen to be excellent. See *Figure 9.28*. The return loss, which is a measure of the impedance matching, is nearly identical. The insertion loss has a very small difference, which can easily be due to the differences in the probe tips' compression. The measurement agreement, even at 20 GHz, is within 25 degrees.

This measurement also shows that the interconnect has a differential return loss of less than –13 dB up to about 6 GHz. This is a typical acceptable specification. The –3 dB insertion loss bandwidth is about 9 GHz.

9.6 Building the Fixture S-Parameter by Fitting a Model to Measurement Data

Figure 9.29 shows the SMA launch to a four-layer circuit board is the fixture to a uniform trace on a circuit board. The SMA is designed with four return pins surrounding one signal pin. The board is designed with plated holes that closely match the pin diameters. The SMA is soldered into the board with the pins connecting to internal ground layers and the signal pin to the top signal layer. The clearance holes in the ground planes have been designed so that the launch is well optimized, but not a perfect match, to 50 Ohms.

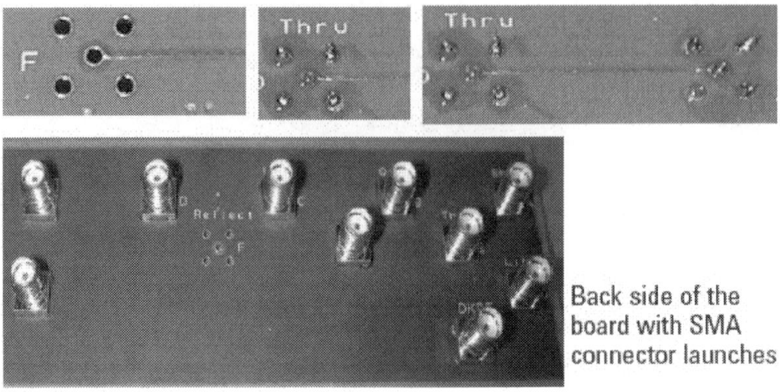

Figure 9.29: *Example No. 2: De-Embedding Circuit Board Traces*

The goal is to generate an S-parameter file for the launch that can be used to de-embed any surface trace from a measurement of the SMA connections on the ends.

The process will be to measure a short length thru with two SMAs on either end and a long thru with two SMAs on the ends. The same SMA model will be used for each end. A topology-based model will be extracted from the measurement and then used to simulate the S-parameters for just one SMA launch. Using this S-parameter file, the intrinsic performance of any structure connected to a launch on the board can be extracted.

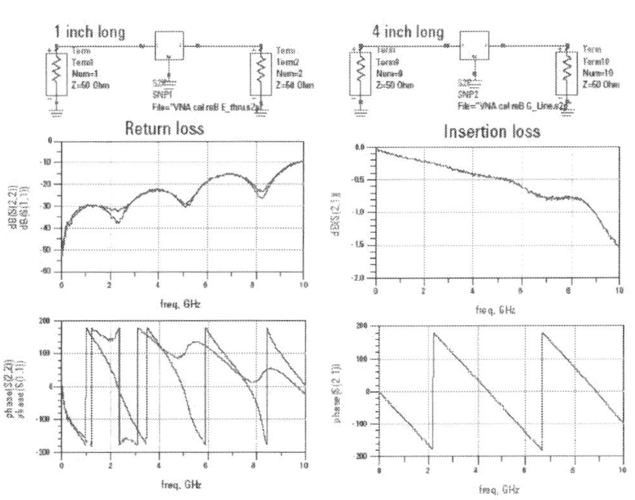

Figure 9.30: Measured Return and Insertion Loss of the 1-Inch-Long Trace

The two sets of measurements taken are shown in *Figure 9.30*: a two-port one-inch-long thru and a four-inch-long thru. As can be seen by comparing the S11 and S22 performance of the one-inch thru, the SMA launches are not identical. They are close, but there are clearly some differences between each end of the line. The same is true about the four-inch-long thru.

This means that there are limits to the accuracy of this de-embedding technique, which are due to variations in the launches from fixture to fixture.

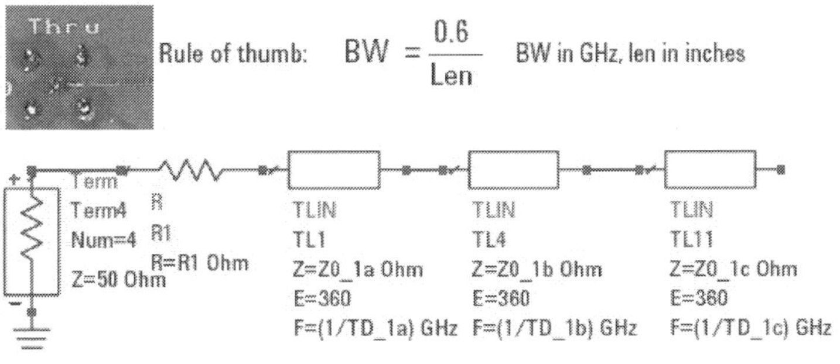

Figure 9.31: Building a Model for an SMA Launch

Any single-ended interconnect can always be approximated by a uniform transmission line. Of course, there is always a bandwidth limitation to how well all models work. The more uniform the interconnect, the higher the bandwidth of the model. When the interconnect is non-uniform, the interconnect can be modeled as a few uniform transmission lines in series. This is always a good model.

The question is always how many sections are needed for a given bandwidth. As a rough rule of thumb, the bandwidth of a single transmission-line section will always be greater than 0.6 divided by the length of the line. If the line is one inch long, the minimum bandwidth of the model will be at least 0.6 GHz.

If you want a model with a bandwidth of at least 10 GHz, this means that for non-uniform regions, you need a new transmission line model for every 60 mils of interconnect. The SMA launch is composed of a uniform barrel about 200 mils long and two shorter regions about 50 and 30 mils long, corresponding to the pin in the board and the excess capacitance at the pin to trace interface.

As a good starting place a 10 GHz model for an SMA launch can be three uniform transmission lines. Each segment has a characteristic impedance and a time delay. In addition, due to the variable quality of the SMA connection, there may be some contact resistance on the order of 100 mOhms, which is added to the model. This is the topology-based model that will be used for the SMA launch. The identical model and parameter values will be used for all launches. To find the parameter values, take the measured results and fit this circuit topology to find the best parameter values.

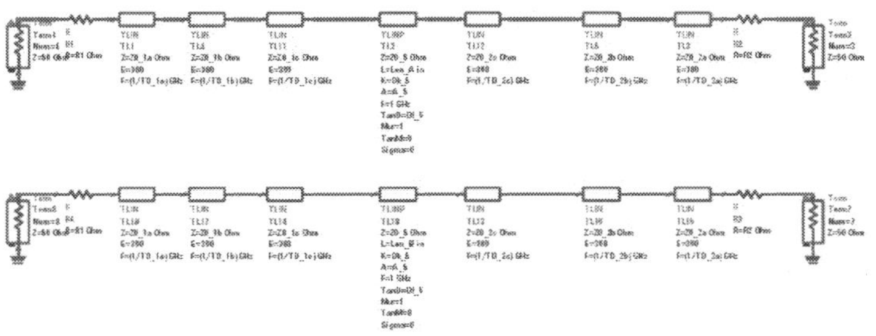

Figure 9.32: *Two Length Thrus*

The two thru interconnects are modeled as identical SMA launches on each end and a uniform transmission line. The only difference between the transmission line models for short and long thrus is their length. They have the same characteristic impedance, effective dielectric constant, and dissipation factor. The ideal, lossy transmission line model in ADS is a scalable model. The variable between the two lines is only their length, one inch and four inches.

In this model, there are 10 parameters: there is a contact resistance for the SMA and two parameters for each of the transmission line models—the characteristic impedance and time delay. This makes seven. With the uniform transmission line thru path, there are three parameters: the characteristic impedance, the effective dielectric constant, and the dissipation factor. This makes 10. Using both sets of measurements as targets, these 10 parameters are simultaneously optimized up to 10 GHz to find the best set of parameter values.

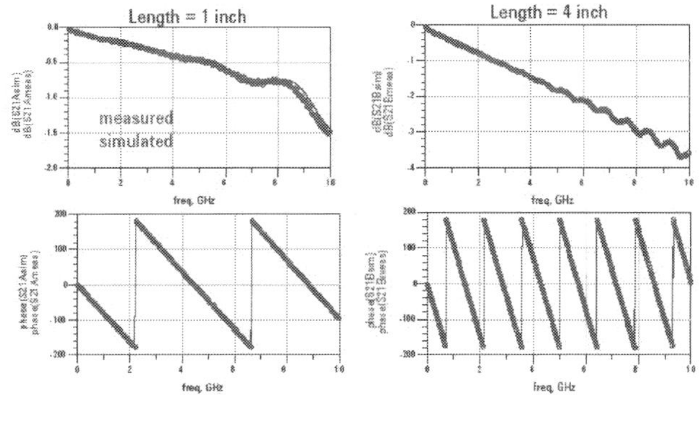

Figure 9.33: *Insertion Loss: Measured and Modeled*

When the best parameter values have been found, the agreement in the simulated and measured performance for both the one-inch and four-inch interconnects is excellent. *Figure 9.33* shows a comparison of the insertion loss for each thru. The agreement to 10 GHz is excellent.

This supports the idea that the model has the right topology and the right parameter values.

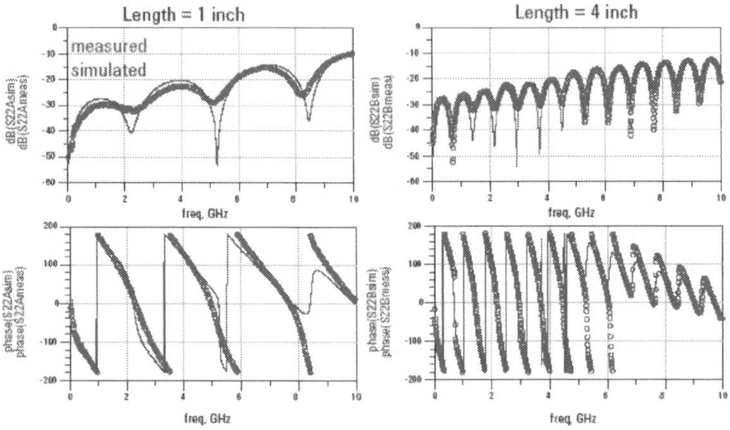

Figure 9.34: *Return Loss: Measured and Modeled*

The return loss measurements and simulations are also excellent. This agreement is all based on assuming the exact same SMA model for each end and a simple, uniform transmission line for the thru path.

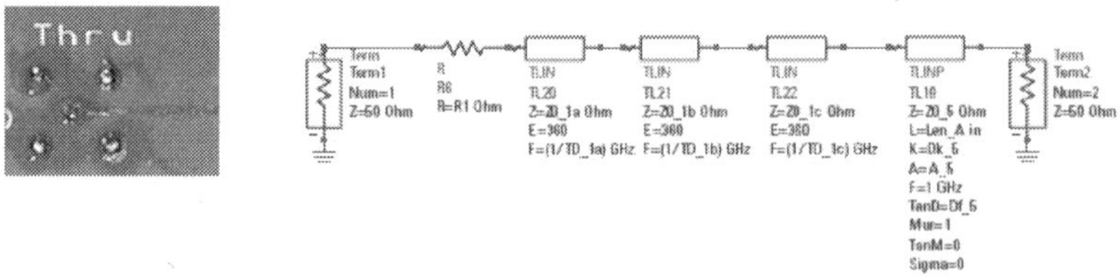

Figure 9.35: *Model for One SMA Launch and Uniform Line*

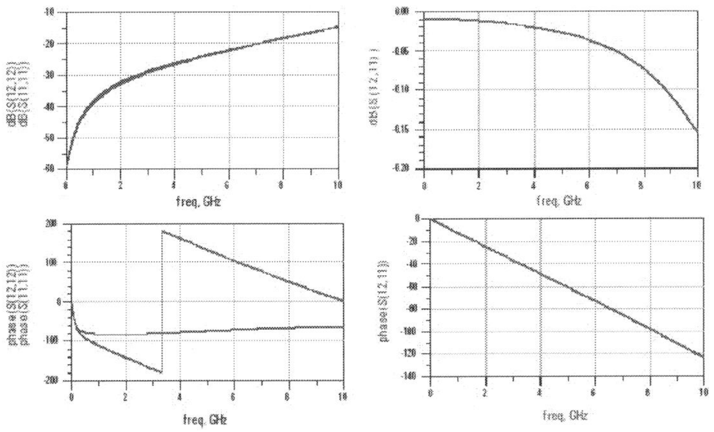

Figure 9.36: *S-Parameter of the SMA Launch with a Zero Length Transmission Line Segment*

After the model has been optimized, the section that is the SMA launch can be pulled out and its S-parameter simulated directly. This two-port S-parameter data set now becomes the SMA launch file, which will be used to de-embed any structure in the path.

In addition, to make it a more generic model, you can add a scaled length of uniform transmission line corresponding to the microstrip line to the three-section transmission line model of the SMA launch. Using the analysis of the two length lines, it is possible to determine the values of the three parameters that define a uniform line, the characteristic impedance, the effective dielectric constant, and the dissipation factor. The length of the line becomes the scaling factor.

The model shown in *Figure 9.37* is the de-embed generator engine, which can produce a two-port S-parameter file for any scaled line segment length. This engine is used to build the .s2p file that connects the SMA into any DUT on the circuit board.

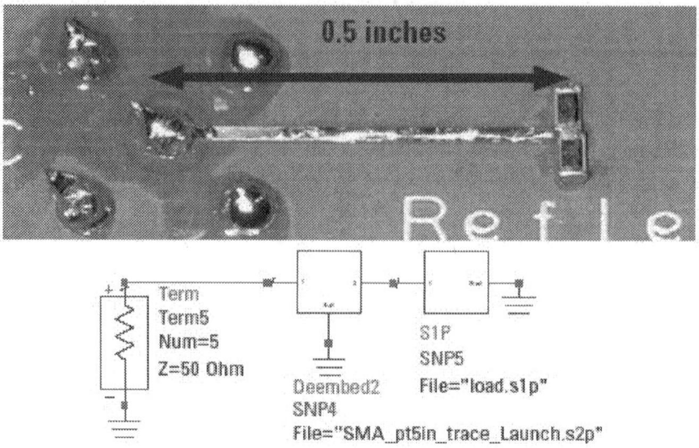

Figure 9.37: *Model for a 50 Ohm Load*

In this example, a standard SMA launch feeds a half-inch surface interconnect trace to a pair of 100 Ohm resistors acting as a termination resistor. At the end of each resistor is a via to the return path, located on the plane directly below the signal layer.

Perform the measurement of the composite structure, and then build a .s2p file consisting of the SMA model and a half-inch length of surface trace. This .s2p file becomes the data set that will be used to de-embed just the terminating resistors from the composite measurement.

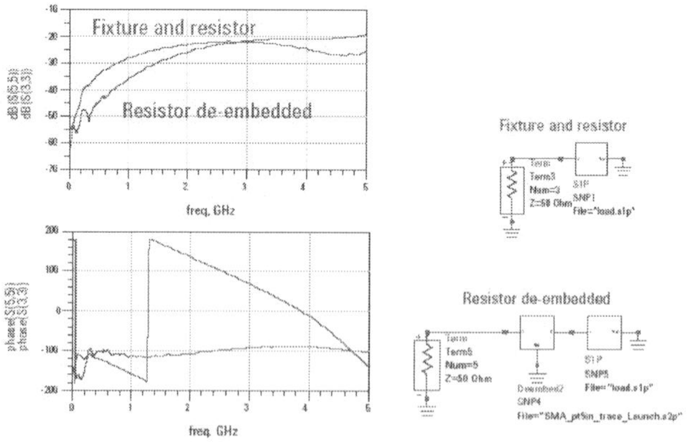

Figure 9.38: *De-Embedded Termination*

The .s2p file generated from the SMA and half-inch trace model is added to the de-embed element in ADS and the composite data of the fixture and resistor is plotted along with the de-embedded data of just the resistor.

The magnitude of the return loss is different and the phase is very far off. Since the SMA launch has been somewhat optimized and the interconnect is roughly 50 Ohms, the magnitude is not strongly affected. The phase is affected since it contributes a half-inch of interconnect.

The de-embedded data can now be used to build a model for the termination resistor. To first order, you might have expected a model to be an ideal resistor. This would give the low return loss. However, the phase is comparable to a capacitor as the data is in the southern hemisphere of the Smith chart. Up to the 5 GHz bandwidth of the measurement, a simple resistor is not enough.

A very good starting place for any component is an RLC model. This model can be used with an optimizer to fit the parameter values.

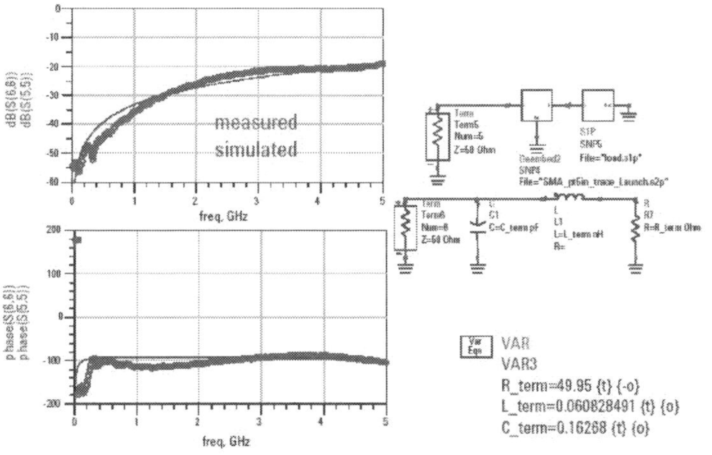

Figure 9.39: *Modeling the Terminating Resistor*

The model for the real terminating resistor is a simple RLC model with the three parameters of the ideal R, L, and C.

These values are adjusted, comparing the simulated return loss of this model to the de-embedded return loss of the actual component. When the optimization is complete, the values of the parameters that best fit the data are R = 49.95 Ohms, L = 0.06 nH, and C = 0.16 pF.

Figure 9.39 shows a comparison of the simulated and measured values; the model is excellent up to 5 GHz.

Now that you understand the de-embedding process, you can look at what you want in an ideal fixture to facilitate its removal from a measurement to de-embed just the DUT.

Optimized Fixture Design

- Minimization of impact on the transfer matrix
- Short
- 50 Ohms
- Low loss
- Allowance of easy obtaining of two-port S-parameters
- S-parameters of fixture when measured should be identical to when fixture connects to DUT

In the design of the fixturing, you want to make sure the fixture has minimal impact on the DUT by designing it to be as transparent as possible. This will make the T matrix values as close to the unity matrix as possible and introduce the minimum numerical error. This means that the fixture should be short, 50 Ohms, and low-loss.

The fixture should be designed so that its S-parameters, and hence its T-parameters, can be easily obtained. Finally, the fixture must be mechanically stable enough so that its S-parameters are the same when connected to the DUT as when they were originally obtained. This will reduce the measurement error in the de-embedding.

When a circuit board is part of the fixture, care should be taken in the design of the fixturing to make it as identical as possible in all instances. The reference structures for the reference and the fixture to the DUT must be as similar as possible.

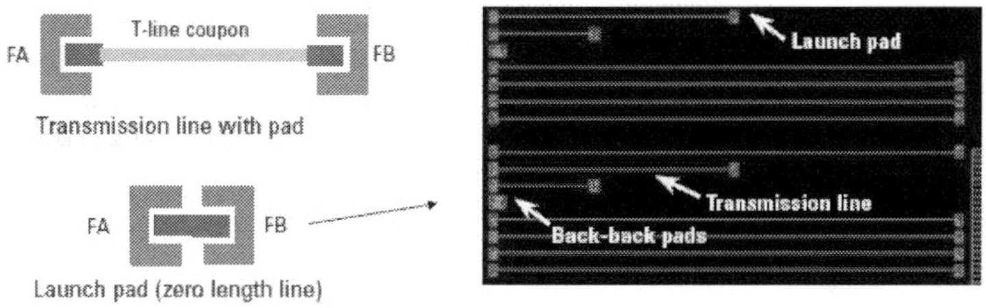

Figure 9.40: *FPGA Test Fixture De-Embedding*

Figure 9.40 shows an example of a fixture design that is optimized for de-embedding. In this example, microstrip traces are fabricated in a substrate used to construct BGA packages. The launch pads are designed for a ground-signal-ground (G-S-G) microprobe with two return pads on either side of the signal pad.

The idea is to perform a two-port measurement of the uniform line and extract the material properties from the substrate, which could then be used as input to a field solver for package design analysis. The design of the fixture uses a launch pad that is effectively a zero length thru. The thru measurement is performed. Then a model is extracted for the two pads and the S-parameters of the launch pad are used to de-embed the S-parameter of the other longer thru lines as ideal transmission lines without the role of the pads in the material measurement.

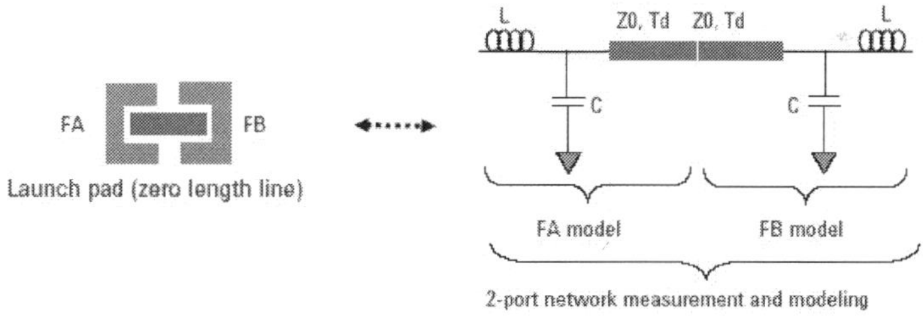

Figure 9.41: *FPGA Test Fixture De-Embedding*

The model for the pads is a symmetric LC model with a single, short-length, uniform transmission line. With symmetry, there are really only four parameters in this model. The parameter values are extracted from the two-port measurement of the thru pads and then the two-port model of just half of this circuit is simulated to create the S-parameter file used to de-embed the longer lines.

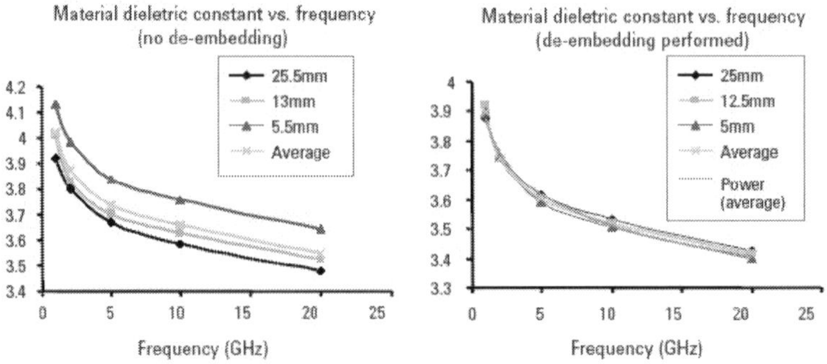

Figure 9.42: *FPGA Test Fixture De-Embedding*

From the de-embedded uniform trace measurements, the dielectric constant as a function of frequency can be extracted based on the phase delay and the physical lengths. In the three lines of different lengths, the dielectric constant should be independent of line length.

Without using de-embedding, but including the effect of the launch area, you can see a wider variation into the extracted dielectric constant and a higher dielectric constant for shorter lines. This is consistent as the capacitance of the launch will skew the phase delay toward higher dielectric constant. This is an artifact.

However, when the launch pads are removed, the skew between the three line lengths is dramatically reduced. The measurements are much more consistent. In this material system, there is some frequency dependence on the dielectric constant.

9.7 Simulating the Fixture S-Parameter with a 3D Field Solver

De-embedding processes can play an important role in simulations as well as measurements. In this example, the performance of a via field is simulated with Momentum, a 3D planar field solver tool integrated into the ADS environment. The problem is to evaluate the performance of a pair of signal vias that connect between the top layer and the bottom layer of a four-layer board. There are two planes between the signal layers acting as return planes, also with vias connecting between them.

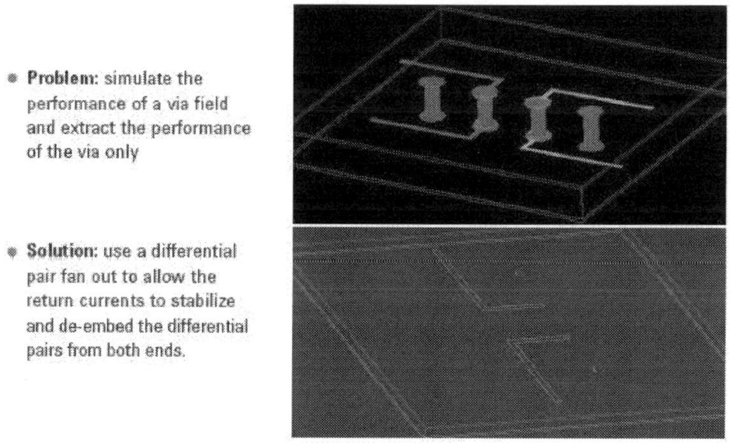

Figure 9.43: *Example No. 3: The Role of De-Embedding in Simulations*

You could attach ports to the top and bottom of the vias with no signal paths. However, this will create an artificial return current path because the signal line path is not present. The signal line feed will affect the return current distribution feeding into the signal via.

However, when you add the signal line feed to the via, you simulate not just the via but also the signal paths. While the magnitude of the insertion and return loss is not much affected by the feed lines, the phase is affected by the feed lines. The phase will affect the circuit model extracted from the simulated S-parameters. De-embedding can be used to take the simulated S-parameters of the composite structure and, by building a model of just the signal feeds, de-embed just the via performance.

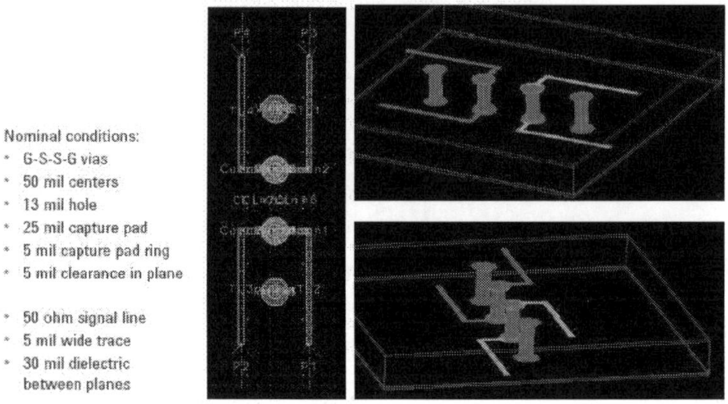

Figure 9.44: *Nominal Via Configuration*

The first step is to build the model of the vias with the signal line feeds. To simulate crosstalk, two adjacent signal vias are used, and the signal feeds separate in opposite directions to minimize the coupling into the surface traces. As an initial setup, the nominal conditions are used of a 50 mil grid for the vias, a 13 mil drilled hole, and a 50 Ohm microstrip. The plane to plane separation is 30 mils, which is very typical.

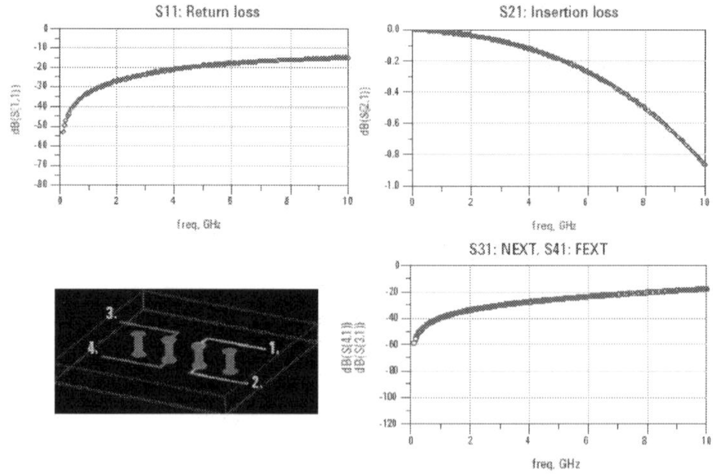

Figure 9.45: *Example of Simulated Return and Insertion Loss of One Signal Via*

Using the Momentum software package, the four-port S-parameter of this via field configuration can be calculated. Since this structure is so small, you should expect to see an insertion loss that is pretty close to 0 dB. The crosstalk, which would be the S31 or S41 terms, should be fairly large, negative dB values, as there should be a small amount of crosstalk between these signal vias. *Figure 9.45* shows the results of this nominal case simulation. The return loss is as expected. It starts with very little, and then creeps up to as high as –20 dB at 4 GHz, a surprisingly large amount. The insertion loss is also fairly large, and the higher the frequency, the worse the insertion loss becomes.

The crosstalk is also larger than might be expected. These values suggest the nominal via built is not so transparent. Of course, another possible explanation is that some of this behavior is due to the fixture, not the via. You can isolate and de-embed just the via from the composite measurement and eliminate this possible artifact by de-embedding it from the measurement.

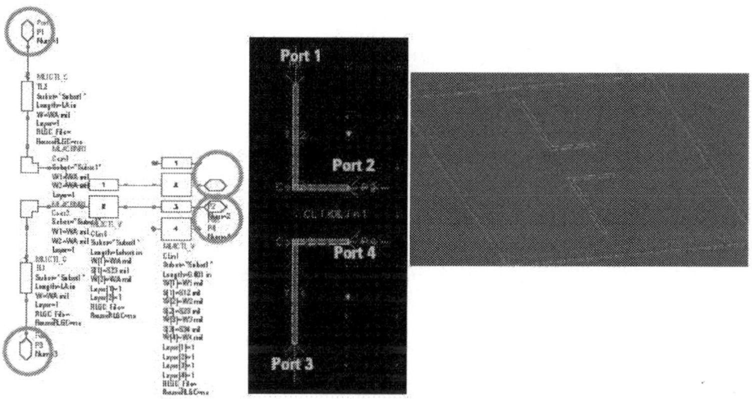

Figure 9.46: *De-Embed File*

The first step in de-embedding the fixture is to create a model for the fixture on each end of the vias. Then simulate the S-parameters of the fixture and use this to mathematically subtract it out or de-embed it from the composite simulation. Break the model off where the signal lines enter the vias. Then place a port at the ends of the signal lines that define the interface to the fixture. There will be one of these on either side of the via field. Since both ends of the fixture are identical, you only need one S-parameter file, just make sure you have the ports connected correctly.

Take this section of the fixture and bring it into Momentum as a 3D structure and simulate the four-port S-parameters from the ends.

This file becomes the part you will remove from the composite simulation of the via and the fixture.

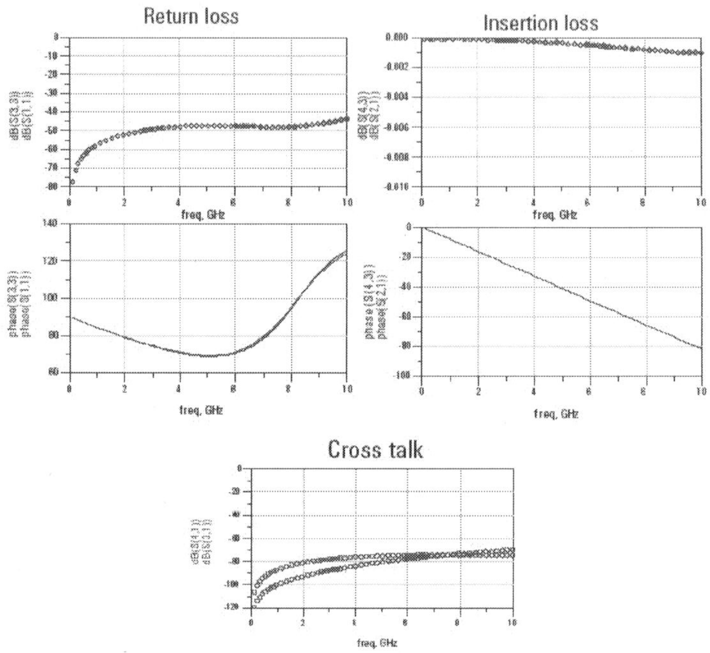

Figure 9.47: *De-Embed Data*

Figure 9.47 shows the simulated file that will be subtracted from the composite simulated S-parameters. It has very low return loss, very low insertion loss, and low crosstalk, but it does contribute phase in each term.

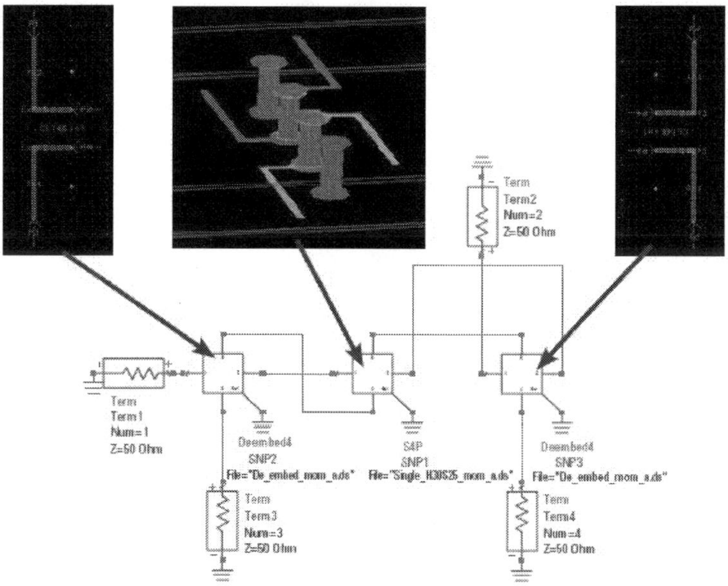

Figure 9.48: *De-Embedding the Four-Port Data*

This is how ADS is set up to de-embed the vias from the composite simulation. You start with the data set file that has the four-port S-parameters of the composite measurement. In ADS, there is a de-embed element that can be used to mathematically de-embed another data set.

Load the fixture only with the correct ports connected as a through path into this de-embed element. Do the same thing on the other end of the composite dataset. Now you simulate the S-parameters of this series combination. The de-embedded S-parameter behavior is really the residual behavior of the complete structure that cannot be fully accounted for by just the transmission line feeds. This would include any return current redistribution effects in the via field.

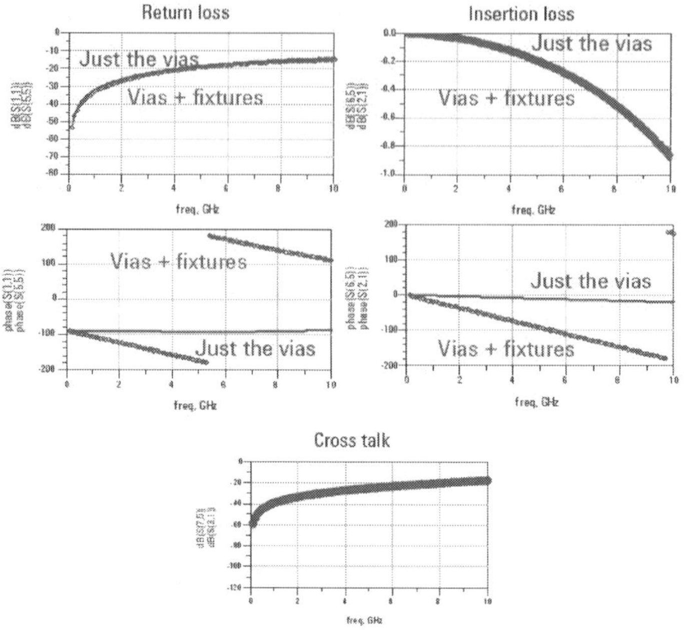

Figure 9.49: *Impact of De-Embedding: De-Embedding Is Critical for Accurate Phase Simulation*

Now you can compare the de-embedded performance of just the via with the composite performance of the via and fixture. As expected, there is little impact on the fixture in the magnitude of the return or insertion loss, but a large impact on the phase. Using de-embedding, you are able to get the S-parameters of just the influence from the vias.

Of course, it might have been possible to get similar results in this case, if the ports were moved closer to where the via started. This would have minimized the impact from the fixture. But, then you could miss any return current redistribution effects when the return vias are moved apart. This technique is a powerful technique for more complex fixture structures that are needed to feed the vias and to evaluate the impact on the region close to the vias. When you do use a long feed fixture, and you do not de-embed, the phase is mostly dominated by the fixture and you may get misleading results.

In addition, while the 3D planar field solver will take into account the return path in the planes that may re-distribute in the presence of the return vias and the planes, when you simulate just the fixture, the return path is well behaved under the signal lines. Any impact on the current re-distribution in the fixture and via case will be left as part of the de-embedded via data set.

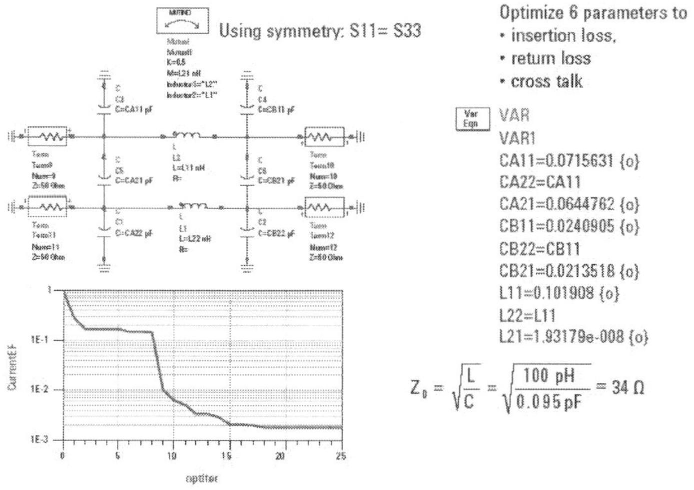

$$Z_0 = \sqrt{\frac{L}{C}} = \sqrt{\frac{100\ \text{pH}}{0.095\ \text{pF}}} = 34\ \Omega$$

Figure 9.50: *Building a Simple C-L-C Model of the Vias*

Using the de-embedded S-parameters of just the vias, you can now build a simple lumped circuit model and begin to analyze the results. The simplest model of a via is a C-L-C topology. When you want to include coupling, you need to add the coupling capacitors and the mutual inductance between the two signal paths.

Figure 9.50 shows the simple model of the two coupled signal vias. The presence of the return path vias is included in the capacitance to ground and the loop inductance of the signal vias. There are nine terms that define this model. However, by using symmetry, you can reduce this to only six terms. Use the built-in optimizer to have ADS find the best set of parameter value that match the simulated S-parameters of the via field.

The convergence is shown in *Figure 9.51*. You can find the best values in less than 20 iterations. Here are the actual model parameter values when the optimizer is done. What do the values mean? What do they tell you about the behavior of the via? As a rough starting place, you can estimate the single-ended impedance of one signal via by taking the square root of the inductance to the capacitance of the via. In this case, the numbers are about 34 Ohms. This says the single-ended impedance of this short via is about 34 ohms—lower than 50 ohms. It looks capacitive, and this is why the return loss is about –25 dB at 2 GHz—it has a mismatch.

But before you read too much into this number, you should check to see how well this simple model matches the actual, simulated S-parameters.

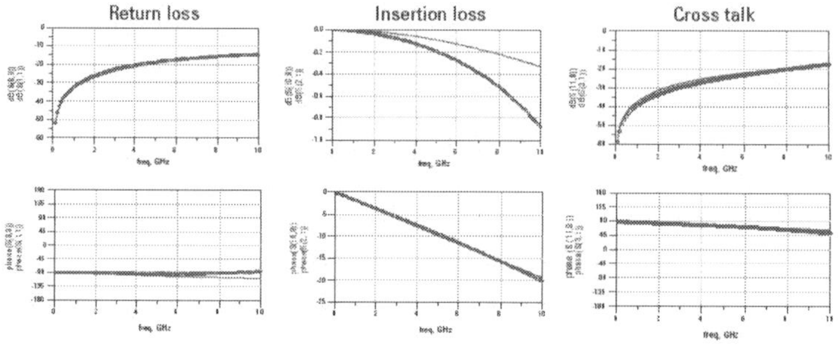

Figure 9.51: *Model Verification*

Figure 9.51 shows the comparison of the 3D planar simulated S-parameters of the two signal vias (thick line) and the simulated s-parameters of the CLC model (thin line) for the return loss, the insertion loss, and the crosstalk.

Observations are as follows:

- Return loss suggests excellent model topology and parameters to 10 GHz
- Crosstalk match suggests excellent modeling of coupling-capacitive dominated
- Insertion loss suggests another loss mechanism not included in the lumped circuit model, but included in the

3D planar field solution: possibly coupling into the plane-to-plane cavity

It is clear that the return loss is a great match, as is the crosstalk. This says that the model is a pretty good model for the behavior of the two signal vias. But the insertion loss is not such a good match. The 3D planar simulation shows more insertion loss than the simple CLC model predicts. The fact that the insertion loss and crosstalk are such a good fit suggests that there is a real effect here. What could it be?

One possible explanation is that as the signal current switches through the region between the planes and the return current flows between the planes capacitively and then to the return vias, there is coupling of energy from the signal into the cavity made up of the planes. The energy that couples into this cavity is lost as radiation into the cavity. This is simulated by the 3D planar tool, but not by the CLC model.

The process of de-embedding the intrinsic via performance from the composite measurement is a generic process you can use to generate a dataset of just the vias of interest, which can then be used for further analysis.

9.8 Summary

De-embedding is a powerful tool that should be in the tool box of every characterization engineer. It requires the S-parameters of the fixture to be known. They can be obtained by direct measurement, fitted with parameterized circuit topology-based models using measured results as the target, or by simulation with a 3D full-wave field solver.

For best results, it is important to minimize the fixture effects by using short sections in the fixture and trying to optimize the design to match to 50 Ohms. Once you have the S-parameter file for the fixture, the built in de-embedding features of the VNA, PLTS, or ADS can be used to separate the true performance of the DUT from the artifacts introduced by its fixturing.

Authors

Eric Bogatin, Signal Integrity Evangelist, Bogatin Enterprises

Mike Resso, Signal Integrity Application Scientist, Component Test Division, Agilent Technologies

Chapter 10

Backplane Differential Channel Microprobe Characterization in Time and Frequency Domains

10.1 Abstract

The chief difficulty with routine characterization of differential channel paths is the current requirement for surface-mount assembly (SMA) connectors to interface with the test equipment. With little extra space available on functional backplanes and the problem of line-loading produced by the SMA stubs, it is not practical to use SMA connectors. This means only specially designed test boards can be routinely characterized. This paper introduces a new methodology for testing the passive interconnects associated with a differential channel in a backplane assembly, which can be used for functional, populated backplane assemblies and shows how the pad layout can be optimized for routine probing without impacting the functional system performance.

10.2 Differential Channels Will Proliferate

We are in the beginning of a revolution in input/output formats. All high-speed interface specifications are migrating toward the use of differential signals implemented with a differential pair of transmission lines. *Figure 10.1* is a partial list of some of these high-speed serial link formats.

All Next Generation High Speed Serial Links will use Differential Signaling

Serial ATA	1.25 Gbps
Hypertransport	1.6 Gbps
AGP8x	2.1 Gbps
Infiniband	2.5 Gbps
PCI Express	2.5 Gbps
Serial ATA II	2.5 Gbps
XAUI	3.125 Gbps
PCI Express II	5.0 Gbps
OC-192	9.953 Gbps
10 GbE	10 Gbps
OC-768	39.81 Gbps

Figure 10.1: *Partial List of Many New High-Speed Serial Link Formats*

In addition to being differential signals, they are all in excess of 1 Gbps. This translates to an analog bandwidth of at least 2 GHz. In the case of 10-gigabit attachment unit interface (XAUI), this is in excess of a 5 GHz analog bandwidth. In order to verify the performance and the quality of models for use in system-level simulation, measurements on the differential channel properties must be at least twice the application bandwidth, or up to a 10 GHz measurement bandwidth.

Characterizing a differential channel can be accomplished in two ways, either by building an equivalent circuit model based on a topology of ideal circuit elements that all simulators can understand, or with a physical layer (PHY) characterization based on performance metrics. The model extracted from PHY characterization is usually called a behavioral model, as it uses the direct measurements as the model itself. A behavioral model is often called a "black box" model because it cannot be determined what specific structures are inside.

A topology-based model can be used in any simulator and provides useful design insight on what physical features influence which electrical features. It requires more work to construct and the effort usually increases exponentially with increasing bandwidth and complexity. Creating a model for a simple structure such as a balanced transmission line (microstrip or stripline) can be difficult at high speeds. Creating a topological model for a large backplane with multiple channels is daunting even for a team of experts.

The alternative characterization approach is to use the direct measurements as the characterization. This behavioral model can then be used directly in a system-level simulation. Careful manipulation of the direct measurements can present the information in a format from which useful performance information can be directly extracted.

In this later approach, the following are a few metrics that describe the performance of the differential channel:

- The insertion loss (quality of the transmitted differential signal)
- The return loss (the differential impedance profile)
- The conversion of differential to common signal
- The location in the interconnect where most of this conversion occurs
- The common impedance profile
- The crosstalk between differential channels
- The eye diagram for a specific bit rate

Each of the preceding figures of merit can be obtained with the Agilent physical layer test system (PLTS) software using a four-port measurement system, whether it is a four-port vector network analyzer (VNA) or four-channel time domain reflectometer (TDR). In theory, the measurements are simple. In practice, they are difficult and constrained by multiple factors such as introduction of excess inductance of probe tips, excess capacitance of test fixture pads, instrument setup, calibration, and de-skewing of differential pairs. There are myriad other subtle process steps that can lead to poor results. The key step in using direct measurements for PHY characterization is to minimize the artifacts introduced by the fixturing used to interface the test equipment with the differential channel under test. An example of a complete four-port VNA–based measurement system is shown in *Figure 10.2*.

Figure 10.2: *Complete Four-Port Characterization System*

10.3 The Bottleneck of SMAs

The most common technique of interfacing a backplane, for example, to a VNA is by designing the board for SMA connectors. An example of a test board instrumented with SMAs is shown in *Figure 10.3*.

Typical SMA Fixtured Test Board

Figure 10.3: *SMA Connectors Used to Interface VNA to Test Board*

The advantage of SMAs is that they are easy to attach a precision 50 Ohm coax cable and, once attached, are robust. While SMAs work, they have three significant drawbacks.

Though the bandwidth of an SMA connector itself may be in excess of 10 GHz, the artifacts it introduces in the test board can often be seen at a much lower frequency. This is usually due to the vias required in the board to plug in the SMA. If this structure is not optimized, the SMA can introduce either an inductive or capacitive artifact.

SMAs are large in size, and with cables attached, there is a limit to how closely spaced they can be mounted to a board. The limit is roughly on a 0.75-inch pitch. This is a density of about 1.8 connections per square inch.

An SMA, in the best case, looks like a 50 Ohm stub about 0.5 inches long. When this is in series with the test line, it may be transparent to the signal. However, if it is placed on an active line, with a driver, it will load the line down and may cause the performance to degrade at Gbps rates.

For these reasons, SMAs are typically used with special, custom characterization boards rather than on the actual product. This is an expensive path and may not give an indication of the performance of the final product boards.

10.4 Advantages of Microprobing

Microprobes typically use precision 50 Ohm micro coax cables with very small tips. They come in the form of coplanar or needle tips. Using a custom calibration substrate, they can be calibrated right down to the tip. The typical parasitics can be less than 100 pH and 100 fF. A close-up of a microprobe in contact with a test pad is shown in *Figure 10.4.*

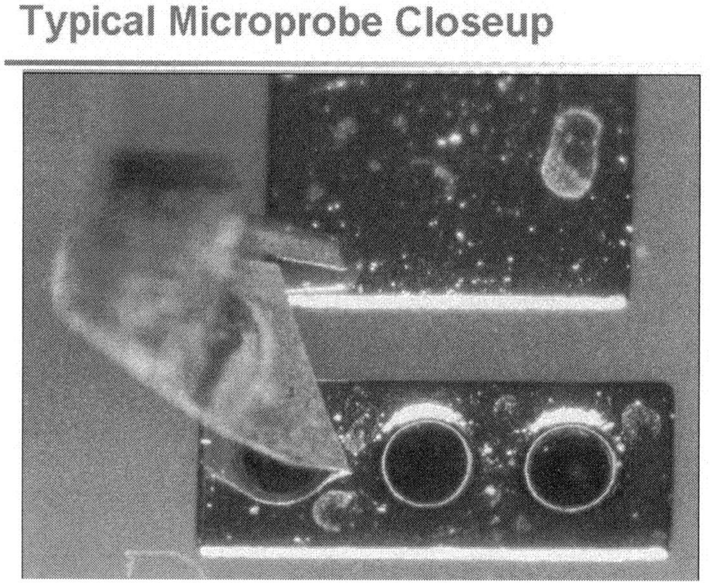

Figure 10.4: *Typical Signal-Ground Microprobe with 450 Micron Pitch*

When the board is designed for microprobing, this approach eliminates the problems with SMA connections and can dramatically increase the value of characterization measurements.

The key to getting the most value from microprobes is to design the board with return paths adjacent to all signal paths. When there is an adjacent return connection to the signal line, the physical probe can be very small and its parasitics very low. This is true whether the probe is a 50 Ohm controlled-impedance probe or a high-impedance active probe.

10.5 Design for Test

As more designs enter the gigahertz regime, designers should think about testability before the design is finished. All it takes to dramatically improve the high-frequency testability of a board is to add a grounded copper fill in the vicinity of the signal vias. This is illustrated in *Figure 10.5*.

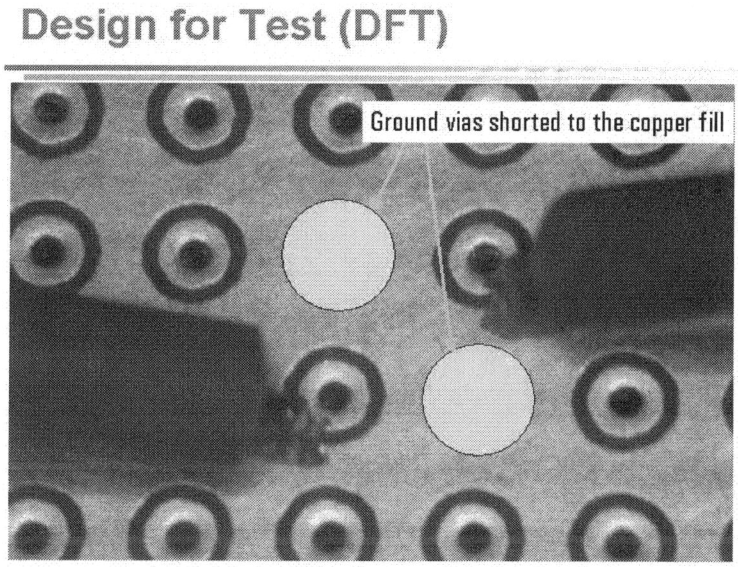

Figure 10.5: *Pad Geometry with Copper Fill Adjacent to Each Signal Via and the Fill Shorted to All Ground Vias*

If this feature is added, a return path will be adjacent to each signal path and microprobe technology can be used. There is a slight increase in capacitance to the signal line, but this can be kept small while still allowing all signal lines to be probed. With this approach, the density of probe points can increase almost 100 times over SMAs, and there is no loading of active signal lines. This approach will dramatically increase the value of instruments, both for passive PHY characterization and active signal monitoring.

10.6 Physical Layer Characterization

All the electrical performance information for a differential channel listed above can be extracted using a four-port VNA. It is actually the single-ended S-parameters that are measured in the frequency domain. *Figure 10.6* shows the format for this information and the port labeling scheme for a differential channel that is measured as two single-ended channels.

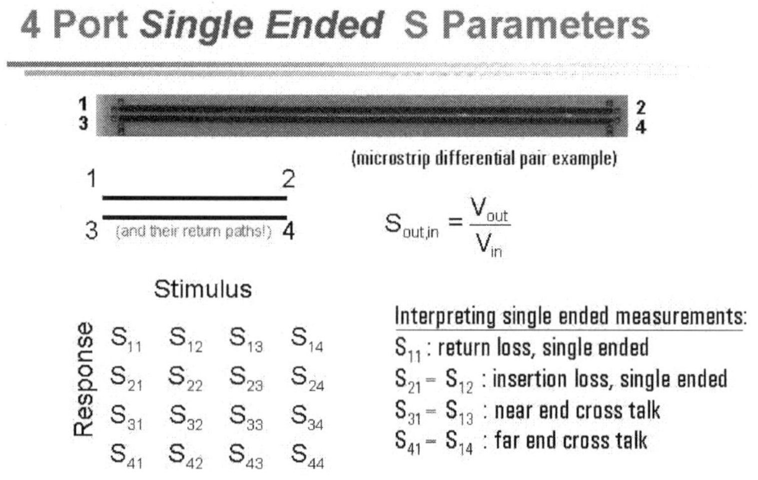

Figure 10.6: *It Is the Four-Port Single-Ended S-Parameters That Are Measured and Then Transformed into Other Forms*

These are mathematically transformed into the balanced, or mixed-mode, or differential S-parameters, also in the frequency domain. These parameters can be used directly to give information about the differential return or insertion loss, or can be transformed into the time domain to give information about the differential impedance profile of the channel or the location of the conversion of differential signal into common signal. The differential parameter matrix is shown in *Figure 10.7*.

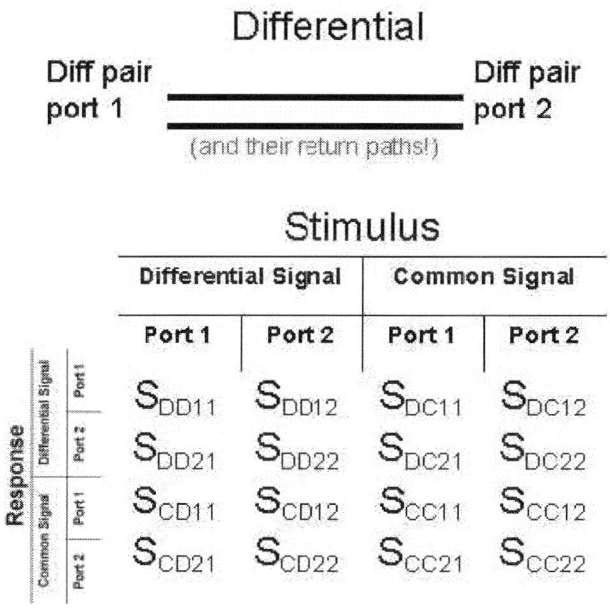

Figure 10.7: *The Result of the Single-Ended to Balanced S-Parameters Transformation Is a Stimulus/Response Matrix of Differential Parameters*

10.7 Understanding Four-Port Mixed-Mode Analysis

In order to interpret the large amount of data in the differential parameter matrix, it is helpful to analyze one quadrant at a time. The first quadrant is the upper left four parameters describing the differential stimulus and differential response characteristics of the device under test (DUT). This is the actual mode of operation for most high-speed differential interconnects, so it is typically the most useful quadrant that is analyzed first.

It includes the input differential return loss from both ends, SDD11 and SDD22, and the differential insertion loss, SDD21 and SDD12. Note the format of the parameter notation SXYab, where S stands for scattering parameter or S-parameter, X is the response type (differential or common), Y is the stimulus type (differential or common), a is the output port, and b is the input port. This is typical nomenclature for frequency domain S-parameters. All 16 differential S-parameters can be transformed into the time domain by performing an inverse fast Fourier transform (IFFT). The matrix representing the time domain will have similar notation, except the "S" will be replaced by a "T" (i.e., TDD11).

The second and third quadrants are the upper right and lower left four parameters, respectively. These are also referred to as the mixed-mode quadrants. This is because they fully characterize any mode conversion occurring in the DUT, whether it is common-to-differential conversion (related to electromagnetic interference [EMI] susceptibility) or differential-to-common conversion (related to EMI radiation). Understanding the magnitude and location of mode conversion is very helpful when trying to optimize the design of interconnects for gigabit data throughput.

The fourth quadrant is the lower right four parameters and describes the performance characteristics of the common signal propagating through the DUT. For most differentially driven systems, the behavior of the common signals is not critically important. The information about how the channel affects common signals, the return and insertion losses, is contained in this quadrant. A summary of the four quadrants in the differential parameter matrix is shown in *Figure 10.8*.

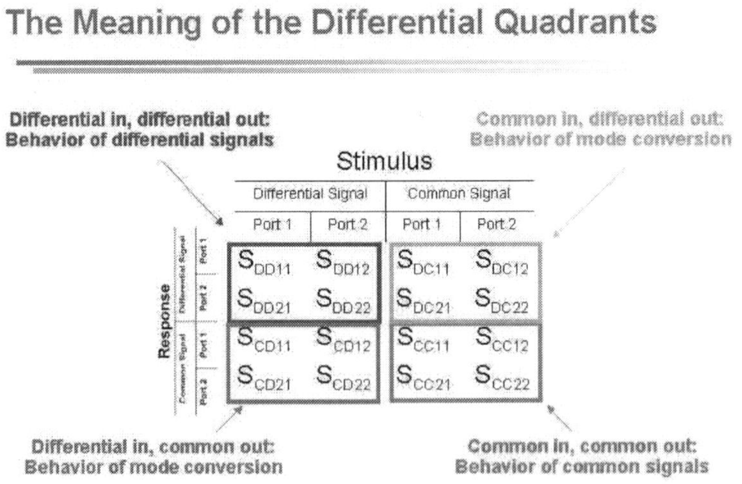

Figure 10.8: *The Result of the Transformation Is a Stimulus/Response Matrix of Differential Parameters*

10.8 Design Case Study: XAUI Backplane

In order to demonstrate the proper measurement technique for characterizing a high-speed, linear, passive interconnect, a popular industry-standard XAUI backplane was used in a design case study. The expected serial data throughput of this device is a minimum of 3.125 Gbps. Daughtercards were used as test fixtures for analyzing the backplane and its connectors, since most test and measurement equipment has 3.5 mm SMA connectors. A picture of the backplane is shown in *Figure 10.9.*

At a few selected locations, the SMA connectors were removed and microprobes were used to probe the differential channels. Because the via and pad design were optimized for SMA and the performance is limited by the pad and vias, there was no improvement with microprobes. Both SMA measurements and microprobe measurements showed the same performance below 14 GHz.

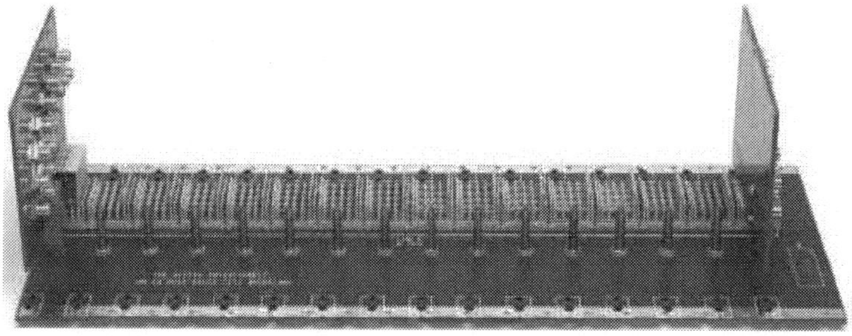

Figure 10.9: *XAUI Backplane Used for Design Case Study (Courtesy of John D'Ambrosia, Tyco Electronics)*

10.9 Comparing Time Domain and Frequency Domain Data

Typically, the first step in analyzing the measurement data is to look at the reflected waveform. In the time domain, this is the differential impedance profile (TDD11). This intuitive waveform will quickly give the designer an idea of where the impedance mismatch of various structures (e.g., printed circuit board [PCB] trace, via, connector) are causing reflections and degrading signal quality.

If a perfectly matched impedance environment is achieved, this waveform will be a flat horizontal line. Any deviation from this flat line can be interpreted as excess inductance (positive deflection from characteristic impedance) or excess capacitance (negative deflection from the characteristic impedance).

With the XAUI backplane, the individual features of the interconnect can be seen in *Figure 10.10*. From left to right, the first feature is excess inductance from the daughtercard SMA connector, then the 110 Ohm daughtercard differential transmission line, then the excess capacitance from the daughtercard via field, then the excess capacitance from the backplane via field, and finally the 100 Ohm differential backplane transmission line. The impedance profile of this complete differential channel can easily be seen in *Figure 10.10*.

Transforming the impedance profile into the frequency domain yields the input differential return loss (SDD11). Measured in decibels, return loss is a negative value that describes how the signal propagating through the DUT is reflected back as a function of frequency. Periodic negative deflections can be seen in this data that are a result of resonant structures within the device.

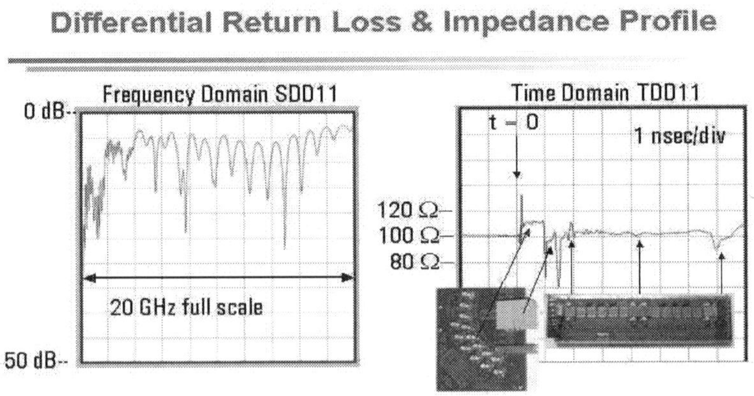

Figure 10.10: *Analyzing Both Frequency and Time Domain Data on a XAUI Backplane*

10.10 Coupling Pulls Down Differential Impedance

In most single-ended high-speed digital interconnects, crosstalk between adjacent transmission lines is undesirable. This is not the case for differential transmission lines. The strong coupling of adjacent PCB traces that make up a differential pair is exactly what contributes to good common-mode noise rejection. When targeting a specific differential impedance of 100 Ohms, this coupling has to be taken into account.

An example of this can bee seen in the impedance profiles in *Figure 10.11*. The single-ended TDR trace shows the daughtercard and motherboard exhibiting around 55–56 ohm of single-ended impedance (one line is driven, the other line is quiet, and the impedance is measured from the driven line to the ground).

The differential TDR trace shows the effect of coupling on the motherboard traces and yields about 100–101 ohms (both lines are driven with equal amplitude/opposite polarity steps and impedance is measured from line 1 to line 2). Notice that the daughtercard differential impedance does not pull down to the target differential impedance of 100 ohms. This indicates weaker coupling on the differential traces due to larger spacing between daughtercard traces.

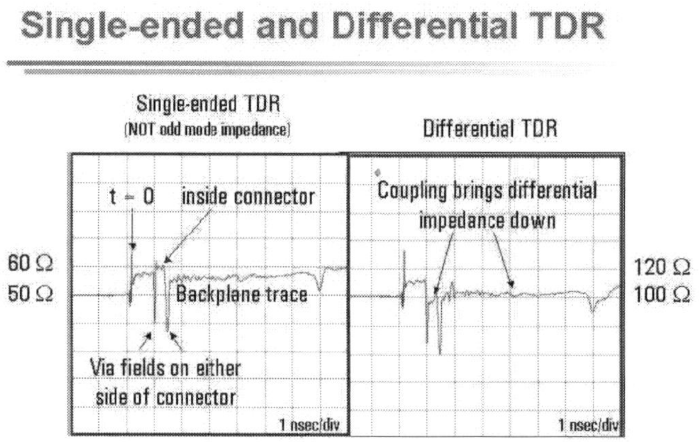

Figure 10.11: *Comparing Single-Ended TDR and Differential TDR Indicates Good Coupling on Backplane and in the Connector*

A figure of merit that has become important for characterizing high-speed differential channels is the input differential insertion loss (SDD21). This is the frequency response of the interconnect for a transmitted signal. It is often insightful to compare various-length channels using SDD21. The two channels in *Figure 10.12* are broadside-coupled stripline traces in a homogeneous dielectric system (both sides of copper embedded in dielectric). The 40-inch-long channel has greater loss, as one would expect. Even though the measurement system had more than 40 GHz of bandwidth, it can be seen from *Figure 10.12* that the device bandwidth falls off dramatically before 10 GHz.

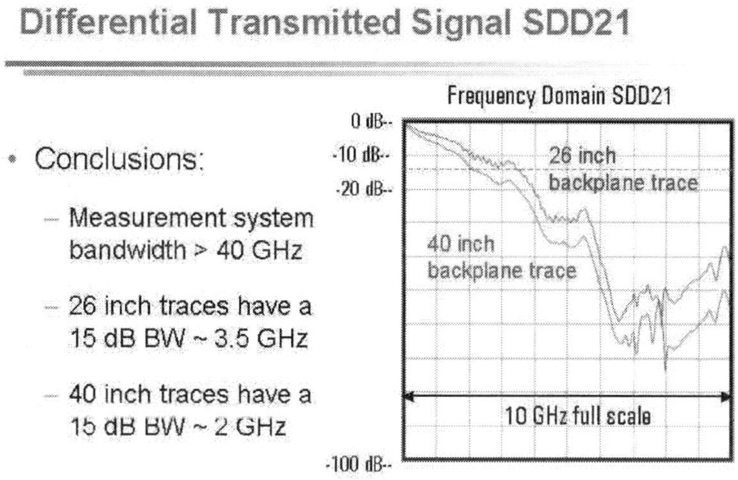

Figure 10.12: *Differential Insertion Loss Is a Key Figure of Merit for Backplane Channels*

Work performed by the High Speed Backplane Initiative (HSBI) includes the use of SDD21 to describe the relative amount of loss in multiple channels as a function of frequency. A typical compliance template would be used in a similar methodology to an eye diagram mask. A proposed SDD21 template is shown in *Figure 10.13*.

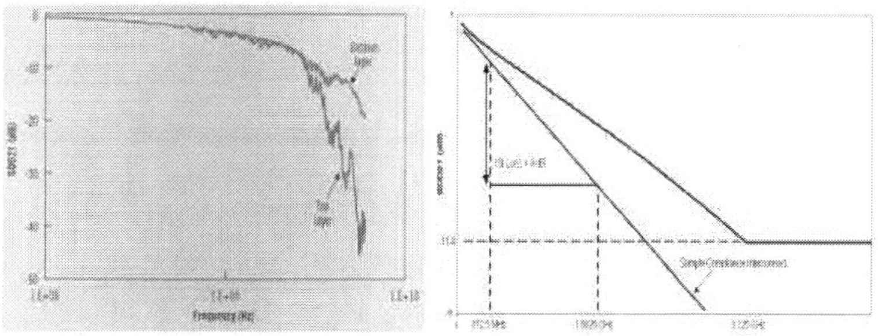

Figure 10.13: *Example of SDD21 Used as a Compliance Specification for 3.125 Gbps Data Transmission*

10.11 Eye Diagram Simulation Using Four-Part S-Parameters

Most high-speed serializer-deserializer (SerDes) chipsets are verified to be compliant to standards by use of eye diagram analysis. A reference receiver used in conjunction with a high-bandwidth equivalent time sampling oscilloscope will display an eye diagram based on the actual bit-stream output from the SerDes. The interconnect between the transmitter and receiver can also be characterized using this test setup. Longer channel lengths between a SerDes chipset will exhibit eye diagrams that close with increasing length due to higher differential insertion loss. There is another way to obtain eye diagram information using S-parameters.

The first step to achieve this simulated eye diagram is to obtain very accurate four-port S-parameters. This can be done using either a four-port VNA or TDR with two differential TDR modules (with PLTS software). Once the S-parameters are obtained, the time domain impulse response of the channel can be derived. Using a novel patent-pending algorithm, the impulse response can then be convolved with an arbitrary binary sequence. By changing the bit rate and rise time of the arbitrary binary sequence, a family-simulated eye diagrams can be developed as shown in *Figure 10.14*.

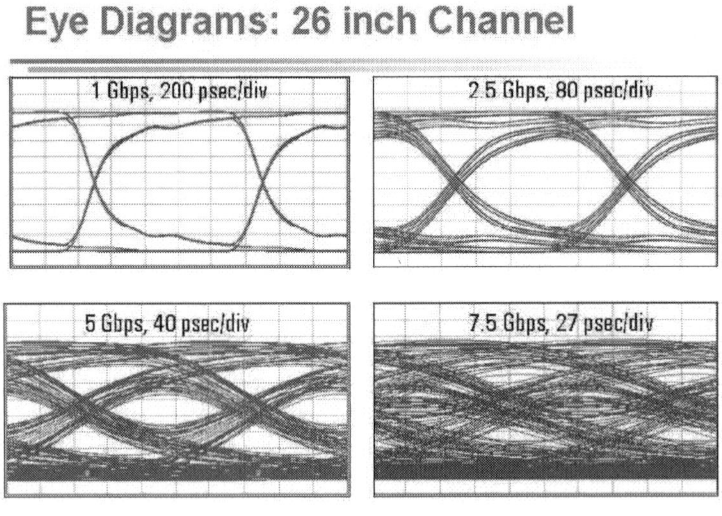

Figure 10.14: *Eye Diagrams Derived from Four-Port S-Parameters*

10.12 Non-Ideal Differential Signaling (AKA Mode Conversion)

When a differential signal propagates through an ideal, symmetric, differential transmission line, no common signal will be generated. In other words, mode conversion should be zero for all frequencies (a flat, zero line). Any asymmetry between the lines that make up the differential pair can potentially convert some of the differential signal into a common signal. We generally call this effect mode conversion.

If no signals get out of the product enclosure and there is a sufficient common-mode rejection ratio in the receivers, mode conversion may not be a problem. However, if a twisted pair connects to the end of the differential channel and any of this is converted to common signal and gets out on the twisted pair, it will contribute to excessive radiated emissions.

Mode conversion can arise from asymmetries in the drivers and asymmetries in the interconnect. An example of active-device mode conversion would be if a differential transmitter or amplifier had one drive signal different from the other supposedly complementary drive signal, either voltage, current, output impedance, or skew. An example of passive-device mode conversion would be if a differential connector had different characteristic impedance lines, different length pins, or different loading on each line such as jags, pads, or ground plane discontinuities. Mixed-mode analysis can be a powerful tool in locating obstacles that limit the highest possible data-rate transmission in interconnects.

The differential-to-common signal conversion can be quantified as a percentage of amplitude by overlaying the time domain differential transmission waveform (TDD21) with the time domain differential to common transmission waveform (TCD21). As seen in *Figure 10.15*, the TDD21 waveform shows the propagation delay and rise-time transition of the degraded differential signal at the output of the DUT.

Shown on a more sensitive scale, TCD21 is the amount of

converted common signal coming out of the channel. This is the mode-conversion waveform showing 7 percent of the original differential signal converted to common signal.

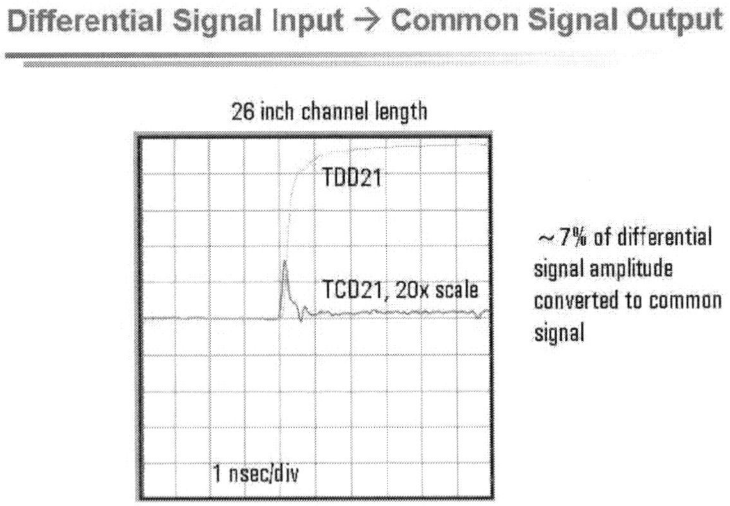

Figure 10.15: *Overlaying Mixed-Mode Waveform TCD21 Shows Mode Conversion of 7 Percent*

After determining that the mode conversion is present in the DUT, the next step is to locate the physical structure that is causing this undesirable effect. The easiest method for doing this is to view the input differential impedance profile (TDD11) and overlay the input time domain reflection mode conversion waveform (TCD11). Since TDD11 is roughly the impedance versus distance, this gives a physical reference to the structure of the daughtercards and backplane. The daughtercard transmission line, daughtercard via field, backplane connector, backplane via field and backplane transmission lines are all clearly discernable from this impedance profile in *Figure 10.16*.

The corresponding location on the horizontal axis for the mode conversion waveform correlates exactly where the mode conversion is happening. This is how the mode conversion source is identified. The maximum deflection (positive or negative) in the mode conversion waveform is the structure that should be addressed first. Interestingly, the largest magnitude of mode conversion was not the backplane connector, but it is the via field layout on the daughtercard and backplane.

In general, the greatest source of mode conversion is the asymmetry of the vias. As they generally represent the largest discontinuity, balancing this effect between both lines of the pair is difficult if attention is not paid. The conversion can be minimized if the via barrel and pads are optimized for 50 Ohms.

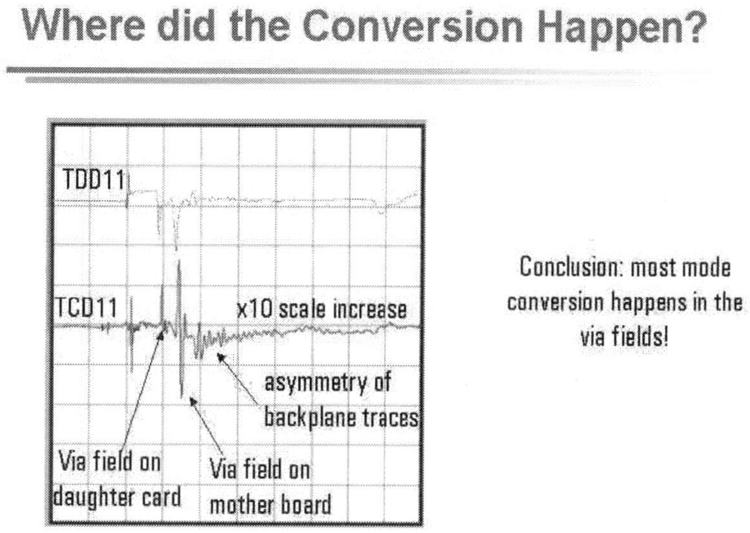

Figure 10.16: *The Location of Mode Conversion to Be Identified Comparing the Differential Return Loss to the Common Signal Coming Back to the Source End*

10.13 Summary

Advanced design tools and test methodologies are now required to address the subtle signal integrity challenges created by microwave transmission-line effects. Proper probing techniques are important to achieving fast design-cycle improvements. In addition, complete measurement characterization of the PHY in both the time domain and frequency domain is crucial to solving tomorrow's signal integrity problems.

References

[1] Eric Bogatin, Signal Integrity-Simplified, Prentice Hall, 2004.
[2] Eric Bogatin, "Getting Your S (Parameters) Together," Printed Circuit Design and Manufacture, February 2003.

Authors

Eric Bogatin, Chief Technological Officer, GigaTest Labs

Mike Resso, Product Manager, Signal Integrity Operation, Agilent Technologies.

Chapter 11

Differential PCB Structures Using Measured TRL Calibration and Simulated Structure De-Embedding

11.1 Abstract

A combined thru-reflect-line (TRL) calibration and de-embedding procedure based on measured and computed data for extracting scattering parameters (S-parameters) of differential structures in multilayer printed circuit boards (PCBs) is presented. The proposed technique starts with measured data using a single-mode TRL calibration to remove the coaxial to planar PCB structure and then simulates the planar single-ended to differential structure transition for de-embedding their effect from the measured data. The result is very accurate mixed-mode S-parameters of the device under test (DUT). This technique is demonstrated on a four-port and 12-port system looking at inter-layer via transitions and shows the benefit of eliminating large multi-port calibration structures.

11.2 Introduction

In modern digital systems based on high-speed interconnection technologies, the correct characterization of discontinuities is mandatory for meaningful signal integrity (SI) analysis [9, 10]. The layout of PCBs involves a large number of discontinuities such as connectors, unusual terminations, bends, presence of packages, via holes, and line crossings (*Figure 11.1*). Their presence can cause signal distortion and functional problems for the mounted digital device. The evaluation of the electrical performance of the discontinuities allows high-speed digital designers to extract equivalent circuit models that can be included as part of more complex circuits representing the board. Such evaluation, conveniently performed in terms of S-parameters, can be done by means of numerical simulations or by measurements.

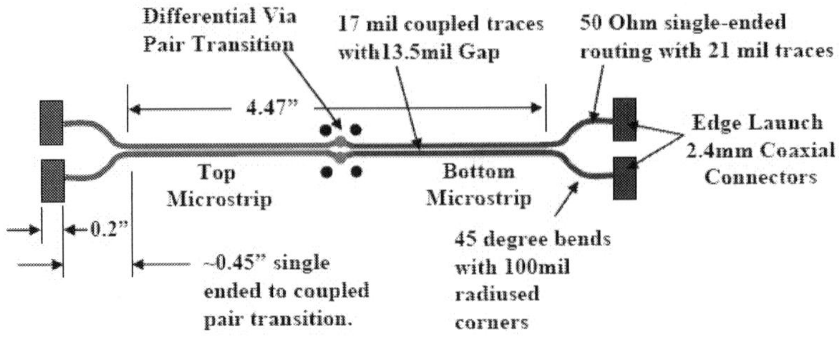

Figure 11.1: *PCB Layout Showing the Characterization Problem*

Numerical simulations must capture the correct distribution of the EM field in order to take into account all the significant effects due to the discontinuity [1]. In multi-gigabit telecommunications systems this means a high level of detail in the description of the geometry that often requires a significant amount of central processing unit (CPU) resources and simulation time. Obtaining measured data provides an excellent calibration point to verify a complex model and ensure that it accurately represents a structure before significant time is spent on optimization for a specific application.

On the other hand, from a measurement point of view, it is generally not possible to gain access to the structure without impacting performance data. An example of this phenomenon is the experimental characterization of a via hole (*Figure 11.1*) in which the presence of feeding parts (e.g., traces, adapters, pads) connecting the instrument's ports to the DUT via must always be present. To overcome the aforementioned difficulties, it has become standard practice to characterize the effects of the test access ports by feeding lines or adapters and then to separate them from the measurement relative to the complete structure. The remaining data are those associated with the electric behavior of the DUT or discontinuity of interest. This procedure is known as de-embedding. In recent years, a number of de-embedding methods have been reported in the literature and their references give useful hints for combining simulations and measurements [2].

A customized single-mode TRL calibration is one method of

providing error correction to remove the launch connector effects when compared to the network analyzer coaxial calibration that just calibrates to the end of the connecting coaxial cable (*Figure 11.2*). However, a simple single-mode two-port TRL calibration does not take out the effects of transitioning from single-ended microstrip to differential microstrip routing for a differential structure. Fabricating a four-port differential mixed-mode TRL structure is feasible [3], but one quickly sees the increase in board area and measurements that it will require. A four-port differential mixed-mode TRL also requires additional mathematics to handle the odd and even modes of propagation found in coupled differential lines when implementing the TRL algorithm. Expanding to a 12-port differential TRL structure has thus far proven to be challenging due to the complexity required.

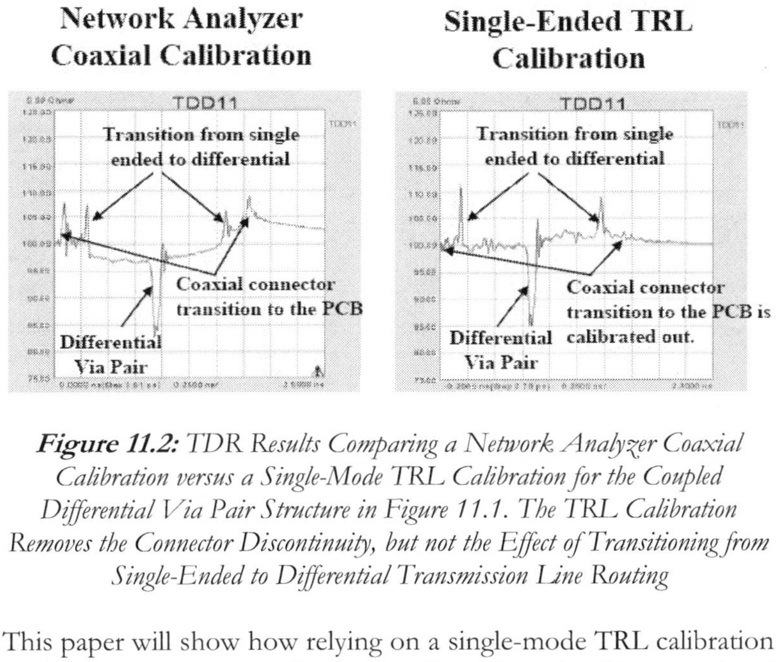

Figure 11.2: *TDR Results Comparing a Network Analyzer Coaxial Calibration versus a Single-Mode TRL Calibration for the Coupled Differential Via Pair Structure in Figure 11.1. The TRL Calibration Removes the Connector Discontinuity, but not the Effect of Transitioning from Single-Ended to Differential Transmission Line Routing*

This paper will show how relying on a single-mode TRL calibration to remove connector effects and then de-embedding simulated single-ended to differential coupled pair transitions is an effective calibration technique for N-port systems. The technique will be demonstrated by obtaining the S-parameters of passive differential inter-layer PCBs via transitions when used in four-port or 12-port routing topologies. Section 11.3 describes the justification for this technique, section 11.4 describes the single-ended TRL, section 11.5 describes the simulated de-embedding structure, section 11.6 looks at a DUT via structure for analysis, section 11.7 looks at verification of this calibration technique with the mixed-mode four-port TRL, and finally section 11.8 concludes with the S-parameters for a 12-port via structure measurement.

11.3 Combined TRL Calibration and Simulated Structure De-Embedding for Multi-Mode N-Port Systems

Obtaining accurate S-parameter data for a two-port, four-port, or N-port system using a network analyzer requires cabling and connectors to launch the signal into the structure under test. The three dimensional (3D) properties of launching from a coaxial cable onto a planar PCB are defined by numerous mechanical and material tolerances on the mating connectors and PCB fabrication that can make accurate 3D–electromagnetic (EM) modeling a difficult task. The utilization of a TRL calibration structure fabricated with the same process as the structure to be measured accurately represents this transition feature and allows the TRL calibration to calibrate out this transition discontinuity from the measurement. Taking a look at the simple single-ended microstrip structure of *Figure 11.3* one can excite the structure with a 10 Gbps PRBS31 signal, but the results will include the whole structure, including the coaxial connector transition to PCB.

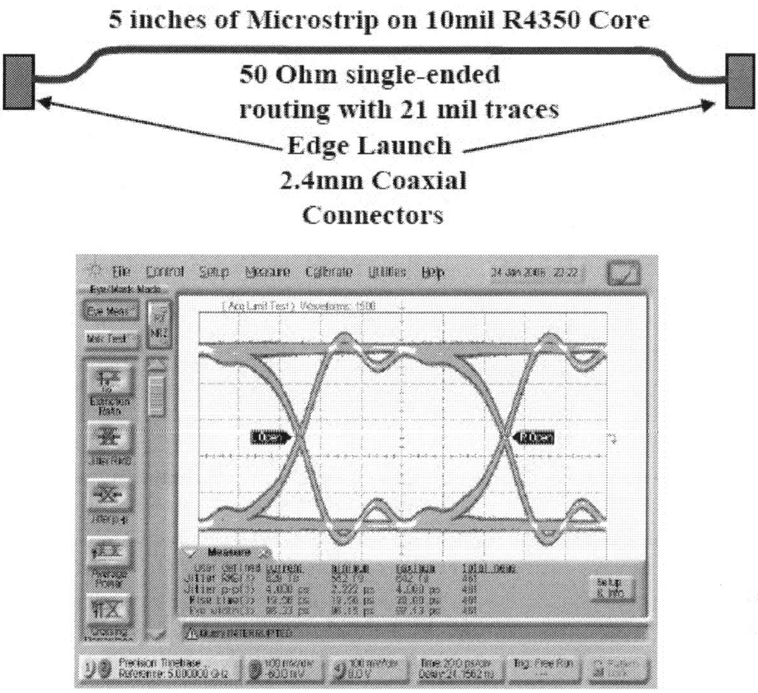

Figure 11.3: *Single-Ended Microstrip Transmission Line and Measured Data Using a 10 Gbps PRBS 31*

Applying a single-ended TRL calibration to remove the coaxial connector transition provides the S-parameters of just the transmission line, which can then be used to generate a resulting eye diagram that does not include the connectors as seen in *Figure 11.4*. Comparing the eye diagram for these two types of calibrations clearly shows that the edge connectors increase the amount of jitter and ripple.

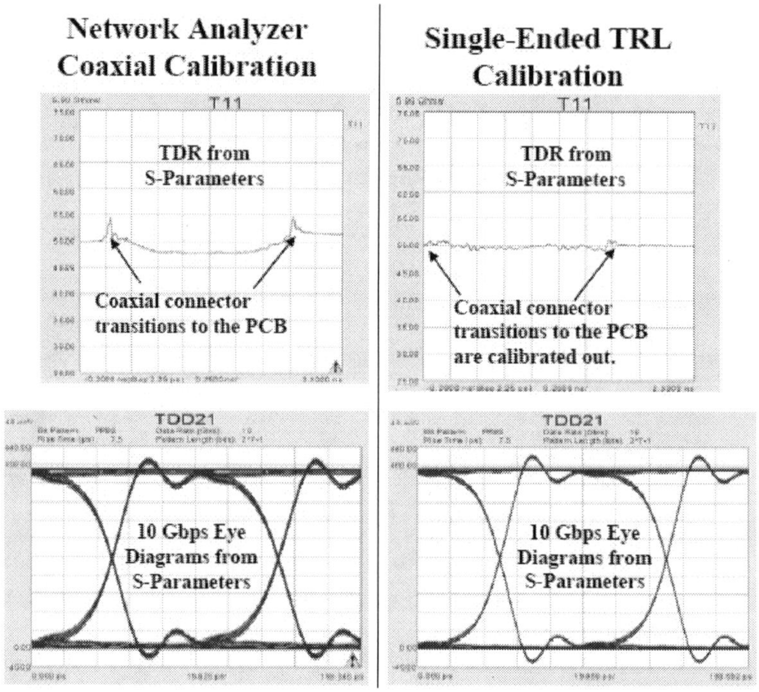

Figure 11.4: *Network Analyzer Coaxial Calibration versus Single-Ended TRL Results for the Microstrip Trace in Figure 12.3*

The TRL calibration is a well-utilized technique for single-ended structures, but as the number of ports increase for N-port systems, the physical space required for the connectors and routing can quickly become cost-prohibitive for layout and materials as well as the increased test time to make all of the connections for an N-port mixed-mode calibration. Faced with this challenge, one looks for a different approach and realizes that the transition from single-ended planar transmission lines to differential or n-port coupled planar transmission lines is a relatively simple structure to model with existing 2D or 3D EM simulators. In this way, accurate S-parameters of the planar transitioning structure can be obtained and used within de-embedding techniques to remove their effect from the measurement, so that one is left with the S-parameters of the desired N-port structure under test.

Applying this technique to a 12-port system quickly shows the benefit. *Figure 11.5* illustrates an example of a structure that routes three stripline differential pairs through a PCB via structure to another layer. The goal is to obtain a full set of 12-port S-parameters so that this component can be used as a part of a model library for PCB design and analysis that will include the effects of crosstalk with neighboring differential pairs. A secondary item of interest is to obtain high-frequency performance of the structure so that a time domain response of the structure can be used for verification of 3D–EM simulations and additional tuning of capacitive and inductive elements.

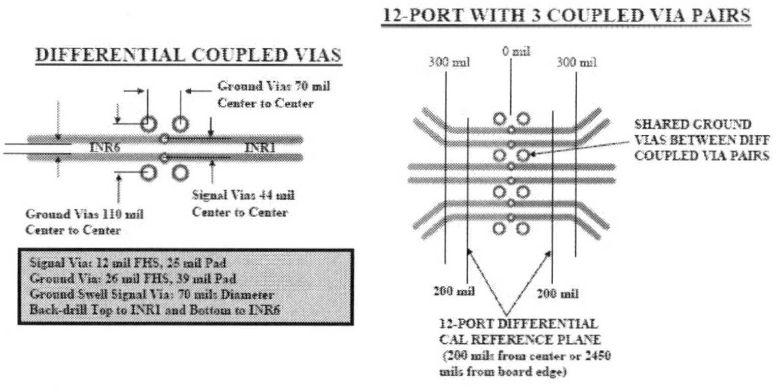

Figure 11.5: *12-Port Differential Via Pairs for the DUT Structure*

To achieve the high-frequency data on this structure, a 2.4 mm coaxial connector is selected for use with a 12-port 50 GHz Agilent physical layer test system so that reliable and accurate coaxial calibration standards can be used. This test system consists of a two-port 50 GHz E8364B vector network analyzer (VNA) and a 10-port U3025AE10 10-port 50 GHz test set. This aforementioned 2.4 mm connector is significantly larger than the via structure under test and thus requires a large area of routing to get from the connectors down to the small structure under test. This feeding line structure is represented in *Figure 11.6*, and it clearly shows the large area required by the connectors as compared to the structure under test.

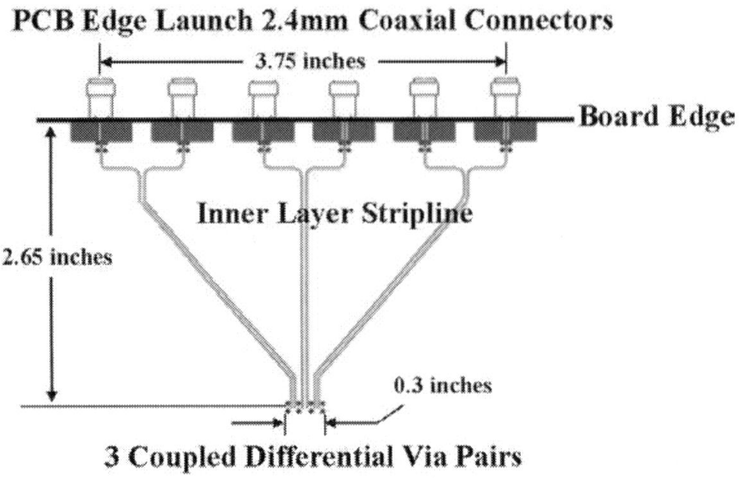

Figure 11.6: *12-Port Feed Line Routing to Get from Coaxial Connectors to the Via Structure under Test*

The justification for utilizing a single-ended two-port TRL with 12-port simulated structure de-embedding quickly becomes clear looking at this structure and comparing the resources that would be required for a 12-port calibration. The edge connectors alone for a 12-port TRL would take close to four feet of board edge space, and if one was to use a two-port network analyzer for the 12-port TRL, it would take 672 coaxial connections just for the calibration. The two-port TRL with simulated structure de-embedding may not be as accurate as a 12-port mixed-mode TRL, but with only 12 connections required for calibration of a 12-port system, it is a very practical approach.

Calibration Technique	2-Port TRL Cal 4-Lines	4-Port TRL Cal 4-Lines	12-Port TRL Cal 4-Lines	12-Port with 2-Port TRL and Simulated Structure De-embed
Number of Connector Footprints	11	22	66	11
2-Port Network Analyzer # of Connections	12	44	672	12
4-Port Network Analyzer # of Connections	12	24	312	12*
12-Port Network Analyzer # of Connections	12	24	72	12*

*Assumes all ports are equal after VNA cal and applies characterized standards for TRL de-embedding on all ports using one set of 2-port TRL Calibration measurements.

Table 11.1: *Simple Comparison of Coaxial Connector Connections to Be Made for Two-, Four-, and 12-Port TRL Calibration with Four Line Lengths versus That of the Combination TRL and Simulated Structure De-Embed Calibration. Assumes Network Analyzer Coaxial Calibration Has Been Performed*

11.4 TRL Calibration Structures and Measurement Technique

The TRL calibration technique has a long history of use with wafer probing for device characterization where it is necessary to calibrate out the effect of the probes using a TRL calibration [5, 6]. The technique provides accurate full two-port calibration of a VNA by employing an eight-term error model for a complete analytical derivation of the error terms. The TRL utilizes three types of standards, starting with the zero-length through line where all the S-parameters are known, an open or short for a high reflect condition, and a non-zero through line with the length chosen so that the frequency of interest has from 20 to 160 degrees of phase rotation over the length of the line. This means that for a wideband calibration, one will need multiple non-zero through lines for complete coverage of the required frequencies. The through lines should also be used in a frequency ratio of 1:8 to allow for enough phase margin in the solution (frequency overlap between bands).

When designing a TRL structure for a PCB, one must consider things such as repeatable connector performance and attach methods to get from the network analyzer coaxial connector to a planar structure, optimized discontinuities to reduce mismatches and reduce calibration errors, selection of routing feedlines that match the DUT of interest, and long enough structures to reduce radiation crosstalk errors. Significant effort has also been put into analyzing microstrip and stripline routing structures, and it has been found that the etching consistency, shielding, and dielectric uniformity of the stripline routing structure has some significant benefits over that of microstrip for the TRL standards. The other reality is that with the existing applications (that use high-density ball-grid arrays [BGAs]), it becomes impossible to route everything on the top and bottom microstrip layers, therefore stripline structure quickly becomes the preferred feedline routing to a DUT.

The 12-port via structure of *Figure 11.5* has been designed to transition from one inner layer to another and thus requires a via transition to get to this inner layer from the coaxial connector. Since the via is also sensitive to numerous board fabrication tolerances, including layer-to-layer alignment, drill placement, and etching, it has been placed next to the edge launch connector so that the TRL calibration can calibrate out these effects. *Figure 11.7* shows the topology for the single-ended launch structure using a very repeatable 2.4 mm edge launch connector to the PCB with an optimized via transition to a 19 mil inner-layer stripline.

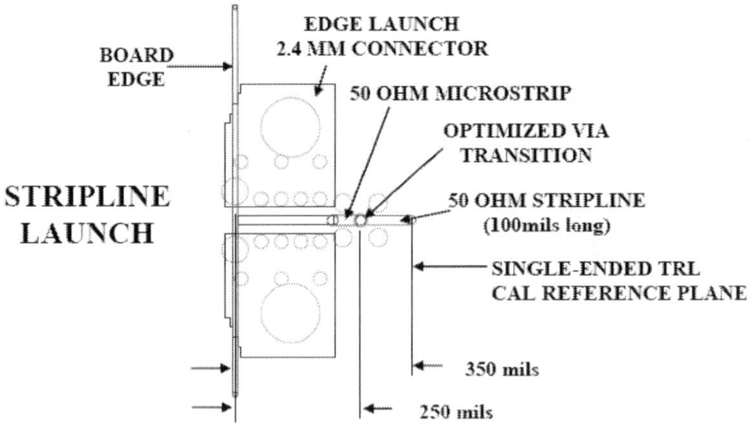

Figure 11.7: *Single-Ended Launch to Inner-Layer Stripline*

Putting this all together, it results in the following line lengths and structures for the single-ended TRL calibration. The thru structure just uses two back-to-back structures of the launch in *Figure 11.7* for a zero-length line. The reflect structure uses an open since this is easier to implement than trying to get low-inductance vias correctly located at the extended-port TRL calibration reference plane. The non-zero line lengths require four lengths to cover the broadband 100 MHz to 40 GHz range needed for detailed time domain analysis of the DUT using measured S-parameters. Additionally, a frequency ratio of 1:7 was selected for the line's start frequency to its stop frequency so that the phase rotation will stay within the required 20 to 160 degrees for the start to stop frequencies of the line. Line lengths are calculated based on the Agilent 8510 Network Analyzer Product Note 8510-8A [7].

TRL Structure	Start Frequency	Stop Frequency	Line Length
Line 1	85 MHz	595 MHz	4,600 mils
Line 2	0.35 GHz	2.45 GHz	1,300 mils
Line 3	1.55 GHz	10.85 GHz	250 mils
Line 4	10 GHz	70 GHz	40 mils

Table 11.2: *Single-Ended TRL Line Calibration Structures with a Frequency Range and Electrical Length Based on the Propagation Velocity in R4350 with a Dielectric Constant of 3.48*

11.5 Simulated Four-Port and 12-Port De-Embedding Structure

The use of 3D–EM simulation tools to obtain the S-parameters for the feed line routing from the single-ended launch (depicted in *Figure 11.7*) to the DUT structure of interest can be a reasonably quick task due to the import capability of most modeling tools. Either Autocad dxf style CAD drawing data or PCB Gerber files can be used to import the exact trace shape that is being used to route from the single-ended launch to a mixed-mode stripline structure. The import feature can minimize the modeling time and avoid errors that can occur when manually drawing complex geometries. The most common error source in full-wave modeling is the incorrect description of the geometry, and this happens if the geometrical dimensions are wrong or some crucial details have been simplified.

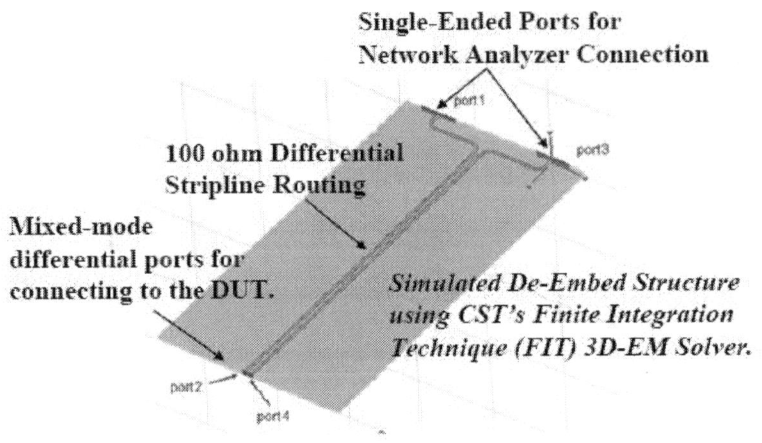

Figure 11.8: *Example 3D-EM Model of a Feedline Structure for Connecting from the Single-Ended TRL Reference Plane to the DUT*

The S-parameters from the modeled structure can then be de-embedded from the measured data to provide full S-parameters for the DUT structure [1, 2]. A simple stripline de-embed structure is illustrated in *Figure 11.8.*

De-embedding is a mathematical process that removes the effects of unwanted portions of the structure that are embedded in the measured data by subtracting out their contribution. The de-embed mathematics relies on the ability of S-parameters to be converted into a T matrix that has the following relationship:

$$T_{TOT} = T_A T_{DUT} T_B \qquad (1)$$

From the previous, by inverting the matrices T_A and T_B, it is straightforward to obtain the following:

$$T_{DUT} = T_A^{-1} T_{TOT} T_B^{-1} \qquad (2)$$

The TDUT matrix can then be converted back to S-parameters for cascading together with other structures in order to predict the performance of a full path with multiple routing structures.

The loss of the simulated transmission line structure will depend on the board fabrication process used and the final structural dimensions. Accurately capturing trace edge features, plating thickness, line widths, dielectric thickness, and conductor surface roughness to name a few is not always a practical task. A simplified method of using the longest TRL calibration through line measurements to establish the impedance and loss of the coupled stripline routing will provide the necessary reference point to tune the loss of the simulation and improve the accuracy of the final de-embed structure.

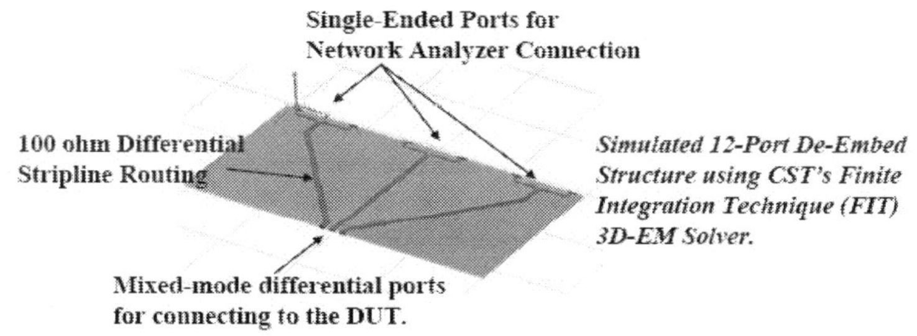

Figure 11.9: *12-Port De-Embed Structure Showing the Flexibility in Routing to the DUT from the Larger Coaxial Connector*

Figure 11.9 shows how the feedlines can be routed from the large coaxial connectors to the differential ports on a DUT structure. One of the main benefits of the simulated 12-port de-embed is that it provides significant flexibility in the routing of the feedlines. A full 12-port TRL would require that the feedlines for all DUT structures being measured use the same routing path to feed into them, but this is not always practical due to layout space constraints or changes in physical size of the DUT structure. The simulated 12-port de-embed feedline routing can easily be adjusted to match the specific DUT structure and allow one single-ended TRL standard calibration to be used for a variety of DUT structures and board layouts.

11.6 Device Structure under Test: Coupled and Un-Coupled via Pairs

Numerous papers have been done on the modeling of via

transitions for PCB layouts to fulfill the growing need for high-frequency performance [1, 8]. Most papers deal with single-ended via transitions, a few with differential coupled via transitions, and very few if any deal with obtaining calibrated mixed-mode S-parameters for the via structure. A poorly designed via can quickly become a well-designed low-pass filter that, if one is not careful, can block the higher frequencies and significantly reduce rise times of a digital signal. Via performance optimization can be tuned using TDR techniques [8] to understand what portions of the structure are capacitive or inductive. Tuning of features can easily be done with a 3D–EM solver, but one must have at least one accurate measured data point on a representative structure so that a 3D–EM model can be verified before extensive time is put into optimization using the simulator.

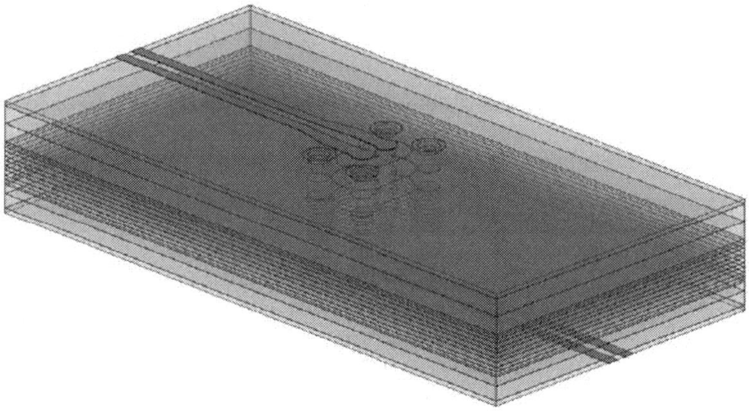

Figure 11.10: *3D-EM Structure Modeling Can Benefit from Validation with Measured Data prior to Extensive Time Spent on Optimizations*

In the case of differential signals, it becomes slightly more complex in that now one has to work with the single-ended and differential impedances. It has been shown that the S-parameters can easily be converted to a reciprocal mixed-mode S-parameter matrix that shows the differential behavior of a pair of transmission lines [11]. Comparing the results (*Figure 11.11*) of the single-ended common-mode performance with that of the differential quickly shows that any tuning or optimization must be done with the differential performance since the benefit of coupling can improve the performance of the differential mode.

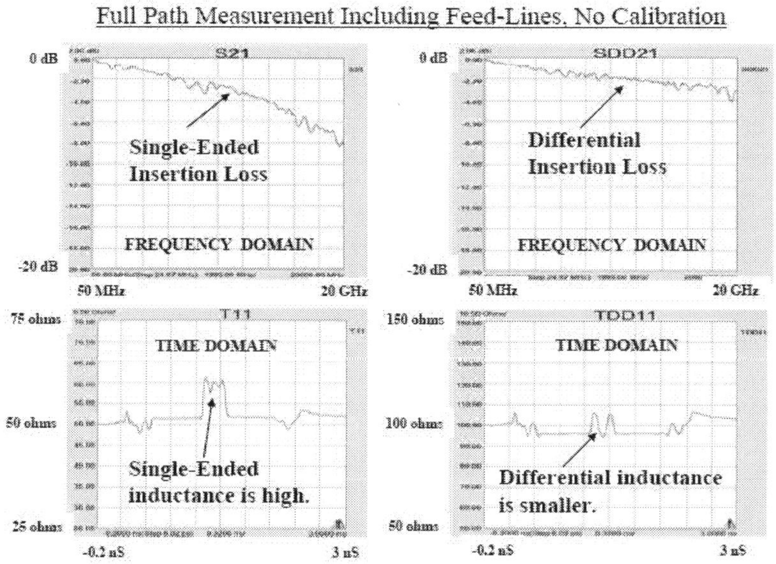

Figure 11.11: *Single-Ended versus Mixed-Mode S-Parameters for the Full-Path Structure of Figure 11.1*

The via structure of *Figure 11.5* with the neighboring differential via pairs for NEXT/FEXT measurements becomes significantly more complex to model with a 3D–EM solver and must rely more on measured data. One approach is to optimize the differential via pairs as separate structures before bringing them together in a NEXT/FEXT coupling proximity. The full-path DUT structure for three via pair topologies has been fabricated on a R4350 test board and measured data exists for these three DUT structures. The first of the three via pair topologies is what it has been shown in *Figure 11.10* and used in the simulated full path of *Figure 11.11*.

11.7 Verification Using Four-Port Multimode TRL Calibration

With the ever-increasing frequencies for digital communications, the need for accurate calibration of mixed-mode differential structures arises. It has been shown that the single-ended TRL can be modified to accommodate the coupling of a mixed-mode structure and provide a calibrated measurement of the DUT [3]. The design of a four-port mixed-mode TRL structure is similar to the technique described for the single-ended TRL in section III with the added requirement that the phase rotation remain between 20 and 160 degrees for all modes. The validity of this four-port mixed-mode TRL has been demonstrated [4] and is an excellent way of checking the performance of the simpler single-mode TRL with simulated structure de-embedding that is presented in this paper.

The fabricated mixed-mode TRL structures will have the topology shown in *Figure 11.12*. This structure uses the same launch as the single-ended TRL structure so that one can use either calibration method when looking at a full-path measurement.

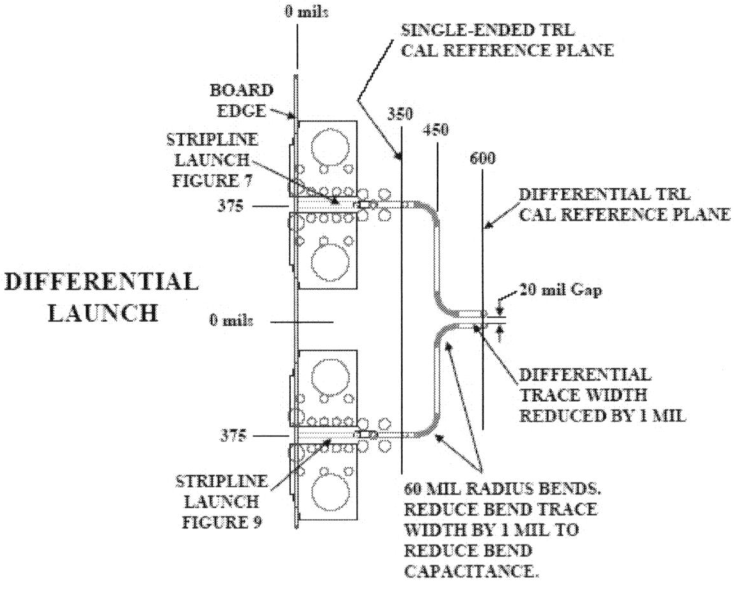

Figure 11.12: *Differential Launch to Coupled Inner-Layer Stripline – Four-Port*

The line lengths for the mixed-mode TRL lines are shown in *Table 11.3*.

TRL Structure	Start Frequency	Stop Frequency	Line Length
Line 1	95 MHz	665 MHz	4,100 mils
Line 2	0.5 GHz	3.5 GHz	800 mils
Line 3	2 GHz	14 GHz	200 mils
Line 4	10 GHz	70 GHz	40 mils

Table 11.3: *Mixed-Mode TRL Line Calibration Structures with Frequency Range and Electrical Length Based on the Propagation Velocity in R4350 with a Dielectric Constant of 3.48*

To insure that the methodology works mathematically, one can start with simulated TRL structures, connector S-parameter data, and modeled DUT via performance to create a full-path structure (*Figure 11.13*). Running the mixed-mode TRL algorithm should result in getting back the via data that one started with, and the same is true for de-embedding the connector and feedlines with the simulated structure de-embedding to get back to the via structure S-parameters.

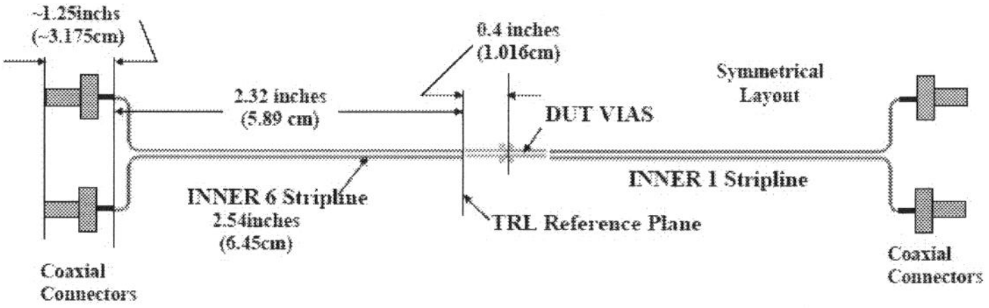

Figure 11.13: *Simulated Full-Path Structure and Data for Verification of the De-Embedding and Mixed-Mode TRL Algorithms*

Running this simulated comparison (*Figure 11.14*) shows that the technique is valid; however, care must be used in verifying the correct characteristic impedances when implementing the algorithms. This comparison quickly shows that the TRL thru line would benefit from being increased in length so that the TRL reference plane is further from the transition from single-ended to differential pair routing. When the transition fields are not allowed to settle, then it results in the thru-line path not having the same characteristic impedance as the longer lines. The final application of comparing measured mixed-mode TRL structures with that of the simulated structure de-embedding will be done once the fabricated structures become available.

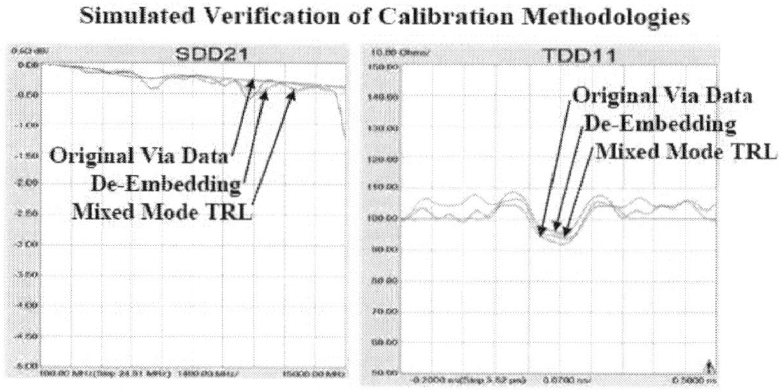

Figure 11.14: *Simulated Differential SDD21 to 15 GHz and Time Domain TDD11 Verification of the Mixed-Mode TRL and Single-Ended TRL with Simulated Feedline De-Embedding. Variations in the Mixed-Mode TRL Show That the TRL Structure thru Line Is not Long Enough for the Single-Ended to Differential Fields to Settle, Resulting in a Characteristic Impedance That Is Different from That of a Longer Line*

11.8 Demonstration of the Combined TRL Calibration and Simulated Structure De-Embedding Technique for Multi-Mode N-Port Systems

The final set of calibration boards are still in progress for the structure of *Figure 11.6* for looking at NEXT/FEXT crosstalk issues, but as mentioned in section V it is still very interesting to look at three separated via pairs to obtain their individual S-parameters.

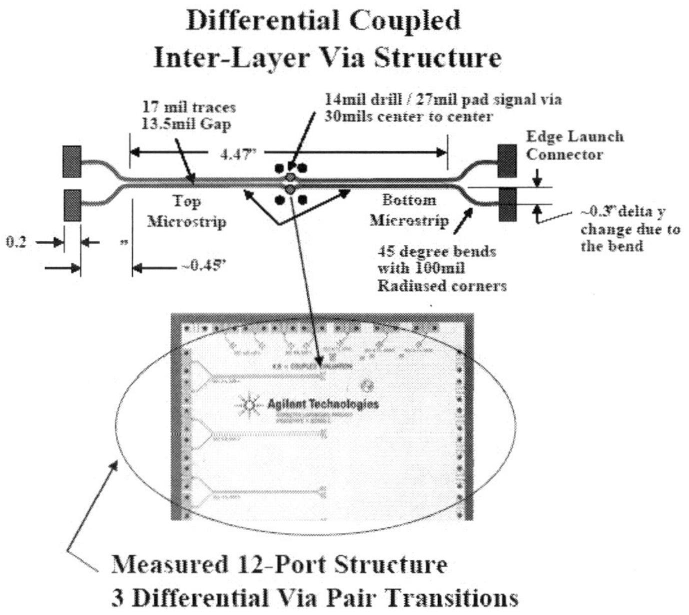

Figure 11.15: *Differential Via Experiments and TRL Calibration*

Although TRL data is not yet available for the single-ended connector topology of this board, one can get close to implementing this same technique by probing a measurement of the connector for use with mathematically de-embedding its contribution from the full-path measurement. This is not as accurate as the single-ended TRL, but it is a very flexible method when one does not have access to TRL structures for the topology of interest. The S-parameters for the connector can be seen in *Figure 11.15*, and the feedline topology for de-embedding is illustrated in *Figure 11.17*.

The simulation of the feedline structure in *Figure 11.17* proved to be a bit more troublesome than expected due to the use of the microstrip routing. The microstrip routing for a large-scale DUT loadboard application utilizes 50 micro inches of Au over 200 micro inches of nickel plating on top of 2 mils of plated-up copper on the outside of the board. The etching variations and the mushroom-shaped edge effects of this style of microstrip resulted in a far more complex modeling problem. The final calibration structures that are in the process of being fabricated will use stripline feedline routing to provide a much simpler structure for modeling and de-embedding. The topside plating issues will be calibrated out in the single-ended TRL calibration to remove the connector transition and the via transition to inner stripline so that modeling of these variations can be avoided.

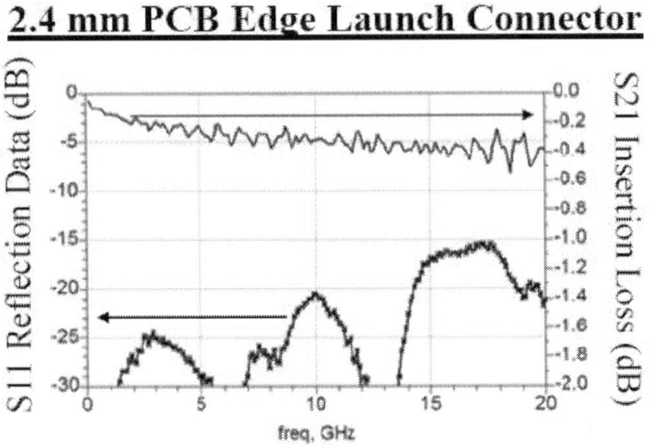

Figure 11.16: S-Parameter Data for the Southwest 2.4 mm PCB Edge Launch Connector Using GigaTest Labs Probe Measurements

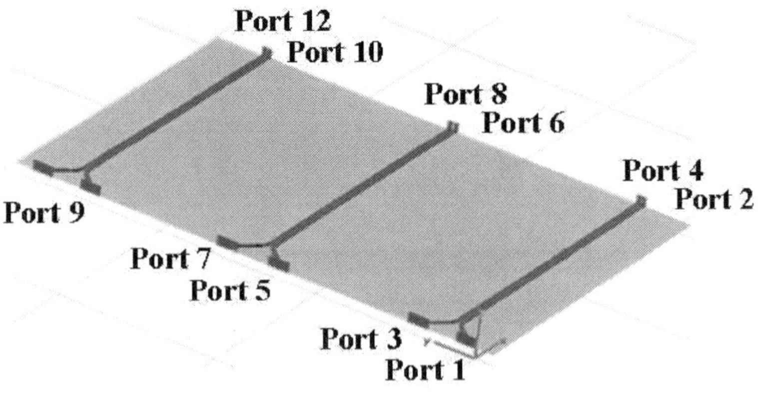

Figure 11.17: Simulated 12-Port De-Embed Microstrip Structure

Applying this technique to the existing microstrip structures of *Figure 11.15* demonstrated the importance of choosing stripline routing paths to implement this single-ended TRL calibration with simulated feedline de-embedding. The use of microstrip feedlines would require additional work to correctly account for plating and etching effects in the simulated structures. The final application of obtaining 12-port S-parameters for coupled via pairs will be done once the fabricated stripline structures become available.

11.9 Summary and Conclusion

The methodology outlined in this paper demonstrates one of the few practical approaches to calibration of mixed-mode differential pairs for 12-port systems. The high number of connectors and connections required for a full 12-port TRL make this a tedious task that quickly benefits from a reduction in the number of connectors and connections at the expense of some loss of precision. The selection of a single-mode TRL structure for de-embedding the coaxial to planer connector launch and any via transitions to inner layers is an efficient way of accounting for PCB fabrication tolerances that are often difficult to determine for an accurate 3D–EM simulation. The use of a simulated N-port planar feedline structure to transition from the single-ended routing to coupled mixed-mode routing is a simple task for a 3D–EM simulator, and it provides the necessary mixed-mode S-parameters for de-embedding this structure from the single-ended TRL calibrated full-path measurement. The single-ended TRL structures also provide the necessary thru line loss information for additional tuning of the simulated N-port de-embed structure for a particular as fabricated PCB.

The increased complexity of multiple transmission lines with the desire to look at NEXT/FEXT crosstalk effects requires accurate S-parameters for use as components in a model library for full-path simulations. This paper has presented a practical solution for obtaining these S-parameters using a combination of single-mode TRL with simulated N-port feedline de-embedding to achieve an N-port calibration for measuring an N-port structure.

Acknowledgments

The authors would like to thank the cooperation between Agilent, Verigy, CST, and Xilinx to make it possible to bring multiple simulation products together to verify the validity of this technique. Key contributors include Jose Moreira of Verigy for the inspiration and funding to make this paper possible, Bob Schaefer of Agilent for the physical layer test system, Sanjeev Gupta of Agilent Eesof EDA division for his technical guidance and assistance with ADS simulations, Yoshio Akashi of Agilent for his multi-mode TRL calculations, Orlando Bell of Gigatest Labs for connector probe measurements and 12-port measurements, Will Burns at Altanova for PCB layout, and Morgan Culver of Agilent for high-quality time domain measurements.

References

[1] G. Antonini, A. Ciccomancini Scogna, and A. Orlandi, "S-Parameters Characterization of through Blind, and Buried Via Hole," IEEE Transactions on Mobile Computing, Vol. 2, No. 2, April–June 2003.

[2] G. Antonini, A. Ciccomancini Scogna, and A. Orlandi, "De-Embedding Procedure Based on Computed/Measured Data Set for PCB Structures Characterization," *IEEE Transactions on Advanced Packaging*, vol. 27, no. 4, November 2004.

[3] C. Seguinot et al., "Multimode TRL – A New Concept in Microwave Measurements Theory and Experimental Verification," *IEEE Transactions on Microwave Theory and Techniques*, vol. 46, no. 5, May 1998.

[4] T. Buber et al., "Multimode TRL and LRL Calibrated Measurements of Differential Devices," ARFTG Microwave Measurements Conference, Fall 2004, 64.

[5] H. Eul and B. Schiek, "Generalized Theory and New Calibration Procedures for Network Analyzer Self-Calibration," *IEEE Transactions on Microwave Theory and Techniques*, vol. 39, no. 4, April 1991.

[6] D. Dunham, V. Duperron, and M. Resso, "Practical Design and Implementation of Stripline TRL Calibration Fixtures for 10-Gigabit Interconnect Analysis," IEC

DesignCon 2006.

[7] Agilent Technologies, "Agilent Network Analysis Applying the 8510 TRL Calibration for Non-Coaxial Measurements," Product Note 8510-8A 2006.

[8] J. Moreira, M. Tsai, J. Kenton, H. Barnes, and D. Faller, "PCB Loadboard Design Challenges for Multi-Gigabit Devices in Automated Test Applications," IEC DesignCon 2006.

[9] E. Bogatin, *Signal Integrity – Simplified*, Prentice Hall, 2004.

[10] H. Johnson and M. Graham, *High Speed Signal Propagation*, Prentice Hall, 2003.

[11] D. Bockelman and W. Eisenstadt, "Combined Differential and Common-Mode Scattering Parameters: Theory and Simulation."

Authors

Heidi Barnes, High-Frequency Device Interface Board Designer, Verigy, Inc.

Antonio Ciccomancini, Application Engineer, CST of America, Inc.

Mike Resso, Signal Integrity Application Scientist, Component Test Division, Agilent Technologies

Ming Tsai, Staff Hardware Development Engineer, Production Technology Division, Xilinx

Chapter 12

Validating Transceiver FPGAs Using Advanced Calibration Techniques

12.1 Abstract

The field programmable gate array (FPGA) is a low-risk solution to implementing silicon architectures in today's fast-paced product development cycle. However, the gigabit transceiver block in the FPGA requires careful design to avoid degrading performance of the high-speed channels. Furthermore, the method of characterizing these transceiver channels in the FPGA is highly dependent on the bandwidth of the printed circuit board (PCB) fixture on which the silicon FPGA die is attached. This paper will address advanced calibration techniques that allow proper and accurate performance analysis of these high-speed channels associated with the FPGA.

12.2 Introduction

Today's high-speed applications need reliable data transfer technology that gets information from the source to the destination fast. FPGAs are a popular solution for integrating transceivers without the cost of hard tooling a fully custom chipset. System architects are challenged with high-end consumer electronics, communications, and mass-storage applications that commonly require 10 or 20 full duplex transceiver channels. In addition, these transceiver channels must be capable of handling 3.125 or 6.25 Gbps data while consuming minimal power. Proper characterization of these gigabit channels requires a multitude of frequency domain and time domain measurements that are typically limited by the bandwidth of the test fixtures to which the silicon package is attached.

Test fixture design demands the most advanced design tools because any limitation of the fixture will directly translate into a measurement that will mask the true performance of the device under test (DUT). At today's current data rates and edge speeds, most of the transceiver channels exhibit microwave transmission-line effects. To complicate matters even more, most serializer-deserializer (SerDes) functionality is implemented with differential circuit topology. This means that all measurements characterizing the critical test fixture transceiver channels must be four-port measurements. Differential insertion loss, differential return loss, and differential eye diagram analysis are mandatory for correctly identifying performance limitations of the test fixture and ultimately of the FPGA.

This paper will present a design case study of an FPGA device that incorporates a gigabit transceiver. The focus will be put on the most challenging FPGA functional block of high-speed channels and how the designer can optimize the testing of the device. Advanced four-port measurements and calibration techniques that enhance the quality of the FPGA characterization will be explored. The breakthrough addressed in this paper is the VNA calibration between two distinct interface types. At one end is an FPGA ball pad that can only be accessed by a probe, and at the other end of the measurement are SMA connectors on the FPGA fixture. Performing full two-port or full four-port short, open, load, thru (SOLT) calibration between them is not possible unless the methodology described in this paper is used. Both frequency domain and time domain examination will provide complementary information yielding unique insights otherwise not realized. A real-world FPGA test fixture will be utilized to demonstrate these methodologies.

12.3 FPGA Applications Overview

Typical FPGA Architecture

FPGA devices can integrate high-speed 3.125 Gbps transceiver SerDes technology with advanced FPGA architecture, as shown in *Figure 12.1*. Historically, designers have used high-speed transceivers strictly in structured line-side applications. Recent breakthroughs in FPGAs have provided gigabit transceiver blocks embedded inside so designers can use transceivers in new demanding system applications. The gigabit transceiver block enables design flexibility, shorter design cycle time, and high performance. The gigabit transceiver block can also simplify the implementation of standard and custom high-speed protocols.

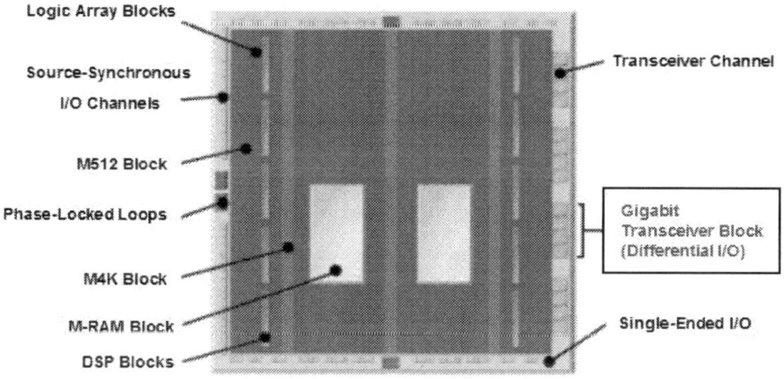

Figure 12.1

FPGA Applications

The main system bottleneck in high-speed communications equipment is data transmission from chip to chip and over backplanes. FPGA devices help designers address this bottleneck by supporting 3.125 Gbps channels and integrating advanced functionality into the device's logic array. The devices are ideal for a variety of applications, including applications bridging, switch fabrics, traffic management functions, wireless, and high-definition television (HDTV) broadcast, as shown in *Figure 12.2*. The use of FPGA devices provides a low-risk path for serial input/output (I/O) applications.

Mobile phone communication is becoming the primary mode of interaction in developed countries. Wireless base stations (BSs) developed on enhanced second generation (2G) technology are a critical link in this system. The transceiver cards within these BSs have FPGAs that receive data via high-speed serial signals (serial RapidIO or proprietary interface) over a backplane from one or more channel cards. Given the high-speed nature of this link, the long trace length that runs across a typically noisy backplane, and the multiple connectors through which the signals must travel, clock-data recovery (CDR)–based implementations are typically used in the FPGA.

Storage-area networks (SANs) have various technical challenges that can be resolved by implementing advanced FPGAs. This includes different storage and networking technologies, advances in storage and infrastructure bandwidth, and information growth. Additionally, users want a virtualized storage repository where they can view and manage all storage assets regardless of technology implementation (network-attached storage [NAS] or SAN), physical location, and various vendor brands. These demands push designers of storage switches to create more flexible and highly integrated systems. Advanced FPGAs can provide up to 20 transceiver channels, enabling designers to use a flexible and integrated solution on the line side and on the backplane with traffic management.

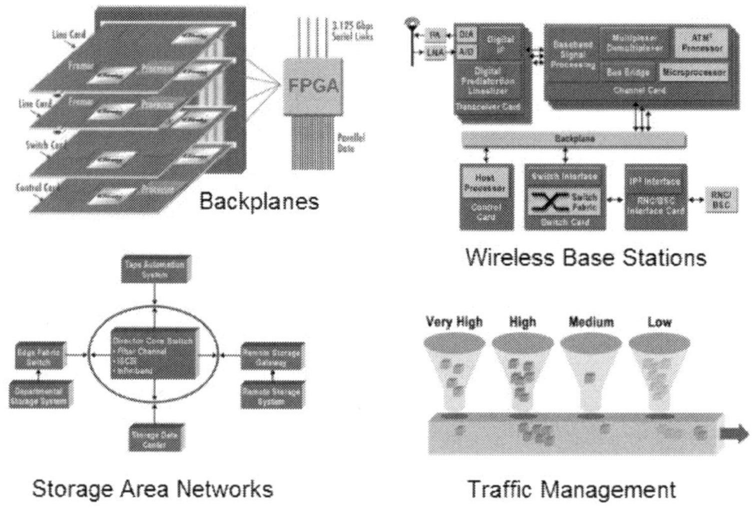

Figure 12.2

Digital Communications Standards

FPGAs support the many emerging protocols in the market that require high-speed differential I/O with CDR. Some examples include high-speed protocols such as RapidIO, 10 gigabit Ethernet support via the 10 Gbps extended attachment unit interface (XAUI), Infiniband, and Fiber Channel. Support for a wide spectrum of applications is shown in *Figure 12.3*.

Each protocol has some unique requirement that demands high levels of signal integrity throughout the physical layer (PHY) channel link. For example, XAUI is designed as an interface extender for the 10-gigabit media-independent interface (XGMII). XAUI can be used in various applications, including 10-gigabit Ethernet line cards, local-area network (LAN)–to–wide-area network (WAN) bridges, backplanes, and chip-to-chip interconnects. The XAUI specification uses four full-duplex serial links operating at 3.125 Gbps in each direction. In aggregate, a total of 12.5 Gbps can be transferred in each direction.

RapidIO technology is a high-performance, packet-switched interconnect technology designed to pass data and control information between microprocessors, digital signal processors, communications and network processors, system memories, and peripheral devices. The new serial link specification from the RapidIO trade association uses the parallel RapidIO protocol from the link layer upward, but with serial rates of 1.25, 2.5, and 3.125 Gbps in the PHY. The Serial RapidIO protocol can be used in backplanes and chip-to-chip applications similar to the XAUI protocol applications. A noteworthy protocol that is quickly becoming popular is PCI Express (formerly 3GIO). PCI Express uses differential CDR signaling to allow transmission of high-speed data while maintaining compatibility with the current PCI software environment. It can be used for chip-to-chip and add-in card applications to provide connectivity for adapter cards, as a graphics I/O attach point for increased graphics bandwidth, and an attach point to other interconnects such as 1394b, USB 2.0, InfiniBand architecture, and Ethernet. Some of the advanced features of the protocol include aggressive power management, quality of service (QoS), and hot plug ability.

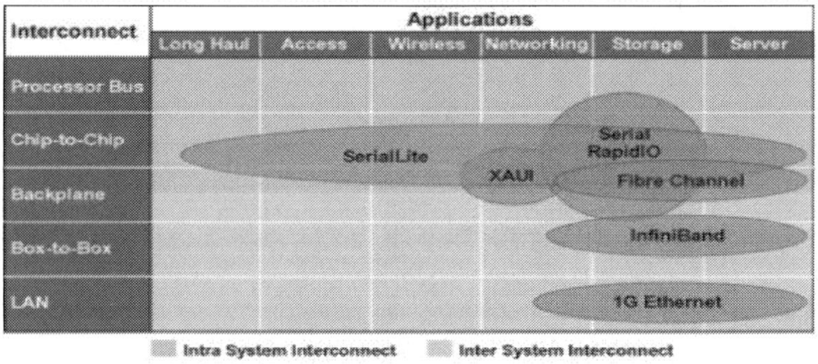

Feature	SerialLite	10 Gigabit Ethernet XAUI	Serial RapidIO	Gigabit Ethernet	Fibre Channel	PCI Express	SMPTE 292M	SONET/SDH OC-48c
Maximum Bandwidth (Gbps)	51	12.5	12.5	1.25	10	80	1.485	2.488
Bus Width	1-256	4	1, 4	1	1	1, 2, 4, 8, 12, 16	1	1
Maximum Single Channel Rate (Gbps)	3.1875	3.125	3.125	1.25	3.1875	2.5	1.485	2.488

Figure 12.3

Test Fixtures Degrade Rise Time

The significant edge rates of today's protocols pose particular challenges for FPGA test engineers. In order to characterize FPGA devices, extremely high-quality test fixtures need to be designed, developed, and fabricated. Advanced simulation, modeling, and measurement tools are required to achieve the proper figures of merit for the silicon. This is especially true for the gigabit IO channel of the FPGA. The goal is to avoid any appreciable rise-time degradation of the silicon output signaling. Otherwise, the test fixture will mask the true performance of the silicon and make it look as if it were a lower-performing device. Another downside is that the test fixture could be eroding the performance margin and causing a good part to be failed. This happens more frequently than one might imagine in the real world today.

The mechanism for high-speed signal degradation due to a test fixture is straightforward, but the solution is not. The generic test fixture in *Figure 12.4* is a good example how rise time is degraded. The mechanism is a combination of two main phenomena: signal amplitude loss due to reflections from impedance discontinuities (e.g., coax connectors, via stubs, solder bumps) and attenuation of high-frequency components from conductor skin effect loss (series) and dielectric loss (shunt). By controlling the impedance environment through the complete test fixture, the true performance of the silicon can be measured. To enhance the accuracy of the gigabit channel even further, systematic error-correction techniques must be employed to remove test equipment error and test probe error.

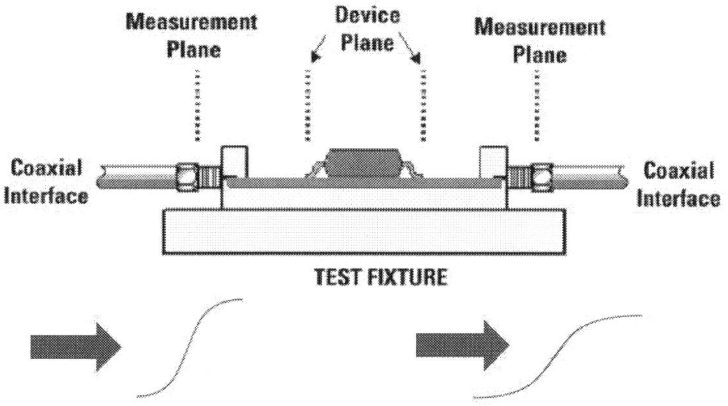

Figure 12.4

12.4 Systematic Error Correction

Fixture Error Correction Techniques

Over the years, many approaches have been developed for removing the effects of the test fixture from the measurement, which fall into two fundamental categories: direct measurement (pre-measurement process) and de-embedding (post-measurement processing). Direct measurement requires specialized calibration standards that are inserted into the test fixture and measured. The accuracy of the device measurement relies on the quality of these physical standards. De-embedding uses a model of the test fixture and mathematically removes the fixture characteristics from the overall measurement. This fixture de-embedding procedure can produce very accurate results for the non-coaxial DUT, without complex non-coaxial calibration standards.

The process of de-embedding a test fixture from the DUT measurement can be performed using scattering transfer parameters (T-parameter) matrices or, in this case, the de-embedded measurements can be post-processed from the measurements made on the test fixture and DUT together. Also modern electronic design automation (EDA) tools have the ability to directly de-embed the test fixture from the vector network analyzer (VNA) measurements using a negation component model in the simulation. An overview of pre-measurement and post-measurement error correction techniques are shown in *Figure 12.5*.

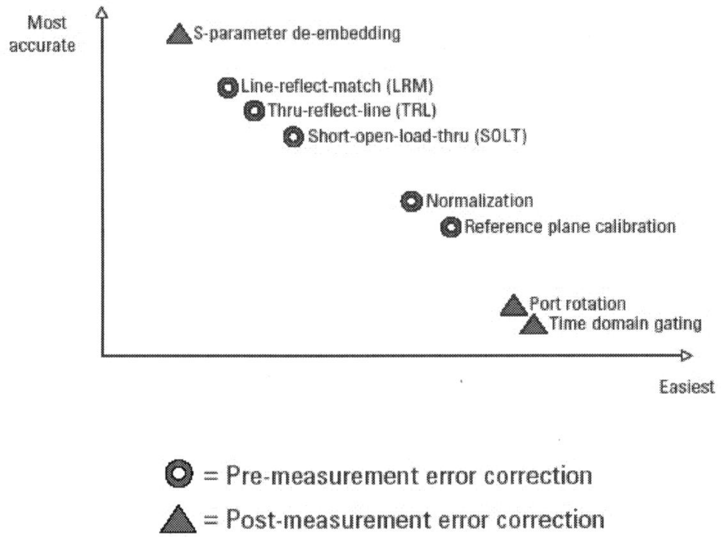

Figure 12.5

TDR and VNA Capabilities Vary

A major problem encountered when making network measurements in microstrip or other non-coaxial media is the need to separate the effects of the transmission medium (in which the device is embedded for testing) from the device characteristics. While it is desired to predict how a device will behave in the environment of its final application, it is difficult to measure this way. The accuracy of this measurement depends on the availability of quality calibration standards. Unlike coaxial measurements, a set of three distinct, well-characterized impedance standards are often impossible to produce for non-coaxial transmission media. For this reason, an alternative calibration approach may be useful for such applications. Example calibrations standards for both time domain reflectometry (TDR) and VNA are shown in *Figure 12.6.*

The thru-reflect-line (TRL) calibration technique relies only on the characteristic impedance of a short transmission line. The unique component of this method is the use of three length lines to cover the complete frequency range of the network analyzer sweep. TRL can be applied in dispersive transmission media such as microstrip, stripline, and waveguide quite easily if the TRL calibration standard is already available. With precision coaxial transmission lines, TRL currently provides the highest accuracy in coaxial measurements available today.

Another useful error correction technique that is very common is SOLT. The well-known examples of these standards are coaxial 3.5 mm connector network analyzer calibration kits. The calibration coefficients are provided with each cal kit and loaded into the network analyzer. This is the caveat with SOLT standards (e.g., the calibration coefficients must be available).

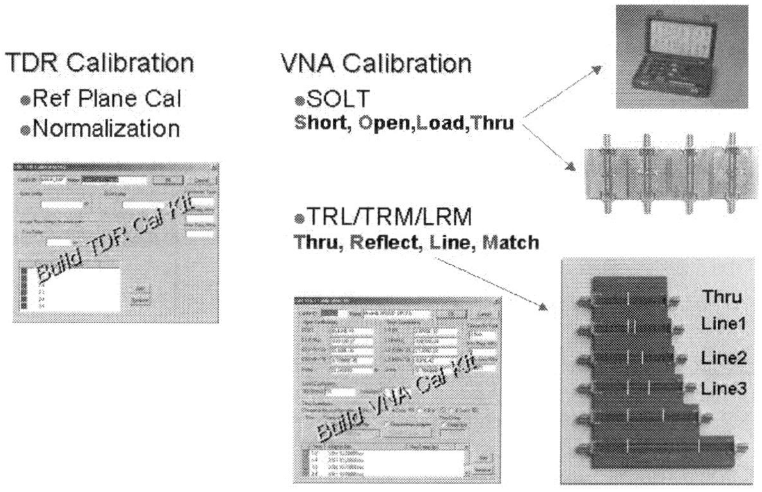

Figure 12.6

Examples of TRL Standards

Some examples of TRL calibration fixtures are shown in *Figure 12.7*. The upper left standard has a strange shape due to the longest line on the board. Most TRL fixtures have three lines (L1, L2, and L3), but some calibration algorithms allow for a fourth line. The longest line covers the lowest-frequency range of the calibration. The longer the line, the closer to DC the calibration has validity. At first glance, this may not seem to be of significance. After all, the higher frequencies are the ones that we are most often concerned with, right? Well, it turns out that accurate low-frequency data is required to assist in the frequency to time domain transformation. So, if cross-domain analysis is desirable (and it usually is), then a longer line in the TRL calibration fixture will help.

The picture in the lower right of *Figure 12.7* shows the example DUT. This happens to be a XAUI backplane with semicircular test fixture cards. The TRL calibration standard previously addressed will remove the semicircular test fixture cards and move the reference plane to the connector.

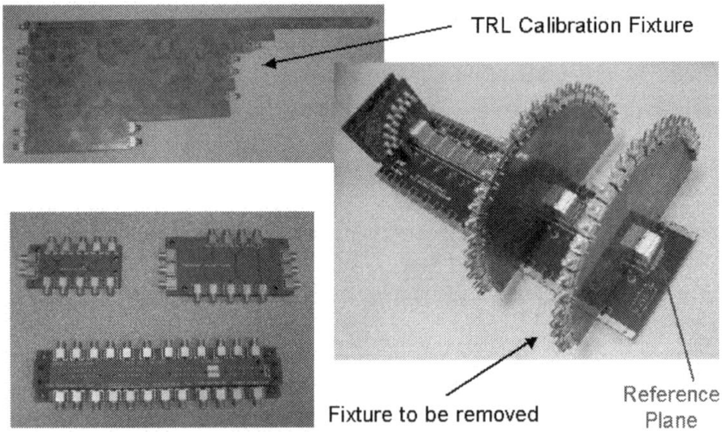

Figure 12.7

Reference Plane Adjustment

A brief description of the user interface used for reference plane adjustment will help explain the process used for this design case study. The dialogue box in *Figure 12.8* shows a step-by-step method for assigning a user-defined reference plane. The first step in using the port reference adjustment is choosing the adjustment method and its options. There are four adjustments: two-port de-embedding, four-port de-embedding, port rotation/extension, and port reference impedance. Each adjustment has its own options that are displayed when the adjustment is selected from the list. Step two is to select the appropriate data file that needs to be de-embedded. In the case of this particular design case study, the authors used vendor supplied two-port Touchstone format files for each of the two single-ended probes used in the measurement setup (courtesy of GGB Industries). Alternately, citifile format could have been used. Step three is to assign the de-embed file to the appropriate port of the test system. The setup in this case required the probes on ports 1 and 3 of the four-port measurement system to be de-embedded. This is where the probes were placed onto the FPGA test fixture in the ball grid array (BGA) area. The last step is to apply the reference plane adjustment by clicking on the "Apply" button in the bottom portion of the dialogue box.

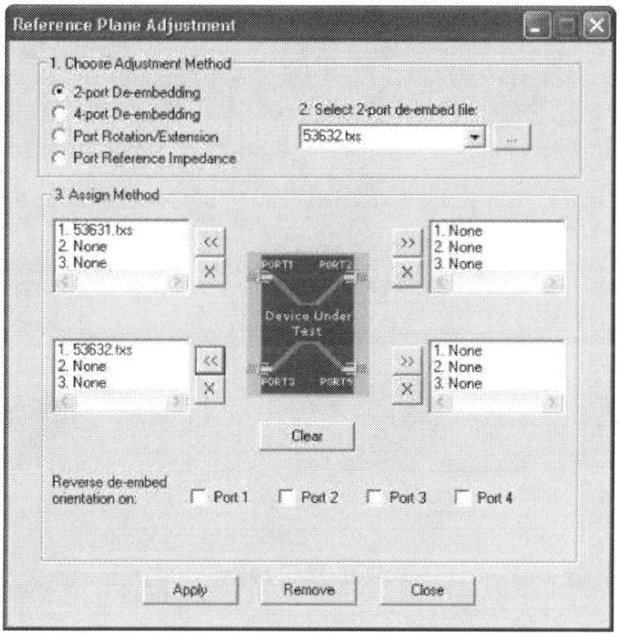

Figure 12.8

12.5 Characterizing Differential Structures

Single-Ended S-Parameters and TDR/TDT

To lay a foundation for understanding how to characterize an FPGA fixture with high-speed differential transmissions lines, a brief discussion of multiport measurements is in order. The four-port device shown in *Figure 12.9* is an example of what a microstrip structure might look like if we had two adjacent PCB traces that are operating in a single-ended fashion. Assume that these two traces are located within relative proximity to each other on a backplane and some small amount of coupling might be present. Since these are two separate single-ended lines in this example, this coupling is an undesirable effect, and we call it crosstalk.

The matrix on the left show the 16 single-ended S-parameters that are associated with these two lines. The matrix on the right shows the 16 single-ended time domain parameters associated with these two lines. Each parameter on the left can be mapped directly into its corresponding parameter on the right through an inverse fast Fourier transform (IFFT). Likewise, the right-hand parameters can be mapped into the left-hand parameters by a fast Fourier transform (FFT). If these two traces were routed very close together as a differential pair, then the coupling would be a desirable effect and would enable good common-mode rejection that provides electromagnetic interference (EMI) benefits.

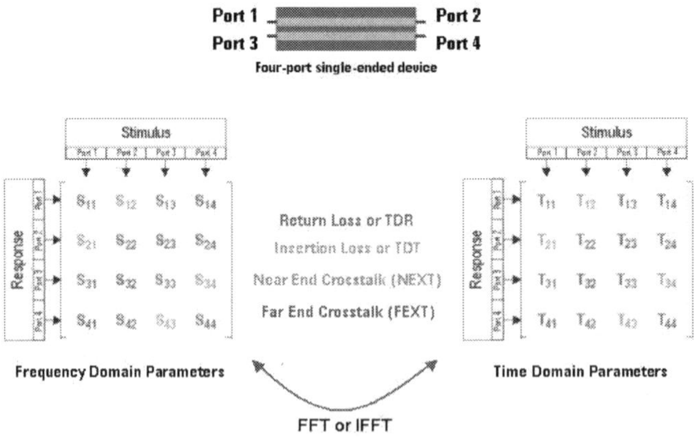

Figure 12.9

Single-Ended to Differential S-Parameters

Once the single-ended S-parameters have been measured, it is desirable to transform these to balanced S-parameters to characterize differential devices. This mathematical transformation is possible because a special condition exists when the DUT is a linear and passive structure. Linear passive structures include PCB traces, backplanes, cables, connectors, and integrated-circuit (IC) packages. Utilizing linear superposition theory, all of the elements in the single-ended S-parameter matrix on the left of *Figure 12.10* are processed and mapped into the differential S-parameter matrix on the right. Much insight into the performance of the differential device can be achieved through the study of this differential S-parameter matrix, including EMI susceptibility and EMI emissions.

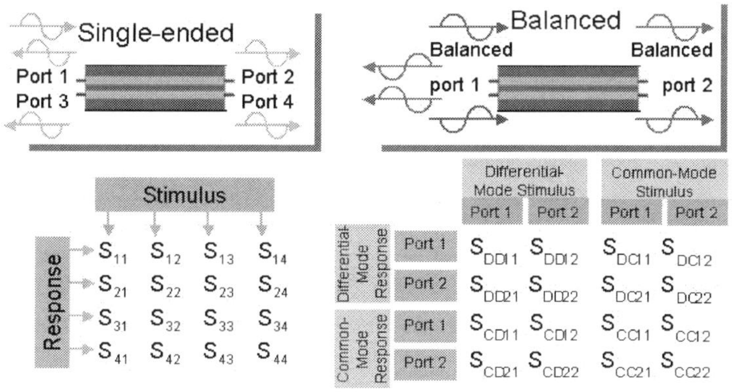

Figure 12.10

Mixed-Mode S-Parameters

Interpreting the large amount of data in the 16-element differential S-parameter matrix is not trivial, so it is helpful to analyze one quadrant at a time. The first quadrant in the upper left of *Figure 12.11* is defined as the four parameters describing the differential stimulus and differential response characteristics of the DUT. This is the actual mode of operation for most high-speed differential interconnects, so it is typically the most useful quadrant that is analyzed first. It includes input differential return loss (SDD11), forward differential insertion loss (SDD21), output differential return loss (SDD22), and reverse differential insertion loss (SDD12). Note the format of the parameter notation SXYab, where S stands for scattering parameter or S-parameter, X is the response mode (differential or common), Y is the stimulus mode (differential or common), a is the output port, and b is the input port. This is typical nomenclature for frequency domain scattering parameters. The matrix representing the 16 time domain parameters will have similar notation, except the "S" will be replaced by a "T" (i.e., TDD11).

The fourth quadrant is located in the lower right and describes the performance characteristics of the common signal propagating through the DUT. If the device is designed properly, there should be minimal mode conversion and the fourth quadrant data is of little concern. However, if any mode conversion is present due to design flaws, then the fourth quadrant will describe how this common signal behaves. The second and third quadrants are located in the upper right and lower left of *Figure 12.11*, respectively. These are also referred to as the mixed-mode quadrants. This is because they fully characterize any mode conversion occurring in the DUT, whether it is common-to-differential conversion (EMI susceptibility) or differential-to-common conversion (EMI radiation). Understanding the magnitude and location of mode conversion is very helpful when trying to optimize the design of interconnects for gigabit data throughput.

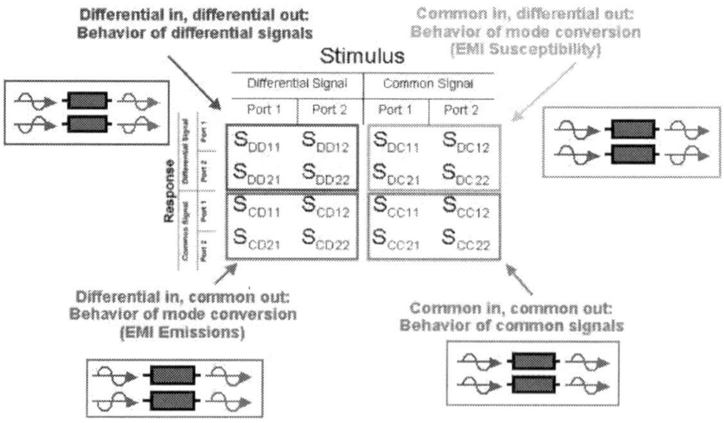

Figure 12.11

FPGA Test Fixture

The FPGA text fixture used in this design case study is shown in *Figure 12.12.* The high-speed gigabit I/O ports for the FPGA can be clearly seen as the wide gold traces on the PCB. Various design features need to be incorporated into this fixture to minimize the impact that it would have on the measurement of the FPGA silicon. A partial list of these are short traces for high-speed signals, use of good quality surface-mount assembly (SMA) connectors (end launch), minimized vias, lower effective dielectric constant, and use of rounded turns. These are all sound objectives for any high-speed board.

The goal of this design case study is to demonstrate an advanced error correction technique that can be used in other applications as well. The most useful and accurate to apply is de-embedding. It is not the simplest methodology due to the fact that the user must have the Touchstone file of the structure to be removed. However, in this case study, the authors were able to obtain the Touchstone file from the probe vendor and then de-embed the probes. If this methodology becomes popular, then perhaps this will encourage more probe vendors to ship Touchstone files with their probes. Also, if this de-embed methodology is used for standardized test fixtures throughout the industry, then applications such as backplane design and validation could benefit tremendously. Complicated probing systems may not need to be used in every instance, thereby reducing the design cycle time.

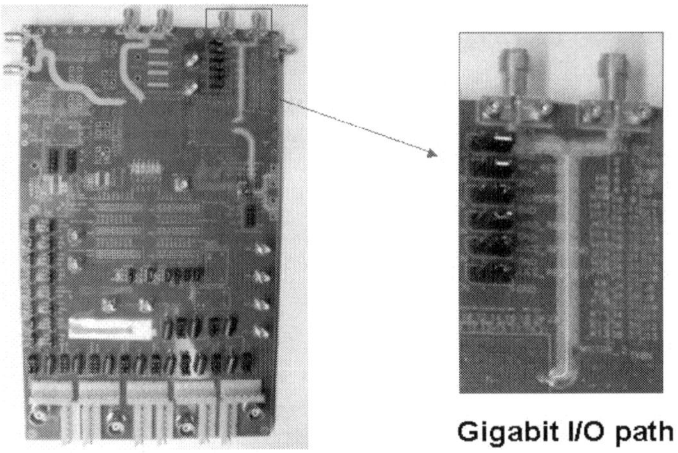

Figure 12.12

12.6 Design Case Study

FPGA Signal Flow Path

The complete signal flow path for a functional FPGA is shown in *Figure 12.13*. The gigabit transceiver block will transmit a gigabit signal starting from the flip-chip IC die, through a controlled collapsed chip connection (C4) bump, through a BGA ball, into a differential transmission line on the FPGA test fixture, out an edge launched co-axial SMA, and finally to the test equipment.

The reference plane for the test setup of the de-embed design case study is shown is red. The graphic has been simplified to show only one side of the differential pair for clarity, but the probing setup is worthy of noting. Two single-ended ground-signal (GS) probes were used to probe the FPGA text fixture at the BGA landing site. The ground and signal configuration of the layout would not easily allow a differential ground-signal-signal-ground (GSSG) probe or ground-signal-ground-signal-ground (GSGSG) probe. Therefore, the single-ended or two-port Touchstone file (.2sp) is what the probe vendor needed to supply for the design case study. In most applications, it is the test setup that dictates the type of Touchstone file and de-embed (i.e., two-port versus four-port) methodology that must be used.

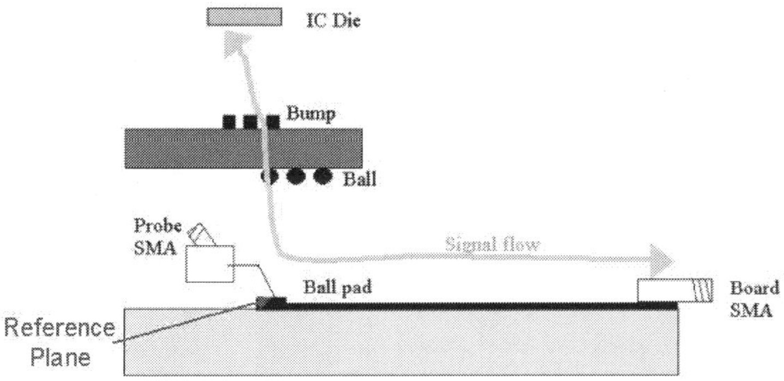

Figure 12.13

FPGA Fixture Layout for Gigabit Channel

The FPGA fixture layout is shown in *Figure 12.14*. The differential transmission line can be easily seen in this view. One of the interesting features that can be noticed is the via field following the outside path of the differential pair. This structure usually indicates a co-planar waveguide topology that allows good control of the impedance environment. This is a good design that minimizes the negative impact that geometric PCB fabrication tolerance can inadvertently have on characteristic impedance.

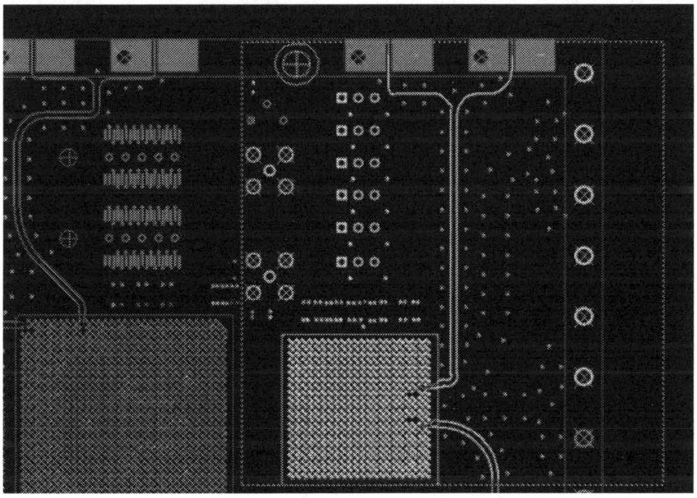

Figure 12.14

Metalization Layout of Ball Grid Array

A close-up of the BGA footprint is shown in *Figure 12.15*. The familiar golden circle for ball attach in combination with the red connecting trace represent what is commonly referred to in the industry as the "dog bone." The physical restrictions due to the tight geometry of the BGA leave little creativity to the PCB layout specialist. There is an unavoidable asymmetry introduced by this geometry, also. The port 1 and port 3 traces will theoretically have mode conversion in the first few microns after the ball attach point due to different width of traces. Practically speaking, the resolution of the test system calibrated to 26.5 GHz will not be able to resolve this mode conversion. The impact of this is minimal. However, as speeds increase, this asymmetry will at some point affect the EMI emissions of the FPGA.

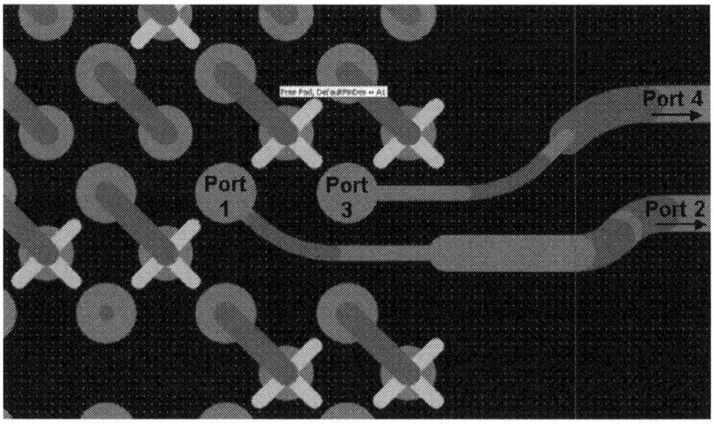

Figure 12.15

Mixed-Mode S-Parameters

Now that we have an intuitive understanding of four-port S-parameters, we can analyze the FPGA test fixture measurements. At first glance, the 16-element mixed mode S-parameter matrix in *Figure 12.16* may seem a bit overwhelming. However, there are certain steps that can be taken to segment the analysis into smaller sections. Typically, the engineer will tend to first analyze the domain that is most familiar, either time or frequency. If the impedance profile is easily recognized by familiar impedance discontinuities on the DUT, then this may be the first step.

If, however, the engineer is more familiar with frequency domain, then the S-parameters are usually viewed first. Whether time domain or frequency domain analysis is done first, the authors tend to view the single-ended parameters first. You may ask why this is the case, since the DUT is a differential device. The answer is that each individual channel can be viewed as a quick "sanity check." If one probe is not coplanar or missing a pad, this will immediately show up in the S11 or T11 term. Likewise, if there is any large impedance discontinuity in an unexpected location, the engineer will know which line within the differential pair to look at for problems. If the differential mode is analyzed first, then an extra step is needed to go back to the single-ended view to get this information. In any case, the benefit of having all four-port data easily accessible in a multitude of formats is obvious.

A clear intuitive understanding of the DUT can be obtained from the user's personal technical preference, then alternate formats and domains can be explored to provide new insight based on familiar knowledge. In the world of high-speed digital, it will become mandatory for top-level designers to be comfortable in both time and frequency domains.

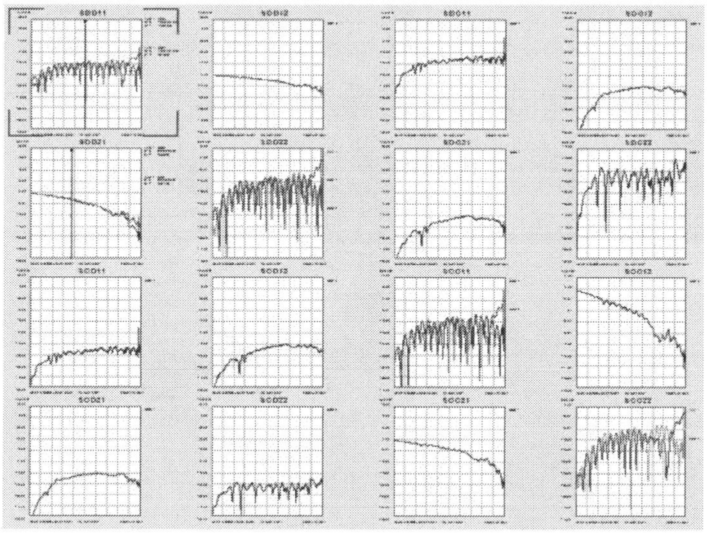

Figure 12.16

Single-Ended Insertion Loss

It is always a good idea to do a brief "sanity check" on the measurements before drawing conclusions and diving deep. Using this strategy, the authors first chose to analyze the forward and reverse single-ended insertion loss of all four ports shown in *Figure 12.17*. The characteristic roll-off at higher frequencies is a familiar sight and easily recognized as a reasonable measurement. A quick marker was placed on each curve for an estimation of the 3 dB bandwidth of each line. This validates good measurements on all four ports and gives the green light to continue analysis. As a note, it is instructive to discuss reciprocity theory at this point. Theoretically, S12 and S21 should be equal because the DUT is linear and passive (same goes for S34 and S43). Since the frequency response of an ideal transmission line is independent of the direction of current flow, we have a reciprocal relationship between the aforementioned parameters. The reciprocity rule of linear and passive devices is nicely validated and further validates measurement success. Proceeding with measurement confidence is always comforting, and this assurance encourages further experimentation.

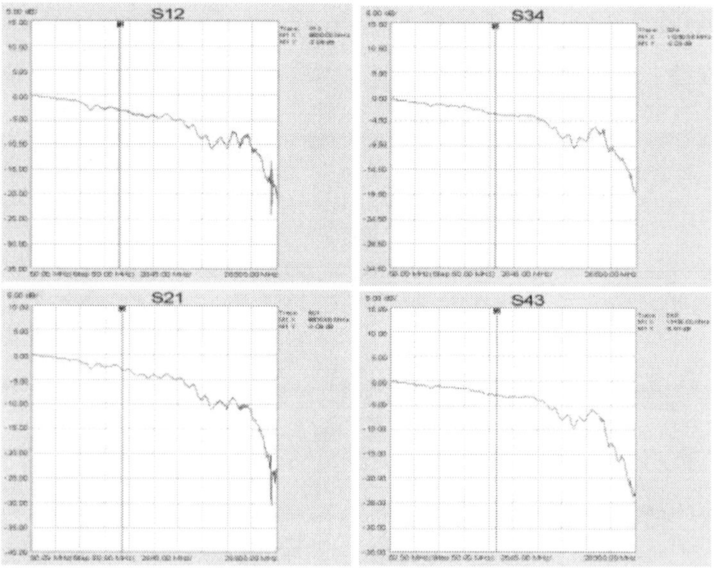

Figure 12.17

Single-Ended Impedance Profile

A similar sanity check strategy works well in the time domain also. This analysis will rely on recognizing PHY features in the DUT. The authors used forward and reverse impedance profile on each of the four ports to gain further insight into the DUT. Being familiar with TDR, this is usually where most digital design engineers will start. Knowing that the measurement was set up with the probes at the FPGA fixture side of the DUT, the most prominent feature in the left-hand side of the upper left graph of *Figure 12.18* is the larger peak hovering around 200 ps, indicating excess inductance. This is usually caused by standard loop inductance of probes and magnitude is dictated mainly by the probe spacing between the signal-carrying conductor and ground pins. A perfect signal launch through probing is a rare occurrence, which is why error correction techniques are employed in advanced measurements.

There is also a smaller peak of excess inductance at around 45 ps. This is very close to the reference plane set by the SOLT co-axial calibration done at the end of the coax cables. This must be the 3.5 mm connector-to-probe body transition. The connector discontinuity should be removed after SOLT, so this must be internal to the probe. The last inductive peak is from the FPGA board to SMA connector transition at the other end of our channel.

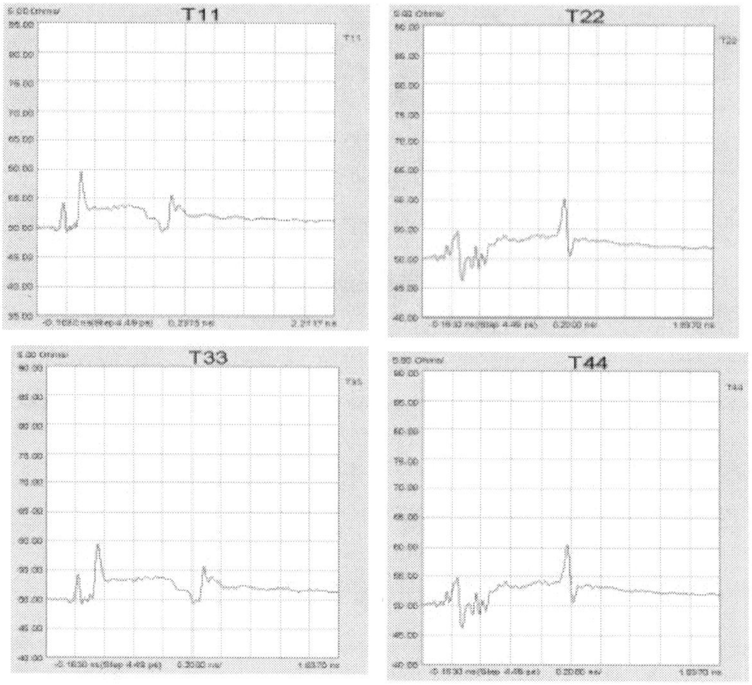

Figure 12.18

De-Embed Probes

In order to validate the S-parameters obtained from the probe vendor, a special test was done. The probes require an SOLT calibration substrate that allows various standards to be probed. This substrate includes a thru line that allows a very short propagation delay between the two probes for a transmission measurement. This thru was probed and measured, and then the Touchstone files from the probe vendor were de-embedded one at a time. Looking at the upper right graph of *Figure 12.19*, the original impedance profile of both probes measuring the thru is shown. A little detective work is needed to uncover what is happening. Viewing this impedance profile, we see the first inductive peak is the SMA–to-probe body transition. The second larger peak should be one of the probes, but only this one additional inductive peak is seen. Should we not see two inductive discontinuities, one for each probe tip? Well, looking at the length of the thru standard and estimating the time of flight through this structure, it is evident that resolving the two probes is not possible with the frequency range used for this measurement (26.5 GHz) and subsequent step size in the frequency domain (19 ps). Therefore, this one peak is actually the excess inductance of the two probes together (remember we are using two probes for this single-ended measurement, one from each side).

The upper right graph in *Figure 12.19* shows the results after the first probe is de-embedded. The first inductive peak from the first SMA–to-probe body transition is removed as expected. It is also noticed that the second peak from the probe tip did not disappear because it is from the combination of the two probe tips in proximity of each other. The interesting note here is that the peak magnitude has changed from 237 mV to 225 mV. This further validates our guesswork that both probes cannot be resolved and the second probe tip still needs to be de-embedded.

The lower left graph in *Figure 12.19* shows the results after the second probe is embedded. This is perhaps the most challenging measurement to decipher. The results are not bad. Anything that has time $t<0$ not perfectly flat is questionable, but the slight non-flatness might be attributed to calibration error, though the amount of non-flatness may not be acceptable for the purist.

However, the lower left graph shows what the authors believe to be less–than-perfect de-embed files for the probes. This is undesirable but indicative of what real-world problems arise in signal integrity labs all the time. The best we can do as engineers is understand the limitation of what we are measuring and try to improve upon it through diligence. At this point in the design case study, it is realized that further de-embed accuracy might be compromised as the experiment continues, but learning more about the process pushed our collective interest.

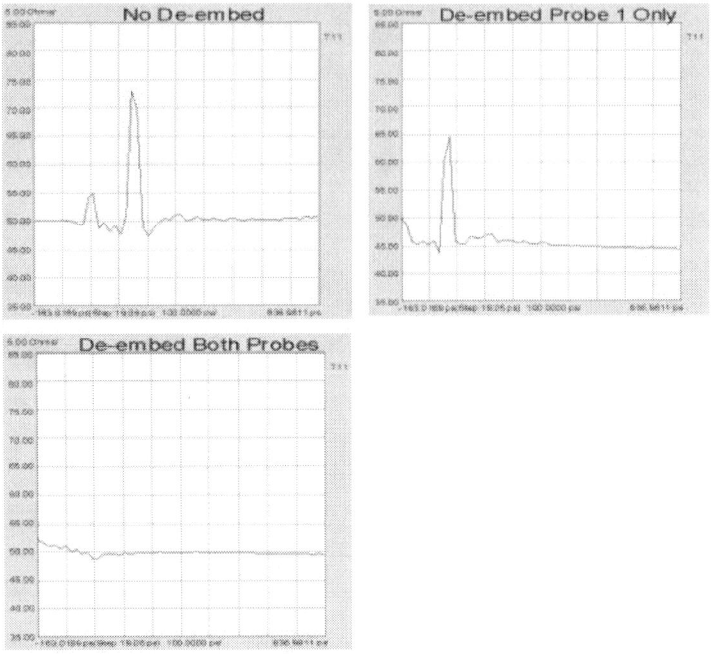

Figure 12.19

Differential Impedance Profile after De-Embed

The next step in the process is to view the corrected differential impedance of the FPGA fixture after the de-embed. Looking at *Figure 12.20*, the "before" graph on the left side shows the extremely well controlled 100 Ohm test cable environment starting at the far left of the impedance profile, then the three inductive discontinuities discussed earlier (SMA connector on probe, probe tip, and FPGA SMA connector at other end). The transmission line on the FPGA fixture is about 104 Ohms throughout and fairly well controlled. After de-embed, the probe is removed and only one inductive discontinuity remains (FPGA SMA connector at other end). Also, notice that there is a slight increase in the amount of ripple in the differential transmission line of the FPGA fixture. This is mostly likely due to the subtle increase of bandwidth due to removing the effect of the probe. The benefit of de-embedding is achieving a higher bandwidth measurement that enables more detailed information about the DUT.

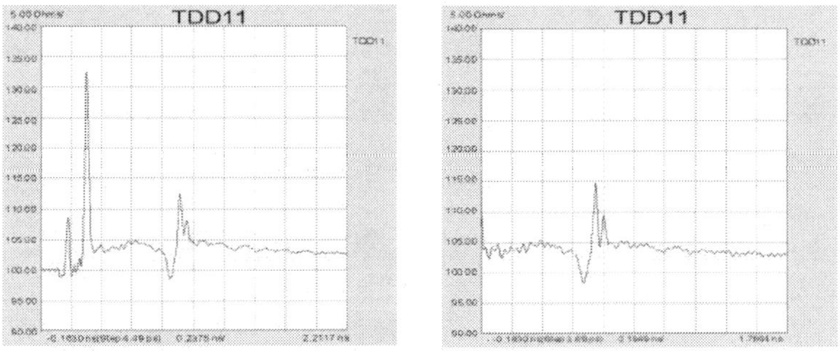

Before **After**

Figure 12.20

Eye Diagram Analysis

Perhaps the last and most familiar step in this process is to look at the performance of the test fixture itself. The eye diagram analysis indicates a very high-performance test fixture. The data rates shown in *Figure 12.21* starting from the upper left and moving clockwise are 3.125 Gbps, 10 Gbps, 30 Gbps, and 20 Gbps. These eye diagrams are synthesized from the S-parameter data of the FPGA fixture. Extracting the impulse response from the S-parameter and then convolving that with a pseudo-random binary sequence (PRBS) achieves an accurate representation of the eye diagram. This algorithm is commonly used in research and development laboratories around the world and will eventually be used as a compliance test for digital standards.

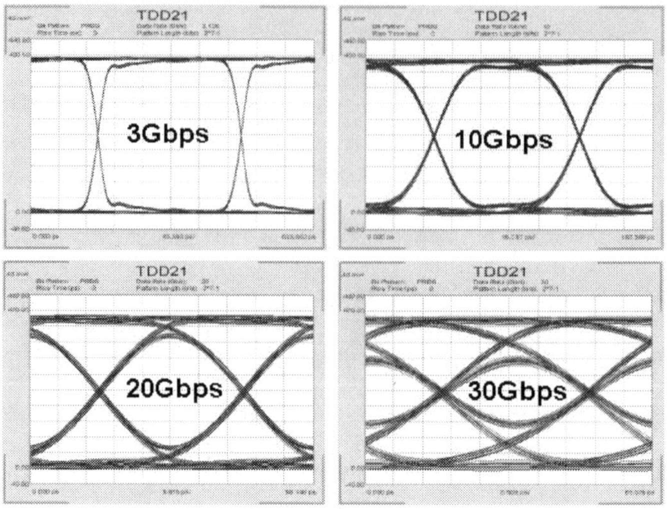

Figure 12.21

12.7 Conclusion

Signal integrity will continue to be a tremendous challenge as data rates maintain their march on to 10 Gbps and beyond. The significant amount of time put into testing digital devices is ultimately a function of how well the device test fixture is designed. The extremely high bandwidth of the DUT requires an even higher-bandwidth test fixture. More sophisticated calibration can enhance measurement accuracy and enable the true device performance to be characterized properly. Since high-speed serial devices are overwhelmingly of differential topology, four-port measurements are required for full understanding. Many error correction techniques can be utilized by the educated designer to improve the confidence of measurement data and the result will ultimately create more margin and higher yield of the end product.

Authors

Mike Resso, Business Development Manager, Signal Integrity Applications, Agilent Technologies

Hong Shi, Member of Technical Staff, Packing Technology, Altera

Chapter 13

Performance at the DUT: Techniques for Evaluating the Performance of an ATE System at the DUT Socket

13.1 Abstract

"Performance at the DUT" is a key phrase in the test and measurement industry and is used to indicate the true electrical signals that exist at the reference plane where the device under test (DUT) connects to the automated test equipment (ATE). The limited specifications provided by test equipment manufacturers and the fact that the test fixture can be a major source of signal degradation create significant challenges in measuring these signals. This paper presents the results of an industry collaboration to address some of these challenges for ATE applications running at 5 Gbps or greater. Topics include probing techniques for measurements at the DUT interface, calibration methods for measuring test fixture effects, source characterization at the DUT, and test fixture de-embedding from the measured device data.

13.2 Introduction

"Performance at the DUT" is the ideal specification for an automated test equipment (ATE) test fixture, but one quickly finds that this simple definition is rather complicated to implement for I/Os running at multi-gigabit data rates. At data rates below 1 Gbps, the propagating edges are on the order of inches in length, long enough that the signal at the test fixture printed circuit board (PCB) socket interface is roughly equal to the signal at the package bump and even further in at the integrated circuit (IC). However, for multi-gigabit ATE systems running with 30 pS rise times, one can no longer make this assumption. The rise-time edge occupies less than 200 mils and requires transmission line theory to understand the performance at each of these interfaces and how they interact with each other as they propagate through the test fixture to the DUT [1]. Depending on what previous data or simulations are available for correlation, the ATE test fixture designer will often struggle to understand whether to put the measurement plane at the IC, at the package bump, or at the transmission line on the test fixture board. It is desirable to have the DUT performance broken down such that the data can be used to evaluate the packaged device performance in the targeted application environment and not skewed by ATE test fixture–induced characteristics.

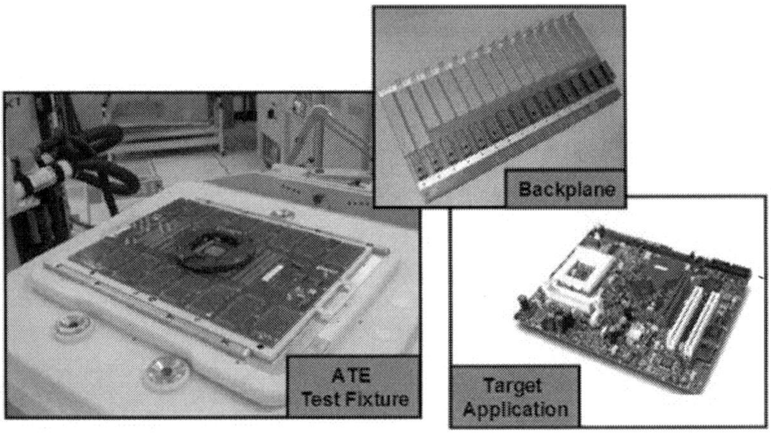

Figure 13.1: *"Performance at the DUT" Specifications Are Not Limited to ATE Applications. Backplane Applications Also Need to Know How the Backplane Interconnect Components Distort the Signal Going into the Device*

This signal integrity challenge is not limited to the ATE test community. Signal integrity issues plague many high-speed digital interconnects, including backplanes, connectors, cables, PCBs, and IC packages. The performance of the device at the device, excluding the electrical effects of the fixturing, is the ideal benchmark. Users of test and measurement equipment always face the challenge that instrumentation is typically specified to the "front panel" or, in the case of an ATE system, the pogo pin interface at the test head. There can easily be significant discontinuities through connectors, cables, and PCB routings to or from the device that degrade the signal from its initial value. To compound this problem, the ATE test fixture designer can be faced with hundreds of multi-gigabit high-speed I/Os with significantly different long trace routings that result in large channel-to-channel variation if it is not calibrated out.

This paper looks carefully at the physical interface structures involved with a DUT on an ATE test fixture to understand the interactions that take place in the ATE environment. Probing and characterization methods will be shown for separating out the performance of the test fixture so that "performance at the DUT" can be accurately obtained from a combination of measurements and 3D–EM simulations. The paper will start with a basic understanding of the benefits of simple test fixture calibration techniques that can provide accurate skew and simple loss models for the ATE test fixture. Interposer design and probing techniques will be demonstrated to provide a standard reference plane for test fixture measurements at the PCB and DUT interface. Advanced calibration techniques such as probe vendor-supplied calibration substrates, custom PCB thru-reflect-line (TRL) calibrations, and 3D–EM simulation de-embedding will be discussed to show the benefits of improved calibration techniques to accurately de-embed the test fixture for multi-gigabit applications.

A design case study will show the results from a test fixture designed for a quad core microprocessor with I/Os running above 5 Gbps on an ATE system (Verigy V93000). The results from this demonstration will clearly highlight the challenges of measuring the "performance at the DUT" and will provide insight into the interactions of the test fixture PCB and the DUT for these multi-gigabit data rates. The paper will conclude with a review of the results of obtaining the synthesized DUT performance based on the various calibration techniques and a discussion of the accuracy of the results. The results will show that the increasing data rates will force one to rely more heavily upon 3D–EM simulations to predict the interaction at an interface.

13.3 Probing Technology, Interposer Design, and Mechanical Challenges

This section addresses the three main technologies needed for fully characterizing a test fixture using probing techniques at the DUT interface. The characterization measurements can be done either isolated on a bench or docked to an ATE system.

Probing Technology

To measure the performance at the socket with enough fidelity for multi-gigabit signals and to obtain accurate S-parameters for de-embedding and model generation it is necessary to use high-performance micro-coaxial probes. *Figure 13.2* shows a picture of the micro-coaxial probes used in this paper together with their individual performance measured by the manufacturer showing that the probes have sufficient characteristics to at least 40 GHz for the needed measurements.

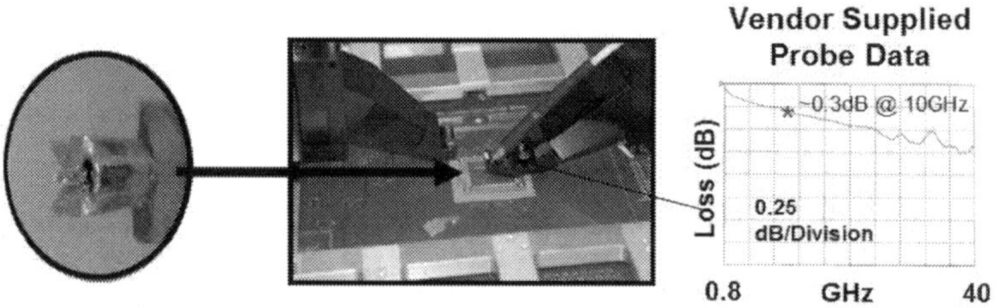

Figure 13.2: *The GGB Picoprobe GSG with 400 um Pitch (Left), Probing an ATE Test Fixture (Middle) and Insertion Loss Performance of the Probe Measured by the Manufacturer (Right)*

Micro-coaxial probes are designed for wafer probing and are fairly delicate as their electrical requirements can only be met by building mechanical compliance into the tips and ground wings. This delicate probe tip is unable to provide the mechanical force required to compress socket pins and, in addition, the ground wings can be damaged if they get caught on the socket contacts. This is an important reason for using a probe interposer board for mating the probes to the DUT interface, which is described in the next sub-section.

The probes come in a variety of configurations with signal only (S), ground-signal (G-S), and ground-signal-ground (G-S-G) contacts at the tip. High-frequency performance is typically limited by the added inductance of the ground connection discontinuity, and one finds that the G-S-G topology provides the best impedance matching by lowering this inductance. The G-S-G probes can be purchased with a variety of ground-to-signal spacings to accommodate the ATE applications ranging from DUT interface types (needles or socket) to tester interface (coax and/or pogo block). The former are usually in the 0.1 mm to 1 mm range, while the latter run in the 1 mm to 3 mm range. Electrically one finds that a tighter G-S spacing that is closer to the dimensions of the ~12 mil 50 Ohm micro-coax used in the fabrication of the probe will improve the impedance matching. This small pitch of the G-S spacing on the probe then requires the use of a probe interposer board to provide this contact spacing and a transition to the DUT pin I/O topology.

Interposer Design

One of the main challenges in probing the test fixture where the DUT resides is the fact that the contact pitch between the signal pins and the reference pins can vary greatly, not only due to the DUT I/O topology but also due to the pin-out where the closest reference pin (e.g., GND in a single-ended measurement) might be far away. Ideally one would like to use a fixed-pitch probe to probe any signal pin without worrying about the pin-out of the device. The solution is to develop a PCB board known as the probe interposer that provides pads on the bottom for connecting to the ATE test fixture DUT interface and pads on the top with a fixed G-S pad spacing of 5 mils for connecting to the probes. The G-S pad spacing is achieved by flooding the topside with a copper pour that directly connects to all ground pins and has a circular clearance around all signal and power pins (*Figure 13.3*). Filled vias with overplating provide a planar surface for the socket contacts and the probe landing. Ideally one would like the probe interposer to be as thin as possible to minimize the impact on the electrical performance of the measurement; however, mechanically the board must be thick enough to avoid significant bending when compressing the high density of socket pogo pins when measuring at the socket pin interface. A thickness of 100 mils was evaluated and determined to be sufficient for the compression of this 3.85 mm x 3.85 mm array of 1,200 pogo pins.

Figure 13.3: *Example of an Interposer Designed for a Specific BGA Ball-Out (Left), Interposer Attached to the Loadboard for Probing Connection (Middle), and 400 um G-S-G Probe Tip Connection to the Probe Interposer Board (Right)*

Mechanical Challenges

Properly probing a test fixture presents several mechanical issues. For example, in an ATE test fixture the DUT socket pins and the ATE pogo vias are usually on opposite sides of the test fixture PCB, requiring simultaneous measurement on two separate faces. Other complexities can arise in trying to accommodate the integral mechanical stiffener attached to the large ATE test fixture, mechanical attachments for temperature forcing environmental systems, and electrical connectors for added test capability. These items can cause mechanical interference with probes that have a limited vertical clearance typically ranging from 3 to 5 mm. Another problem of ATE test fixtures is that they are usually large and heavy, which adds to the complexity of maintaining a rigid probing connection. *Figure 13.4* shows the mechanical solutions developed for the Verigy V93000 ATE system that can be used for probing the test fixture on the bench or with the test fixture docked to the ATE system.

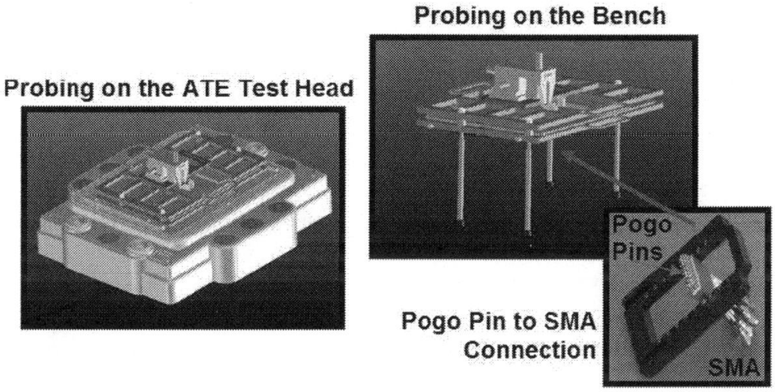

Figure 13.4: *Mechanical Solution for Probing an ATE Test Fixture. Left: Mechanical Fixture with Probe Positioner Attached to the Loadboard That Is in Turn Docked to the ATE System. Right: Mechanical Fixture with Probe Attached to the ATE Test Fixture Including Legs for Bench Measurement and Pogo Pin to SMA Block Assembly for Thru Path Measurements*

13.4 Calibration Techniques

The first issue that arises in obtaining calibrated electrical performance of the ATE test fixture is that the ATE tester interface with the pogo pins or coaxial connections is not the same as the probing interface where the DUT socket resides. Therefore, a special calibration technique must be used to place the measurement reference plane at the coaxial connectors on one end and at the probes on the other. At DC this is a simple correction of subtracting out the losses of the cables and adapters used by the measurement system. At higher frequencies, the calibration increases in complexity with reflection, radiation, and phase changes as well as the resistive losses being considered. Techniques have been developed with frequency domain network analyzers to provide National Institute of Standards and Technology (NIST) traceable calibration methods for moving the electrical reference plane to the end of the measurement cables for accurate characterization of the electrical performance of a DUT [6].

The simplest calibration technique, called an "insertable calibration," places reference standards on each port or cable end and then measures the thru path by connecting the two measurement cable ends together. In the case of a DUT such as the ATE test fixture where one end is a probe interposer and the other is a coaxial connector, then an adapter is needed for connecting the two measurement cables together for the through path calibration. This requires a "non-insertable" calibration technique such as a "defined thru," "unknown thru" or two-tier "adapter removal" calibration. Calibration standards come in the form of open, short, load, thru, and multiple thru line lengths for connecting to the measurement cables, and one must identify which combination of standards provides the best accuracy versus measurement simplicity for a given application. A very common selection of coaxial standards is the short, open, load, through (SOLT) calibration. Calibration standards can easily be purchased for a variety of coaxial connector types, and even the probe manufacturers sell characterized thin-film calibration substrates. However, the probe interposer board presents a challenge, since it requires the design of custom standards. Experience has shown that the thru-reflect-multiple lines (TRL) calibration standards can easily be fabricated on a PCB and provide the ability to move the electrical reference plane onto the PCB test fixture or to the bottom of the interposer board [5, 7].

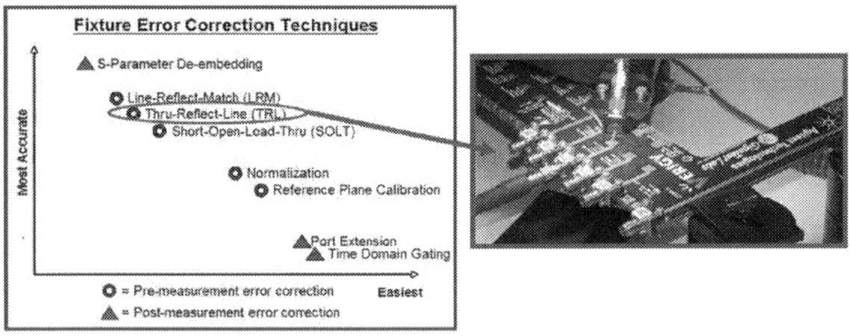

Figure 13.5: Qualitative Comparison of Fixture Error Reduction Techniques and the PCB Implementation of a TRL Calibration for the Probe and Probe Interposer. Note That the Accuracy of the S-Parameter De-Embedding Assumes a Perfect Data Set in This Graph and That Actual Accuracy Will Depend on which Calibration Method Is Used to Obtain the De-Embedding S-Parameters

The theory of this calibration technique works quite well (*Figure 13.5*). However, implementing this technique for an ATE test fixture application poses a considerable challenge. The non-insertable calibration using customized standards requires a significant number of connections to be made, as is the case for connecting the probes to as many as seven standards in some TRL calibration kits. Increasing the number of network analyzer ports or cable connections from two to four compounds the problem, and consideration of the case of 12 or 16 ports with a TRL calibration in this fashion becomes prohibitively time-consuming [5]. The other issue is that as the number of required calibration standards increases, so does the probability of an operator error, and one can easily get an erroneous calibration.

A practical solution to this problem is to identify a way to use an automated electronic calibration module to the ends of the network analyzer cables to minimize operator connection errors and provide a NIST traceable reference plane. Then using post-processing tools, one can de-embed the effects of the adapters required to connect from the NIST traceable reference plane at the end of the network analyzer cables to the desired reference plane on the ATE test fixture [8, 9]. This method as shown in *Figure 13.6* also provides the advantage of checking the data before and after the de-embedding of the adapter to make sure that the calibration method used to remove the adapter is providing the desired accuracy. Now the issue becomes one of accurately measuring or simulating the electrical data for the adapters that are used to get to the desired reference plane such as the probe and probe interposer board in the case of the ATE test fixture.

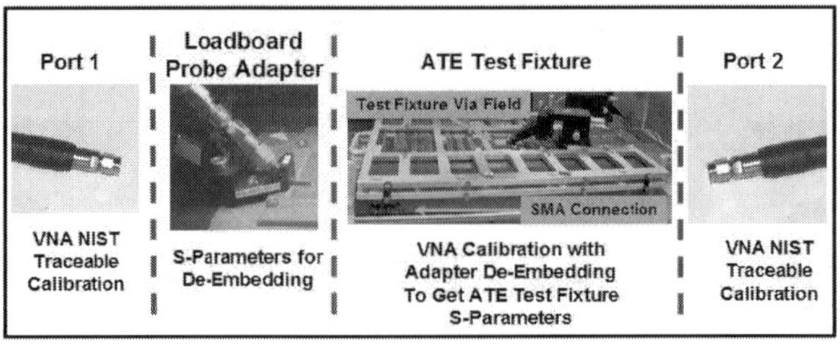

Figure 13.6: *NIST Traceable Coaxial Calibration to the End of the Network Analyzer Cables Combined with Loadboard Probe Adapter De-Embedding to Get ATE Test Fixture S-Parameters*

As we mentioned before, the "non-insertable" calibration method of adapter removal with a TRL calibration will allow placement of a coaxial reference plane on one cable end and a PCB reference plane on the other. This is precisely what is needed for measuring the adapter going from the network analyzer coaxial cable connector to the ATE test fixture reference plane (*Figure 13.7*). This adapter removal with TRL calibration can be quite tedious for the case of the probe connection, but with this technique it only needs to be done for two ports, it does not have to be done at the same time or place as the ATE test fixture measurements, and the method can be repeated over time to evaluate the electrical repeatability of the adapter.

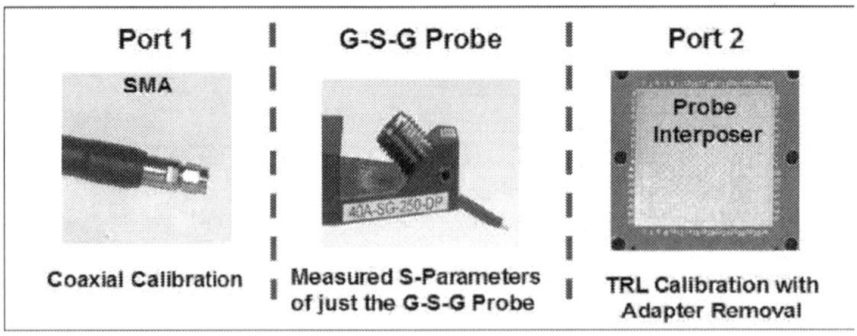

Figure 13.7: *Utilization of TRL Calibration Standards and Adapter Removal Methods to Move the Measurement Plane to Either Side of the G-S-G Probe so That the Electrical Performance Can Be Measured for the Desired Application*

In theory the TRL calibration structures work quite well, but here again the practical implementation does not always work as well as one would like. A large challenge comes in matching the ground reference topology for the ATE test fixture measurement with that used by the calibration structures. Probes are essentially point sources that can only launch signals at specific locations. By contrast, many DUT and tester interfaces use several grounds and have other signals in proximity, therefore the flow of the ground currents at the reference plane may not match those measured with the calibration structures. In some cases this can lead to inconsistencies in the launch response. The use of an interposer adds to this grounding complexity since now the calibration structure needs to account for the probe and the interposer connection to the ATE test fixture. In an ideal world one would fabricate separate calibration structures for every signal pin to be measured so that the grounding topology can be replicated in the calibration. However, this would be time- and cost-prohibitive.–

To understand the effect of the neighboring ground vias on the measurement path, one can run a simple experiment of probing a signal via pad with one, two, three, or four neighboring ground vias in the DUT via field being probed.

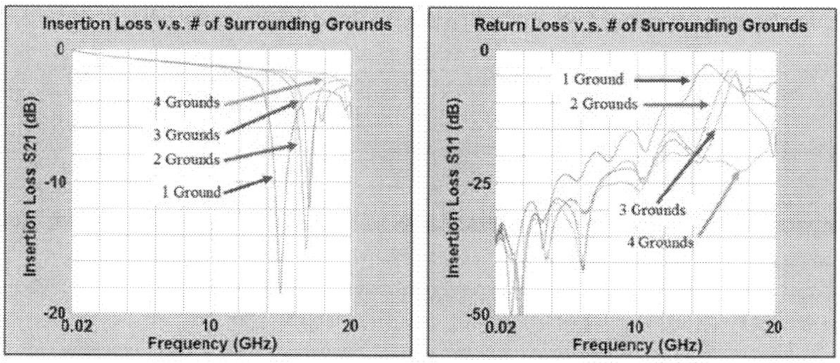

Figure 13.8: *Experiment to See the Effect of Ground Vias Next to the Signal Via*

The insertion loss and reflection data in *Figure 13.8* show that for frequencies below about 10 GHz the use of one ground via versus four ground vias results in minimal differences for frequencies below 10 GHz. The TDR reflections from the probe end (*Figure 13.9*) also show that the best match is achieved with four surrounding ground vias at the DUT BGA pin-out locations.

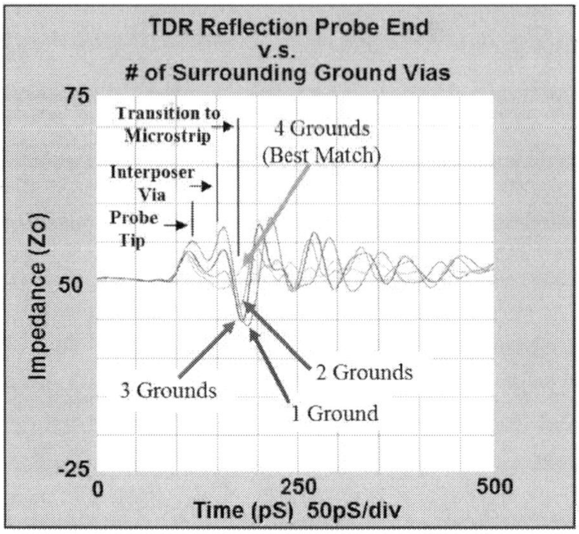

Figure 13.9: *TDR at the Probe End Showing How the Impedance Discontinuity and Resonant Ringing Decreases When the Number of Surrounding Ground Vias Is Increased*

Simulations of the interposer (*Figure 13.10*) show that the resonance is coming from the structure of the neighboring via field as the signal transitions through the interposer. In the case of the interposer calibration structures, these neighboring pins are coupling with the signal pins as frequency increases and the design of the interposer above 10 GHz becomes a challenge.

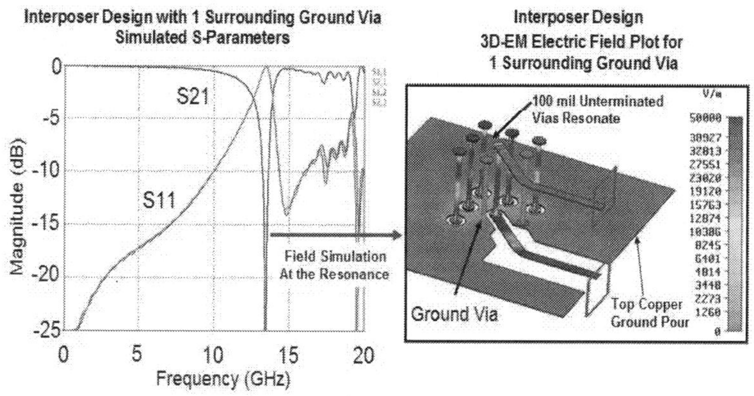

Figure 13.10: *Simulations of the Probe Interposer Design Show That with Only One Neighboring Ground Via, the Unterminated Adjacent Vias Can Start to Resonate. The Grounding Topology of the Neighboring Vias Will Vary When Attached to the Loadboard, and This Resonance Will Shift*

13.5 Measuring the Probe and Probe Interposer Adapter

Now that a method of calibration and an understanding of the challenges in making the adapter from the network analyzer coaxial cable to the desired reference plane on the ATE test fixture have been established, the TRL calibration structures can be fabricated.

TRL calibration and coaxial SOLT calibration with adapter removal will then allow the ability to measure the performance of this adapter for future de-embedding on test fixture measurements (*Figure 13.11*).

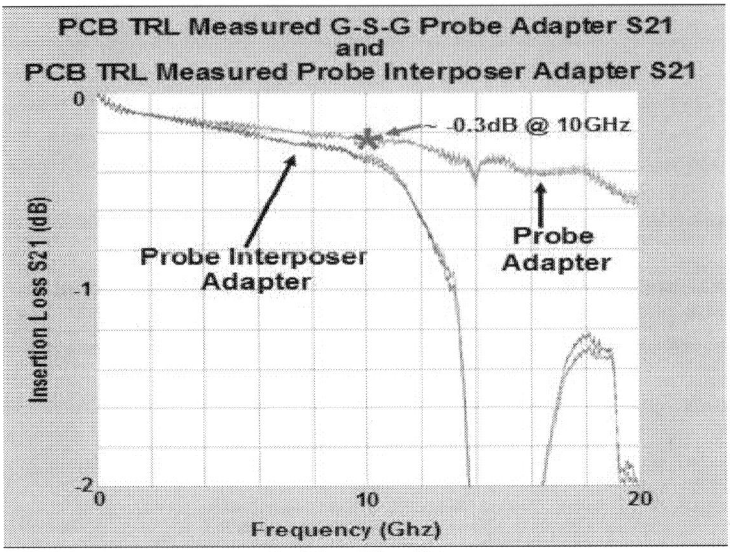

Figure 13.11: *Adapter Removal Calibration with PCB TRL Standards and Coaxial SOLT Standards Provides Measured S-Parameters of the G-S-G Probe and the Probe with Interposer. This Allows One to De-Embed This Data and Move the Reference Plane to the End of the Probe Tip or to the Bottom of the Interposer Board*

Measuring just the probe connecting to the top of the interposer with the TRL adapter removal calibration shows data that is very similar to the calibration data provided by the vendor (*Figure 13.2*) and indicates that the interface of the interposer to the probe is working well. Measuring the probe and the interposer together (probe interposer adapter) with the reference plane at the bottom of the interposer shows that it matches well with the 3D–EM simulation. *Figure 13.11* illustrates the resonant roll-off in the probe interposer adapter insertion loss similar to interposer simulations with only one adjacent ground via.

Looking at the bandwidth of the measured data for an ATE test fixture path of 17 inches of stripline prior to de-embedding, the probe interposer adapter can provide some useful insights into how well the de-embedding process will work. The filter roll-off resonance location has moved further out in frequency than that measured for one neighboring ground via on the interposer, indicating that the grounding topology and connections of the interposer to the real ATE test fixture is closer to a neighboring via topology of three grounds and should work well with the adapter de-embedding for frequencies of 10 GHz and below.

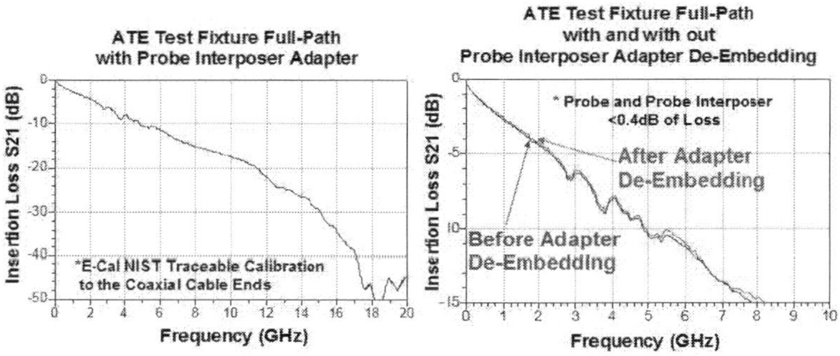

Figure 13.12: *ATE 17-Inch Stripline Full-Path Measurements and the Effects of De-Embedding the Probe Interposer Adapter*

At frequencies above ~10 GHz, one will need to look at improved calibration structures to better match the reference plane location on the ATE test structure. A simplistic way of looking at this reference plane issue is to consider two perfect 50 Ohm coaxial cables of different dimensions (*Figure 13.13*).

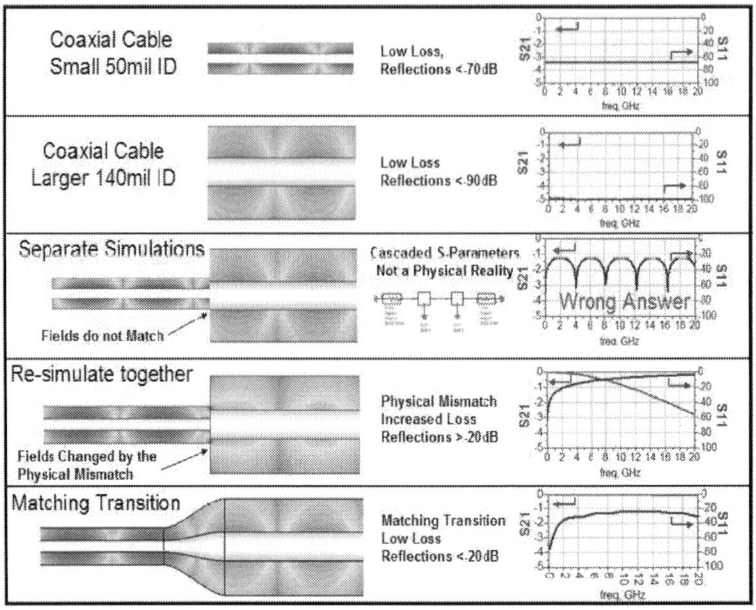

Figure 13.13: *Obtaining S-Parameter Data on Individual Components Does Not Guarantee That They Will Give the Correct Answer When Cascaded Together in a Simulation (3D-EM Simulations Using CST Microwave Studio Time Domain Solver)*

Both of these 50 Ohm cables would have S-parameters with very low loss and extremely small reflections, and if one cascaded the S-parameters together in a simulator tool the result would be a low-loss, low-reflection cable. However, in the real world when one tries to connect these two different cable sizes together, there is a physical discontinuity that can cause significant reflections or low-pass filtering of the data. Placing a reference plane at such a location makes it difficult to build the standards in a way that the measurement technique does not significantly change the physical topology. For the case of a socket or a PCB via field, this would mean creating standards that expand in the vertical direction at the reference plane requiring sockets or PCBs with varying height. Varying the height of the socket or the PCB can add significant cost and time to a project, and one may still question the ability to accurately fabricate the appropriate structures.

A better solution is to look at the use of 3D–EM simulation to provide a flexible tool for evaluating the interaction of two materials at a reference plane. The 3D–EM tool will also provide insight into the benefits of coupling for differential probing applications. The previous analysis assumes that the coupling is low and relies on single-ended calibration techniques to avoid the more complex structures and calculations required for a multi-mode, four-port TRL [5]. The other advantage to using a 3D–EM simulation tool is that it will also provide insight into how one can optimize or improve the probing adapter to ATE test fixture for higher–data-rate applications.

13.6 Test Fixture Performance Measurement

Now that a method has been established to de-embed the effects of the loadboard probe adapter and move the electrical reference plane to the DUT via field pads on the ATE test fixture, we can compare the measured results with more traditional methods. A very common way to obtain the electrical data for a signal path on an ATE test fixture is to fabricate a test coupon with traces routing to coaxial connectors that simulate the best- and worst-case routing topologies (*Figure 13.14*).

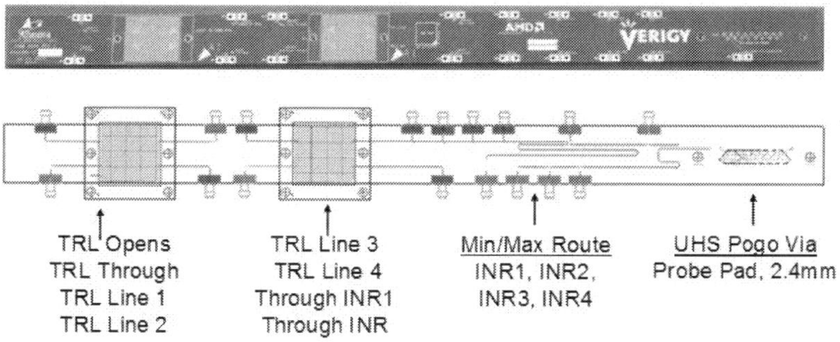

Figure 13.14: *Example ATE Test Coupon Fabricated on the Same Panel as the ATE Test Fixture. The Test Coupon Will Typically Include Minimum and Maximum Trace Routings for Each Signal Layer to Quantify the Losses and Assist in the Analysis of the DUT Data Measured with the Test Fixture*

Plotting the data for the minimum and maximum trace lengths for four signal layers shows how there can be as much as 4 dB of loss between the best- and worst-case traces on an ATE test fixture at 10 GHz. Transforming this data to loss per inch by subtracting the maximum and minimum trace length losses for a given layer shows that some of the differences are also coming from variations in the stripline dielectric losses. This data clearly shows the benefit of correcting the data to remove the effects of the ATE test fixture.

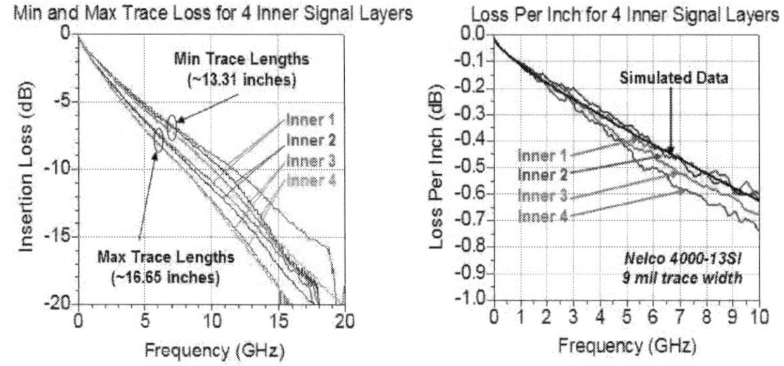

Figure 13.15: *Variation in Loss for Different Signal Layers and Different Routing Lengths. Total Loss Is Shown on the Left and Then a Loss per Inch Is Shown on the Right Based on Subtracting the Minimum and Maximum Trace Losses for a Given Layer*

Measuring the corresponding minimum 13-inch trace routing on the ATE test fixture by probing at the DUT via field with the probe interposer adapter shows a higher loss then the data from the test coupon (*Figure 13.16*). The difference is more than 1 dB at 10 GHz, which is more than a 10 percent difference in voltage levels.

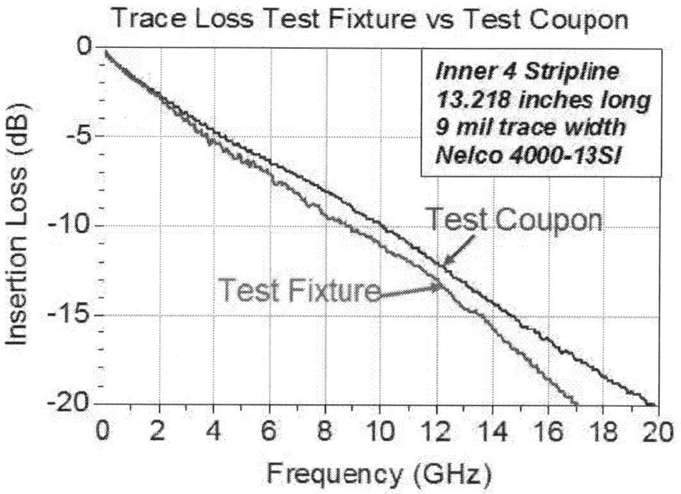

Figure 13.16: *Trace Loss for a 13.218-Inch Trace on the ATE Test Fixture versus the Loss for the Same Length Trace on a Test Coupon*

Other options exist such as consulting a probing house that can perform full S-parameter characterization. The probing house typically has neither the luxury of a coaxial pogo-pin to PCB adapter for the ATE interface nor a custom probe interposer for the DUT interface end. Without the custom interfaces, one must select from a variety of probe spacings to find the best fit for the via field topologies connecting to the signal path. *Figure 13.17* illustrates the type of setup required for making these measurements. The physical size of the test sample and the two-sided probing requirement dictate the need for a custom probing system with four positioners on a dual-platen system with remote optics. The positioners are able to swivel ±45 degrees on the platens, enabling the user to access the ground pins in multiple orientations. This capability is the key to probing directly on the pogo vias and on the device footprint.

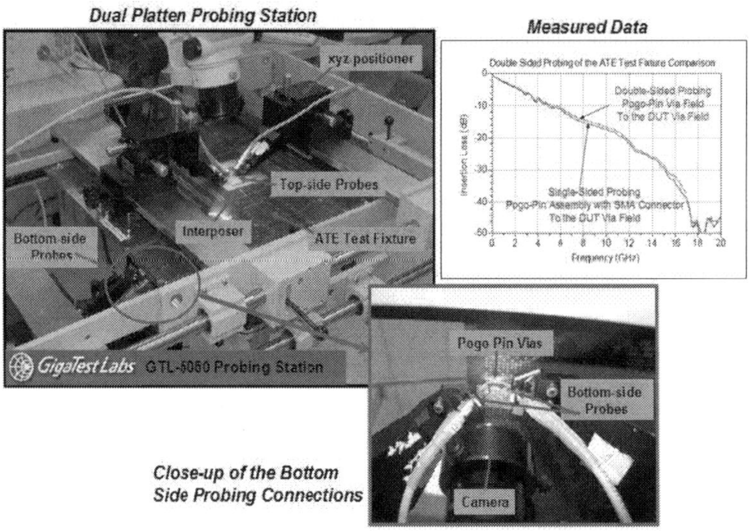

Figure 13.17: *Measurement of ATE Test Fixture by Probing Directly at the Pogo Vias on the ATE Interface Location and with the Probe Interposer Adapter at the DUT Location*

The measurements using the pogo-pin assembly compare well with the double-sided probe-based measurements, the difference being the signal-loss of the pogo-pin assembly, which is on the order of 1 dB at 10 GHz.

13.7 Focus Calibration on an ATE System: Measuring "at the DUT"

The calibration examples in the previous section were all done by taking bench measurements of the test fixture, and the resulting data provides an in-depth understanding of the losses of the test fixture and the accuracies involved in using a probe interposer adapter to measure "at the DUT" performance. The frequency-dependent losses measured on the ATE test fixture clearly show that at multi-gigabit data rates, the long 30+ cm traces typical for a dense microprocessor application can significantly degrade the signals to and from the DUT. The increasing loss with frequency causes data-dependent level and timing jitter in addition to degrading the signal slew rate [2, 3]. The standard ATE calibration to the pin electronics does not take into account any of the test fixture losses, and this results in measured device data with less performance margin than what is expected. Applying the probe interposer technique to the in-situ focus calibration of the ATE system will allow the measurement of the "performance at the DUT" for the ATE–transmitted signals going to the DUT and for the DUT signals being received at the ATE pin electronics.

One typical approach for determining the "performance-at-the-DUT" is to use a time domain transmisometry (TDT) or a vector network analyzer (VNA) to obtain trace loss data for each channel by employing either the simple test coupon approach or the higher-accuracy measurement of each channel using the probe interposer. This data can then be used to de-embed the effects of the test fixture on the measured signal (e.g., by filtering the signal through an appropriate software filter that compensates for the test fixture effect) for a DUT transmitter eye-height measurement or to define how much level compensation is needed for the ATE driver to get the needed data eye opening at the DUT for a receiver tolerance test. The accuracy of this method will depend on the ability to obtain accurate ATE pin electronic models over the desired frequency range.

A specific example of this is to "focus calibrate" the data eye height that is provided to a DUT receiver for a "receiver tolerance" test. Since the test fixture will add data-dependent level and timing jitter, it is expected that the programmed level on the ATE software will not correspond to the eye height seen by the DUT receiver. The proposed measurement-based modeling approach is to simulate the data eye at the DUT using a model of the pin electronics, pogo assembly, and measured test fixture S-parameters. In this example of a Verigy V93000 PinScale HX card, the pin electronics already contains an integrated equalizer [3, 4] that compensates for part of the loss and also needs to be included in the simulation. *Figure 13.18* shows the simulation setup implemented using Agilent ADS.

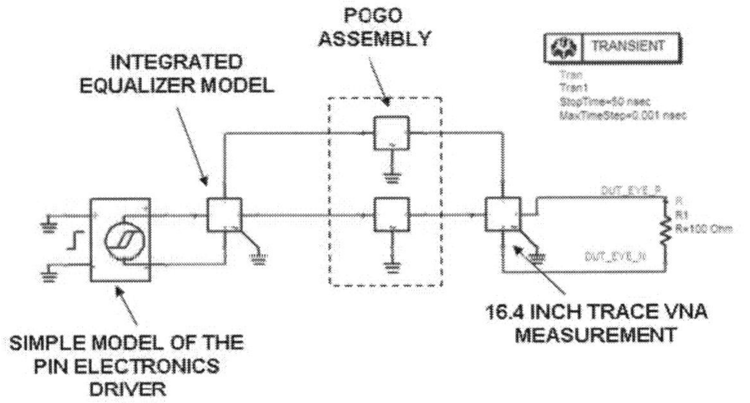

Figure 13.18: *ADS Simulation Setup for Evaluating the Needed Focus Calibration Factor for the Receiver Sensitivity Test*

Note that the simulation uses a very simplistic model of the pin electronics. *Figure 13.19* left shows the simulated data eye at greater than 5 Gbps with a PRBS7 data pattern. The pin electronics driver levels were set to a 350 mV differential swing with the objective that the DUT receiver sees a 350 mV eye opening at approximately the middle of the data eye. From the simulated eye opening in *Figure 13.19* left, a focus calibration scaling factor of 1.64 was derived for achieving the 350 mV eye opening at the DUT for this data pattern. *Figure 13.19* right now shows the date eye at the DUT receiver with the programmed level swing at the ATE pin electronics calibrated by the 1.64 factor and obtaining the 350 mV eye opening.

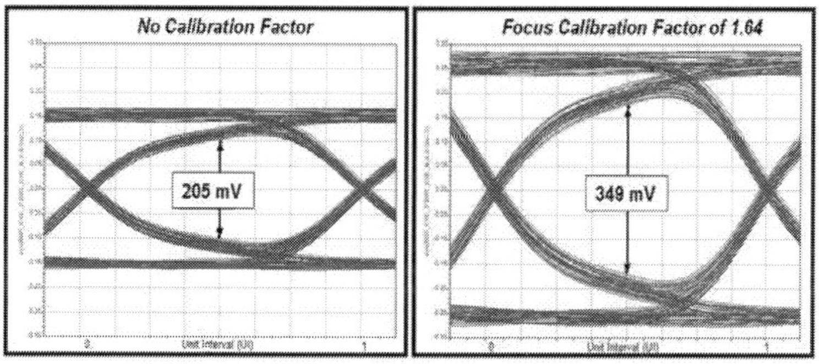

Figure 13.19: *Simulated Data Eye without Any Focus Calibration Factor (Left) and with the Focus Calibration Factor Inferred from the Simulation (Right)*

If measured data is unavailable for the test fixture, then one could use trace geometry, length, and dielectric material to simulate the loss of the test fixture for use in determining the focus calibration factor. However, as shown in *Figure 13.15* the dielectric losses can vary from layer to layer and the losses of the via transitions are not insignificant and thus a simple transmission line model will have limited accuracy.

The measurement based modeling approach with VNA measured S-parameters makes it easy to model a wide variety of data patterns and data rates. However, it is generally useful to verify the simulations with an in-situ focus calibration measurement to cross-check the results. The in-situ probing measurement at the DUT has the advantage of using the active ATE pin electronics source signal with its inherent jitter characteristics. *Figure 13.20* shows a picture of a manual focus calibration system (prototype) docked to a test fixture on the ATE system.

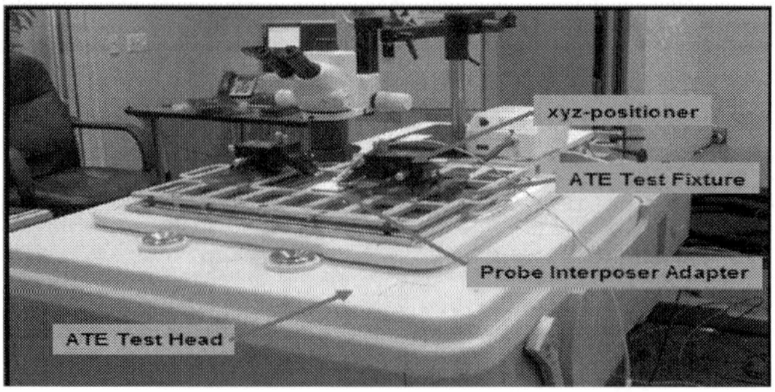

Figure 13.20: *Manual Focus Calibration System (Prototype) Docked to the Test Fixture and ATE System*

The idea is to measure the stimulus signals from the ATE pin electronics at the DUT and in the other direction be able to inject a stimulus at the DUT that can be measured by the ATE pin electronics, as shown in *Figure 13.21*.

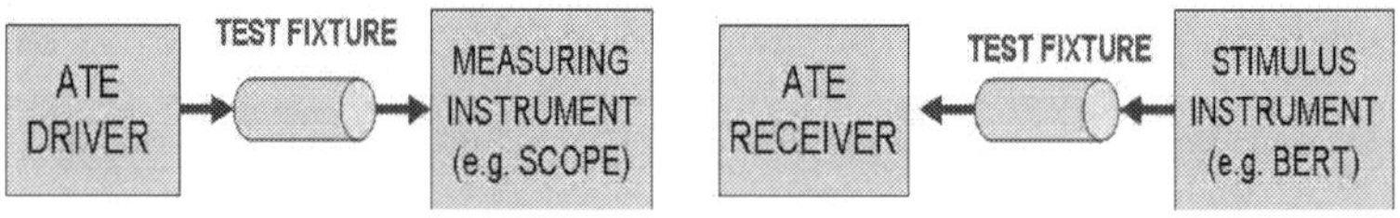

Figure 13.21: *ATE "at the DUT" Focus Calibration Approaches*

The same process implemented in *Figure 13.18* is used to determine the calibration factor needed to correct the eye height at the DUT, but this time the focus calibration setup shown in *Figure 13.20* will provide the data for derivation of the focus calibration factor. *Figure 13.22* shows a comparison of the differential data eye using an optimal test fixture for the integrated pin electronics equalization (as mentioned before the ATE pin electronics used in this example already include an integrated equalizer) for a 350 mV single-ended greater than 5 Gbps PRBS7 data signal and the same signal measured with the interposer probing setup for the 16.4-inch 10 mil signal trace.

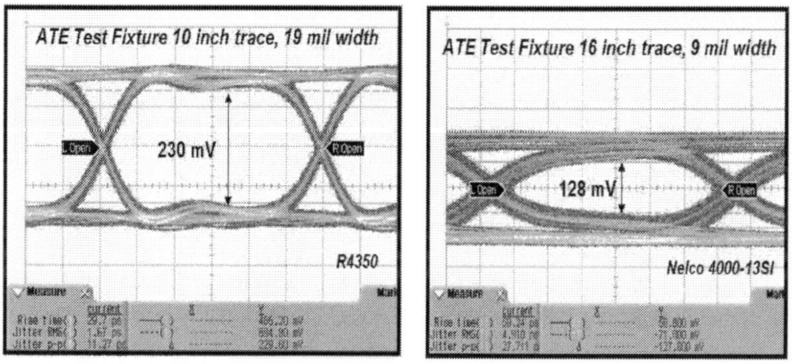

Figure 13.22: *Comparison of the Single-Ended Data Eye at Greater than 5 Gbps with a PRBS7 Data Pattern Using an Optimal Test Fixture for the Pin Electronics Equalization (Left) and Using the Longest Trace (16 Inches) in the ATE Test Fixture for a Microprocessor Characterization Application (Right). Note That Vertical Scales Are Not Equal*

From the measured results it is clear that the inner eye height (defined by the markers) is significantly reduced compared to the optimal test fixture trace, and more important, both do not provide the exact 350 mV data eye height at the DUT. This is expected given that the trace is longer and thinner. It is then necessary to compensate for this reduction in the eye height by using a higher programmed voltage swing from the tester pin electronics. *Figure 13.23* shows the results comparing the data eye with the ATE driver programmed to 350 mV and with the ATE driver programmed to 800 mV. This means that a calibration factor of approximately 2.28 is needed for this measurement point.

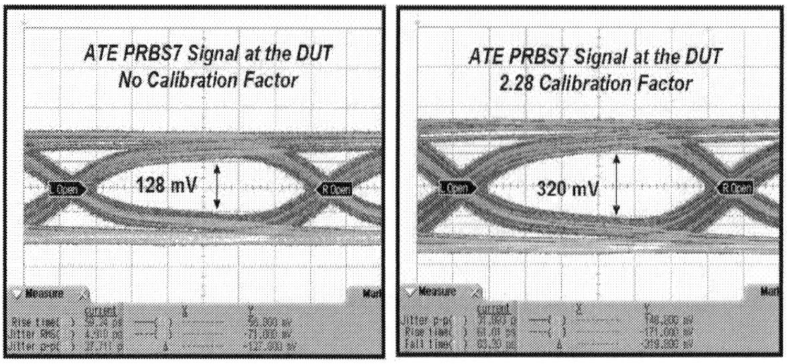

Figure 13.23: *Comparison of the Single-Ended Data Eye at Greater than 5 Gbps with a PRBS7 Data Pattern without Any Calibration Factor (Left) and with the Calibration Factor (Right). (Note: The Vertical Scales Are Not Equal)*

The difference in the calibration factor when compared with the simulation-based results is expected since the simulation model of the pin electronics does not perfectly model the ATE driver and more work needs to be invested in refining the simulation. It is expected that the time domain simulation would provide more optimistic results and that the in-situ calibration is pessimistic in that it also includes the probe interposer adapter and cabling to the measurement instruments.

The previous examples only dealt with focus calibrating the stimulus signal from the ATE system at the DUT, but as shown in *Figure 13.21*, it is also important to calibrate or de-embed the signals measured by the ATE receiver for the test-fixture loss. *Figure 13.24* shows a real example of de-embedding the test fixture effects from the measured waveform and data eye using the test fixture characterization data to develop an inverse filter for convolving with the measured data pattern.

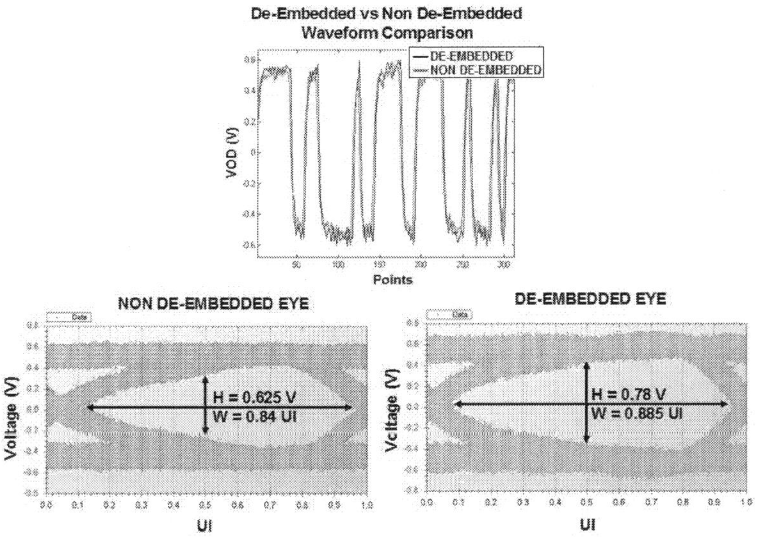

Figure 13.24: *De-Embedding the Test Fixture Effect from the ATE Measured Data*

All of the approaches described in this section can be repeated for other ATE–focused calibrations—for example, the transmitter eye-height measurement or the jitter tolerance test. It is important to notice that items such as data-dependent jitter (DDJ) due to inter-symbol interference are not that easy to focus calibrate or de-embed. Therefore the simplest approach to reduce it is to compensate for it by equalization embedded on the pin electronics or on the test fixture [4].

13.8 Conclusion

Accurate measurement of DUT performance is needed for understanding how a device will perform in its target environment. This paper clearly shows that achieving "performance at the DUT" characterization on an ATE system is not a simple task as data rates enter the multi-gigabit domain. Long trace routings that are typical in high-density microprocessor ATE test fixtures can easily degrade the signals, and calibration techniques are required to remove these test fixture effects.

A probe interposer technique has been described that allows one to obtain accurate electrical performance of the test fixture up to 10 GHz and 3D–EM modeling techniques have been suggested for going higher in frequency. The probe interposer technique was compared with less accurate test coupon trace measurements and trace simulations to show that above 3 GHz the more sophisticated direct probing of the actual test fixture trace does make a difference. The probe interposer technique has the added benefit that it can enable VNA calibrations that use custom PCB TRL standards to move the reference plane for the S-parameter measurements to the interface of the DUT with the ATE test fixture.

"Focus calibration" of the ATE system refers to a variety of techniques that are used for in-situ calibration of the ATE pin electronics to get an "at the DUT" signal characterization. Simulating ATE pin electronic models with the measured S-parameters of the test fixture signal trace can provide a powerful tool for synthesizing the quality of the eye reaching the DUT and for de-embedding the test fixture effects from the measured data at the pin electronics. Accurate models for the ATE pin electronics that include jitter- and frequency-dependent source and receiver effects are not easy to obtain, so one must be careful to check the simulations with in-situ measurements at the DUT to ATE test fixture interface by using the probe interposer. Bench instrumentation can then be used to measure the signal coming from the ATE pin electronics or to inject a known source signal into the tester.

The test fixture will clearly be the bottleneck as the pin count and data rates continue to increase on ATE applications. Advancements in equalization techniques are coming that will allow much stronger integrated and programmable equalization on future ATE pin electronics [10] to compensate for test fixture losses. However, the challenge of how to correctly program this equalization remains, and it will be necessary to measure and stimulate the signal at the DUT and feedback this information to the ATE system to correctly program the pin electronics equalization. This is the only way to assure the highest levels of measurement accuracy.

This paper cannot address every type of ATE "focus calibration" measurement, but the methodology used in evaluating the validity of the probe interposer technique demonstrates a systematic approach that can be applied in general for qualifying a multi-gigabit measurement at a complicated 3D interface. The presented techniques of moving the measurement reference plane through de-embedding will hopefully lead to the ability to separate out the performance of the IC, the IC and the package, and the full combination of IC, package, and socket from a single set of measured ATE data for improved correlation between the die performance and the final packaged performance in the target environment of the end application.

Acknowledgments

We would like to acknowledge Vishal Mittal and Neerav Mehta from AMD as the original developers of the channel de-embedding algorithms used in part to generate the waveforms presented on the "measuring at-the-DUT" section, and thank them for their valuable contributions. We would like to thank Kevin White from Verigy for all the great mechanical work on the socket probing solution, Will Burns at Altanova for his design layout skills for the interposer boards, Mehdi Mechaik of AMD for his discussion of probe de-embedding, Bob Schaefer and Dave Blackham for their discussion of VNA calibration methods, and Gary Otanari for probing support at Gigatest.

References

[1] Moreira, M. Tsai, J. Kenton, H. Barnes, and D. Faller, "PCB Loadboard Design Challenges for Multi-Gigabit Devices in Automated Test Applications," IEC DesignCon 2006.

[2] M. Shimanouchi, "New Paradigm for Signal Paths in ATE Pin Electronics are Needed for Serialcom Device Testing," Proceedings of the International Test Conference, 2002.

[3] W. Humann, "Compensation of Transmission Line Loss for Gbit/s Test on ATEs," Proceedings of the International Test Conference, 2002.

[4] J. Moreira et al., "Passive Equalization of Test Fixture for

High-Speed Digital Measurements with Automated Test Equipment," Proceedings of the 2006 International Design and Test Workshop.

[5] H. Barnes, A. Ciccomancini, M. Tsai, and M. Resso, "Differential PCB Structures using Measured TRL Calibration and Simulated Structure De-Embedding," IEC DesignCon 2007.

[6] H. Shi, G. Liu, and A. Liu, "Accurate Calibration and Measurement of Non-Insertable Fixtures in FPGA and ASIC Device Characterization," IEC DesignCon 2006.

[7] D. Dunham, V. Duperron, and M. Resso, "Practical Design and Implementation of Stripline TRL Calibration Fixtures for 10-Gigabit Interconnect Analysis," IEC DesignCon 2006.

[8] Agilent Technologies, "Agilent Network Analysis Applying the 8510 TRL Calibration for Non-Coaxial Measurements," Product Note 8510-8A 2006.

[9] G. Antonini, A. Ciccomancini Scogna, and A. Orlandi, "De-Embedding Procedure Based on Computed/Measured Data Set for PCB Structures Characterization," IEEE Transactions on Advanced Packaging, vol. 27, no. 4, November 2004.

[10] J. Moreira, H. Barnes, H. Kaga, M. Comai, B. Roth, and M. Culver, "Beyond 10Gb/s? Challenges of Characterizing Future I/O Interfaces with Automated Test Equipment," Presented at the first IEEE International Workshop on Automated Test Equipment: VISION ATE 2020.

Authors

Heidi Barnes, Senior Application Consultant, Verigy

José Moreira, Senior Application Consultant, Verigy

Michael Comai, Senior Product Engineer, AMD

Abraham Islas, Senior Product Engineer, AMD

Francisco Tamayo-Broes, Product Development Engineer, AMD

Mike Resso, Signal Integrity Measurement Specialist, Component Test Division, Agilent Technologies

Antonio Ciccomancini, Application Engineer, CST

Orlando Bell, Vice President, Engineering, GigaTest Labs

Ming Tsai, Principal Engineer, RF Design Group, Amalfi Semiconductor

Chapter 14

Frequency Domain Calibration: A Practical Approach for the Serial Data Designer

14.1 Abstract

The proliferation of high-speed serial links means an ever-growing number of engineers is concerned with multi–Gbps data rates on interconnects that span a few inches to tens of meters. The characterization of interconnects in this frequency regime is defined in terms of scattering parameters (S-parameters) and typically requires a network analyzer. The steep learning curve and detailed calibration procedures of a vector network analyzer (VNA) place this technique beyond the reach of many of the serial data engineers that need the measurements the most. When only one S-parameter is really needed for a compliance test, the user still has to pay the overhead of obtaining all the S-parameters when doing a frequency domain measurement. An alternative approach for high-bandwidth characterization of interconnects leverages the simplicity of time-domain reflectometry (TDR) measurements, yet provides an accuracy level well suited to the requirements for high-speed serial data systems.

14.2 Characterization of Serial Channels

Interconnects are not transparent in the multi-gigahertz frequency range where high-speed serial links operate. This means characterizing interconnects at these high frequencies is a critical step in verifying compliance for operation in applications such as PCI Express, HDMI, SATA, and Infiniband.

Historically the VNA has been the workhorse of the RF engineer with operating bandwidths well above the 20 GHz range, and accuracy of a small fraction of a dB. Even though digital performance is always measured in the time domain, the popular use of VNAs has pushed standards organizations to begin to define compliance of channels in the frequency domain as well.

Whether the measurement is performed in the time or frequency domains, it can be mathematically transformed from one to the other using the Fourier transform or inverse Fourier transform. This makes the source of the measurement transparent. *Figure 14.1* is an example of the measured return and insertion loss of the same interconnect measured with a VNA in the frequency domain and with a TDR, measured in the time domain.

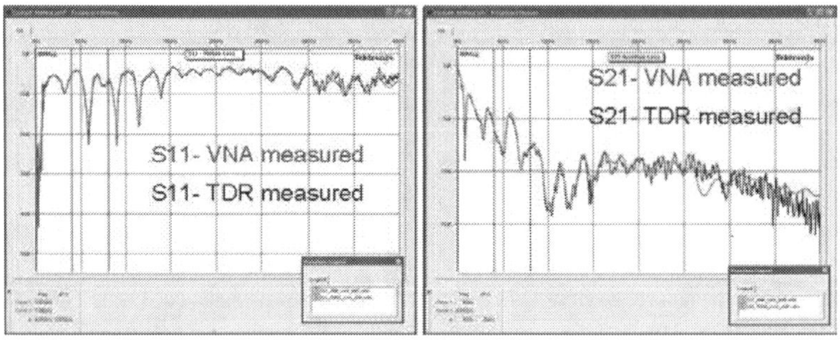

Figure 14.1: *Comparison of Measured Return and Insertion Loss of an Infiniband Cable, Independently Measured with a TDR and VNA*

With the ability to move seamlessly back and forth between the time and frequency domains, it is possible to select the optimum domain to display the data based on the sort of question being asked.

In a four-port measurement of a differential channel, there are really 10 unique S-parameter terms. When converted into differential form, these elements describe 10 qualities of the channel. Some terms are more useful than others.

For example, to evaluate the quality of the signal at the receiver, the most important S-parameter term is the differential insertion loss or the SDD21 term. The frequency domain form of this term is often used by compliance organizations in specifications.

However, to evaluate the performance of the interconnect and what a received eye might look like, the time domain impulse response is actually used. From the transmitted behavior of the impulse response, the transmitted performance of any arbitrary waveform can be simulated using convolution integral methods. For example, the convolution of a pseudo-random bit sequence (PRBS) and the impulse response of the channel is the real-time signal the receiver might see. This can be sliced synchronously with the clock, and an eye diagram can be created. An example is shown in *Figure 14.2*.

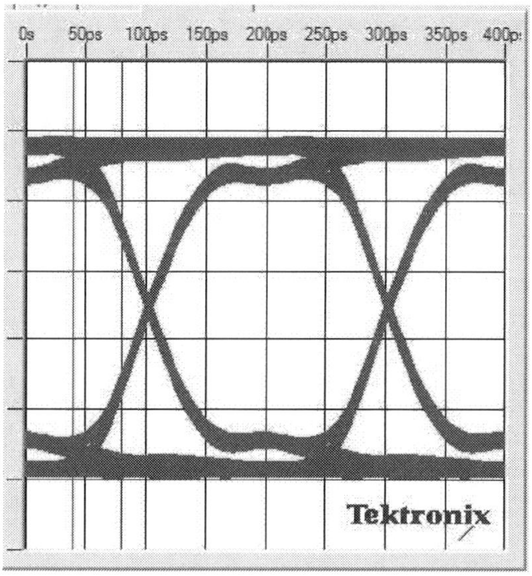

Figure 14.2: *Synthesized Eye Diagram for a 5 Gbps PRBS Signal through a Cable Using the One-Step Thru Calibration*

It is not possible to tell by looking at the eye diagram alone whether the original data was taken in the frequency domain or the time domain.

14.3 Calibration Methods

The primary distinction between the domains in which the measurement is taken is the instrument used. In the frequency domain, the instrument is exclusively a VNA, while in a time domain measurement, the instrument is exclusively a TDR, capable of also measuring a transmitted voltage response.

There are two important distinctions between the instruments. Because of the nature of the sampling for the TDR measurements, there are some constraints on the frequency steps and the highest-frequency range. In addition, the method of calibration is radically different.

A common calibration method with a VNA is the short-open-load-thru-isolation, or SOLT I, method. In this technique, measurements are performed on each port when a precision reference standard composed of a short, open, 50 Ohm load is connected. Pairs of ports are connected together and a thru is measured. Finally, all ports are disconnected and the isolation between ports is measured. In a four-port calibration, there are 17 calibration measurements performed.

The measured S-parameters of these references are combined to create correction factors applied to subsequent measurements to de-embed the internal sources, receivers, switches, cables, and connectors to the end of the cables.

After this calibration procedure, all 16 S-parameter terms, both as single-ended and as differential or balanced S-parameters, are available. Even if only one term such as SDD21 is required for a test, the same calibration process must be used. The cost for the one measurement pays the overhead for all the unused S-parameter terms.

This is in contrast to a time domain measurement and calibration. Each specific S-parameter term can use its own calibration step so if only one term, such as SDD21 is required, only the calibration step for that term needs to be performed.

For example, an SDD21 measurement through a device under test (DUT) can be completed with a one-step thru calibration process. A reference thru is initially measured. This can be as simple as a direct connection between the sources and receivers of the two lines that make up the differential pair. Or, it can include any fixturing on the printed circuit board (PCB) launch. A direct connection between the transmitter and receiver of one line in a differential measurement is shown in *Figure 14.3*.

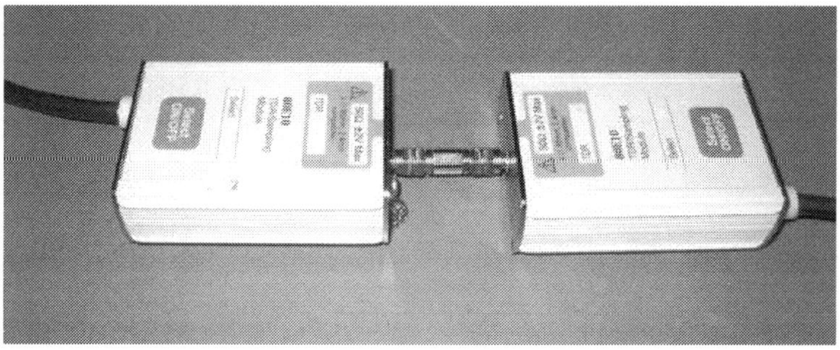

Figure 14.3: *Photo of the Transmit and Receive Channel of One Line in a Pair, Connected in a Thru Reference Configuration*

The measured received voltage signal, VDD21-reference, becomes the reference standard. Next, the DUT and its fixturing is connected in a thru and the transmitted voltage signal is measured, VDD21.

These time domain measurements are converted into the frequency domain with an FFT and their ratio, at each frequency, is calculated. This becomes the differential insertion loss, SDD21. This process is shown graphically in *Figure 14.4*.

$$SDD_{21} = \frac{FFT\{VDD_{21}(t)\}}{FFT\{VDD_{21\text{-reference}}(t)\}} =$$

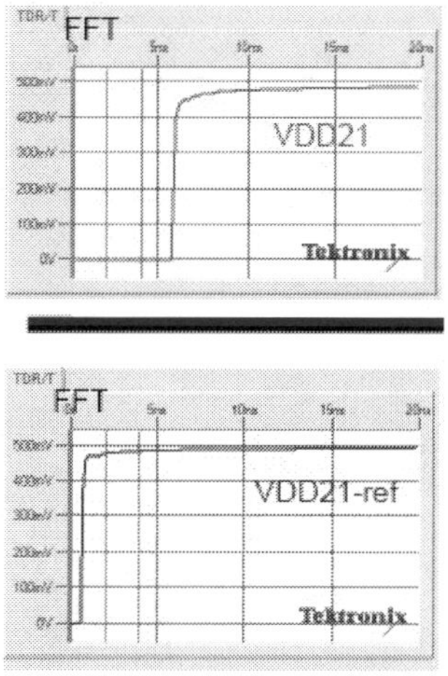

Figure 14.4: *Illustration of the One-Step Thru Calibration Process in a Time Domain Measurement of the Differential Insertion Loss of a DUT*

The error in performing this simplified calibration is proportional to the square of the SDD11 value of the fixturing. If the cables and connectors of each line are well matched to 50 Ohms, this calibration process can have an absolute accuracy of better than 1 percent.

In addition to simply removing the reference thru, in dB, the phase of the reference thru is also removed from the phase of the combination of DUT and fixturing.

An example of the resulting differential insertion loss of a SATA cable is shown in *Figure 14.5*.

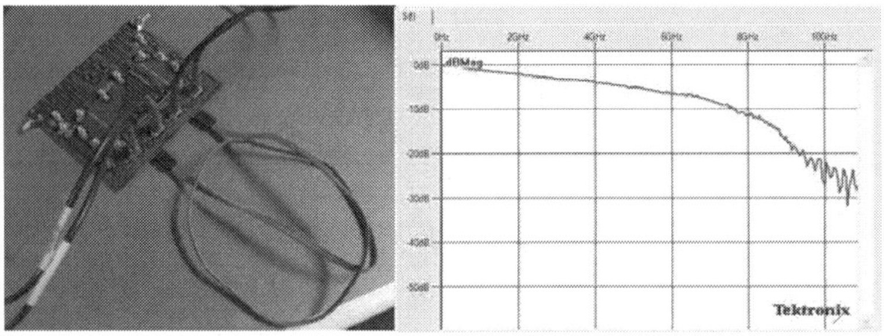

Figure 14.5: *Measured Differential Insertion Loss, SDD21, of a SATA Cable, with the TDR Cabling and PCB Trace Fixture Thru Removed from the Measurement Using the One-Step Thru Calibration Process, as Described in the Text*

This one-step thru calibration can dramatically simplify differential insertion loss measurements and increase productivity.

14.4 Frequency Limits from Time Domain Measurements

A second distinction between frequency domain and time domain measurements is the frequency range spanned by the instruments. In a network analyzer, the lower-frequency range is usually set by the quality of the directional coupler. As the higher-end frequency increases, the low-end frequency usually goes up as well. This is usually in the 10 to 50 MHz range. The high end in a VNA can exceed 50 GHz.

In a time domain measurement, the low-end frequency is much closer to DC. Rather than being set by the time window, which might be 10 MHz, the actual DC limit in a TDR response is the period of the repetition rate of the pulse generator, which can be on the order of 50 kHz. This enables a better extrapolated value to DC than a VNA measurement.

The frequency resolution is set by the time window of the measurement. If the time window is 100 ns, the frequency interval is 1/100 ns or 10 MHz.

In order to unambiguously measure the total time delay of an interconnect, there must be at least two frequency measurements within a cycle, otherwise aliasing can occur. Since a cycle is 1/time delay, the minimum frequency interval must be less than ½ x 1/TD. This corresponds to the rough rule of thumb that the time window in the equivalent time domain measurement of a VNA, should be at least twice the time delay of the cable.

This will assure receiving the transmitted signal through the interconnect and enough time to detect the settling of the signal and some of the low-frequency properties of the losses. A longer time window will display finer frequency resolution in the lower-frequency range.

With a wiring delay of about 5 ns/m, a 10 m long cable has a time delay of about 50 ns, requiring an equivalent time window of at least 100 ns. The required frequency interval of 10 MHz sets 10 m as the boundary of the longest-length cable that can be measured unambiguously with a VNA.

This is not a limitation in TDR–based measurements, where the measured time interval can be extended to many times the time delay of the interconnect. This is especially important in long cable characterization. Cables longer than 10 m can be easily measured in the time domain.

The high-frequency limit of the displayed data in a time domain measurement is ultimately set by the Nyquist limit. This is the highest frequency that can be measured based on the sampling interval for 2 measurements per cycle. Given the frequency interval, the Nyquist frequency is N/2 x the frequency interval, where N is the number of sampled time domain points.

For example, if the time base is 100 ns, the frequency interval is 10 MHz. If 4,000 points are used, then the Nyquist limit is 4,000/2 x 10 MHz = 20 GHz. Increasing the number of points that are included in the FFT will push the Nyquist limit to higher frequency.

This can be accomplished by recording consecutive blocks of transmitted signal, all in the same time window, but shifted in time. *Figure 14.6* shows an example of four consecutive time windows that increase the number of points to 16,000. With a time window of 100 ns, these four sets of measurements result in a Nyquist limit of more than 80 GHz.

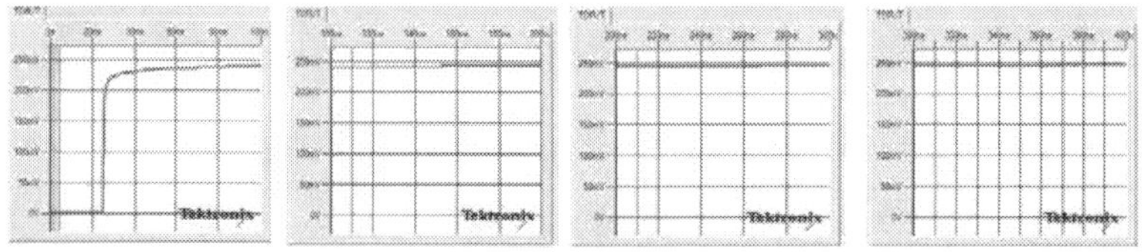

Figure 14.6: *Four Consecutive Measurements of the Transmitted Signal through a Long Cable, Increasing the Number of Collected Points to 16,000 and Increasing the Nyquist Upper-Frequency Limit to 80 GHz. Tektronix's Iconnect Software Addresses This Requirement and Does the Waveform Stitching Automatically, Enabling up to 1,000,000 Points or More to Be Captured*

Before the 80 GHz Nyquist limit is reached, another factor limits the upper-frequency limit from the time domain measurements. This is due to a reduced signal-to-noise ratio (SNR) of the higher-frequency components. In a step-edge response, the amplitude of the higher-frequency components drops off as 1/f. This is illustrated in *Figure 14.7.*

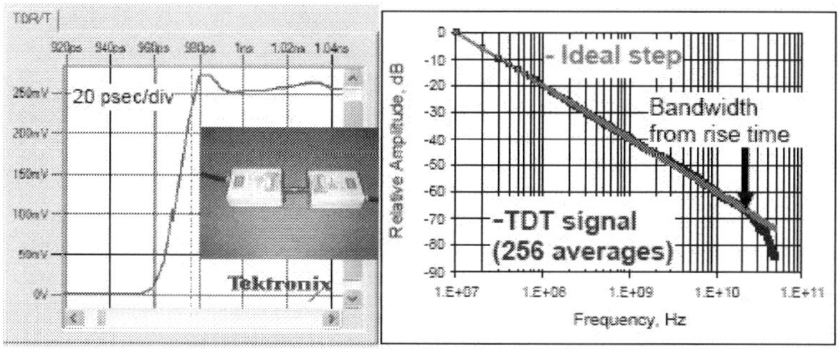

Figure 14.7: *Time Domain Received Transmitted Signal with 16 ps 10–90 Rise Time and Frequency Domain Spectrum Compared to an Ideal 0 ps Rise-Time Step*

The amplitude of the ideal step edge, with a 0 ps rise time, drops off at the rate of 20 dB/decade. The rise time of the measured step edge is about 16 ps. Its bandwidth is about 0.35/16 ps = 22 GHz. As seen in *Figure 14.7*, at slightly beyond 22 GHz, the amplitude of the received signal drops off faster than the ideal case. The practical high-frequency limit from a time domain measurements is set when the amplitude of the received frequency component approaches the noise floor of the wideband receiver. In this example, it is above 50 GHz and can best be explored by measuring specific structures in the time domain and converting to the frequency domain.

14.5 Intrinsic Performance Limits

The ultimate limit to the quality of the measurements in the time domain and displayed in the frequency domain is set by the noise of the receiver in the time domain. An isolation measurement, when the receiver is disconnected from the thru connection and the receiver measures only noise, clearly shows this limit. *Figure 14.8* demonstrates the measured noise in the time domain being about 0.1 mV in amplitude.

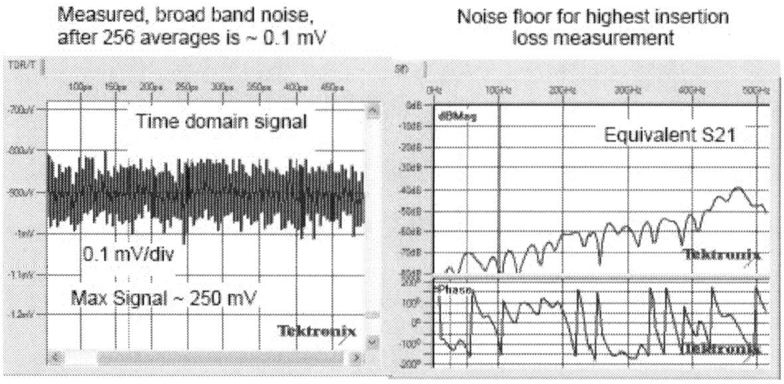

Figure 14.8: *Measured Noise in the Time Domain and the Corresponding S21 Response in the Frequency Domain with No Connection between the Receiver and the Transmitter*

This noise is converted to the frequency domain by using the reference thru as the signal reference in a one-step calibration. With a noise amplitude of roughly 0.1 mV and a signal amplitude of 250 mV, the noise is about -68 dB down. This is roughly the noise floor for the time domain measurement with 256 averages.

In the frequency domain, the noise floor for the insertion loss measurement starts at about -80 dB and increases to about -70 dB at 10 GHz and approaches -40 dB near 50 GHz. This sets the limit to the lowest insertion loss or the most loss that can be measured using 256 averages.

The second limit is set by the smallest change in insertion loss that can be measured in the frequency domain. This can be obtained by measuring a thru connection repeatedly and using one of the measurements as a reference. In effect, it is a measure of the impact on insertion loss from the random noise of the wideband amplifier.

Figure 14.9 is an example of the time domain measurement of a thru measurement, with two averaging times of one average and 256 averages and the corresponding S21 measurement in the frequency domain.

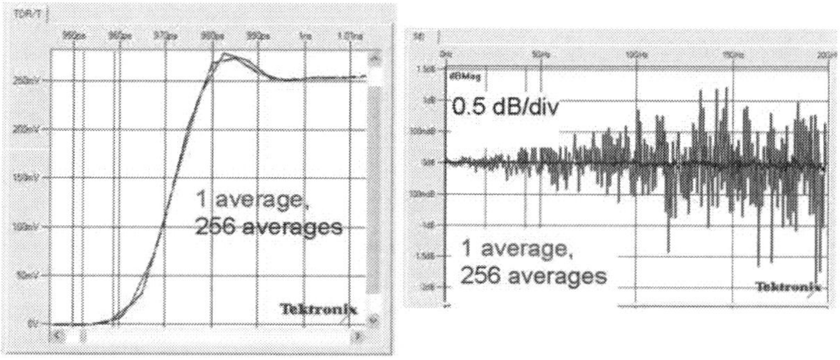

Figure 14.9: *Measured Thru Reference with One Average and 256 Averages in the Time Domain and the Frequency Domain*

Using one of the thru measurements as a reference, the S21 response of the other thru measurements can be displayed in the frequency domain using the one-step thru calibration. This is a direct measure of the impact the noise in the time domain has on the noise in the frequency domain. With 256 averages, the noise amplitude is about 0.05 dB at 10 GHz, and much smaller below 10 GHz. *Figure 14.10* shows four thru measurements and the intrinsic measurement limit to the smallest change in insertion loss that is significant.

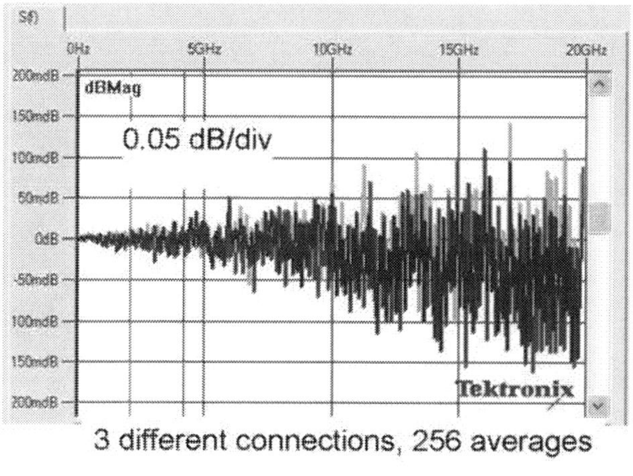

Figure 14.10: *Repeated Thru Measurements Showing the Reproducibility Limit to Be about 0.05 dB at 10 GHz for a Time Domain Measurement*

These measurements set the sensitivity limits to be expected for a time domain measurement using 256 averages. At 10 GHz, the largest insertion loss measurable is about -70 dB and the smallest insertion loss measurable is about -0.05 dB.

14.6 Variation in Typical Fixtures

Very few high-speed serial link interconnects use coaxial connectors in their typical applications. To enable a clean interface between the interconnect and the coaxial connectors of the VNA or TDR, an interface board, which acts as a geometry transformer, is usually used. A coaxial connection, usually an SMA, is at one end of the board and the mass termination connector for the interconnect is on the other end. This is the general structure for cable interconnects as well as backplane interconnects.

Any SDD21 measurement of the DUT includes the SMA launch, the short length of trace on the board, and the standard connector. When just the measurement of the interconnect is desired, various methods of removing the fixture's impact from the measurement can be used. These include de-embedding, using reference SOLT structures on the fixture board, TRL calibration, and simple 2x thru measurement.

In practice, only experienced users go through the added task of a TRL or de-embedding procedure. The most common solution is to either ignore the impact of the board fixture, or subtract the 2x measurement from the total measurement to get the DUT only.

An implicit assumption with all of these techniques is that the reference used to create the calibration is exactly the same structure with the same high-frequency performance as the specific element that is in series with the DUT. Any variations between the calibration standard measured and the fixture to the DUT will not be compensated by the calibration process and will be included in the DUT measurement.

For example, just the simple process of connecting cables and an SMA thru barrel of typical lab quality and not calibration quality has considerable variation from connection to connection. *Figure*

14.11 shows an example of the repeated measurements of adaptors, cables, and SMA thrus, measured for four nominally identical connections.

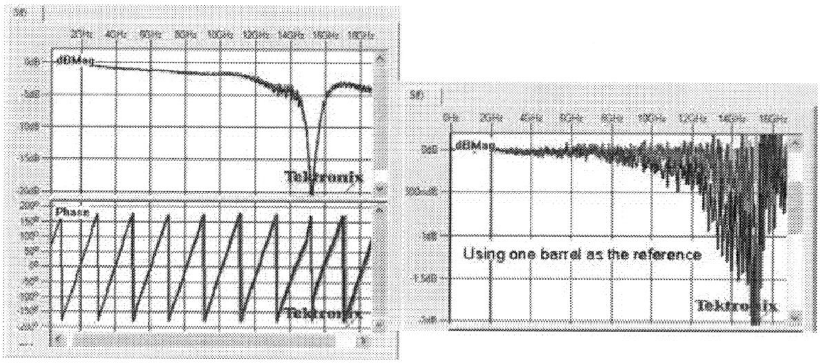

Figure 14.11: *Left: Measurement of the Adaptors, Cables, Connectors, and SMA Barrel Using a Direct Connection as the Reference Thru. Right: Using One of the SMA Barrel Thrus as the Reference in a One-Step thru Calibration, Comparing the Variation between Four Other Measurements*

The variation between different, but nominally the same, connections at 10 GHz is about 0.15 dB. This is more than three times above the intrinsic limit of the time domain measurement. It also increases very rapidly above 10 GHz, an indication of the sensitivity of high-frequency performance to the quality and normal variation of coaxial connections.

At 10 GHz, the insertion loss of the adaptors, cables, and SMA connection is about -2 dB. The variation of 0.15 dB around this nominal value is about 7 percent variation from connection to connection.

The large, narrowband resonance dip at about 15 GHz is due to a transverse resonance in one of the couplers that cuts off frequency components with a half wavelength that matches its transverse mode. This is close to the normally reported limit to low-cost SMA connections of 18 GHz.

If just lab-quality cables and SMA connections were used in the fixture path, the reproducibility from measurement to measurement would be about 0.15 dB. On top of this is the variation in the fixture boards typically used as geometry transformers. *Figure 14.12* is an example of a 2x thru reference connection in a six-layer circuit board used to interface to a SATA cable with a standard connector.

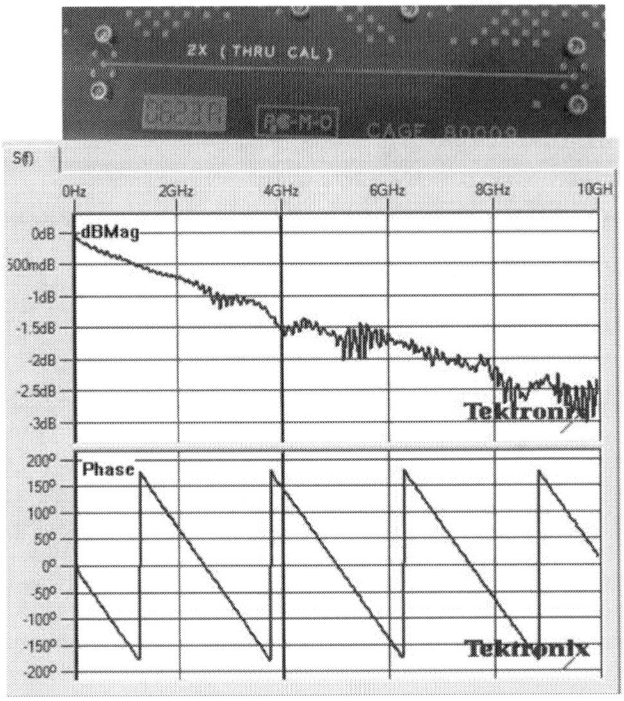

Figure 14.12: *Photo of the 2x Ref Calibration Structure Used to Connect from an SMA Connection to a SATA Cable. The Measured Insertion Loss of the 2x Reference Trace Is Shown as Roughly -2.5 dB at 10 GHz*

As a rough measure of the variation of the reference 2x thru to the actual connection between the SMA and the connector, the 2x reference thru on eight boards was measured. The comparison of these eight reference thrus is shown in *Figure 14.13*, when one of them is used as the reference for the others in a one-step thru calibration.

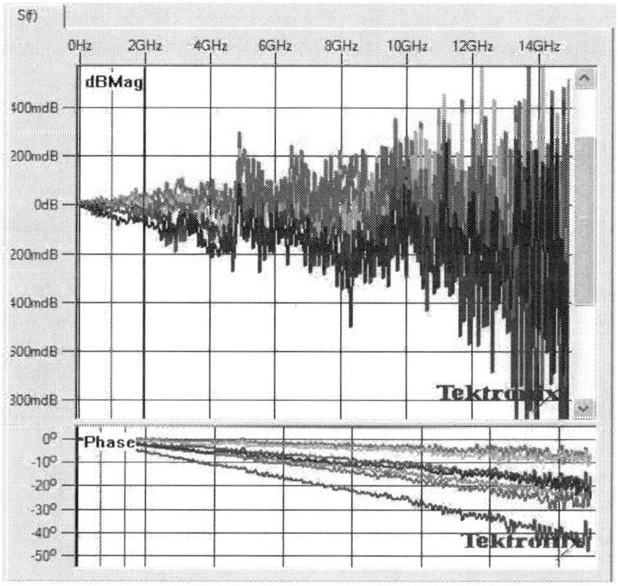

Figure 14.13: *Measured Insertion Loss of Eight 2x Reference Thrus on Eight Interface Boards*

These are nominally identical traces on the boards, yet show a variation of as much as +/- 0.3 dB at 10 GHz in their insertion loss. This is out of a nominal value of about -2.5 dB or about 10 percent in the insertion loss. This could be from variation in the dissipation factor from board to board, or how the SMA was mounted to the board.

The phase variation of about +/- 22 degrees at 15 GHz, corresponds to about +/- 4 ps, out of a time delay of about 320 ps, or about 1 percent. This could easily be due to dielectric constant or glass weave effects.

These variations are a good measure of what might be expected as the reproducibility of interface fixtures. Even if these 2x reference thrus are used to calibrate out the roughly -2.5 dB of insertion loss from the composite measurement to get the behavior of just the DUT, the difference between the reference standard that is measured to what is between the coaxial connector and the DUT may vary by at least 0.3 dB.

This variation is at least six times larger than the measurement limit in a time domain measurement.

14.7 Conclusion

The intrinsic measurement capabilities of state-of-the-art TDR–based systems provide equivalent S-parameter measurements from near DC to bandwidths in excess of 50 GHz. The dynamic range at the low end is 70–80 dB, with at least 40 dB dynamic range at 50 GHz. The smallest, significant, measurable change in the insertion loss can be as small as 0.05 dB.

Using a one-step thru calibration process, a time domain measurement of an S-parameter such as differential insertion loss, does not have to carry the calibration overhead of all 16 S-parameter measurements. The same 2x reference standard on the typical interface card is the only standard needed to calibrate out the impact of the cables, connector, and circuit board connection from a composite measurement to reveal just the behavior of the DUT.

The typical variation in a reference standard from board to board or between the reference standard and the connection to the DUT is more than 0.3 dB. This is well above the intrinsic limit of 0.05 dB in time domain measurements. The one-step thru calibration's ease of use, intrinsic accuracy, and reproducibility make this technique well suited to high-speed serial data interconnect characterization tasks.

Authors

Steven Corey, Principal Engineer, Electro-Optical Product Line, Tektronix

Eric Bogatin, Signal Integrity Evangelist, Bogatin Enterprises

Dima Smolyansky, Product Marketing Manager, Tektronix

Chapter 15

Practical Design and Implementation of Stripline TRL Calibration Fixtures for 10-Gigabit Interconnect Analysis

15.1 Abstract

The design of today's gigabit interconnects require sophisticated measurements; however, error correction techniques described to date have been overly theoretical. This paper will illustrate the practical steps required to create a stripline thru-reflect-line (TRL) calibration kit for a vector network analyzer (VNA). The creation of a real-world stripline TRL cal kit will result in discovering more interconnect performance margin than originally expected. Example elements will be illustrated for a six-layer Rogers 4350 printed circuit board (PCB) and an eight-layer Rogers 4350 PCB.

15.2 Introduction

As communication speeds push beyond 10 Gbps, the need for accurate measurements of components of the physical layer become critical. Backplanes, PCBs, and connectors must be characterized with precision in order to gain the performance margin required for the industry's highest data rates. A number of pre-measurement and post-measurement error correction techniques can be utilized to obtain the appropriate figures of merit for a specific component. Time domain gating, port extension, reference plane calibration, normalization, short-open-load-thru (SOLT), TRL, load-reflect-match (LRM) and de-embedding are a few of the most popular techniques used today. Of the most useful calibration techniques that can yield accurate measurements with less effort than most is the TRL calibration. TRL is a pre-measurement error correction that is primarily used in non-coaxial environments such as testing waveguides, using test fixtures, or making on-wafer measurements with probes. TRL uses the same 12-term error model as a SOLT calibration, although with different calibration standards. The standard SOLT calibration standards are provided by the equipment manufacturer, whereas the TRL standards must be designed, developed, fabricated, and characterized by the signal integrity engineer. Since these TRL standards often consist of stripline PCB fixtures, the need for via structures and connectors creates design challenges that can degrade the calibration accuracy.

The relationship between the calibration elements and the TRL fixture will be discussed. To accomplish a good measurement, the fixture and the calibration kit must share several common elements, which include the SMA to stripline interface (the launch) and a specified length of transmission line. Minimizing variation between the fixture elements and calibration kit elements should be a design goal for proper measurement, as well as careful attention to vias, yielding gains in calibration kit and fixture design efficacy. If proper design methods are used, the load can be used from DC to several GHz and the SMA launch can be used to approximately 20 GHz. This allows fewer lines, simplifying the calibration kit. It will be shown how it is possible to calibrate a VNA from DC to 24 GHz using a single line element. Fewer connections to accomplish a calibration will result in fewer mistakes and better calibrations. It is our goal to present a case study that will emphasize practical tools and techniques to help ease the burden of creating a TRL calibration kit using common stripline PCB processing methodologies.

15.3 Why Calibrate?

Why do we need to calibrate a network analyzer? Isn't this expensive equipment good as is? To answer these questions, we need to examine the key building blocks of a network analyzer, what it measures, and the major contributors of measurement errors. Only perfect test equipment would not need correction. Imperfections exist in even the finest test equipment and cause less-than-ideal measurement results. Some of the factors that contribute to measurement errors are repeatable and predictable over time and temperature and can be removed, while others are random and cannot be removed. The basis of network analyzer error correction is the measurement of known electrical standards such as thrus, open circuits, short circuits, and precision load impedances. Typical VNA measurement errors are shown in *Figure 15.1*.

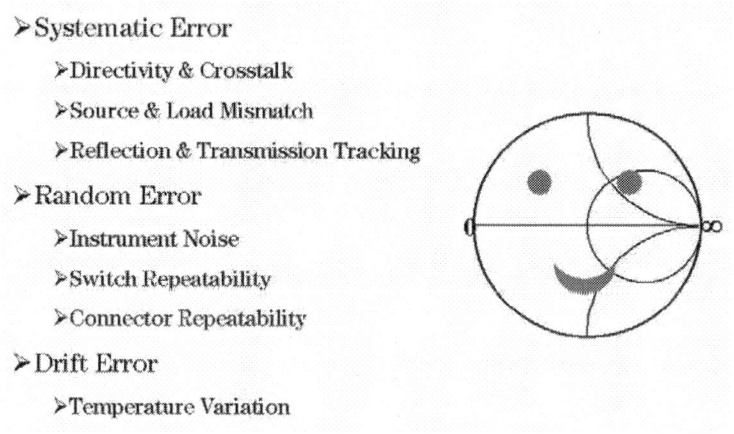

Figure 15.1: *Calibration Considerations in VNAs*

15.4 Linear Two-Port Network Analyzer Measurements

The foundation for understanding VNA measurement errors lies within understanding the general architecture of this test instrumentation. The most basic network analyzer shown in *Figure 15.2* consists of an accurate sine wave signal source and a high-frequency switch that routs the signal to the forward measurement direction or the reverse measurement direction. A signal separation device called a coupler is used to sample the incident signal and the reflected signal at the input port of a DUT. Another coupler is used in a similar fashion to separate the signal at the output port of the DUT. The sampled signals ao, bo, a3, and b3 can be processed to obtain the input reflection and forward transmission characteristics of the DUT.

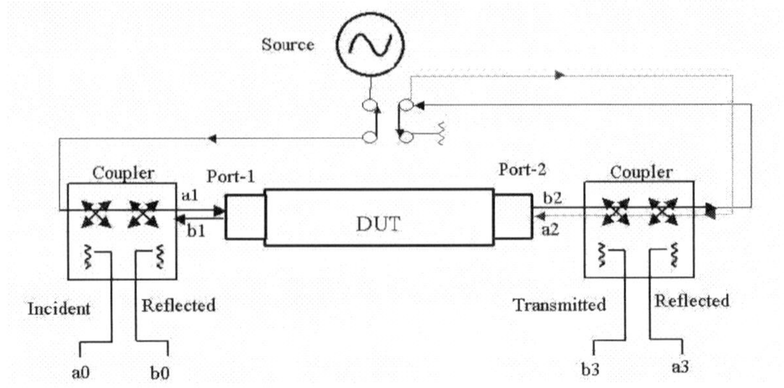

Figure 15.2: *Linear Network Analyzer Measurements*

Two Port S-Parameters Defined

These input and output signals can be represented by a signal flow graph and expressed mathematically. The signal flow graph shown in *Figure 15.3* is a good picture of what happens in a stimulus/response type of measurement. The formulas for relating the measured quantities to the S-parameters of a DUT are also shown. The definitions of terms are as follows: S = scattering, a1 = incident wave at port 1, a2 = incident wave at port 2, b1 = reflected/transmitted wave at port 1, and b2 = reflected/transmitted wave at port 2. The relationship between these parameters can be expressed in a matrix math equation.

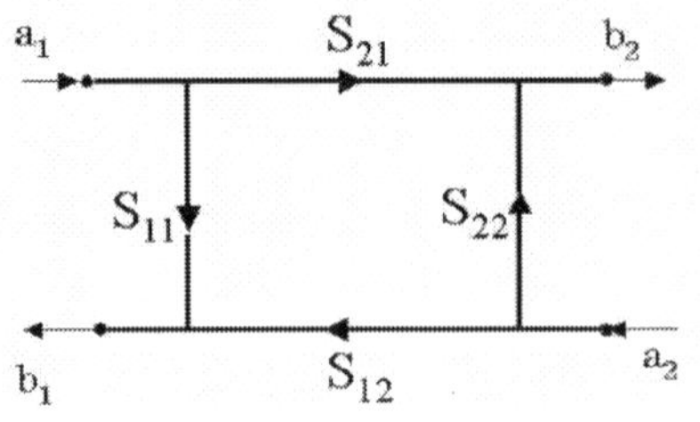

$$b_1 = a_1 S_{11} + a_2 S_{12}$$
$$b_2 = a_1 S_{21} + a_2 S_{22}$$

$$\begin{bmatrix} b_1 \\ b_2 \end{bmatrix} = \begin{bmatrix} S_{11} & S_{12} \\ S_{21} & S_{22} \end{bmatrix} * \begin{bmatrix} a_1 \\ a_2 \end{bmatrix}$$

Figure 15.3: *Two-Port S-Parameters Defined*

15.5 VNA Measurement Errors

All measurement systems, including those employing network analyzers, can be plagued by three types of measurement errors: systematic errors, random errors, and drift errors.

Systematic errors are caused by imperfections in the test equipment and test setup components such as cabling. If these errors do not vary over time, they can be characterized through calibration and mathematically removed during the measurement process. There are six types of systematic errors: directivity and crosstalk errors relating to signal leakage, source and load impedance mismatches relating to reflections, and frequency response errors caused by reflection and transmission tracking within the test receivers.

Random errors vary randomly as a function of time. Since they are not predictable, they cannot be removed by calibration. The main contributors to random errors are instrument noise (e.g., sampler noise, the IF noise floor), switch repeatability, and connector repeatability. When using network analyzers, noise errors can often be reduced by increasing source power, narrowing the IF bandwidth, or using trace averaging over multiple sweeps.

Drift errors occur when a test system's performance changes after a calibration has been performed. They are primarily caused by temperature variation and can be removed by additional calibration. The rate of drift determines how frequently additional calibrations are needed. However, by constructing a test environment with stable ambient temperature, drift errors can usually be minimized. While test equipment may be specified to operate over a temperature range of 0°C to +55°C, a more controlled temperature range such as +25°C ± 5°C can improve measurement accuracy (and reduce or eliminate the need for periodic recalibration) by minimizing drift errors. *Figure 15.4* shows the block diagram of various components that are susceptible to these errors.

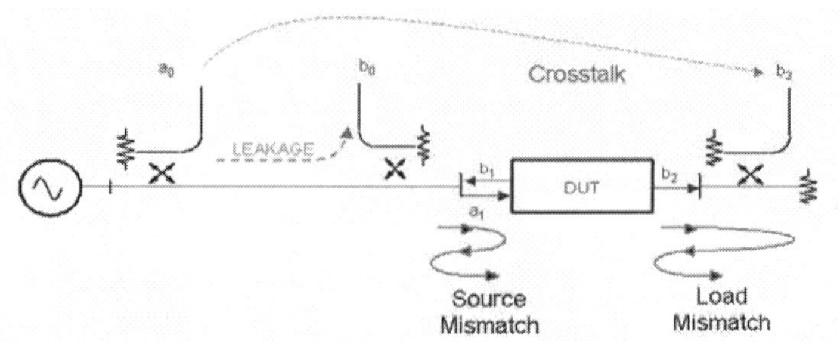

Figure 15.4: VNA Measurement Errors

15.6 Vector Network Analyzer with Four Ports

The four-port VNA shown in *Figure 15.5* has a single swept frequency source that is switched to each port to make a reflection and transmission measurement. The source is sampled by the reference receiver. The switches are set to route the incident signal through the directional coupler and to the desired test port. The directional coupler separates the reflected signal from incident signal and switches route the reflected signal to the "A" sampler. The S11 measurement is the ratio of A/R, which is equivalent to TDR measurement in the time domain. Transmission measurements are the ratio of B/R and are equivalent to the TDT measurement. The source, reflected, and transmitted signals are appropriately routed to complete the set of 16 S-parameter measurements for a four-port DUT.

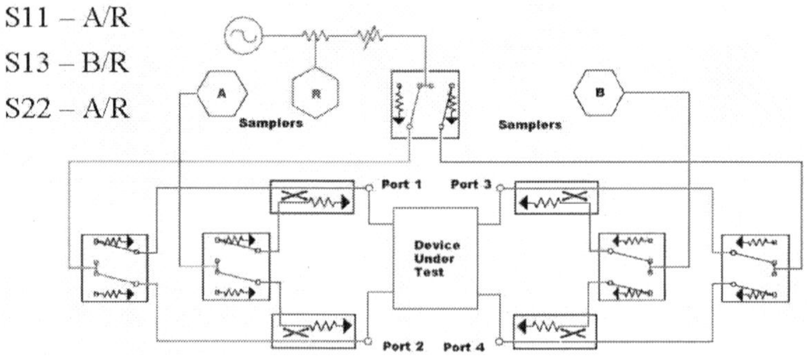

Figure 15.5: Vector Network Analyzer Block Diagram with Four Ports

N-Port S-Parameters Defined

As shown in *Figure 15.6*, the S-parameter definition can be expanded to N-ports for multiport applications. Many new high-speed digital protocols now recommend measurements that require the use of more than four ports, including differential crosstalk on serial ATA, PCI–X, high density multimedia interface (HDMI) and RapidIO. The most efficient way to make these measurements is with a physical layer test system (PLTS) that has more than four ports. By implementing this type of measurement technique, all device ports are connected at once for multiple differential channels. For example, a 12-port measurement system can make connection to three adjacent differential channels. This facilitates all near-end and far-end crosstalk combinations by eliminating the need to perform multiple calibrations and manage all the S-parameter data. Without the appropriate data management software, this can be an overwhelming task, because a 12-port system will produce 144 S-parameters.

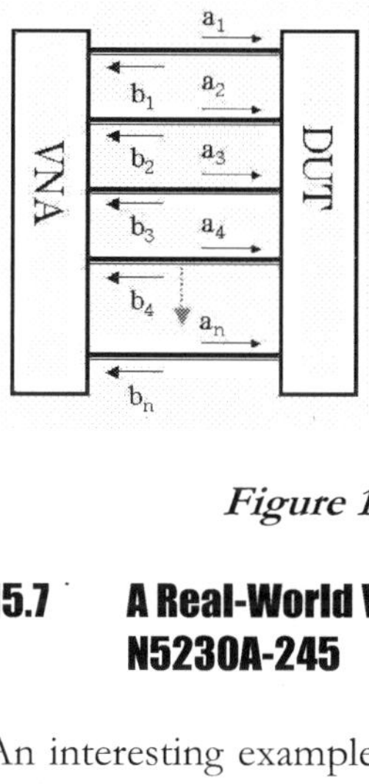

$$\begin{bmatrix} b_1 \\ b_2 \\ b_3 \\ \downarrow \\ b_n \end{bmatrix} = \begin{bmatrix} S_{11} & S_{12} & S_{13} & \rightarrow & S_{1n} \\ S_{21} & S_{22} & S_{23} & \rightarrow & S_{2n} \\ S_{31} & S_{32} & S_{33} & \rightarrow & S_{3n} \\ \downarrow & \downarrow & \downarrow & \square & \downarrow \\ S_{n1} & S_{n1} & S_{n1} & \rightarrow & S_{nn} \end{bmatrix} * \begin{bmatrix} a_1 \\ a_2 \\ a_3 \\ \downarrow \\ a_n \end{bmatrix}$$

Figure 15.6: *N-Port S-Parameters Defined*

15.7 A Real-World VNA Block Diagram Example – The Agilent N5230A-245

An interesting example of a real-world multiport network analyzer is the Agilent N5230A-245. High-speed digital data transmission is composed of differential signals, so four ports are now a requirement for measuring important performance parameters such as differential insertion loss in interconnects. This four-port VNA, shown in *Figures 15.7 and 15.8,* has all of the standard microwave components mentioned previously in the two-port VNAs.

However, it has an inherently lower system noise floor (trace noise of 0.006 dB rms at 100 kHz bandwidth) and higher dynamic range (up to 120 dB at 2 GHz). By utilizing advanced oversampling techniques, the system architecture enabled a large improvement in stability and drift errors over previous generations of VNAs. High-quality couplers and switches allow better error correction to achieve the lowest possible errors in measurements.

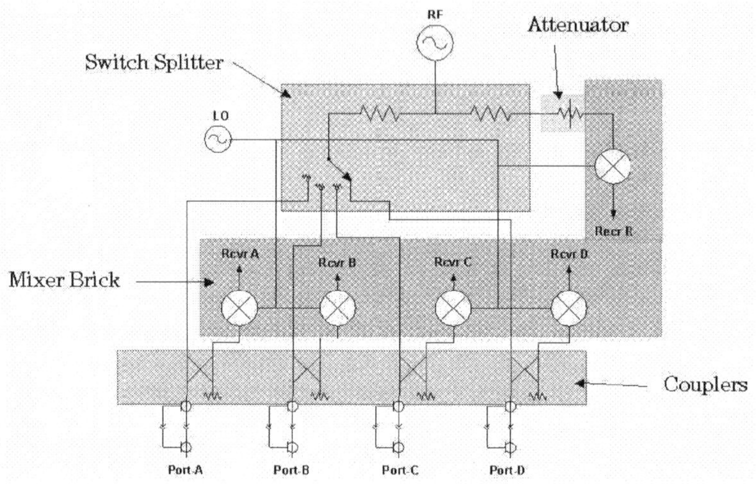

Figure 15.7: *Functional Block Diagram of Four-Port Vector Network Analyzers*

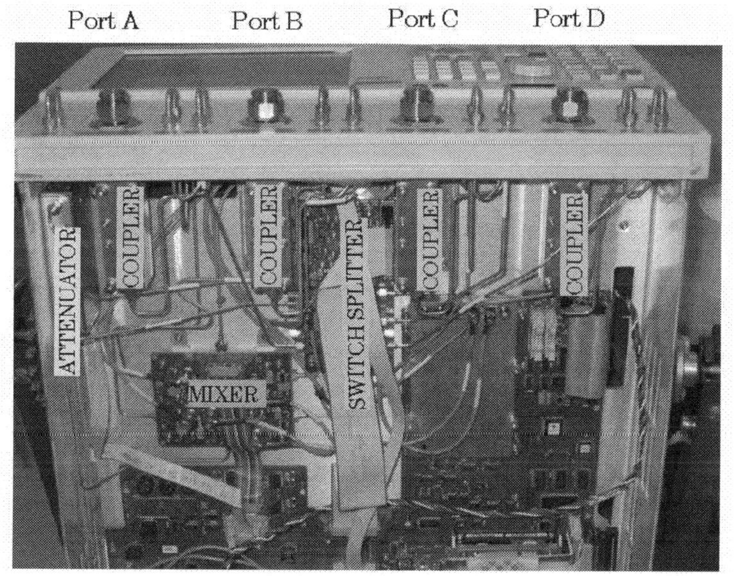

Figure 15.8: *Hardware Layout of Four-Port Vector Network Analyzer*

15.8 TRL Calibration Types

A major problem encountered when making network measurements in microstrip or other non-coaxial media is the need to separate the effects of the transmission medium (in which the device is embedded for testing) from the device characteristics. For example, testing a high-speed backplane connector requires the use of PCB test fixtures that adapt the test equipment 3.5 mm connectors to the mated connector pair. While it is desired to predict how the connector will behave in the environment of its final application, it is difficult to measure without the appropriate test fixture. The accuracy of this measurement depends on the availability of quality test fixtures. Unlike standard 3.5 mm connectorized coaxial measurements, a set of three distinct well-characterized impedance standards are often impossible to produce for non-coaxial transmission media (like the connector). For this reason, an alternative calibration approach may be useful for such applications. The TRL calibration technique relies only on the characteristic impedance of a short transmission line. From two sets of two-port measurements that differ by this short length of transmission line and two reflection measurements, the full 12-term error model can be determined. Due to the simplicity of the calibration standards, TRL can be applied in dispersive transmission media such as microstrip, stripline, and waveguide. With precision coaxial transmission lines, TRL currently provides the highest accuracy in coaxial measurements available today. Many names have been given to this overall approach, including self-calibration, thru-short-delay, thru-reflect-line, thru-reflect-match, line-reflect-line, line-reflect-match, quick-short-open-load-thru, and short-open-load-reference line. These techniques are all variations on the same basic approach and are shown in *Figure 15.9*.

Cal Type	2-port Std 1	2-port Std 2	2-port Std 3	Port 1 Std 1	Port 1 Std 2	Port 1 Std 3	Port 2 Std 1	Port 2 Std 2	Port 2 Std 3
TRL/LRL	Thru line		Delay line	Unkn# reflect			Unkn# reflect		
TRM/LRM	Thru line			Unkn# reflect	Match		Unkn# reflect	Match	
TRA/LRA	Thru line		Attenuator	Unkn# reflect			Unkn# reflect		
TNA/LNA	Thru line	network	attenuator						
QSOLT 1	Thru line			short	open	load			
QSOLT 2	Thru line			short		load	short		
QSOLT 3	Thru line				open	load			open
QSOLT 4	Thru line			short		load			load
LRRM	Thru line			Unkn# short	Unkn# open	load	Unkn# short	Unkn# open	
unkn thru (SOLR)	Unknown thru			short	open	load	short	open	load

Figure 15.9: *Various Types of TRL Calibration Variations Are Used Today by Microwave Engineers around the World*

15.9 A Stripline TRL Fixture – A Design Case Study

Many electrical interconnect manufacturers have the need to accurately measure the performance of their devices. The challenge for high-speed interconnect measurements is to not let the test fixturing itself interfere with obtaining accurate data. Very often, poorly designed test fixtures have poor signal integrity and display excessive impedance discontinuities, high series loss due to conductor skin effect, and high shunt loss due to dielectric materials. This all translates into a low-bandwidth measurement that makes the interconnect look much different than it is in reality.

This design case study will step the reader through the process of creating a TRL standard for avoiding these measurement errors. The authors have purposefully chosen a simple and straightforward tone when describing the methods used, including a highlighting of the pitfalls. However, great detail is provided in many specific areas. Hopefully, this practical approach will encourage more engineers to experiment with this calibration method and implement it in their next project. Design for testability is a discipline that will reap large rewards if the time investment is made early in the design cycle.

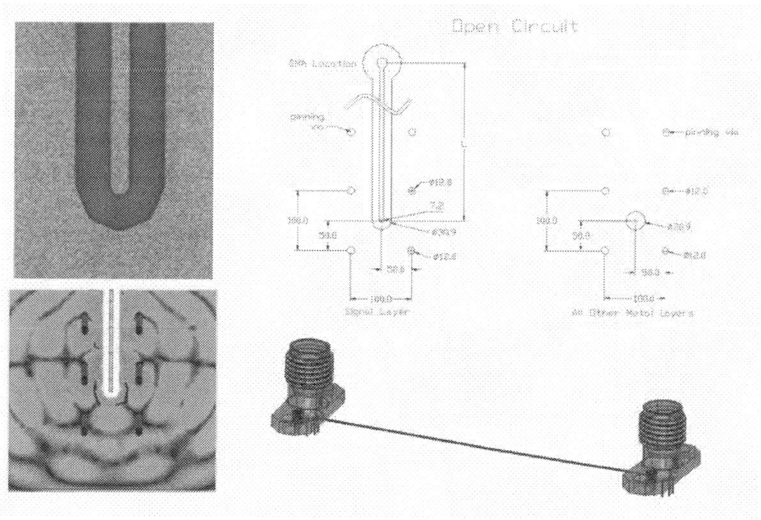

Figure 15.10: *A TRL Fixture Is Designed with the Assistance of a Three-Dimensional Electromagnetic Field Solver*

It may be best to consider a TRL design in two parts, macro and micro. In the macro consideration we will look at the design of the PCB, the fixture in general, and the TRL portion in particular. In the micro portion, we will consider the details that are needed for stripline. Most of the details revolve around the via and how to create a good launch and calibration standards in spite of this feature. First, a little philosophy.

TRL Least Common Denominator, an Open Circuit

We just discussed how calibration removes many types of errors. We can choose to introduce an error and remove it on purpose, as long as the error is systematic and repeatable. One thing we can do with this idea is change the location along the transmission line where the instrument stops and the DUT begins. To do this, we simply place our calibration standards at the location where we want to create the instrument–DUT interface. We could, in principle, use SOLT standards, but they are difficult to manufacture in coax with a machine shop. In stripline they will be exceedingly difficult to manufacture to the degree of accuracy needed to allow for a good calibration. Use the TRL standards instead, because the electrical characteristics of TRL standards do not need to be precisely known, they do not need the tight manufacturing controls of a machine shop. We can subject the TRL cal kit to the variation inherent in PCB manufacturing and still achieve a good calibration.

The least common denominator of the fixture and the cal kit is the open circuit. Every element contains at least this. For our purposes here, the open circuit includes a specific length of stripline, a launch, an SMA, a length of coax, and the VNA instrument itself. Because we are going to use the same instrument and length of coax every time, we can forget about these during the fixture design. As long as a torque wrench is used to mate the SMAs, the coax and instrument can be regarded as repeatable. To create the rest of the fixture and cal kit, we are going to keep the electrical characteristics of the stripline, launch, and SMA the same as much as we can. To a large degree this translates into keeping the mechanical characteristics of this collection of parts the same each time we create it. This collection of parts can be thought of like a time domain scope probe. Better still, like a probe station probe. Like a scope probe, this open circuit probe can measure whatever it touches, the cal features, or a DUT. Unlike a scope probe, we do not have the luxury of using the same device every time. A good, high-bandwidth, repeatable design is a must. And unlike a probe, our open circuit probe cannot be placed in a random location after the fixture is constructed. It must be placed during the PCB design and layout. Make every open circuit probe the same, as much as possible. This is the part we want to remove from the measurement, either through calibration or through de-embedding. In calibration, we will place the collection of standards at the stripline open circuit probe tip. For de-embedding, this is the collection of parts to model using a 3D field solver. Another name for the open circuit probe in the context of measuring the DUT is "the fixture." See *Figure 15.11*.

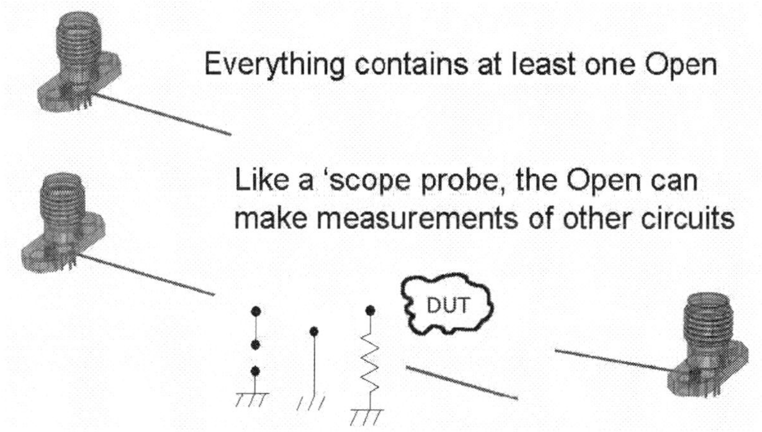

Figure 15.11: *The TRL Calibration Uses the Open Circuit Standard as a Probe*

The Macro Half of a TRL Design

We will consider the entire PCB for a moment and ask the following questions:

- What experiments should be conducted? Make sure the fixture allows for the experiments that are required. What non–TRL features need to be included? Make sure that they are included. How many signal layers? How thick should the PCB be? Will a full panel or a partial panel be used? Will the PCB lay flat on the lab bench, or will it be placed in a stand? Fixtures are costly, are time-consuming to design and manufacture, and do not forgive missing features.
- What material (DK) will be used to construct the fixture? The dielectric constant of the material to be used should be known, because time of flight delays will need to be

calculated. A material representative of the finished product is a common choice. Because this division of Molex, Inc., does not sell PCBs, we often choose Rogers 4350 to minimize rise times at the DUT. Rogers 4350 is a good material for VNA fixture dielectric use as well. Other dielectrics can be used. If your goal includes system-level testing, a dielectric representative of the one used by the system may be a good choice. If the goal is component-level testing, a high-performance material is a superior choice because it offers more consistent performance and better repeatability.

- How far away is the SMA interface from the DUT? Should the board "look" like something else for time domain tests? Are there mechanical considerations that limit SMA location? Generally speaking, the closer the SMA–PCB interface is to the intended DUT, the better, regardless of which calibration method will be used. Mechanical considerations caused us to place the SMAs about 3.5 inches from a low-loss DUT that is only about an inch long. Even so, we believe we obtained good measurement results.
- What level of performance does the cal kit and fixture need to deliver? The expected performance from the cal kit and the bandwidth of the desired data set will drive some compromises into the design. Needless to say, the care and attention to detail needed for a 1 Gbps design is different than that needed for a 10 Gbps design. If the measurements will be used for simulations (e.g., S-parameters used in SPICE), a wider bandwidth measurement than what is needed to confirm adherence to a specification may be required.
- How close to the DUT should the reference plane be? Determine where the reference plane will be placed. The reference plane is a largely arbitrary location along the transmission line from the VNA to the intended DUT. At the reference plane the measurement phase is zero, the gain is zero, and the return loss is the same as the noise floor of the measurement. It can be, and often is, placed very close to the intended DUT. The reference plane should be placed along the transverse electromagnetic

(TEM) mode transmission line; do not place it in a discontinuity (e.g., a via, a pin, an antipad). It may be necessary to leave some of the fixture transmission line in the measurement. As long as the transmission line that is included is understood and recorded, it can be accounted for. Once this location is determined, the actual DUT becomes that thing that is between the reference planes.

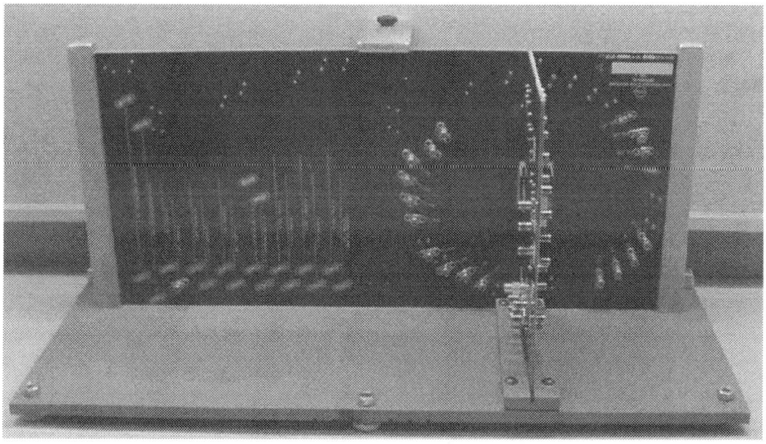

Figure 15.12: *Test Fixture with a TRL Calibration Kit (Left)*

15.10 The Macro Element View

We know what experiments and measurements we want to conduct; we know something about the mechanical constraints of the fixture. Now we need a calibration kit to assist with the measurements. Because we just considered the overall fixture constraints, we should know where to set the reference plane and therefore how much transmission line to assign to the open circuit standard.

A TRL calibration kit has the following four basic elements:

- Thru—This is just a pair of opens, tip of stripline to tip of stripline. A zero-length thru is discussed here. This element defines the zero-loss, zero-phase point. Ultimately, the DUT will be centered in the thru.

- Reflect—An open or a short, this element has to maintain its polarity (sign of the reflection coefficient), but the magnitude of the reflection need not be known. Use the same reflect on each port during calibration.
- Line—This is just like the thru, with an extra piece of transmission line inserted in the center. This extra transmission line needs to be 90° long at the center of the frequency band to be covered. The propagation delay of the extra transmission line in each line must be known as well. This element establishes the reference impedance.
- Load (sometimes called match)—Technically this is not a necessary element, but practically it is for broadband measurements. It serves to cover the bottommost frequency band and relieve the fixture of the need for very long lines. Two are required and each should deliver the same impedance. This element also serves to establish the reference impedance.

The elements previously listed are the minimum necessary to construct a broadband, TRL cal kit. A minimal kit, like the one illustrated in *Figure 15.13*, might allow calibration from DC to 20 GHz, if the load is carefully designed. Because it is difficult to design a load that works over a broad enough frequency range (DC to 2.5 GHz), it is common practice to employ more than one line in a kit. The lines need to be a quarter wave length (90°) long, plus or minus a lot, in the dielectric used for the stripline cavity. It may be better to think of them as lengths of transmission line that are NOT integer multiples of a half wave (180°) long. It is recommend that we stay 20° or more from the half wave length point.

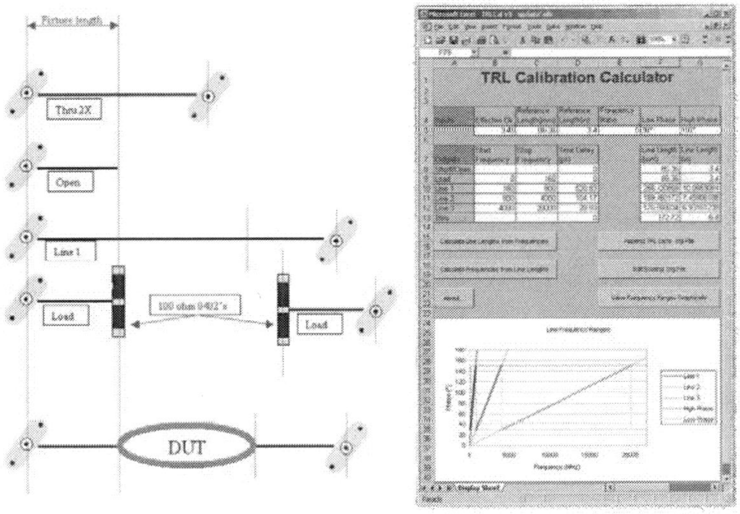

Figure 15.13: *TRL Calibration Fixture Layout with Line-Length Calculator. The Kit Illustrated Is the Bare Minimum Needed for a Broadband Calibration. Often Additional Lines Are Required to Extend Frequency Range*

It turns out that an engineering margin of 20° from a half wave is the same as a factor of eight in frequency range. We might design line-1 to work from 200 MHz to 1,600 MHz. The next line, line-2, will need to pick up where the last one left off, 1,600 MHz to 12,800 MHz. Line 1 would need to be a quarter wave longer in the dielectric than the thru at the center frequency (900 MHz), or about 1.639 inches longer if the system is in a DK=4 material. If all this seems complicated, do not worry, an example calculation can be found below, and there is a calculator available at the Web sites referenced in the appendix.

The loads are treated like a very long line by VNA during a TRL calibration. There has to be a pair of them, and they are connected instead of a very long line. The first line has to take over where the loads stop working properly. The best way to determine this is to use a full-wave field solver and design a high-bandwidth load. It will be apparent where the load starts to show a high return loss, or stops acting like a simple 50 Ohm resistor during the design process. Our observation has been that a pair of 0402 thin film resistors mounted to a PCB will stop working well between 100–200 MHz depending on the system. Turning them on their edge, to reduce capacitance, might allow them to work well to about 1 GHz. Turning 0402 resistors on their side is tedious work, and what's worse, it will yield mixed results. In the appendix, there is a load design to get you started. If a full wave solver is not in your tool kit, measure the load performance after the board is realized and adjust the design for the next board accordingly. Baring a good load design, do your best and aim low. Even if the load is just a pair of 0402's on a circuit board, it should be possible to reach 20 GHz or more with the three additional lines PLTS allows.

15.11 Putting It Together

The following is an example design.

For this example we will want to test a mated backplane connector pair. The connectors will rest on a PCB constructed with an FR4 core and Rogers 4350 outers. The PCB will be 0.093" thick with the calibrated transmission lines suspended in a Rogers 4350 stripline cavity. We will place the SMAs 3.500 inches from the intended DUT. Mechanical considerations force this as the minimum distance, if we want to maintain mostly straight transmission lines. This decision results in an arc of SMAs around the intended DUT. Ultimately, we want to test the DUT only, we will place the reference plane very close to the connector pins, 100 mil. This means that the open circuit standard will be 3400 mil long. The actual DUT will be a pair of mated connectors with a via and a 100 mil piece of stripline on each side.

The thru is easy to design. Just take two open circuits and connect them, stripline tip to stripline tip. This will define a circuit with two SMA connectors, one at each end of a 6800 mil stripline. The time to propagate a signal through the thru should be exactly the same length of time as it takes to propagate a signal to the end of the reflect and back. In stripline, just flipping the two opens, tip to tip, gets very close. It is likely that no further effort will be required here.

On to the lines and the load. Each line will be incrementally longer that the thru, and it will cover some defined frequency band. To avoid the need for a really long line, a load will be substituted. The first line has to take over where the load stops working well. The length of the line is inversely proportional to the band it covers. Lower-frequency bands require longer lines that consume more board real estate. Ideally, the load and the lines should be at the same reference impedance, our goal here is 50 Ohms. A pair of 100 Ohm 0402's placed on the top of a circuit board can stop acting load-like around 100 MHz. This could happen sooner if the via connecting the load has a stub. Some time spent pushing the bandwidth of the loads up will be rewarded with shorter lines and possibly fewer of them. For this example, assume we did not optimize the load and that it stops working well at 160 MHz.

Because the load stops at 160 MHz, this first line needs to take over at this point. We will start our design with a frequency factor of eight. Using this criteria, the first line (line-1) would work from 160 MHz to 1,280 MHz. The next one (line-2) would work from 1,280 MHz to 10,240 MHz. And the third line will work from 10,240 MHz to 81,920 MHz. This is probably a much higher frequency than needed and higher than most VNAs can cover. Try a factor of 5 instead. The lines now cover the following bands: line-1 covers 160 MHz to 800 MHz, line-2 covers 800 MHz to 4,000 MHz, and line-3 covers 4,000 MHz to 20,000 MHz. This will give us more engineering margin without increasing the number of calibration structures that we need to measure. The system set up with a factor of eight will probably work, but there is less margin for error. If a DK comes out wrong, a transmission line does not come in at exactly the length we wanted, or something else happens, the factor of eight kit may run into trouble.

The lines need to be a quarter wavelength long in the material in which they are constructed. Assume DK=4, because it is a common value. Recall, light travels at 299,792,458 m/s in a vacuum. The center frequency of line-1 would be ((160E6+800E6)/2) = 480E6. A quarter wavelength in DK=4 dielectric at 480 MHz is, c/f * ¼ * 1/SQRT(DK) * 39.37 inches/meter = 3.074 inches. Thus, the length of line-1 is 3.074 inches longer than the thru, or 9.874 inches. The VNA will need to know the delay for each line. For our line-1 example this works out to be, (length/c)*SQRT(DK) = 3.074 inch * 84.7 ps/inch*SQRT(4) = 520.7 ps. Make a similar calculation for each of the lines and layout the lines accordingly. See *Figure 15.14* for a complete order, including some extra features for non–TRL calibrations.

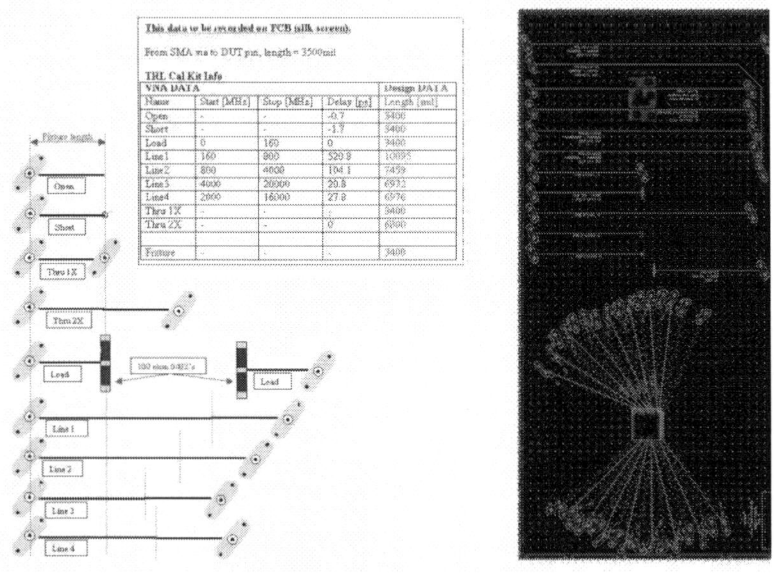

Figure 15.14: *TRL Calibration Fixture Layout Order with Line Lengths (Left). CAD Illustration of PCB (Right). Note That the 1X Thru Is Not Needed for the TRL Method, It Was Included for Time Domain Work. Line-4 Is Part of an Experiment to Calibrate the VNA Using the Load up to 2 GHz. To Calculate Lengths and Delays in the Table, DK=3.48 Was Used*

15.12 The Micro Half of a TRL Design

From a macro point of view, we are done. The length of all the necessary parts has been calculated and we can move on to layout. If this were coax, we really would be done. Alas, this is not coax, and we are really only about half done. It is time to consider the micro half of the problem. This was alluded to in the discussion above about the load and how its performance changes the length, and possibly the number, of lines needed to complete a calibration. Stripline is a pretty good transmission line, but it has a significant problem in that it is usually accessed by the use of a drilled, plated thru hole (PTH) or via. If you are designing for high speed, you are already familiar with the problem. Via stubs create resonances and the location and depth of the resonance changes with the length of the via, the length of the stub, the barrel diameter, the number of pads on the via, the diameter of the pads, the size of the antipads, and probably some things we do not keep very good track of, including the wear on the drill and the amount of etch. Vias are a mess and they are almost always necessary. What to do? The same things we do for any high-speed circuit, only more so if we can. If possible, use a full wave simulator to assist with this (e.g., HFSS, CST). Consider blind, or backdrilled, launches and loads. If you want to measure a component placed on the PCB, consider high-performance materials and optimized stackups. Yes, it adds cost, but it improves the performance of a PCB that is likely to see a very short run. Remember, one of our goals is to be able to subtract the fixture from the measurement. Removing sources of variability from the fixture will be rewarded by more accurate and repeatable measurements, regardless of the calibration technique used.

PCB Launch Characteristics

The SMA to PCB interface, the launch, should be electrically transparent as can be managed and easily reproduced. If the signal is sufficiently attenuated or altered by the launch, no useful measurement will result. If the launch does not provide consistent performance, it cannot be subtracted through either TRL calibration or de-embedding. There are several papers available on good PCB launch design. Ask your SMA vendor for help if you need more detail about launch design. If you have time to optimize the performance of only one feature, make it this one. See *Figure 15.15*.

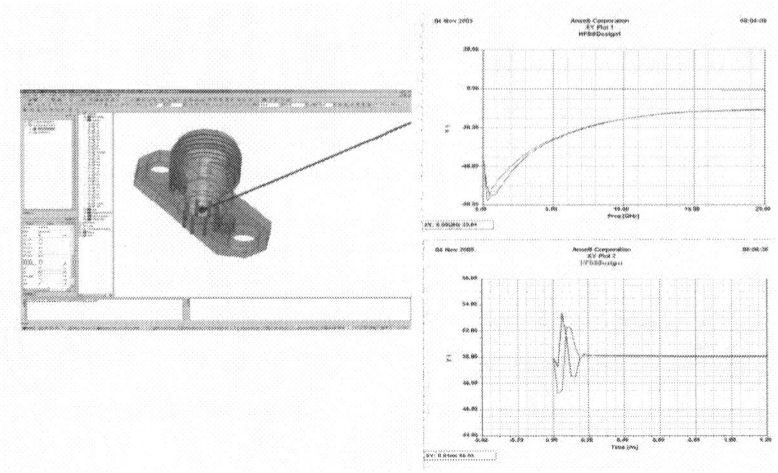

Figure 15.15: *Optimizing the PCB Launch Is a Critical Step to Successful Stripline VNA Measurements*

If broadband measurements are desired, an economical way to achieve them is to include a broadband load design. Consider a system with no load. The first line will need to cover 10 MHz – 80 MHz, thus it will need to be about 42 inches long in a DK=3.5 material. The kit will need three additional lines, all of which take up space and all of which need to be measured to perform the calibration. This is a lot of transmission line to place on an 18x24-inch panel. Worse, the topmost frequency that can be achieved with a TRL cal kit based on a no-load design is 5 GHz—not adequate for many of today's designs. An extreme example, yes. Consider another extreme example, a load design that works to 3 GHz. Now the first, and only, line need be only about 7 inches long in the same DK=3.5 material. The kit becomes very compact, it is easier to calibrate because fewer connections are needed, and the top bandwidth of such a kit is 24 GHz—Ample for many of today's designs and near the limit of APC 3.5 mm connectors. There is a load design in the appendix.

Consider the open or short. Is the discontinuity well defined? This is relatively easy to do, but it is worth considering anyway. The location of the reflection from the reflect standard should be well defined, for ease of design and layout. The mechanical length of the reflect should be one half the length of the thru. The reflect must maintain its polarity (sign of the reflection coefficient) throughout the desired bandwidth of the measurement. There is an open and a short design in the appendix.

Check (CAD), check (Gerbers), check (finished product)! Most CAD packages for PCB layout do not do well with some of the TRL stripline features. The short seems to cause the most grief. It is common for the short to lose its length and become shorted right at the SMA interface. The CAD software apparently believes that there is no difference between a short at the end of a transmission line several inches long and one that has zero length. For this reason, and because the open seems to perform better in simulation, I usually use opens as the reflect standard for stripline. The short is not the only place where CAD problems arise. Check the Gerbers and the finished product as well. Remember, the goal is to use identical open circuit probes to measure the DUT and the cal kit. If this happens, the open circuit probe can be reliably subtracted from the measurement and the actual DUT can be measured. See *Figure 15.17*.

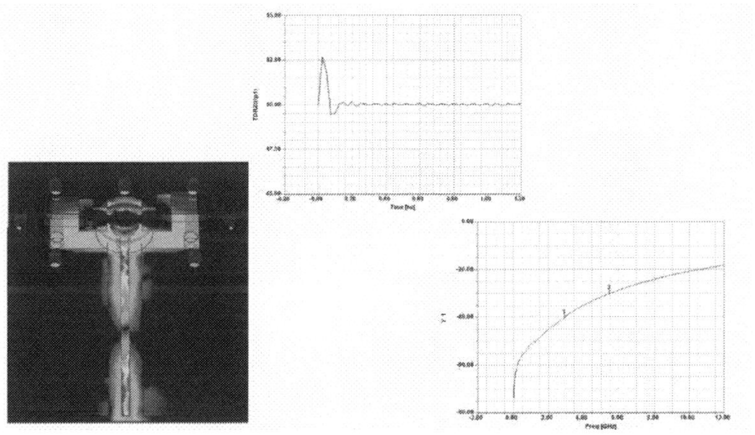

Figure 15.16: *High-Frequency Structure Simulator (HFSS) Is a Powerful Tool for Load Analysis*

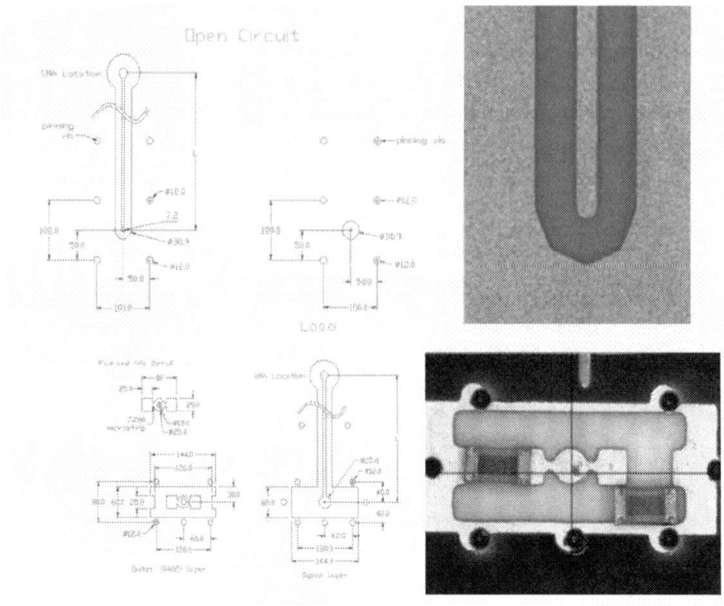

Figure 15.17: *Comparing Design Intent to Realized Prototype TRL Fixture*

15.13 Validation of TRL Fixtures

We will compare some measurements using a stripline TRL calibration to similar measurements made using the gold standard, a SOLT calibration plus de-embedding. For this comparison, a fixture such as the one described in the previous example was employed and both a TRL calibration and a SOLT calibration was applied and measurements taken. In this case, the connector is differential and the TRL calibration reference plane is placed near the point where the transmission lines get close enough together to create a differential mode. This is about an inch from the desired DUT, a mated pair of connectors. For the SOLT calibration, the reference plane was placed at the end of the coaxial cable and then moved to the same place using de-embedding. To de-embed, an S-parameter set of the item to be de-embedded must be created. To do this, an HFSS simulation of the open circuit probe was created and an S-parameter set extracted. Now we should have two comparable data sets.

The first pass does not look good (*Figure 15.18*). At 1 GHz, the delta between a TRL calibration and an SOLT plus de-embedding is 0.29 dB; at 7 GHz, it is 1.17 dB. Where did we go wrong? The piece to be de-embedded, our open circuit probe, was carefully modeled using HFSS. Dimensions were double-checked and confirmed correct. Upon measuring the dielectric properties of the material used, we realized that the DK and DF varied from the advertised values by a fairly large margin. DK was off by 10 percent, and DF was off by more than 100 percent. To obtain the actual DK and DF values, two lines of different length (the thru and line-1) were measured using an SOLT calibration. The group delay and loss values were subtracted to obtain time of flight and loss per unit length. Using these values, new DK and DF parameters were found. Applying a TRL calibration and measuring the delay and loss in line-1 directly confirmed this. The lesson here is, do not count on the vendor for in-situ DK and DF values. Measure them yourself.

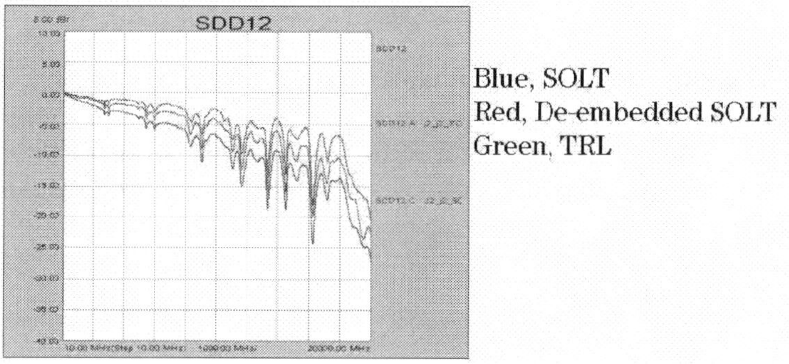

Figure 15.18: *First Pass Does Not Compare TRL and SOLT plus De-Embedding Well*

15.14 Using the Corrected Material Properties

The new material properties are entered into the HFSS simulation for our open circuit probe and S-parameters were extracted for entry into PLTS. This resulted in curves that correlated very well in the forward direction and pretty well in the reflected direction. The insertion measurement shows the de-embedded curve resting just below the TRL derived curve. A closer examination of material properties might help to bring these even closer together. We used a single value for DK and DF, even though the measurements made it clear that they are frequency-dependent. Copper parameters were assumed to be textbook. See *Figure 15.19.*

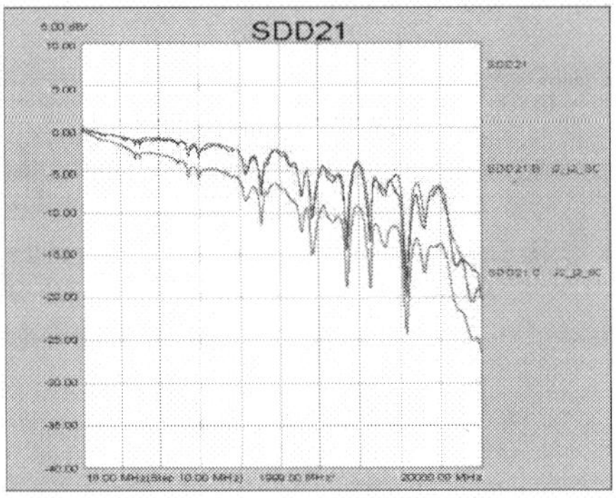

Blue, De-embedded SOLT
Green, TRL
Red, SOLT

Figure 15.19: *Applying Correct Dielectric Constant of PCB Material Shows Correlation between De-Embedded SOLT and TRL*

Near-End Crosstalk Characterization

What about near-end crosstalk (NEXT)? This is a more difficult measurement to make than insertion loss. An adjacent pair of pins was selected and a NEXT measurement was obtained. These curves match up very well in insertion loss, in group delay, and in the time domain. See *Figure 15.20.*

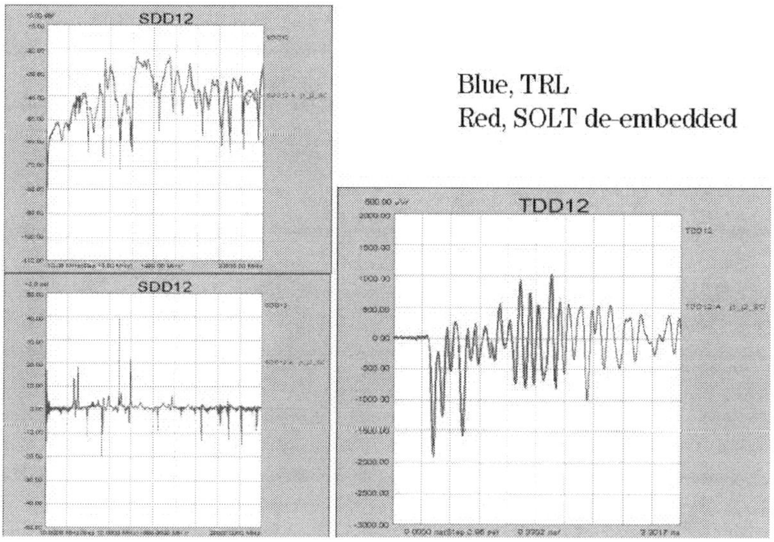

Figure 15.20: *Near-End Crosstalk Analysis Shows Good Correlation*

Far-End Crosstalk Characterization

How far can we push the forward direction? A pair of pins on the diagonal was selected and far-end crosstalk (FEXT) was measured. We see correlation in the S-parameters all the way down to the noise floor. Insertion loss, group delay, and time domain all correlate. See *Figure 15.21*.

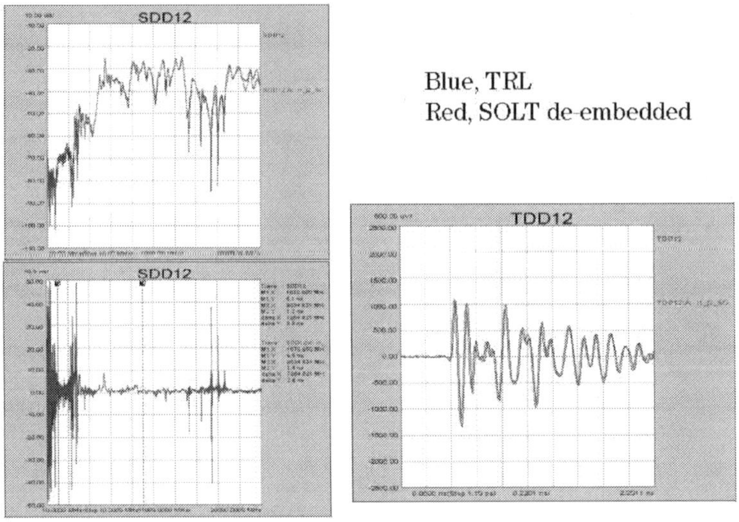

Figure 15.21: *Far-End Crosstalk Analysis Shows Good Correlation*

Finally, we will compare the reflections. A pair of pins was selected and the reflect direction was examined. Correlation is good, but not as good as in the forward direction (*Figure 15.22*). Impedance measurements of the actual DUT varied by 2–3 Ohms and the return loss curves do not show the nearly perfect overlap that the forward direction curves do. The cause of this variation is not fully understood, but it could be due to inaccuracies in the open circuit probe model that was de-embedded. Or it could be due to imperfections in the TRL cal kit or fixture. Another source of variation is the impedance standard. For the SOLT calibration, it was a broadband coaxial standard. For the TRL calibration, it was a resistive load and the characteristic impedance of the lines. In any case, the correlation between these two methods is quite good.

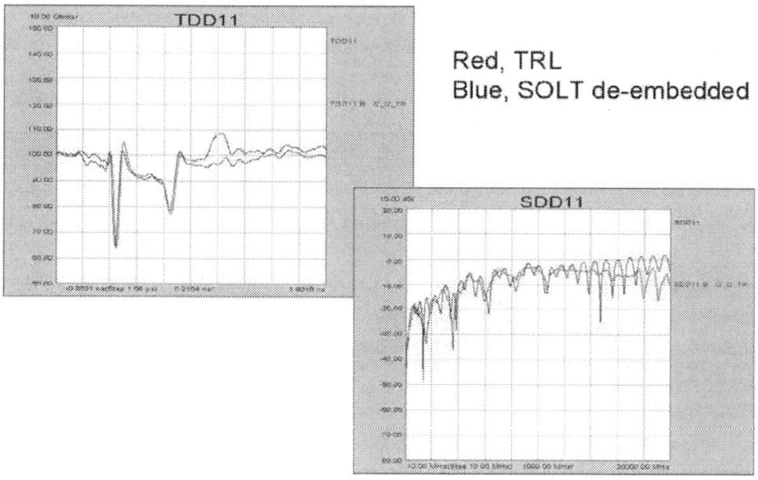

Figure 15.22: *Anomalies in the Reflect Direction*

15.15 Conclusion

Both techniques showed very good correlation in the forward direction, from 0 dB down into the noise floor of our techniques (~-50 dB to -80 dB). The reflect direction was not as satisfactory, but correlation was still good. To design the TRL cal kit, it was necessary to spend some time with a full wave solver and design a launch and a load. Because this was done, S-parameters for the open circuit probe could be obtained and de-embedding could be done. To obtain correct material properties for the SOLT de-embedding, it was necessary to include and measure some of the TRL cal kit features (the thru and line-1 were measured). A well-designed and -implemented TRL cal kit will allow for both de-embedding and TRL to be easily used. Ideally, TRL and SOLT de-embedding reinforce each other. Either measurement would result in a very good set of data for compliance testing or SPICE simulations.

It is worth noting that the TRL cal kit was designed using the same erroneous DK value that caused de-embedding to fail. We did not need to know much about DF to design the kit, only that it is low. The extra margin we gave ourselves by using a factor of five instead of a factor of eight was sufficient to ensure that the lines portion kit still worked in a different dielectric than was planned for. By measuring the TRL standards, the loss and phase of the fixture (the open circuit probe) was taken into account with no further effort on our part. Conversely, if we did not see the disparity between the SOLT plus de-embedding measurement and the TRL measurement, the dielectric properties might not have been measured and an erroneous measurement could have been reported. It is easier to obtain a good measurement with TRL than by de-embedding an SOLT measurement. The flexibility of reference plane placement is a very welcome feature as well.

Appendix

Additional material, including detailed drawings of various components, a TRL calculator, and HFSS simulations, can be found at the following:

www.agilent.com/find/plts
www.sun60.com

References

[1] Martin Vogel, Suresh Subramaniam, and Brad Cole, "Method for Optimizing a 10Gb/s PCB Signal Launch," DesignCon, 2004.

[2] Scott McMorrow and Alfred Neves, "A Hybrid Measurement and Field Solver Approach for the Design of High-Performance Interconnects," DesignCon, 2004

[3] "Network Analysis Applying the 8510 TRL Calibration for Non-Coaxial Measurements," Agilent Application Note 8510-8A.

Authors

Vince Duperron, Design Engineer, Molex

Dave Dunham, Electrical Engineering Manager, Molex

Mike Resso, Product Manager, Signal Integrity Applications, Agilent Technologies

Part IV

Jitter and Active Signal Analysis

Chapter 16

Channel Compliance Testing Utilizing Novel Statistical Eye Methodology

16.1 Abstract

New growth demands in the bandwidth requirements for short- and long-reach electrical channels have pushed the performance limitations of traditional linear passive interconnects such as high-speed backplanes and connectors. Most chip and system vendors regard 3 Gbps data rates over copper as mature technology today and are rapidly prototyping 6 Gbps and 10 Gbps hardware. Original standardization work for high-speed digital interfaces focused on defining the transmitter and receiver compliance in terms of eye diagram analysis utilizing masks and jitter tolerance. The remaining work of defining complete channel requirements is less accurately defined. As the bandwidth and crosstalk limits of the electrical interface are approached, the ultimate result is a degradation of bit-error rate (BER) over the physical layer. This undesirable effect mandates the need to fully characterize the capability of the electrical channel from end to end.

Traditional methods to simulate an emphasized transmitter, physical channel, and equalizing receiver in the time domain or frequency domain are not sufficient. Low probability events cannot be accurately modeled without long simulation times. To optimize the efficiency of simulations, a novel methodology based on statistical methods is introduced that allows fast and accurate compliance testing of differential channels. The following paper describes the algorithm and compares measurement results to a Matlab implementation of the algorithm using a design case study of 3 Gbps and 6 Gbps transmitter technology in 0.13 um complementary metal oxide semiconductor (CMOS).

16.2 Introduction

Channel compliancy is a key issue in the development of any electrical signaling specification that must contend with the interrelations between the transmitter, channel, and receiver. Typically two of these three elements will be defined, with the third element being a function of these two definitions. Furthermore, high-speed serial links (HSSLs) employ signal condition techniques at the transmitter and equalization techniques at the receiver. Thus, HSSL specifications must deal with the situation represented in *Figure 16.1*.

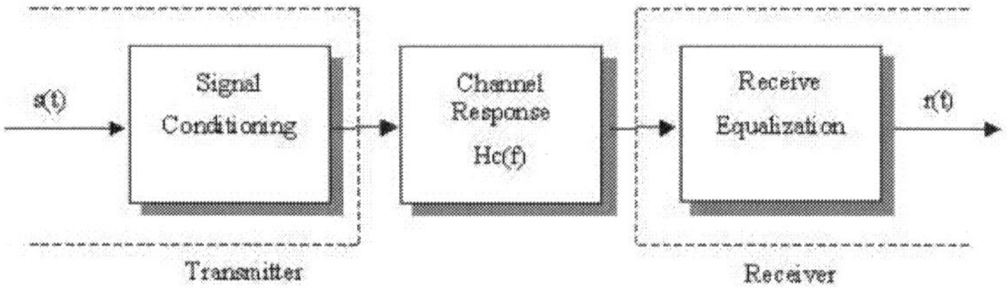

Figure 16.1: *The Interconnect*

In defining the performance of a transmitter or receiver, eye masks are specified. These eye masks are so-called worst-case definitions for a BER of definition 10^{-b}. This implies that if the signal were sampled in real time, it would only violate the mask 1 bit every 10^{b} bits. The transmit eye effectively limits the maximum jitter for a given edge rate and the minimum amplitude for an ideal termination impedance. The receiver eye mask defines a worst-case stressed eye. This implies that if a jittered and attenuated signal violates the eye mask once every 10^{b} bits in real time, then a receiver must receive it with a BER better than 10^{-b}.

In defining the performance of a channel, scattering parameters (S-parameters) are used. Vector network analyzers (VNAs), e.g., Agilent Technologies' N1900-Series Physical Layer Test System (PLTS), can be used to provide mixed-mode frequency domain measurements. The forward differential channel response, SDD21, is a natural starting point for defining a channel compliancy model. The use of SDD21 to develop a channel compliancy model poses a very significant challenge when applied to a backplane environment. The associated "layer connection effect" creates an environment where the performance of one layer can significantly vary from another, with the top layer representing the worst-case layer and the bottom layer representing the best-case performance. *Figure 16.2* demonstrates the impact of layer connection on 20-inch channels where the layer connection was varied in a 0.200" thick backplane. Furthermore, the amount of variation will grow with the associated board thickness of the backplane and daughtercards.

During the development of XAUI in the 10G Ethernet specification, the decision was made to define a SDD21 channel compliancy model, based on the median performance of observed channel data up to a frequency range of 3.125 GHz[1]—"The compliance interconnect limit of 47.4.1 represents the median performance of a range of interconnect designs. The range included designs from 46 to 56 cm in total length, having trace widths of 0.125 to 0.300 mm, and using different grades and thicknesses of FR4."[2] This model is shown in *Figure 16.3.* When combined with either a compliant transmitter or receiver, channels with a SDD21 response above the compliancy model would meet the required BER of 10^{-12}.

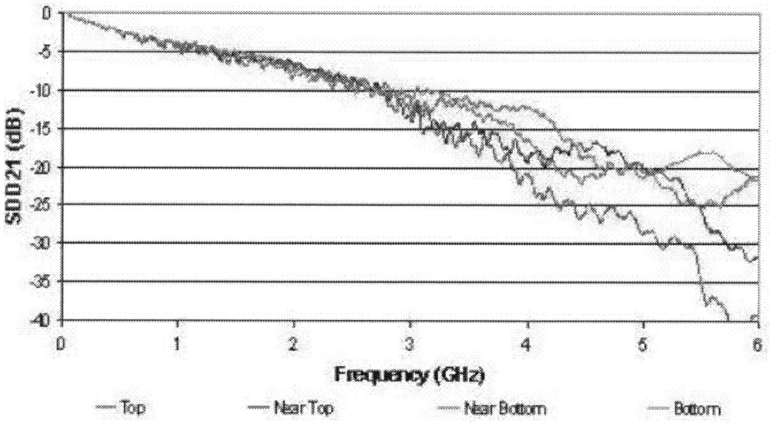

Figure 16.2: *Impact of Layer Connection*

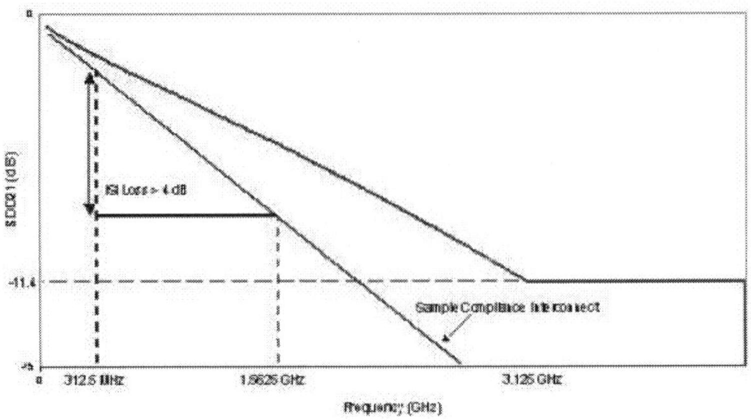

Figure 16.3: *XAUI Channel Model*

Looking at the channel response in the frequency domain is fundamental to gaining insight into the performance of the channel and how it affects the signal feeding into the receiver using either Fourier transforms $r(t) = ifft\{fft(s(t))\cdot H(f)\}$ or standard time domain simulators capable of accepting the S-parameters. The XAUI channel model, however, is truly only informative, as the impact of discontinuities and group delay are not represented. It is smooth in comparison to the real channels shown in *Figure 16.2*, with no apparent ripple. These ripples can easily result in a "dipping" below the channel compliancy model at various frequencies. This leads to concern regarding whether the channel will operate as intended since the channel compliancy model has not been met.

So from a specification perspective the limitations of this type of channel compliancy model make it more useful for informative purposes than normative purposes. Thus as equalization schemes such as transmit emphasis and receiver equalization are introduced to push a given channel's BER to its theoretical limit a new channel compliancy methodology is needed by the user. This methodology needs to look at the channel as part of the system, illustrated in *Figure 16.1*, rather than on its own merits. This methodology needs to utilize all the information provided by four-port S-parameter characterization combination with the total jitter present at the receiver sampler to give insight into the BER performance of the total system.

Total sampling jitter is contributed from the transmitter, channel, and receiver. However, the jitter introduced by the channel is inherently contained within the S-parameters and, as will be explained, is a function of the transmitter. Jitter is best defined by understanding the bathtub representation, i.e., the measurement BER or inverse error function (Q) versus the sampling offset. Gaussian jitter (GJ), otherwise known as random jitter, is defined as the gradient of the linearized portion of the bathtub at the BER of interest. High-probability jitter (HPJ), otherwise known as deterministic jitter, is the intersection of the linearization at the x-axis where Q=0.

The model shown in *Figure 16.4* is formally known as a dual dirac model and allows the system designer to add all HPJ terms in the link linearly, and to root-mean-square all GJ terms. The GJ terms are then multiplied by twice the Q required by the link to give the total GJ. And the total jitter is calculated as the linear sum of the total HPJ and total GJ, which must be less than 1 unit interval for the link to work.

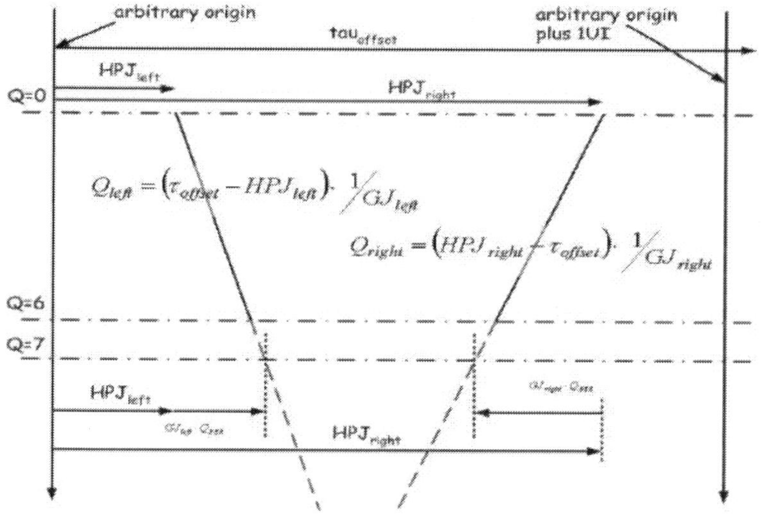

Figure 16.4: *Dual Dirac Model*

It is necessary to fully characterize the interconnect capability from the transmitter through the channel to the receiver. It is also necessary to employ a statistical methodology that will capture the impact of low-probability events. The S-parameter characterization of the channel and system will provide the building blocks for further simulation and analysis that will provide the quantitative answers. This will allow the inclusion of the following:

- Real channel data, including phase data and mode conversions
- In-band crosstalk resulting from similar switching signals
- Out-of-band or alien crosstalk resulting from other signaling
- Transmit jitter
- Receiver jitter

The inclusion of all of these parameters can then be applied statistically to the problem at hand to determine the channel's expected BER in the system and its compliancy to the targeted specification. Infineon Technologies has developed such a technique referred to as "StatEye." The focus of this paper will be to explain this technique utilizing backplane channels from Tyco Electronics' HM-Zd legacy backplane. Validation testing of the technique will be provided by Agilent Technologies.

16.3 StatEye Methodology

Pulse Response Theory

The pulse response of a channel can be understood as the signal resulting when an non-return-to-zero (NRZ) pulse, $tx(t)$, is transmitted through the channel under investigation. The NRZ pulse has a width equal to the period of the bit rate of interest, and therefore the pulse response is only valid for a specific baud rate. Given that the transfer function of the channel, $S(\omega)$, is known in the frequency domain, the pulse response is best calculated by multiplying it with the fast Fourier transform (FFT) of the NRZ pulse, $tx(\omega)$.

$$tx(t) = H(t) \cdot H(t_{period} - t)$$
$$tx(\omega) = FFT(tx(t))$$
$$rx(\omega) = tx(\omega) \cdot S(\omega)$$
$$rx(t) = iFFT(rx(\omega))$$

As the time domain and frequency domain functions are actually discrete measurements, great care must be taken in the generation of the transmit pulse to ensure the period accuracy. The pulse response of the channel should not be confused with the impulse response, which is equivalent to a transmit pulse of infinitely small period but a total area of one, aka dirac.

The pulse response allows an understanding of how inter-symbol interference (ISI) occurs through superposition (see *Figure 16.5)*. Given a stream of either positive or negative pulses (dashed), the superposition of these pulses (solid) causes an ever-worsening signal integrity as the width of the pulse response increases. This ISI is very similar to wander encountered with AC coupled signals, except that the pulse response of electric channels is asymmetrical. The ISI due to the variation of the initial signal amplitude at the beginning of each period is at the zero crossing time converted into a time jitter. At the point where the ISI is of the same amplitude as the signal, the eye becomes closed and the jitter more than a complete bit period.

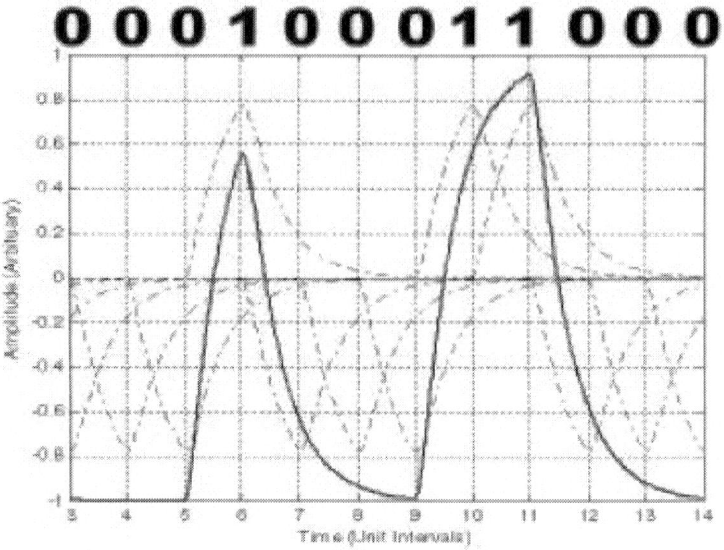

Figure 16.5: *Inter-Symbol Interference*

Equalization

Bandwidth limitations can be equalized by cascading the channel with a linear time continuous filter, i.e., one that can be represented in both the time and frequency domain, which is the exact inverse pulse response of the channel. Given this theoretical filter response, the resulting transfer function would be a linear response with constant gain over frequency. As this type of filter is both theoretical and impractical to build, an approximation is usually used. Working at lower bit rates, where accurate signal processing can be performed, a resulting raised cosine spectrum or partial response can be achieved[3]. For higher bit rates where only simple equalization can be implemented, the time continuous filter is usually only a finite number of zeros and poles. By optimal placement of these zeros and poles, an enhancement of the bandwidth of the channel can be achieved, which to a certain degree approaches a simplified partial response. *Figure 16.6* is an example of how the frequency response of the channel is increased in bandwidth and the resulting pulse response is narrowed, which results in a reduction of ISI.

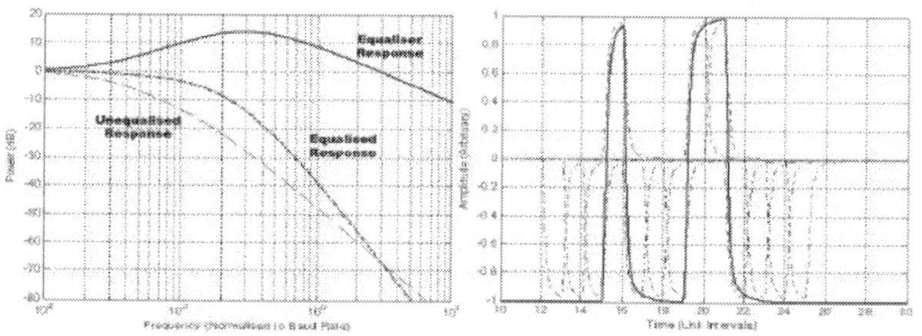

Figure 16.6: *Equalized Frequency and Pulse Response*

The disadvantage of linear-time continuous equalizers can be understood by considering the crosstalk frequency response. The crosstalk frequency response is the transfer function of a near-end crosstalk (NEXT) aggressor Tx pair or far-end crosstalk (FEXT) aggressor Tx pair to the receiver, as shown in *Figure 16.7*. The previously described equalizer, for example, will enhance this crosstalk frequency response in the same way, increasing the crosstalk energy at the receiver and thus decreasing the resulting total received signal.

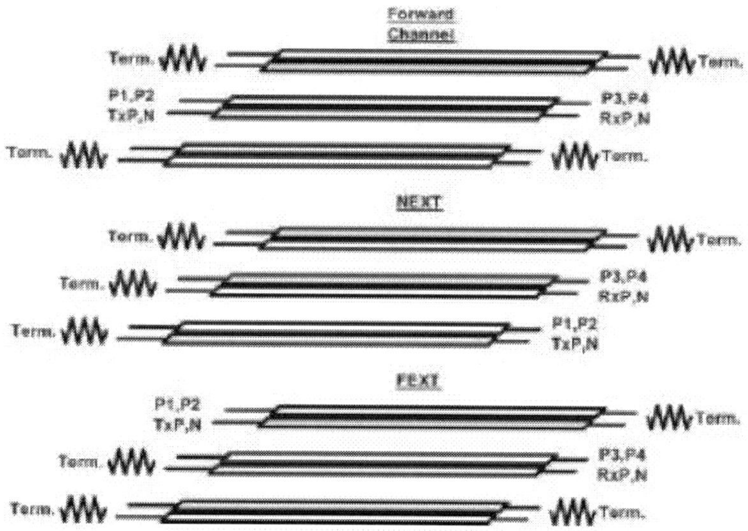

Figure 16.7: *Crosstalk Definitions*

A decision feedback equalizer (DFE) is a nonlinear equalizer insofar that it cannot be represented in the frequency domain. The concept of a DFE lies in its ability to cancel the post-cursors of the channel pulse response[4]. If we define a pulse response in terms of its amplitude at baud spaced samples, shown in *Figure 16.8*, where r_n with $n<0$ are called pre-cursors and r_n with $n>0$ are called post-cursors, the DFE can cancel the ISI caused by post-cursors. Given that the channel response is known and an equalized signal has been correctly received as a 1 or 0, then referring to *Figure 16.9,* the influence of the post-cursors can be removed by setting $k_n=r_n$. Since the DFE only feedbacks a decision, any noise present on the signal is not amplified. In the case, however, where the noise causes a decision error, this error is then fed back to the receiver and can cause error propagation[5].

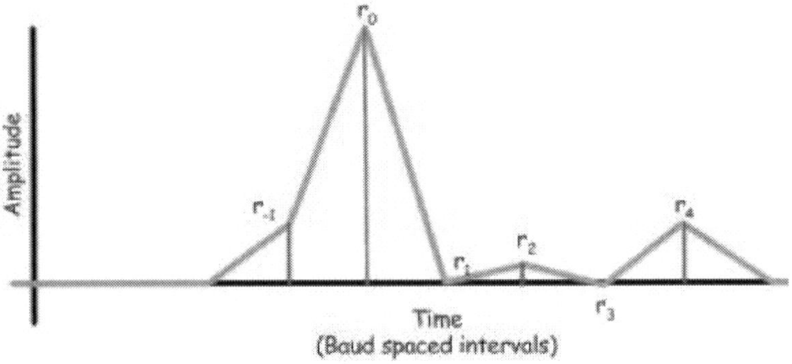

Figure 16.8: *Pulse Response Definition*

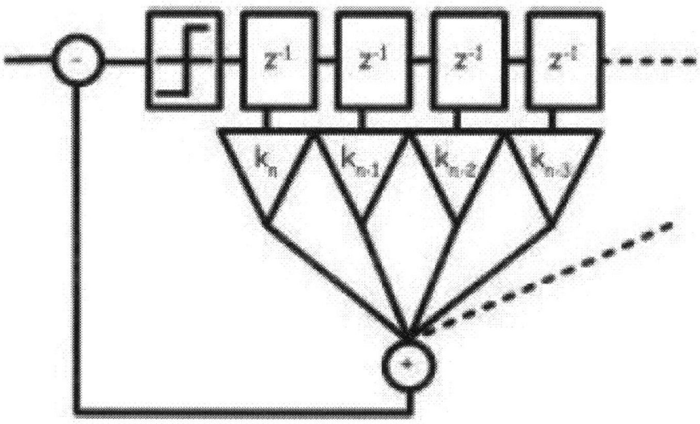

Figure 16.9: *DFE Architecture*

Channel Characterization

All the electrical performance information for a differential channel can be extracted using a four-port VNA. It is actually the single-ended S-parameters that are measured in the frequency domain. *Figure 16.7* shows the format for this information and the port-labeling scheme for a differential channel that is measured as two single-ended channels.

These are mathematically transformed into the frequency domain balanced, mixed-mode, or differential S-parameters (see *Figure 16.11*). These parameters can be used directly to give information about the differential return or insertion loss. Further transformation into the time domain gives information about the differential impedance profile of the channel or the location of the conversion of differential signal into common signal.

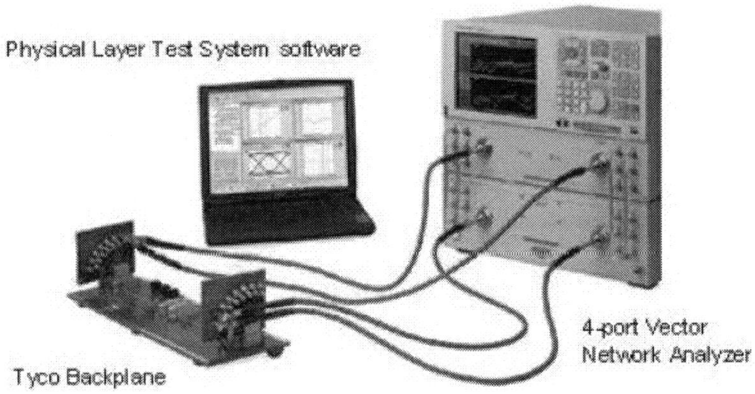

Figure 16.10: *Lab Setup*

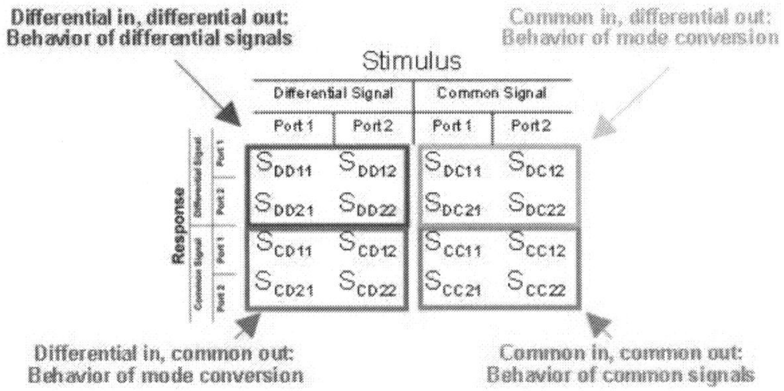

Figure 16.11: *S-Parameter Quadrants*

In order to interpret the large amount of data in the differential parameter matrix, it is helpful to analyze one quadrant at a time. The first quadrant is defined as the upper left four parameters describing the differential stimulus and differential response characteristics of the device under test (DUT). This is the actual mode of operation for most high-speed differential interconnects, so it is typically the most useful quadrant that is analyzed first. It includes input differential return loss (S_{DD11}), input differential insertion loss (S_{DD21}), output differential return loss (S_{DD22}), and output differential insertion loss (S_{DD12}). Note the format of the parameter notation S_{XYab}, where "S" stands for S-parameter, "X" is the response mode (differential or common), "Y" is the stimulus mode (differential or common), "a" is the output port, and "b" is the input port. This is typical nomenclature for frequency domain S-parameters. All 16 differential S-parameters can be transformed into the time domain by performing inverse fast Fourier transform (IFFT). The matrix representing the time domain will have similar notation, except the "S" will be replaced by a "T" (i.e., T_{DD11}).

The second and third quadrants are the upper right and lower left four parameters, respectively. These are also referred to as the mixed-mode quadrants. This is because they fully characterize any mode conversion occurring in the DUT, whether it is common-to-differential conversion (electromagnetic interference [EMI] susceptibility) or differential-to-common conversion (EMI radiation). Understanding the magnitude and location of mode conversion is very helpful when trying to optimize the design of interconnects for gigabit data throughput.

The fourth quadrant is the lower right four parameters and describes the performance characteristics of the common signal propagating through the DUT. For a properly designed device, there should be minimal mode conversion and the fourth quadrant data is of little concern. However, if any mode conversion is present due to design flaws, then the fourth quadrant will describe how this common signal behaves.

16.4 Cascading of Channel with Transmitter and Receiver Return Loss Model

After the full S-parameters of a channel have been measured, the effect of reflections can be modeled using a worst-case transmitter and receiver return loss. After converting the S-parameters into the transmission matrix[6], performing a matrix multiplication, and converting back to the S-parameters, the overall transfer function is obtained.

Given the following:

$S_{m\,n}$ is the measured four-port differential data of the channel
Tx is the transmitter return loss

Tx is a single pole filter
Rx is the receiver return loss
Tr(ω) is the channel response transfer function as defined

then

$$Tr(\omega) = \begin{bmatrix} Tx\ (\omega) \\ Tx\ (\omega) \end{bmatrix} \otimes \begin{bmatrix} S\ (\omega) & S\ (\omega) \\ S\ (\omega) & S\ (\omega) \end{bmatrix} \otimes \begin{bmatrix} Rx\ (\omega) \\ \end{bmatrix}$$

In addition to the return loss, a typical transmitter is not capable of generating an ideal NRZ pulse, so the transfer function is additionally multiplied by a single pole filter with a corner frequency of 0.75xfbaud, typical for current mode logic (CML) style outputs.

Theoretical Analysis of Receiver Pulse Response

The superposition of post-cursors and pre-cursors to form ISI is statistical in nature. Given a full random binary data stream and a finite number of cursors, each possible combination of cursors can superimpose each with equal probability.

A simple example shown in *Figure 16.12*, with an arbitrary sample time, has one pre-cursor and two post-cursors and shows how the eight possible combinations of the cursors can combine to cause ISI. In this example, the ISI is as large as the signal itself and closes the eye. The probability of each amplitude can be represented by a conditional probability distribution function, which for this example is simply eight diracs or deltas, δ, of equal probability, i.e., 1/8. As more cursors are taken into account the number of possible combination increases and the probability distribution becomes more detailed and can be represented mathematically, taking into account the effect of the DFE cancellation of the post-cursors.

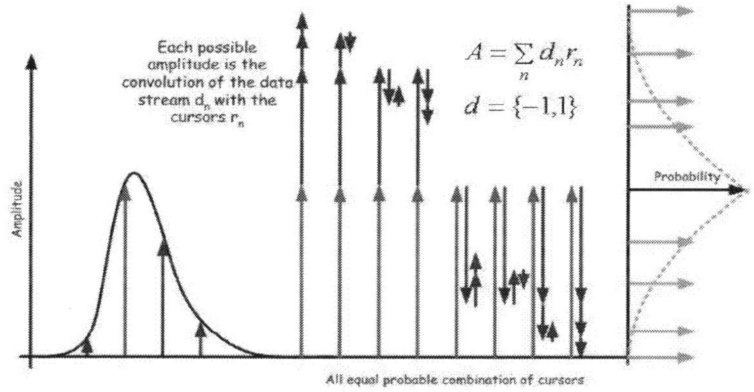

Figure 16.12: *ISI PDF*

Given the following:

$r_n(\tau)$ are the cursors of the pulse response at sampling time τ
e_b is the ideal static equalization coefficients of the b tap DFE
$c(\tau)$ is the set of equalized cursors at sampling time τ
$\delta(\tau) = \lim_{\varepsilon \to 0} \varepsilon |x|^{\varepsilon - 1}$ is the dirac or delta function

then

$$c(\tau) = \begin{bmatrix} r_{-m/2}(\tau) & \cdots & r_{-1}(\tau) & r_1(\tau) - e_1 & \cdots & r_b(\tau) - e_b & r_{b+1}(\tau) & \cdots & r_{m/2}(\tau) \end{bmatrix}$$

$$d_n = \begin{bmatrix} d_n^1 & d_n^2 & d_n^{..} & d_n^{m-1} & d_n^m \end{bmatrix}$$

$$n = \left[\sum_{b=1..m} d_n^b \cdot 2^{(b-1)} \right] + 1$$

$$n = 1..2^m$$

$$p(ISI, \tau) = \frac{1}{2^m} \sum_{n=1..2^m} \delta\left(c(\tau) \cdot (2d_n - 1) - ISI\right)$$

where,

d_n are the possible combinations of the data stream and is either 1 or 0
$p(ISI, \tau)$ is the probability of a given ISI occurring

It would seem from the methodology that to calculate the PDF of a pulse response with a large number of cursors, >10, all possible combination would have to be calculated, e.g., for 30 cursors, 2^{30} combinations would have to be calculated. This sledgehammer approach is not necessary, as the problem can be broken down into small problems.

Given the following:

c is an example set of four cursors
d_n are the possible combination of bits in the data stream

$$c = \lfloor c_1 \quad c_2 \quad c_3 \quad c_4 \rfloor$$
$$d_n = \left[d_n^1 \quad d_n^2 \quad d_n^3 \quad d_n^4 \right]$$
$$n = \left[\sum_{b=1..4} d_n^b \cdot 2^{(b-1)} \right] + 1$$
$$n = 1..2^4$$
$$p(ISI) = \frac{1}{2^4} \sum_{n=1..2^4} \delta(c \cdot (2d_n - 1) - ISI)$$
$$= \left[\frac{1}{2^2} \sum_{n=1..2^2} \delta(c_{1..2} \cdot (2d_n - 1) - ISI) \right] * \left[\frac{1}{2^2} \sum_{n=1..2^2} \delta(c_{3..4} \cdot (2d_n - 1) - ISI) \right]$$

Given an example of four cursors, the PDF can either be calculated from 2^4 diracs or the convolution of two PDFs calculated from 2^2 diracs. In this way a pulse response of 40 cursors could be calculated by convoluting 20 smaller PDFs calculated from 2^2 diracs. The savings in required processing steps can clearly be seen.

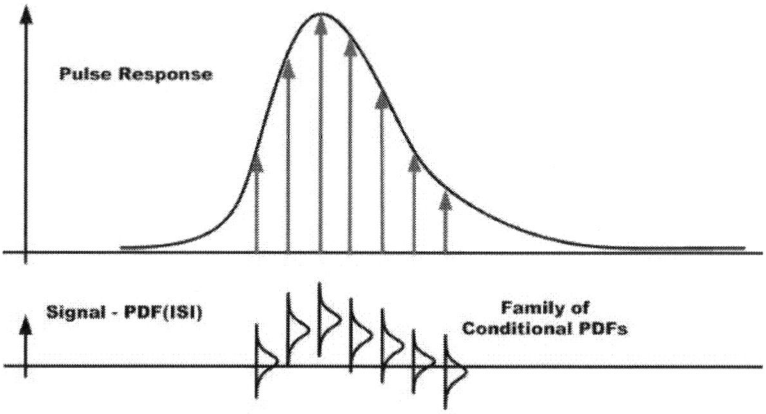

Figure 16.13: *Family of Conditional PDFs*

For each arbitrary sample point within the pulse response, a conditional PDF can be calculated, forming a family of conditional PDFs (see *Figure 16.13*). Due to the discretization when storing and building a PDF, an accuracy error occurs that can be defined (see *Figure 16.14*). Given that the discrete PDF array is defined from –AMAX to AMAX and has m bins, then the error of a single entry has a defined bin error, ε

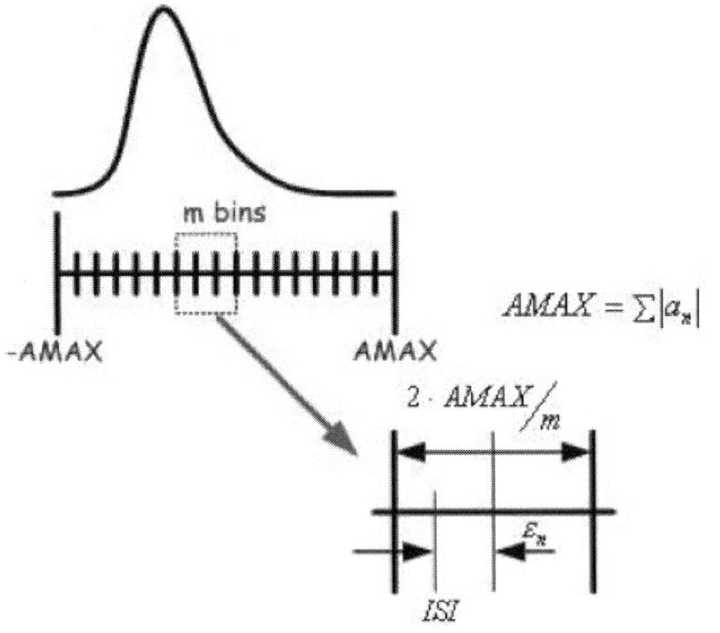

Figure 16.14: *Binning of PDFs*

where $-\dfrac{AMAX}{m} > \varepsilon > \dfrac{AMAX}{m}$

For each convolution, the error accumulates and can be represented in the simple aforementioned example.

$$
\begin{aligned}
&p(ISI_{c1..c4}) = p(ISI_{c1..c2}) * p(ISI_{c3..c4}) \\
&= \left[\frac{1}{2^2}\sum_{n=1..2^2}\delta(c_n\cdot(2d_n-1)-ISI)\right] * \left[\frac{1}{2^2}\sum_{n=1..2^2}\delta(c_{n+2}\cdot(2d_n-1)-ISI)\right] \\
&= \frac{1}{2^2}\left[\delta\left(c_1\cdot d_1^1 + c_2\cdot d_1^2 + \varepsilon_1 - ISI\right) \ \ldots \ \delta\left(c_1\cdot d_4^1 + c_2\cdot d_4^2 + \varepsilon_2 - ISI\right)\right] * \\
&\frac{1}{2^2}\left[\delta\left(c_3\cdot d_1^1 + c_4\cdot d_1^2 + \varepsilon_3 - ISI\right) \ \ldots \ \delta\left(c_3\cdot d_4^1 + c_4\cdot d_4^2 + \varepsilon_4 - ISI\right)\right] \\
&= \frac{1}{2^4}\left[\delta\left(c_1\cdot d_1^1 + c_2\cdot d_1^2 + c_3\cdot d_1^3 + c_4\cdot d_1^4 + \varepsilon_1 + \varepsilon_2\right) \ \ldots \ \ldots \ \ldots\right]
\end{aligned}
$$

Given that the bin error associated with a PDF is evenly distributed with a zero mean and variance

$$\mu_2 = \sigma^2 = \frac{\varepsilon_n^{\ 2}}{12}$$

then N convoluted PDFs will also have a bin error with zero mean but a variance of

$$\mu_2 = \sigma^2 = N\frac{\varepsilon_n^{\ 2}}{12}.$$

The peak error given a probability equal to the BER of interest is then $Q \cdot \sigma$, where Q is the inverse error function of the probability of interest, e.g.,

$$AMAX = 1.0, m = 1000, \varepsilon_{\max} = 0.001, N = 50$$

$$\mu_2 = 50 \cdot \frac{0.001^2}{12} \approx 4.2 \cdot 10^{-6}$$

$$\varepsilon_{\max}\Big|_{N=50}^{BER=10^{-12}} = 7 \cdot \sqrt{\mu_2} \approx 0.0143 \approx 1.5\%$$

Given an arbitrary receiver sampling point with no jitter, then the associated PDF sampled is simply the already generated conditional PDF from the family of PDFs. If the sampling point is jittered with a known distribution then the sampling sees an average conditional PDFs (see *Figure 16.15*).

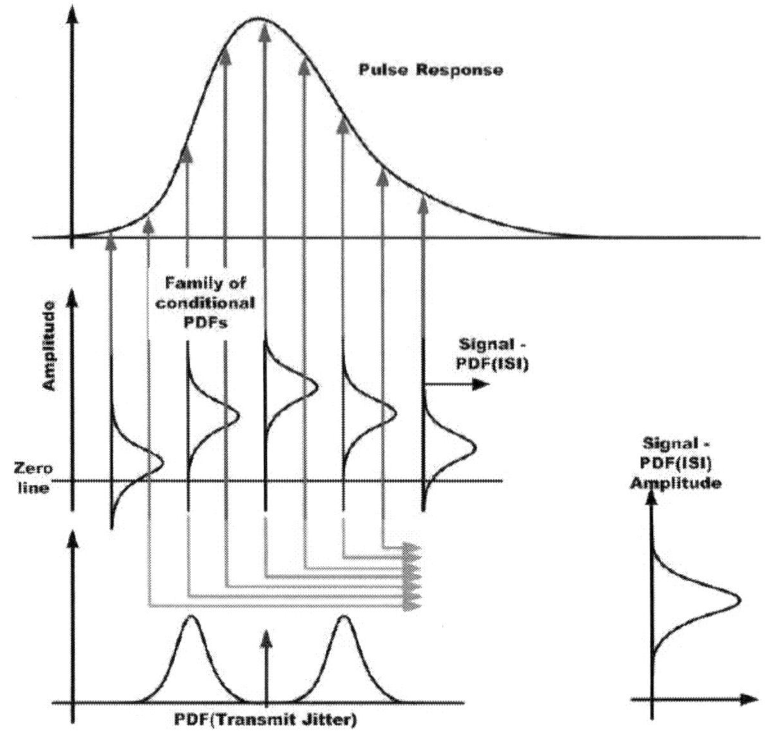

Figure 16.15: *Average Conditional PDF*

Given a jitter distribution

$$P_{jitter}(\tau', w, \sigma) = \frac{1}{2\sqrt{2\pi}} \cdot \frac{1}{\sigma} \cdot \left[e^{-\left[\frac{(\tau' - w/2)^2}{2\sigma^2}\right]} + e^{-\left[\frac{(\tau' + w/2)^2}{2\sigma^2}\right]} \right]$$

the average conditional PDF is then

$$p_{average}(ISI, \tau) = \int_{-\infty}^{\infty} p(ISI, \tau + \upsilon) \cdot P_{jitter}(\upsilon, w, \sigma) \cdot d\upsilon$$

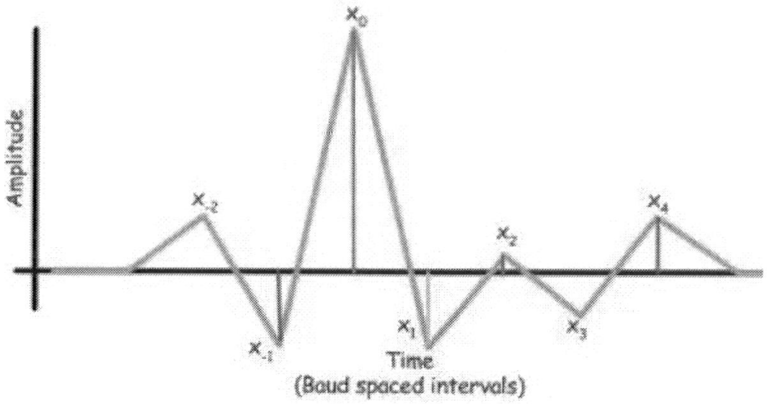

Figure 16.16: *Crosstalk Pulse Response*

To account for all the effects in the channel, the transfer function (i.e., including return-loss effects) and pulse response is calculated for all possible sources, including crosstalk (see *Figure 16.16*):

1. Forward transmitter differential mode to differential mode
2. Each crosstalk aggressor differential mode to differential mode
3. Each crosstalk aggressor common mode to differential mode
4. Forward transmitter common mode to differential mode

For each of the additional transfer functions, i.e., 2, 3, and 4, a set of cursors is defined

$$c(\tau)=\left[x_{-m/2}(\tau)\ \ \dots\ \ x_{-1}(\tau)\ \ x_0\ \ x_1(\tau)\ \ \dots\ \ x_{m/2}\right]$$

with $n = 1\ldots2^{m+1}$ and an associated family of conditional PDFs, respectively:

1. $p_{DD12}(ISI,\tau)$
2. $p_{DDx2}(ISI,\tau)$
3. $p_{CDx2}(ISI,\tau)$
4. $p_{CD12}(ISI,\tau)$

The overall average conditional PDF can then be calculated

$$p_{average}(ISI,\tau)=$$
$$\int_{-\infty}^{\infty}\left[p_{DD12}(ISI,\tau+\upsilon)*p_{CD12}(ISI,\tau+\upsilon+w)*p_{DDx2}(ISI,\tau+\upsilon+w)*p_{CDx2}(ISI,\tau+\upsilon+w)\right]\cdot$$
$$P_{jtter}(\upsilon,w,\sigma)\cdot d\upsilon$$

where each of crosstalk conditional PDFs are convoluted in turn.

The arbitrary position of $p_{average}(ISI,\tau)$ can now be swept over the complete pulse to give a family of average conditional PDFs, *Figure 16.17a/b*. Using a contour algorithm the points of equal probability of the integrated PDFs can be drawn, creating a "so-called" StatEye, *Figure 16.17c*. This StatEye shows the probability of receiving a specific amplitude for a given arbitrary receiver sampling point.

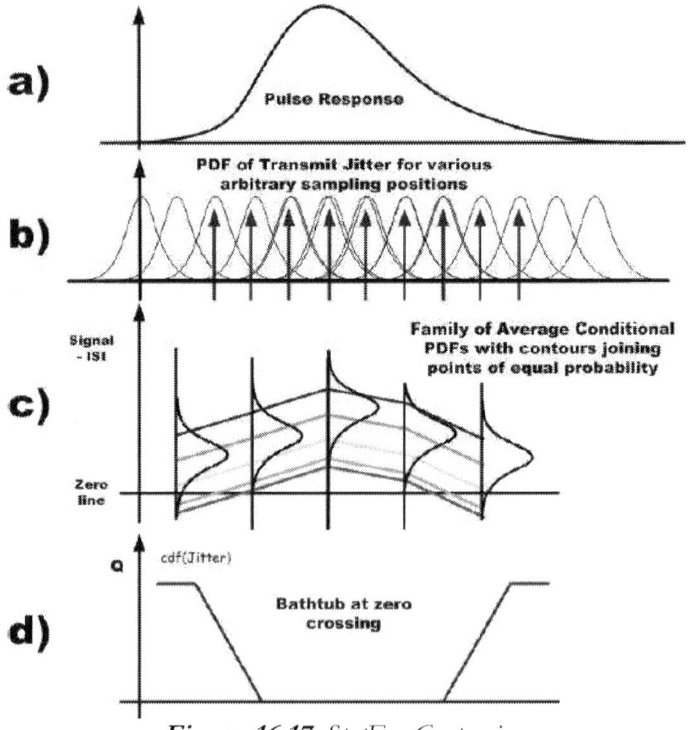

Figure 16.17: *StatEye Contouring*

By plotting the probability at the zero-amplitude crossing for the family of CDFs against the receiver sampling point, a bathtub, *Figure 16.17d* is generated from which the GJ and HPJ can be extracted.

Implementation

The implementation of such an algorithm is feasible on a typical >1 GHz processor within interpreting languages such as Matlab. Matlab provides a user-friendly and fast prototyping interface that allows the algorithm to be implemented and debugged.

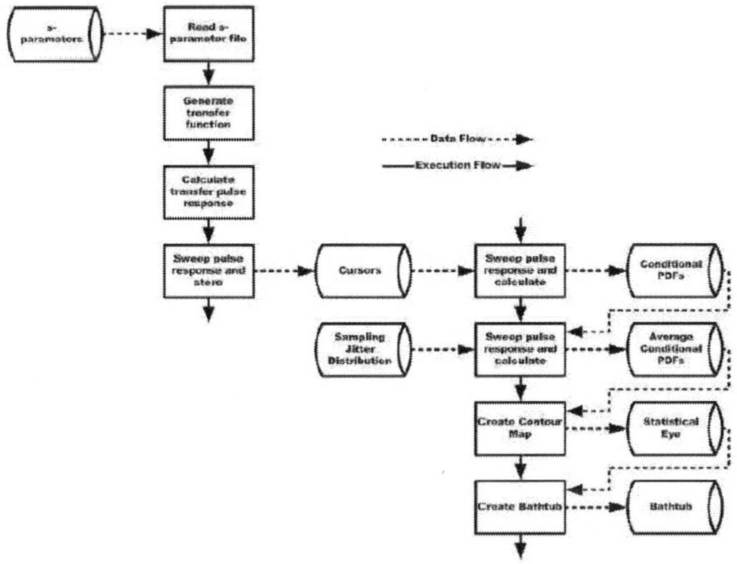

Figure 16.18: *Implementation Execution and Data Flow*

A recommended flow, shown in *Figure 16.18*, consists of three non-nested loops. Loop #1 scans the pulse response, initially storing the cursor values. Loop #2 then calculates the conditional PDFs, and loop #3 scans the pulse again and calculates the average conditional PDFs. Using time-set intervals of 0.01 unit intervals and typically 1,000 bins for the PDFs, a good compromise between accuracy and execution time (~30 seconds) is found. If faster execution times are necessary, then it is recommended to move to a C implementation, where execution times can be increased by a factor of four. However it is recommended that this step only be taken after an initial implementation in Matlab. A version of this algorithm is currently under development, which will interface directly to Agilent's PLTS software.

System Budgeting and Interoperability

The "StatEye" methodology systems can be budgeted to incorporate both transmit and receiver equalization and utilize adaptive algorithms.

To ensure interoperability, the transmitter is controlled by defining an eye mask that limits the GJ and HPJ and ensures a certain signal amplitude.

Defining the channel as the combined forward response and "all" significant crosstalk responses, the cascaded channel with a representative transmitter and receiver return loss is defined as being compliant if the resultant "StatEye" for a worst-case transmit signal and an ideal receiver DFE filter meets a worst-case eye mask, i.e., it has an amplitude and a GJ and HPJ better than that defined.

The receiver should be capable of receiving a worst-case stressed[7] eye with a BER better than that required of the system. Given that the "StatEye" analysis is performed with an ideal DFE and no receiver sampling jitter, this allows a standard to not impose any ideas or assumptions concerning the penalties associated with the implementation.

A designer of such a receiver should do the following through simulation:

- Estimate the sampling jitter and add this to the transmitter jitter
- Estimate the DFE penalties and adjust the cursor set
- Recalculate the "StatEye" using the total sampling jitter is then verify that the receiver's sampler is capable of receiving it

The "StatEye" therefore uses the dual dirac model only as a means of quantifying the jitter present at certain points in the system without relying on the invalid dual dirac "addition" of GJ and HPJ over a band-limited channel. Because the "StatEye" is not defining an exact implementation of the channel but merely an overall performance of the channel, it allows the PCB implementer to trade off certain parameters against each other, e.g., channel length can be decreased, in turn decreasing attenuation and allowing crosstalk to increase.

16.5 Design Example Results

Channel Description

Tyco Electronics announced the availability of its HM-Zd Legacy interoperability platform in August 2003. This platform provides the industry with a backplane that is representative of the conditions seen in typical system vendor environments. The Legacy platform builds on Tyco Electronics' HM-Zd XAUI interoperability platform, which the 10Gigabit Ethernet Consortium selected as a common platform for interoperability testing in early 2001. The HM-Zd Legacy backplane platform is conceptually shown in *Figure 16.19.* It consists of two line cards that provide SMA access and the Z-PACK HM-Zd–based backplane. Each line card is 0.093" (nominal) thick, consists of 14 layers, and is fabricated using Nelco 4000-2 material. There are four signal layers distributed throughout the entire stackup, where the 100W differential geometries are based on 0.006" (nominal) wide traces. The trace length from the SMA to the Z-PACK HM-Zd connector is 2". The backplane is 0.200" (nominal) thick, consists of 18 layers, and is fabricated using 4000-6 material. There are six signal layers distributed throughout the entire stackup, where the 100W differential geometries are based on 0.0055" (nominal) wide traces. On the backplane there are three sets of trace lengths—1", 16", and 30". Thus, for the platform there are overall system lengths of 5", 20", and 34".

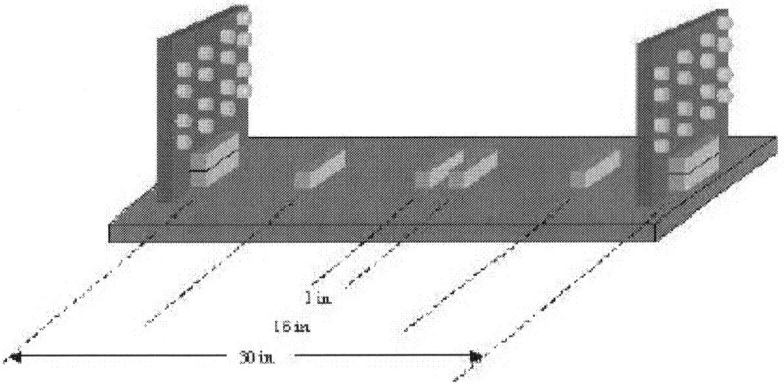

Figure 16.19: *XAUI HM-Zd Interoperability Platform*

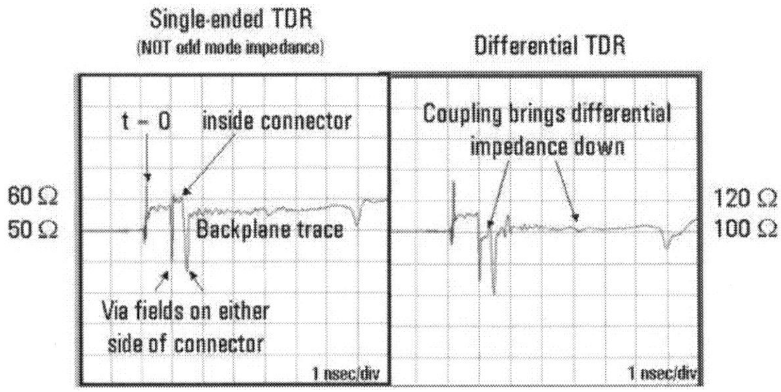

Figure 16.20: *Differential Coupling on XAUI Channel Pulls Impedance Down*

Differential Impedance Analysis

In high-speed digital interconnects, crosstalk between adjacent transmission lines is usually undesirable. However, there is one notable exception to this case—that is when we are designing differential transmission lines. The strong coupling of adjacent PCB traces that make up a differential pair is exactly what is needed to achieve good common-mode noise rejection. So when targeting a specific differential impedance of 100W, this coupling has to be taken into account. An example of this can be seen in the impedance profiles in *Figure 16.20*. The single-ended TDR trace shows that the daughtercard and motherboard exhibit around 55–56W of single-ended impedance (only one line is driven, the other line is quiet, and the impedance is measured from the driven line to the ground). The differential TDR trace shows the effect of coupling on the motherboard traces and yields about 100–101W (both lines are driven with equal amplitude/opposite polarity steps, and impedance is measured from line 1 to line 2). Notice that the daughtercard differential impedance does not pull down to the target differential impedance of 100W. This indicates weaker coupling on the differential traces due to larger spacing between daughtercard traces.

Mode-Conversion Analysis

When a differential signal is propagating through an ideal differential transmission line, there theoretically should be no other mode transmitting. If there is anything that is different on one line of the differential channel from the other line of the differential channel, then we have something called mode conversion. This will result in EMI problems, the severity of which depends on the type of mode conversion and application.

Mode conversion relating to high-speed digital channels can be grouped into two categories: mode conversion caused by active devices and mode conversion caused by passive devices. An example of active-device mode conversion would be if a differential transmitter or amplifier had one drive signal different from the other supposedly complementary drive signal—voltage, current, or skew. An example of a passive-device mode conversion would be if a differential connector had different characteristic impedance lines, different length pins, or different loading on each line such as ground plane discontinuities. In any case, mixed-mode analysis can be a powerful tool in locating obstacles that limit the highest possible data rate transmission in interconnects.

As seen in *Figure 16.21*, the differential-to-common mode conversion can be quantified as a percentage of amplitude by overlaying the forward time domain transmission waveform (TDD21) with the forward time domain transmission mode conversion waveform (TCD21). As expected, TDD21 shows the propagation delay and rise-time transition of the degraded signal at the output of the DUT. The resulting measurement is the mode conversion waveform showing 7 percent of the original signal.

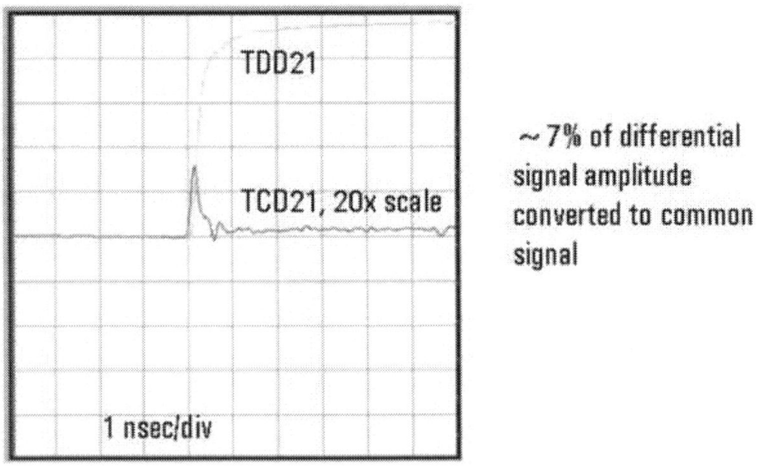

Figure 16.21: *Measuring Forward Time Domain Transmission (TDD21) and Superimposing Differential-to-Common Transmission Mode Conversion Yields an Indication of EMI Radiation*

CMOS 0.13 um Design Case

A typical 0.13 um pure CMOS design is capable of implementing multiple transceiver channels on one piece of silicon. This typically can lead to problems of crosstalk, which limit the maximum performance achievable. For the purpose of comparison, the following two links will be analyzed:

- A 3.125 Gbps XAUI transceiver with emphasized transmitter and traditional receiver
- A 6.25 Gbps transceiver with five-tap DFE receiver

For the purpose of channel analysis, the electrical characteristics defined in *Table 16.1* will be assumed.

Baud Rate	3.125	6.250	Gbps	
Return Loss @ fbaud*0.75	-8.000	-8.000	dB	
Transmit Jitter				
Unbounded Gaussian	0.012	0.011	UIrms	Unit Intervals Root Mean Square
Bounded High Probability	0.180	0.150	UIpp	Unit Intervals peak peak
Total	0.265	0.235		
Transmit Amplitude	400.000	400.000	mVsepp	Single Ended peak peak
Maximum Emphasis	6.000	0.000	dB	
Receiver Jitter				
Unbounded Gaussian	0.012	0.011	UIrms	
Bounded High Probability	0.150	0.100	UIpp	
Total	0.235	0.185		
Sampling Error Jitter				
Unbounded Gaussian	0.017	0.015		
Bounded High Probability	0.330	0.250		
Total	0.451	0.370		
BER	12.000	15.000	10^(-x)	
Q	7.040	7.940		

Table 16.1: *Assumed Electrical Characteristics*

StatEye Analysis

The return loss model of the transmitter, *Figure 16.22*, and receiver is a smooth response defined by a simple parallel resistor with 60W and a capacitor such that the return is just violating the specification at three-quarters the baud rate. The return loss of the channel returns a large amount of energy that increases the reflections and resonances of the forward transfer function. The pulse response of the forward channel, *Figure 16.23*, can be seen to contain post-cursors out to 53 UI (+7 UI) for 6.25 Gbps compared to the 3.125 Gbps, highlighting the need for additional receiver equalization. The influence of reflections at 6.25 Gbps can be seen in the tail at 56 UI, which is enhanced due to the non-ideal return loss of the transmitter and receiver.

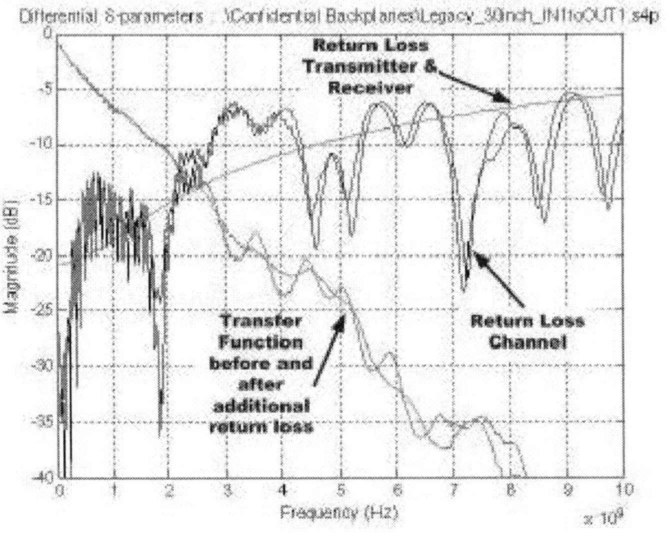

Figure 16.22: SDD21 with Return Loss of 6.25 Gbps Link

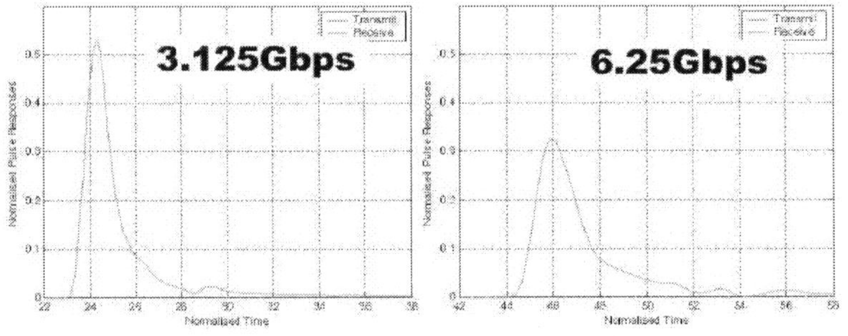

Figure 16.23: Pulse Response for 3.125 Gbps and 6.25 Gbps NRZ Transmit Pulse

Although no crosstalk measurements were available, two crosstalk models were generated from the SDD11, *Figure 16.24*, to show the influence of return loss, transmit emphasis, and DFE. The crossover of the return loss and forward channel for the -30 dB crosstalk example is already at three-quarters the baud rate for 6.25 Gbps, demonstrating that a simple analog bandwidth enhancement filter would additionally boost the crosstalk.

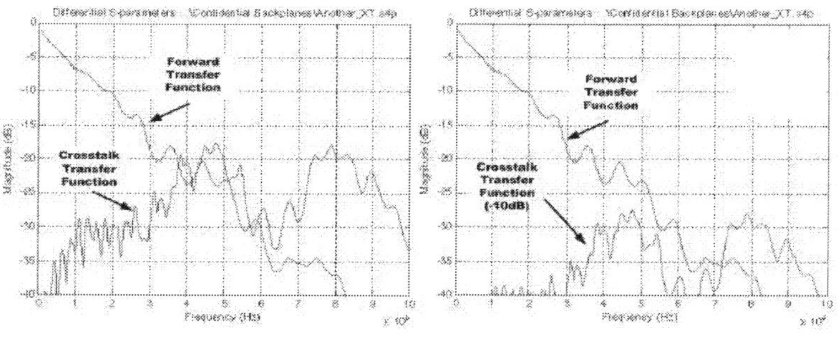

Figure 16.24: -30 and -40 dB Generated Crosstalk Transfer Function

After allowing the StatEye script to optimize the transmit emphasis coefficients[8], the resulting eye and bathtub, *Figure 16.25*, for the required Q has an amplitude of 88 mVsepp (0.22 normalized to the transmit amplitude) with a GJ of 0.014 UIrms and HPJ of 0.265 UIpp. When using the total sampling jitter for the calculation of the StatEye (see *Figure 16.26*), the effect of the receiver jitter can be seen to decrease the amplitude of the signal to 52 mV (0.13) and increase the jitter to 0.020 UIrms and 0. 397 UIpp. It can be seen that the additional receiver jitter does not add classically using dual dirac theory again, clearly demonstrating the need for the StatEye methodology.

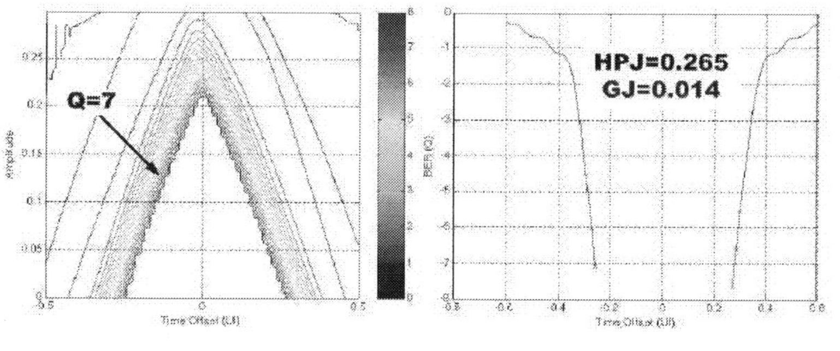

Figure 16.25: -3.125 Gbps StatEye and Bathtub with only Transmit Jitter

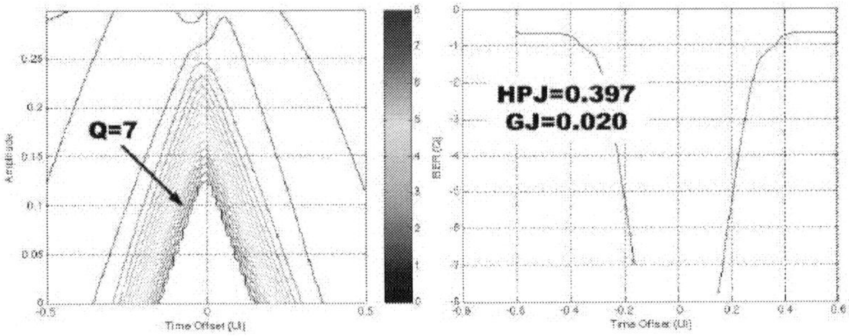

Figure 16.26: *3.125 Gbps StatEye and Bathtub with Total Sampling Jitter*

As the baud rate is increased to 6.25 Gbps, the increased width of the pulse response causes the StatEye without additional equalization to close. Using again an optimization script to find the coefficients of the DFE, the StatEye is calculated for the total sampling jitter and -40 dB crosstalk. The resultant eye, *Figure 16.27*, for the required Q has an amplitude of 50 mV (0.125) with 0.022 UIrms and 0.196 UIpp jitter, which corresponds to the requirements of the sampler. It should be noted that the worst-case StatEye does not represent a single worst-case signal, as this would correspond to a bit period of 49 ps. As the crosstalk is increased to -30 dB, a loss in eye opening is incurred (see *Figure 16.28*); however, unlike a time-continuous filter, the loss of amplitude is linear and not emphasized and would still allow a workable system.

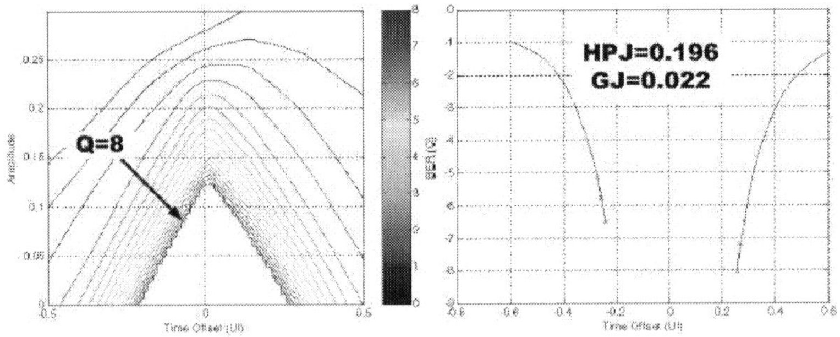

Figure 16.27: *6.25 Gbps StatEye and Bathtub with Transmit Jitter and -40 dB Crosstalk*

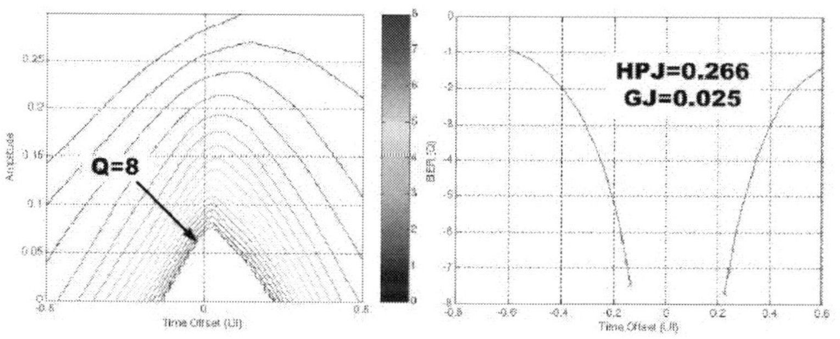

Figure 16.28: *6.25 Gbps StatEye and Bathtub with Transmit Jitter and -30 dB Crosstalk*

16.6 Conclusion

This paper outlines the "Legacy" methodology used by XAUI and explains under what conditions limitations become unacceptable as the channel is pushed to its limit either in terms of speed or BER. The StatEye methodology was then introduced, explaining how the statistical methodology, in combination with the use of the full S-parameter information of the forward channel and crosstalk aggressors, enables the limitations to be overcome. A small explanation of the way the StatEye fits into the entire standards and interoperability picture showed how no additional engineer requirements are needed in terms of hardware and how the problems of measuring "closed" eyes are circumvented. Finally an example of how a typical legacy channel can be measured with available equipment was demonstrated, pointing out typical problems using frequency and time domain representation and ending with the StatEye results showing that for 6.25 Gbps operation, such a channel, given the appropriate circuitry, would be sufficient for BER well below 10–15.

The StatEye methodology appears initially to be quite complex; however, in comparison to any typical simulation tool, it is quite straightforward, demonstrated by its simple implementation in Matlab. Alternative methods using time or frequency domain analysis would seem to be able to give a similar result, but care must be taken, as the very low-probability events are not possible to represent due to the very long simulation times required, e.g., $100 \cdot 10^{15}$ bits for a BER of $1 \cdot 10^{-15}$.

References

[1] The channel model is defined by

$$|S_{21}(f)| \leq -20\log(e)\left[a_1 f^{0.5} + a_2 f + a_3 f^2\right]$$

where f is frequency in Hz, $a_1 = 6.5 \cdot 10^{-6}$, $a_2 = 2.0 \cdot 10^{-10}$, and $a_3 = 3.3 \cdot 10^{-20}$.

[2] "IEEE802.3 2000," IEEE, Clause 47.3.5, 2000.
J. G. Proakis and M. Salehi, "Communication Systems Engineering," Prentice-Hall, pp. 555, 1994.

[3] J. G. Proakis and M. Salehi, "Communication Systems Engineering," Prentice-Hall, pp. 590, 1994.

[4] J. E. Smee and N. C. Beaulieu, "Error-Rate Evaluation of Linear Equalization and Decision Feedback Equalization with Error Propagation," IEEE Trans. Commun, pp. 656, vol. 46, May 1998.

[5] H. Johnson, "Scattering Parameters," High-Speed Digital Design, vol. 6 issue 03, June 2003, http://signalintegrity.com/Pubs\news\6_03.htm.

[6] The generation and calibration of the stressed eye is not the subject of this paper.

[7] The optimization uses a LMS–gradient–based algorithm, which is not the subject of this paper.

Authors

Anthony Sanders, Principal Engineer, Infineon

Mike Resso, Signal Integrity Application Scientist, Component Test Division, Agilent Technologies

John D'Ambrosia, Manager, Semiconductor Relations, Tyco Electronics

Chapter 17

Characterizing Jitter Performance on High-Speed Digital Devices Using Innovative Sampling Technology

17.1 Abstract

The proliferation of gigabit serial links has exposed signal integrity issues not typically encountered in the design laboratory. Predicting and measuring performance parameters such as differential impedance, crosstalk, skew, attenuation, and jitter have become crucial to assure success of component manufacturers. Understanding the various types of jitter and properly locating the jitter source are the first steps to achieving breakthrough performance of data transmission systems based solely on copper. This paper will provide a brief tutorial on jitter definitions, present jitter measurement techniques used in the past, and address a newer technology that will allow advanced designers to measure sub-picosecond jitter and beyond.

17.2 Introduction

Today's leading-edge high-speed digital designers are pushing the performance limits of copper as a transmission media. The proliferation of gigabit serial links has exposed signal integrity issues not typically encountered in the design laboratory. Predicting and measuring performance parameters such as differential impedance, crosstalk, skew, attenuation, and jitter have become crucial to assure success of component manufacturers. Accurately measuring jitter performance of copper interconnects over a very broad frequency range continues to be a challenging task. Serial data can have jitter imposed upon it from various sources, including passive devices such as stripline, microstrip, backplane connectors, cables, and flexible interconnects. Significant knowledge of jitter phenomena has been obtained during the development of optical fiber transceivers, and this has been leveraged to achieve breakthrough performance of data transmission systems based solely on copper. Understanding the various types of jitter and properly locating the jitter source are the first steps of this process.

Jitter is defined as the deviation in timing of an ideal event. By its very nature, jitter comprises many components. The two major types of jitter in transmission systems are random jitter (RJ) and deterministic jitter (DJ). RJ is caused by random behavior of components and devices such as thermal noise. Because RJ is unbounded, statistical analysis is required to characterize it. On the other hand, DJ includes timing fluctuations that are bounded in nature and can be characterized by a peak-to-peak value.

DJ is usually classified in various subcategories depending on its phenomenological cause: data dependent jitter (DDJ) and inter-symbol interference (ISI) includes bandwidth limitations in the transmitters and receivers as well as dispersion in the transmission lines; duty cycle distortion (DCD) and pulsewidth distortion (PD) includes timing errors in clocks and data modulators; periodic jitter (PJ) includes timing fluctuations with a periodicity other than the bit rate, such as the ones generated by a multiplexer with uneven input delays; uncorrelated bounded jitter (UBJ) includes less common effects such as gain crosstalk in amplifiers. In most applications, the total jitter (TJ) has both RJ and DJ components. In addition, very low-frequency jitter (usually less than 10 Hz) is also classified as wander or drift. In some cases, including packet transmission, wander is not as relevant as high-frequency jitter.

17.3 Jitter Measurement

There are several approaches to characterize jitter, depending on the type of signal that is output from the device under test (DUT). Repetitive signals provided by devices such as clocks have available both time and frequency domain techniques. Time interval analysis measures fluctuations of the clock period with respect to an ideal clock (time interval error/period jitter) or between adjacent cycles (cycle-to-cycle jitter). *Figure 17.1* shows an example of a time interval error measurement.

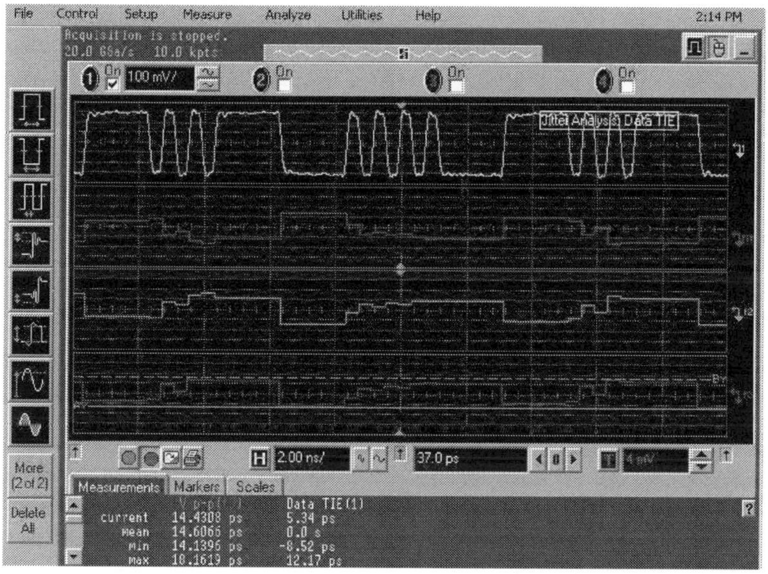

Figure 17.1: *Time Interval Error Analysis Using a Real-Time Oscilloscope*

In addition, jitter can be calculated from phase noise measurements by integrating the curve under the noise spectrum. In this type of measurement, DDJ can be extracted since it is manifested as spikes in the spectrum (*Figure 17.2*).

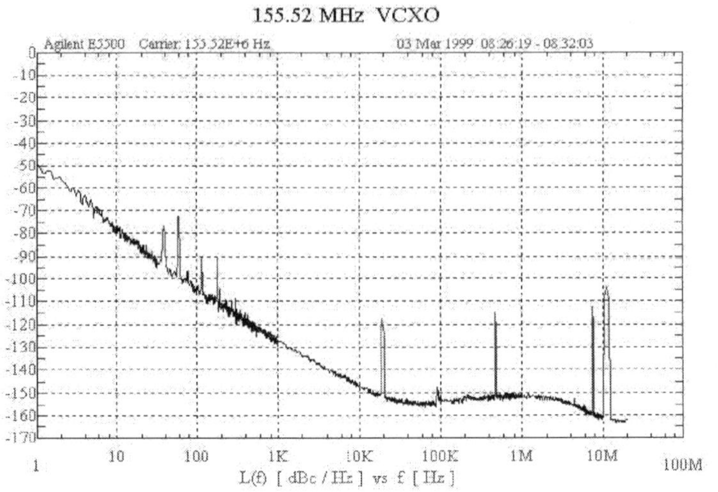

Figure 17.2: *Noise Spectrum of a 155.52 MHz VCXO*

In non-repetitive signals such as data transmissions, other techniques are usually used. Bit-error rate (BER) tests and eye diagram analysis are very common measurement methodologies. The former does not measure jitter directly, but characterizes the total jitter that a transmission system is able to tolerate for a specific BER. In the following paragraphs, we will concentrate on jitter characterization based on eye diagram analysis, which measures jitter from the spread of data transitions.

Because jitter has random components, a histogram is built and the standard deviation is used as the value of the RMS jitter. A Gaussian distribution is usually assumed if RJ is dominant. In the presence of DJ, it is necessary to separate the bounded and unbounded components for an accurate jitter measurement. The latest advances in jitter modeling can make this separation possible. However, the arsenal of tools to measure jitter is based on the fundamental principle of accurate waveform timing. Waveform timing is heavily influenced by how much error the test instrument is introducing. The first parameter to consider is the instrument frequency response. This should be flat and have a high corner frequency to guarantee that the edges are reproduced accurately. Instruments with poor frequency response could introduce ISI.

However, the most important factor in achieving accurate jitter measurements is the jitter introduced by the instrument itself. Instrument jitter not only increases the absolute value of the jitter measurement, but also masks DJ and makes it difficult to estimate. In low-frequency transmission systems, the errors introduced by these parameters are usually negligible. However, as transmissions frequencies get to 10 GHz or higher, measurements hit the instrument limitations (dictated by the instrument hardware). In this paper we present an innovative solution to overcome current hardware limitations in jitter measurements.

17.4 Random Sampling and Precision Time Base

Most of the jitter contribution in instruments is in the synchronization circuitry (trigger). In a traditional, high-bandwidth, equivalent-time, sampling-oscilloscope time base, the time axis is generated with a stable oscillator, which acts as a reference for data sampling. Since the data rate and the oscillator frequency are independent of each other, a trigger is required to produce the appropriate time equivalent sampling. The trigger starts the oscillator, which generates the required delays to sample the data at the right time interval. This time interval is determined before the actual sample is taken. Unfortunately, there is a significant amount of jitter associated with this triggering process. To overcome this fundamental limitation, a different sampling architecture was developed. This new random sampling architecture utilizes a free running oscillator internal to the test equipment. The time values for each sample are determined using accurate time estimator models. The amplitude information and time information are then combined into a data/time stamp pair for further processing by the mainframe.

The concept of using an external clock as a time reference is based on converting an amplitude value to a time value from a known waveform (e.g., sinusoidal). To determine the time of a data sample, both data and clock are sampled simultaneously. From the amplitude of the clock and its quadrature, the angle ωt between 0 and 2π is obtained. Time is then calculated using the clock frequency entered by the user. The combination of random sampling and a clock-based time base gives rise to a precision time base. This time base is not suitable for pattern display, but it is excellent for eye diagram analysis and repetitive signal measurements.

Communication signals have an important advantage over other type of waveforms. They are generated from a stable, accurate clock, and this clock can be used as a time reference. In this case, no triggering or synchronization is required and hence very low jitter can be obtained. In addition, improving the clock quality can increase time accuracy and time-base linearity. Another advantage is that high-frequency clocks, up to 45 GHz, can be used to display data in an oscilloscope. This is in sharp contrast with the ~10 GHz trigger limit in a traditional time base.

Figure 17.3 shows the advantage of the precision time base versus a standard time base. The first obstacle to overcome to display the 40 GHz signal is the requirement of a clock divider to reduce the trigger signal to 10 GHz. When the signal is displayed, the measured jitter is about 840 fs. This is a very good value, but compared to the period of the signal, it is significant enough that the signal trace cannot be determined precisely (*Figure 17.3(a)*). When the precision time base is used, the 40 GHz signal can be used as the clock and the signal jitter is reduced to 158 fs.

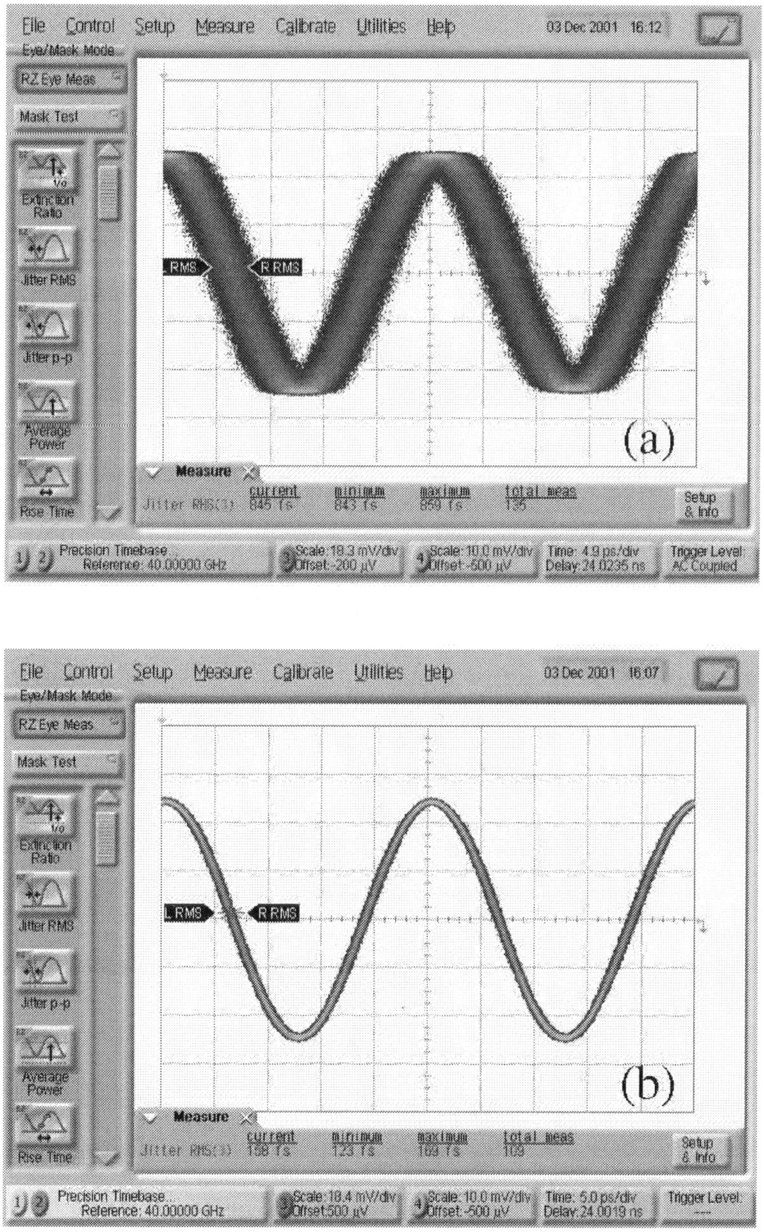

Figure 17.3: *40 GHz Signal Measurement Using Standard Triggering (a) and Precision Time Base (b)*

An immediate application in eye diagram analysis is visualized in *Figure 17.4* for a 40 Gbps optical signal. In *Figure 17.4(a)*, there is some DJ component in the eye diagram, but the instrument jitter masks it in a way that makes it impossible to deconvolve it accurately from the histogram. In *Figure 17.4(b)*, the lower jitter of the precision time base makes it possible to use histogram analysis to separate RJ from DJ.

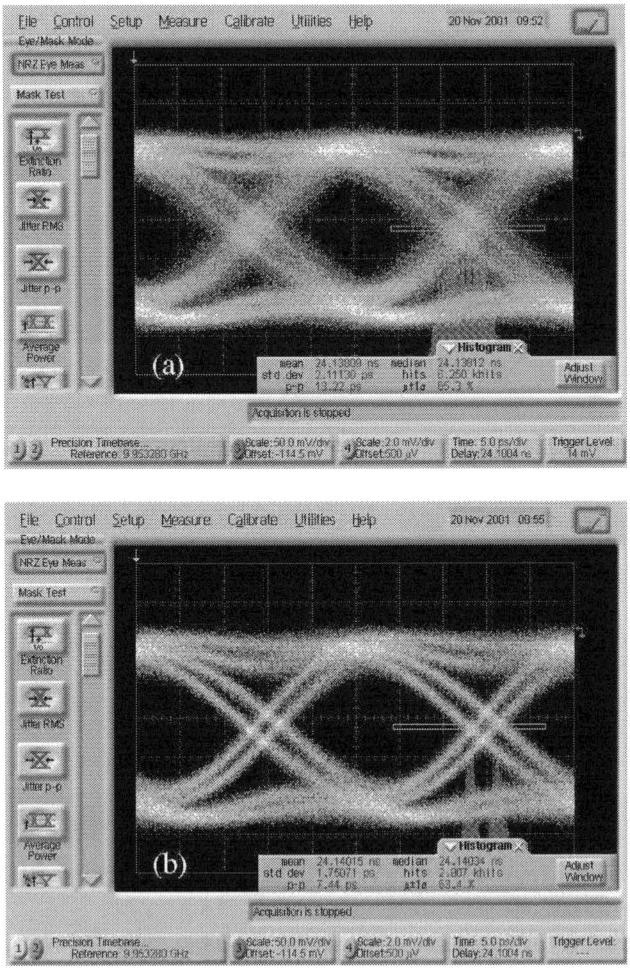

Figure 17.4: *Comparison of a 40 Gbps Eye Diagram Using an External Trigger (a) and Using the Precision Time Base (b)*

It is not necessary to go to frequencies as high as 40 GHz to see the influence of instrument jitter in eye diagrams. Even at lower frequencies, instrument limitations can affect jitter measurements. *Figure 17.5* shows an eye diagram of a synchronous optical network (SONET) OC-192 electrical signal. The signal looks very clean until jitter analysis is attempted.

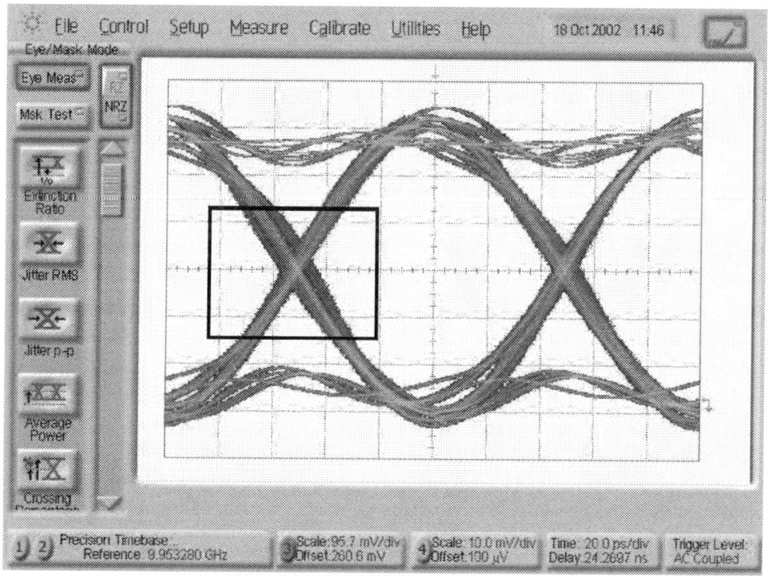

Figure 17.5: *SONET OC-192 Eye Diagram*

Zooming into the data transitions (rectangular box in *Figure 17.5*), we can see the instrument limitations. *Figure 17.6* shows the presence of deterministic jitter, but again an accurate measurement is not possible.

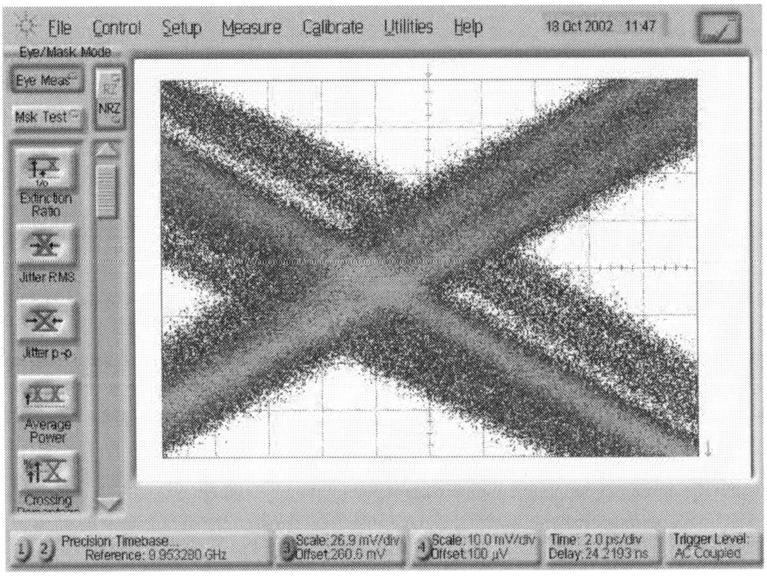

Figure 17.6: *Detail of OC192 Eye Diagram with Standard Triggering*

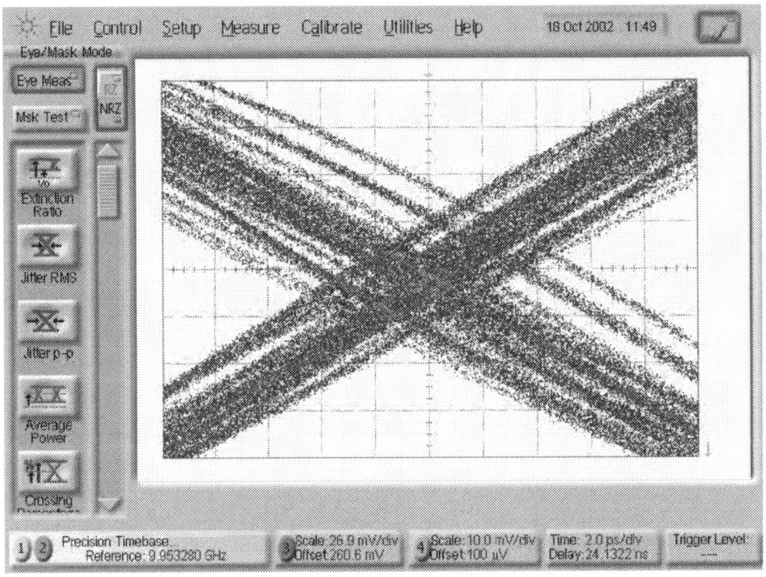

Figure 17.7: *Detail of OC192 Eye Diagram with Precision Time Base*

In contrast, *Figure 17.7* exhibits a much cleaner trace that shows DJ is dominant. Selecting specific patterns to generate the eye diagram can complete the characterization. *Figure 17.8* shows the transitions when a short pattern (seven bits) is used. The pattern dependency can then be traced to a specific component in the transmission system.

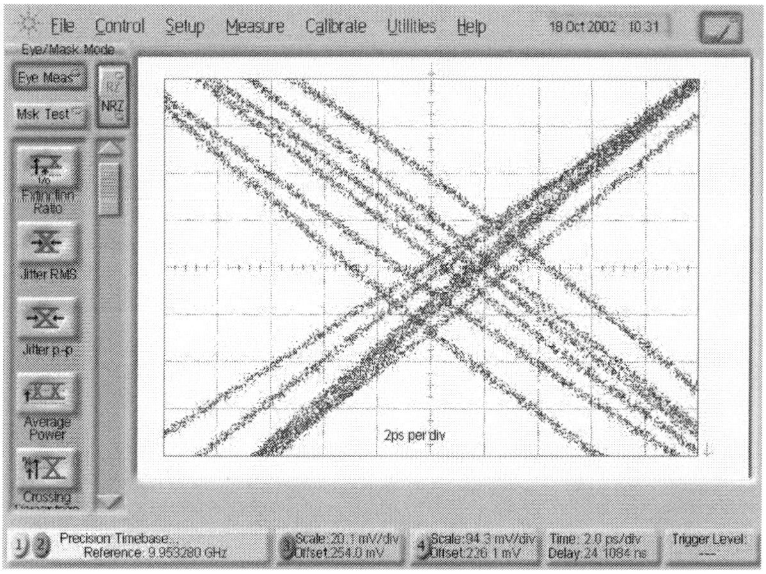

Figure 17.8: *Detail of OC192 Eye Diagram with Precision Time Base*

The previous examples showed the power of the new time-base approach in jitter measurements. The ultimate limitation of the system is given by signal-to-noise ratio of the clock signal. Optimization of the hardware parameters was able to produce less than 50 fs jitter measurements, as shown in *Figure 17.9*.

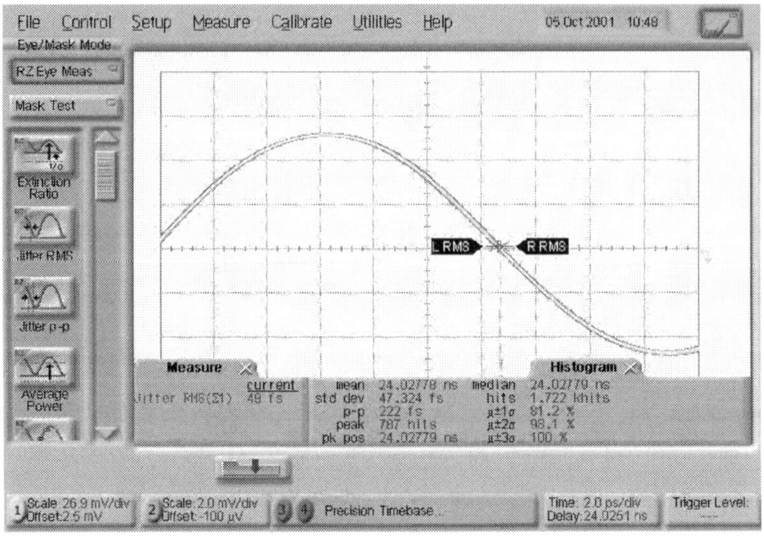

Figure 17.9: *Parameters Can Be Optimized to Produce RMS Jitter below 50 fs*

17.5 Future Trends: Optical Sampling

As digital data rates increase through traditional copper networks, there is a natural migration to build optical fiber networks. Fiber has some distinct advantages, but eliminating jitter is not one of them. There are many sources that can generate jitter in both optical and electrical systems. These include clock and data recovery circuits, transceivers, loop filters in phase-locked loops, and serializer/deserializer circuits. In addition to a diminished jitter budget, the problem of test equipment receiver bandwidth grows more serious at data rates increase. Creating and characterizing error-free gigabit serial data transmission is a manageable task. However, when the data rate makes a quantum leap into the OC-768 realm of 40 Gbps, the design engineering skills and tools take on different requirements. The data modulation format changes from standard non-return-to-zero (NRZ) to return-to-zero (RZ). These new and demanding ultra-high-speed components and systems require next-generation tools to accurately characterize performance. One of the critical tools of designing, developing, and manufacturing these OC-768 systems will be the optical sampling oscilloscope. An example of measuring a 350 Gbps signal with an optical sampling oscilloscope is shown in *Figure 17.10.* Without sub-picosecond instrument-jitter values, the picosecond pulses of this high-frequency rate-transmission system will not be displayed properly.

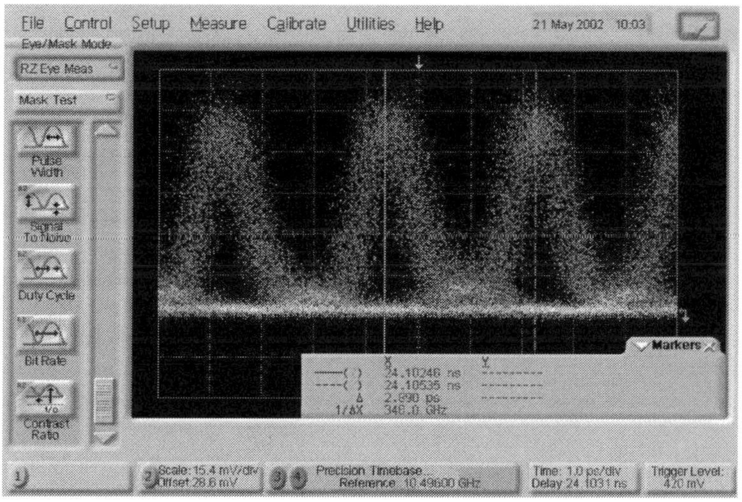

Figure 17.10: *Characterizing a 350 Gbps Signal with Optical Sampling*

17.6 Summary

It is critical that high-speed digital designers understand signal integrity issues when developing state-of-the-art systems. Margin for timing error is decreasing in digital transmission systems due to the push for higher serial throughput. As data rates increase, jitter measurements become more of a challenge and test equipment error becomes significant. To minimize test equipment error, an innovative random sampling technique has been employed to create a precision time base for equivalent time oscilloscopes. It has been shown that accurate jitter measurements can be made on signals up to 350 Gbps and that jitter below 50 fs can be accurately measured. While this technology was originally developed to characterize high-speed optical devices, it has been shown to be very useful in eliminating traditional limitations of jitter measurements on electrical signals.

Authors

Osvaldo Buccafusca, Development Scientist, Lightwave Division, Agilent Technologies

Mike Resso, Signal Integrity Application Scientist, Component Test Division, Agilent Technologies

Chapter 18

Signal Integrity Concerns When Modulating Laser Transmitters at Gigabit Rates

This article will describe the challenges facing high-speed digital designers when trying to optimize optical transmitter performance. The paper begins by giving a brief overview of the evolving optical network. It will lay a foundation of semiconductor laser fundamentals, then address the relationship between electrical signal integrity of laser driver circuits and the resulting impact on optical performance of distributed feedback (DFB) and electro-absorptive modulated lasers (EMLs). Laser modulator package architectures using flexible interconnects will be investigated to determine the effect on eye diagram quality at 2.5 and 10 Gbps. Finally, the paper will present state-of-the-art optical laser transmitter measurement techniques for 40 Gbps return-to-zero (RZ) data signals.

- The Evolving Optical Network
- Semiconductor Laser Fundamentals
- Common Laser Driver Circuit Designs
- Signal Integrity Challenges
- Flexible Interconnect Application
- Future Trends

Figure 18.1: *Agenda*

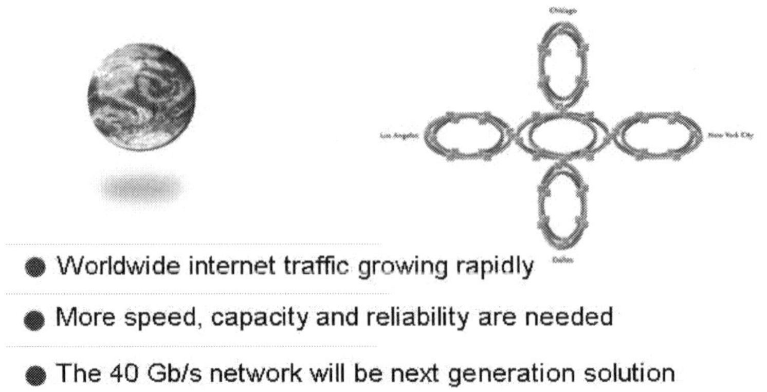

Figure 18.2: *The Evolving Optical Network*

In the last 20 years, optical fiber transmission networks have progressed tremendously in terms of link distance and information-handling capability. Major advances in the electronic and optical technology of laser transmitters have made this possible. The single-mode photonic emission from a semiconductor DFB laser can be coupled efficiently into today's single-mode fiber. In addition, these lasers can be switched on and off with transition times on the order of 5–10 ps to provide data rates up to 40 Gbps. The electrical circuits required driving the related laser modulators and multiplexers are so demanding that manufacturers are developing new material technologies to meet the challenge. Indium phosphide (InP), silicon germanium (SiGe), and gallium arsenide (GaAs) are materials that will be needed as data rates reach 160 Gbps in development laboratories in 2002. The outstanding performance of semiconductor lasers has in fact provoked considerable effort to improve the capabilities of test equipment in order to fully characterize the capabilities of these optical components. A technological breakthrough in test instrumentation is needed to expand the time domain bandwidth for all digital communications analyzers to analyze the modulation characteristics of tomorrow's optical networking laser transmitters.

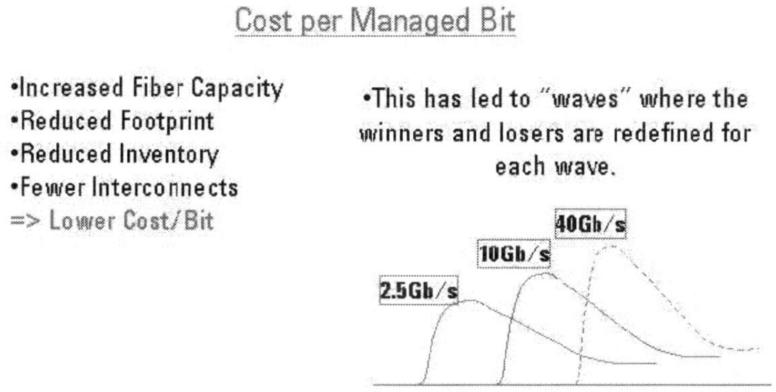

Figure 18.3: *Why Gigabit Rates?*

Historically, each successive wave of higher speeds has led to a lower cost per bit for network equipment. Once a new technology becomes mature, it has typically delivered four times the bit rate at only two and a half times the cost. A "managed" bit includes all the indirect costs associated with data transport. Increased fiber capacity has prevented the laying of new fiber in areas where there is fiber exhaust and has avoided the cost of additional repeater stations. Higher rates usually lead to smaller form factor equipment, smaller backplanes with shorter transmission lines, and reduced power requirements. These can be a major cost of a central office or other point of presence. Also, by increasing the capacity of each wavelength by a factor of four, the number of wavelengths can be reduced. This technology is called wavelength division multiplexing (WDM). Since each data-channel wavelength will need at least one spare circuit pack for that wavelength, this can reduce the inventory of spare equipment for repair by a factor of four. These cost dynamics have led to technology waves, where the winners and losers are redefined for each wave. We saw this happen at 2.5 Gbps, when Lucent took leadership market share in 2.5G dense WDM (DWDM). Then we saw the 10 Gbps wave take over 2.5 Gbps for DWDM equipment. Nortel, the first to reliably deploy 10 Gbps, took enormous market share, moving from 13 to 43 percent in six months. So, the industry expects the same thing to happen with 40 Gbps once it becomes a mature technology and delivers a true cost-per-bit advantage. This is why the stakes are so high and so many investments are being made.

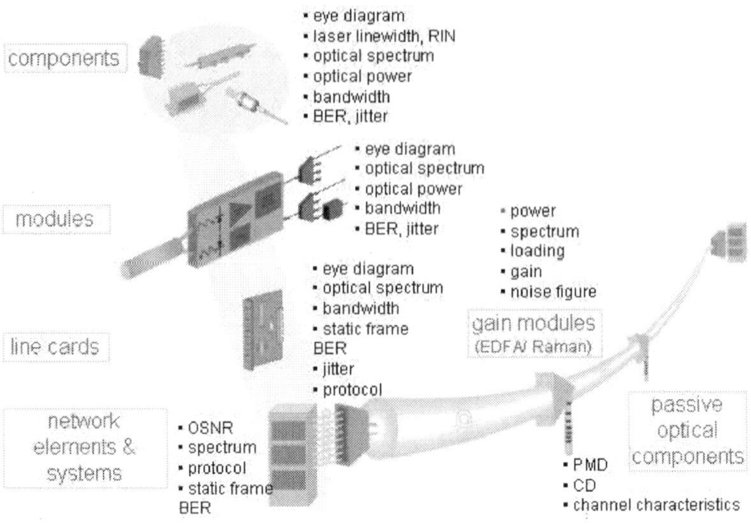

Figure 18.4: *Laser Transmitters Used in the Optical Network*

This slide lists some of the key test and measurement challenges in developing 40G components and systems. At the equipment level, there is the network terminal equipment, such as line amplifiers, and passive optical equipment, such as optical multiplexers and dispersion compensators. The equipment that is in a central office, often referred to as network elements, is composed of line cards, which are composed of modules, which are composed of basic electrical and optical components. Gigabit data rates challenge today's technology, so characterizing and testing each point in the supply chain is essential. Listed besides each component are some of the critical measurements that must be made to ensure reliable products.

- Measurement tools
  - Electronic design automation (ADS, IC-CAP)
  - Waveform analysis (DCA)
  - Active optical components (DCA,LSA)
  - Raman amplifier test (OSA, TLS)
  - Passive optical components (OSA,CD,PMD)
  - BER test (BERT)

Figure 18.5: *Optical Network Physical Layer Test Instrumentation*

There is a fairly large suite of measurement solutions available today to address the 40 Gbps industry. The majority of the measurements that we will focus on today is the waveform analysis section of this suite that incorporates the use of the digital communications analyzer. Hopefully, this paper will stimulate your interest in this 40G arena such that you will be interested in learning about more 40G test equipment.

Figure 18.6: *Signal Integrity Concerns*

When designing high-speed digital electronics, it becomes more difficult to transmit ones and zeros as the data rates increase. This is because high data rates usually translate into a faster rise-time transition between a logic low level and a logic high level. As this rise time becomes shorter, the subsequent electrical length of a PCB trace that becomes a transmission line is much shorter. Design of transmission lines require significantly more care and rigorous understanding of underlying microwave phenomena. To enhance signal integrity of the optical network physical layer (e.g., electrical drivers of optical transmitters, router backplanes, gigabit interconnects), designers must follow good high-speed practices. Minimizing reflections by reducing impedance mismatch between components and lowering crosstalk between adjacent electrical lines is important. These techniques will avoid problems such as signal attenuation, skew, and jitter.

- Direct Modulation of the Laser Current ($\leq$ 10GB/s)
  - Fabry-Perot Semiconductor Lasers (FP)
  - Vertical-Cavity Surface-Emitting Lasers (VCSEL)
  - Distributed Feedback Lasers (DFB)
- External Modulation of Laser Light ($\geq$ 2.5Gb/s)
  - Electro-Absorptive Modulator (EAM)
    - Electroabsorptive Modulated Laser (EML)
  - Mach-Zehnder Interferometer Modulator(MZ)

Figure 18.7: *Fiber-Optic Transmitter Technologies*

Telecommunications and data networks require more bandwidth over longer distances at a lower cost. As a result, fiber optics is evolving into a diverse array of technologies that can accommodate these factors depending on the applications. For lower-cost, short-distance networks, direct modulation of the current of a semiconductor laser is the best choice. There are three main types of lasers used for direct modulation: Fabry-Perot (FP) lasers, vertical-cavity surface-emitting lasers (VCSEL) and DFB lasers.

For higher-quality, long-distance networks, external modulation of the laser light is the preferred approach. This maintains the spectral purity of the laser light. Two main technologies compete for this space: electro-absorptive modulators (EAMs) and Mach-Zehnder (MZ) interferometers. EAMs are constructed of semiconductor materials such as InP. EAMs modulate light by absorbing the optical energy when a negative voltage is applied. The main advantage of EAMs is the ability to be integrated with a DFB laser on the same die. When the EAM and DFB laser are monolithic, the assembly is commonly referred to as an EML. Although EMLs are an improvement over directly modulated lasers, they still have some technical disadvantages when compared to MZ interferometer modulators.

Designers use lithium niobate (LiNbO3) MZ interferometer modulators to achieve the highest-quality optical transmitter. More recent technologies use GaAs, InP, and polymers to create MZ modulators. Each material has technical advantages; however, LiNbO3 remains the main technology for MZ modulators. This is due to the proven reliability, low insertion loss, and broad operating wavelength range of LiNbO3. This makes this material ideal for tunable lasers and WDM.

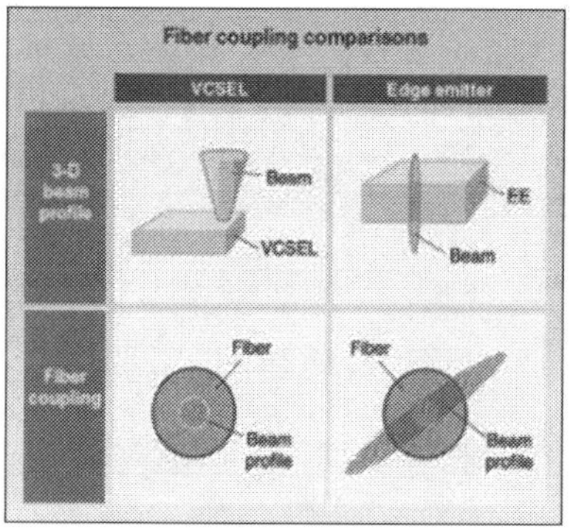

	FP	VCSEL	DFB
Common Wavelengths	1310 nm, 1550nm	850nm,1310nm	1310 nm, 1550nm
Materials	InGaAsP / InP	GaAs, InGaAsP	InGaAsP / InP
Output Power	≤ 10 mW	≤ 5 mW	≤ 40 mW
Spectral Width	3 nm	0.05 - 0.2 nm	0.0001nm
Threshold Current @25C	5mA – 15mA	2 – 4mA	5 – 15mA
Transmission Distances (at 2.5Gb/s)	< 20 km	< 40 km	<200 km
Coupling Efficiency	30% – 50%	85% – 90%	30% – 50%
Cost	$25-$500	$5-$100	$100-$3,000

Figure 18.8: *Semiconductor Laser Comparison*

Semiconductor laser technology has been advancing at an ever-accelerating pace. FP lasers are the simplest laser structure and have been around the longest time, forming the foundation for the more recent technologies. At 1,310 and 1,550 nm, FP lasers are fabricated on InP, with the active laser layers made of InGaAsP. The waveguide structure of the laser is grown horizontally on the semiconductor wafer. Hence in order for the light to be emitted, there is a post-process of cleaving, polishing, and thin-film–coating that adds cost and time to manufacturing. The lasing action is simply created by two semi-reflecting surfaces with the semiconducting material in between, setting the gain and wavelength of the laser device. Thus, there are multiple modes of light creating a picket-fence–type spectral wavelength map, making the effective spectral width of the laser of up to 3 nm. Due to the large spectral width of FP lasers, they cannot be used for long distances, since chromatic dispersion would limit their performance.

DFB lasers were developed next by adding a Bragg grating structure inside the laser waveguide between the reflecting surfaces of a FB laser. This isolates only on the mode inside the laser, making its spectral width extremely thin on the order of 5 MHz. Hence, DFB lasers are used for much longer-distance transmissions. DFBs, however, still require the extra post-processing steps, since it is an edge-emitting laser. Furthermore, the output beam is elliptical due to the rectangular waveguide structure of the laser.

In order to bring the laser cost down, much research has gone into the development of VCSELs. VCSELs are radically different structures from FPs and DFBs. Their laser cavity is built vertically on the semiconductor substrate during the doping and etching processes, and there is no need to dice the chips up or to polish the output surfaces. Hence, VCSELs can be manufactured at a reduced cost, since their manufacturing process requires fewer steps. Furthermore, since they emit light from their top surface, their output beam is designed to be round, making optical coupling to a fiber much more efficient and easier. VCSELs also have more stable output over temperature, hence they do not require a monitor photodiode. VCSELs also have lower threshold currents, which means they require less power. Although VCSELs have advanced significantly in the last couple of years, 5 MHz line widths and high-power 1,550 nm operation have not been demonstrated yet with VCSELs; therefore, DFB lasers still dominate long-range applications.

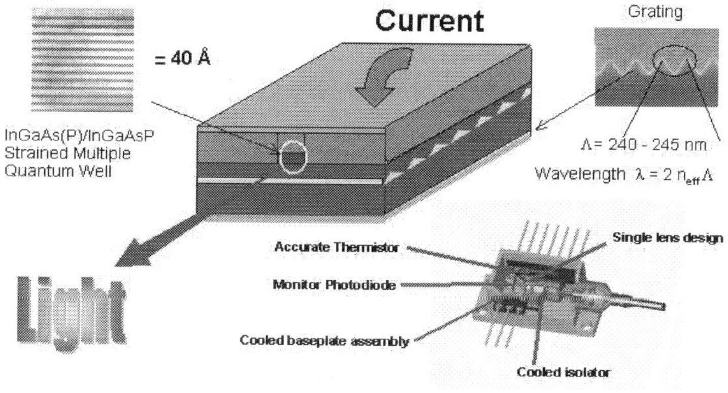

Figure 18.9: *Distributed Feedback Laser for WDMs*

This slide illustrates the fundamental constructions of an indium gallium arsenic phosphate (InGaAsP) multiple quantum well DFB laser used for WDM applications. The multiple quantum well structure provides the gain medium for the laser cavity. The grating structure underneath and perpendicular to the laser waveguide selects the operating wavelength. The wavelength of the laser has a linear relationship with change in temperature of approximately 0.7 nm/C, so the laser must be placed on a thermoelectric cooler in order to maintain the correct operating wavelength over time and temperature. Thus, a thermistor is placed in thermal contact with the laser so that it can be used in a control loop that maintains the laser at a specific operating temperature. The laser is then placed into a butterfly package for thermal heat-sinking purposes. The laser light exits on both sides of the laser, with a majority of the light emitting out of the front face to the lens and a small amount exiting on the back side toward the monitor photodiode. The monitor photodiode is used in a control loop that maintains the output power constant over temperature and time. The light exiting the front face goes through a lens and through an optical isolator, which keeps light reflecting back into the laser cavity and disrupting the lasing process.

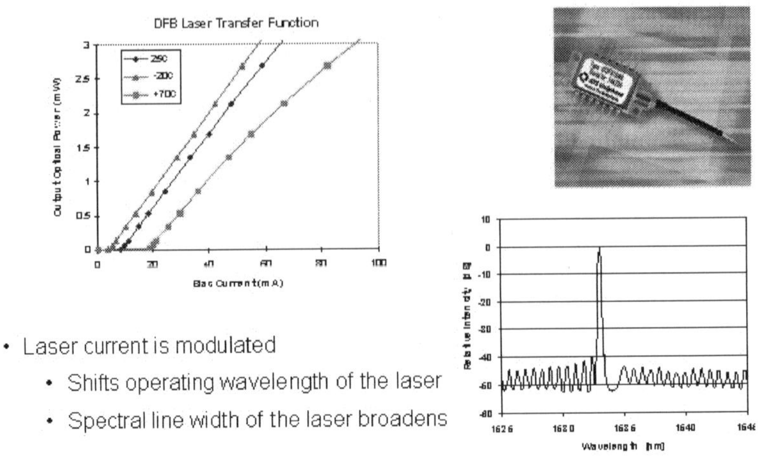

Figure 18.10: *Semiconductor Laser Fundamentals*

The transfer function of a DFB laser is very temperature-sensitive. As illustrated in *Figure 18.10*, the threshold current and the slope efficiency degrade at higher temperatures. Of course, with thermo-electrically controlled (TEC) lasers, as illustrated in *Figure 18.10*, the DFB should remain at a stable operating temperature and see much fewer variations with temperature. The spectral image of the laser shown demonstrates the thin line width of a DFB laser. When the laser is directly modulated at 2.5 Gbps, the spectral width of the laser broadens, which will limit the distance the light can travel over standard single-mode fiber due to the fiber's chromatic dispersion. This spectral broadening of the laser, when directly modulated, is commonly referred to as laser chirp performance. The laser actually shifts in operating wavelength as the current through the device turns the laser on and off. Typically, the laser has a positive chirp (a shift toward shorter wavelengths) when being turned on and a negative chirp (a shift toward longer wavelengths) when it is turned off. The shift in operating wavelength occurs more severely when the laser first turns on at the threshold current of the laser and can cause a new shorter-wavelength peak in intensity, as illustrated in *Figure 18.10*.

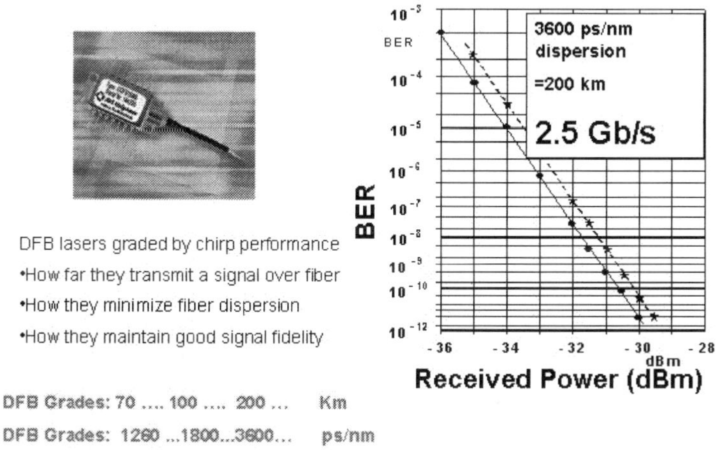

Figure 18.11: *DFB Laser Chirp*

In the 1550 nm range, DFB lasers are graded on the amount of chirp they produce when modulated directly. To separate the low chirp lasers from the high chirp lasers, DFB lasers are graded by the amount of dispersion that the laser can handle before there are significant errors in received signal at the end of the fiber link. Bit-error rate (BER) curves at different lengths of fiber are used to illustrate how the fiber chromatic dispersion has caused the chirp of the DFB laser to limit the transmission distance. The common metric used is the dispersion penalty, which is the measure in dB of optical power between the back-to-back BER curve and the BER curve over some fixed fiber length. Typically, if the dispersion penalty is greater than 2 dB, then the DFB cannot be used for that length of fiber. The DFB laser manufactures would then specify that a laser could handle a fixed amount of dispersion in ps/nm and guarantee that it could handle this dispersion with less than a 2 dB dispersion penalty. Recent advances in DFB technologies have produced lasers with fewer chirps, allowing them to reach distances of 200 km. In the above graph, the BER curves for a 3,600 ps/nm DFB laser is plotted first for the case of a the transmitter connected directly to the receiver, then with 200 km of standard single-mode fiber inserted between them. Standard single-mode fiber has a dispersion coefficient of up to 18 ps/(nm km) in the 1,550 nm range, hence the dispersion would equate to the fiber length times the dispersion coefficient.

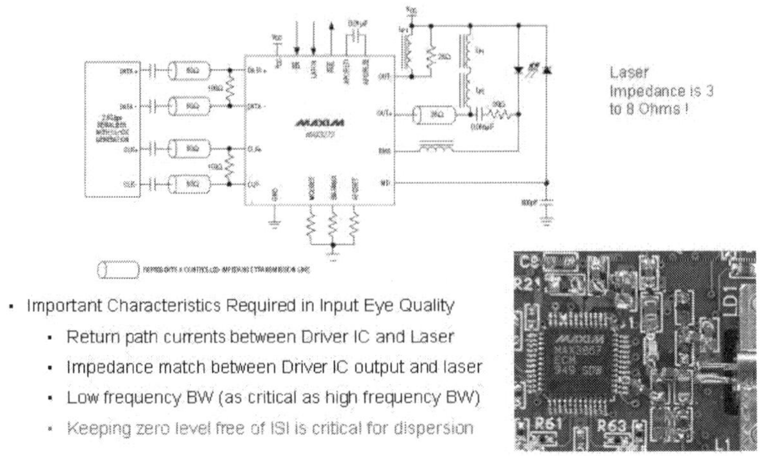

Figure 18.12: *Sample Driver Circuits for a Laser at 2.5 Gbps*

Figure 18.12 is a schematic of a MAXIM laser driver connected to a semiconducting laser. This integrated circuit (IC) takes the 2.5 Gbps data stream and 2.5 Gbps clock and re-times the data, providing a low-jitter data stream to the laser. The input impedance of the drive is 50 Ohms for the input differential signals, and 25 Ohms for the output differential signals . The 25 Ohm transmission impedance is used because it is closer to the input impedance of the laser, which is between 3 and 8 Ohms, depending on the bias and the laser. A series 20 Ohm resistor is commonly used in series as the most straightforward way to match the impedance of the laser to the driver output impedance. Ideally, the laser and driver are very close together, but in some cases, there is significant distance between the laser and drive circuit. This requires a controlled transmission line microstrip between the laser and the driver. It is critical that the return path for the laser, the driver, and the microstrip have a common AC ground with as little resistance or inductance between them.

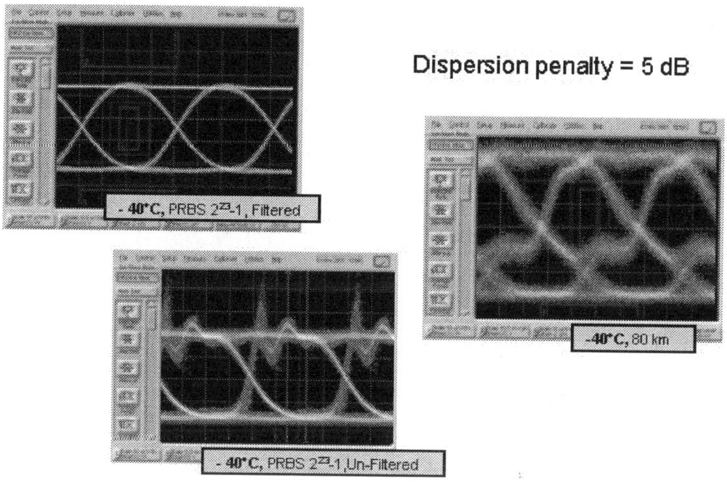

Figure 18.13: *1,550 nm DFB Performance at 2.5 Gbps*

Figure 18.13 shows three eye diagrams of an uncooled 1,550 nm DFB laser. The top eye diagram is the output of the laser directly into the Agilent 86100A digital communications analyzer using a plug-in with an OC-48 filter. This filter removes the laser ringing that occurs in the eye diagram. The filter is used as a standard for all extinction ratio, jitter, and eye quality measurements. This filter is a fourth-order Bessel-Thompson filter that approximates the receiver bandwidth and it is used for compliance measurements. Its 3 dB bandwidth is typically 75 percent of the data rate.

The unfiltered eye diagram can observe key parameters such as the laser relaxation frequency and magnitude. The filtered eye measurement may cover up the relaxation frequency ringing and may not properly characterize possible problems when tested in an actual link. This holds true in *Figure 18.13* when viewing the eye diagram after 80 km of standard SMF-28 fiber. The large extinction ratio (~11dB) was causing the laser to chirp significantly during turn-on, which would also cause excessive ring of the laser. In this case, the dispersed eye diagram caused a 5 dB dispersion penalty. The correct way to characterize relaxation frequency ringing is to make an unfiltered measurement.

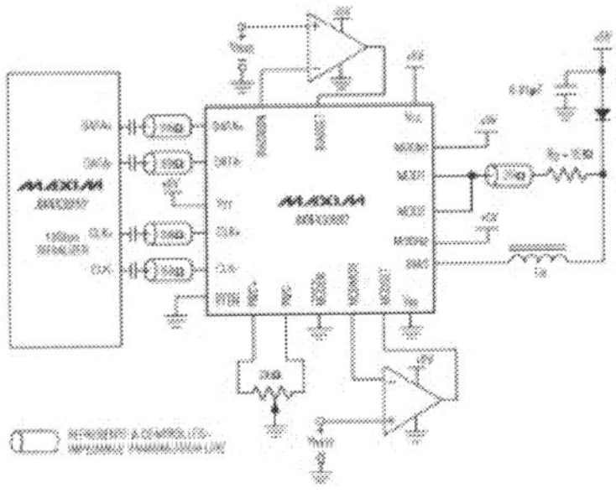

- **Important Characteristics Required in Input Eye Quality**
  - Return path currents between Driver IC and Laser are critical
  - Impedance match between Driver IC output and laser is critical
  - Low frequency BW is just as critical as high frequency BW

Figure 18.14: *Sample Driver Circuit for Laser at 10 Gbps*

Direct modulation of lasers at 10 Gb has been shown to work very effectively at 1,310 nm due to the zero-dispersion point of standard single-mode fiber. Thus, the spectral broadening of the laser at 1,310 nm is not as critical as it was at 1,550 nm. Again, as with 2.5 Gbps, return path currents between driver IC and laser are critical, impedance match between driver IC output and laser is critical, and low-frequency bandwidth (BW) is just as critical as high-frequency BW. However, the signal integrity is four times more challenging due to the parasitic and laser limitations. At 10 GHz, the laser leads become inductive and the internal bond wires to the laser chip radiate and capacitively couple to the laser case, which is typically AC ground. Hence, packaging the lasers becomes challenging from a radio frequency (RF) perspective, severely limiting the rise and fall times. Hence, industry trends have been pushing integration of the driver with the laser in a hybrid package to improve performance and yields in production.

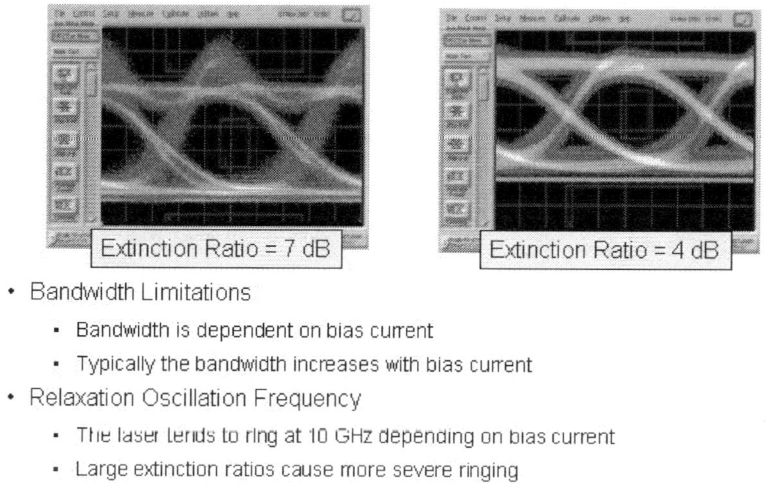

Figure 18.15: *1,310 nm DFB Laser Performance at 10 Gbps*

Operating semiconducting lasers at 10 Gbps pushes the technology to its limits, squeezing every ounce of performance out of the device. Typically, biasing the laser for more output power increases the bandwidth of the laser but, at the same time, requires more drive level to create the same extinction ratio. Higher bias levels mean shorter life or higher-power lasers. Lower bias levels mean a decrease in laser relaxation oscillation frequency, along with an increase in the magnitude of oscillation. This sometimes occurs below 10 GHz, causing the ringing to dip into the middle of the eye diagram when trying to reach the required minimum 6 dB extinction ratio for short-fiber links (< 2Km) at OC-192. This means that there is this tradeoff between laser extinction ratio and eye quality.

More recent DFB lasers have reduced the magnitude of the relaxation oscillation and have pushed relaxation frequency past 17 GHz, but still most 10 Gb/s DFB lasers used today have significant ringing that can be masked by the OC-192 filter, causing interoperability problems between receivers. Fiber-optic receivers from different manufacturers use different photodiodes with different bandwidth and group delays. Receivers that work great with EML and MZ transmitters sometimes produce errors across the optical input range due to direct modulated DFB lasers, which have severe ringing in the middle of the eye diagram. The ringing passes through the receiver bandwidth and into the error detector, creating BER floors in the BER curve. In other words, the receiver always produces at least a fixed number of errors, no matter what optical input power is placed into it.

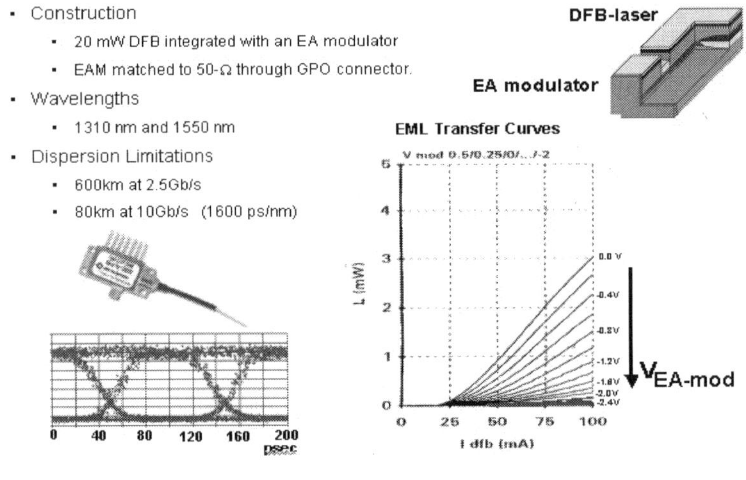

Figure 18.16: *EML Laser Fundamentals*

Due to the limitations of DFB lasers at 2.5 Gbps and 10 Gbps, electro-absorptive lasers (EMLs) have become the standard laser for medium- to long-reach applications. At 2.5 Gbps, the transmission distance moved from 200 km maximum for directly modulated lasers to 600 km with EMLs. EMLs are fabricated together on the same die with a DFB laser, so their transfer function is dependent on the DFB bias current and the bias on the electro-absorptive section. Although not as severe as direct modulation, the chirp performance of EMLs is dependent on the bias level and the drive level of the EA section. The EA section is relatively efficient at about 5 dB/V. The drive levels required for EAMs are typically around 2.5 Vp-p in order to obtain extinction ratios greater than 8.2 dB. EAMs also require a negative bias adjust between –0.7 and –1.8V in order to optimize the output waveform extinction ratio and insertion loss.

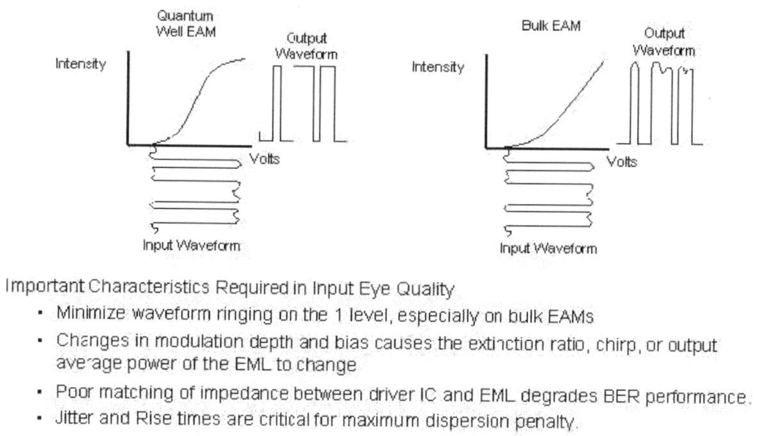

Figure 18.17: *Signal Integrity Concerns with EML Lasers*

There are basically two types of EAMs: bulk EAMs and quantum well EAMs. Bulk EAMs work over a broader wavelength range and require slightly higher drive levels than quantum well EAMs do. One advantage of some quantum well EAMs is that the one and zero levels, when driven optimally, have flattened transfer functions that saturate, so pattern-dependent noise on the one and zero levels can be cleaned up. On the other hand, bulk EAMs remain linear at the one level, so any pattern-dependent noise on the one level is magnified on the optical output. However, bulk EAMs have a zero level that will flatten out without having to worry about it folding over like quantum well EAMs. This means that with bulk EAMs, the optimum eye diagram for best dispersion penalty performance has a crossing that is skewed lower than 50 percent. For both types of EAMs, it is critical not to overdrive them or bias them positively above ground. Not only will this degrade the eye diagram, but also, it could eventually damage the device.

- 50 Ohm Impedance used at Driver output

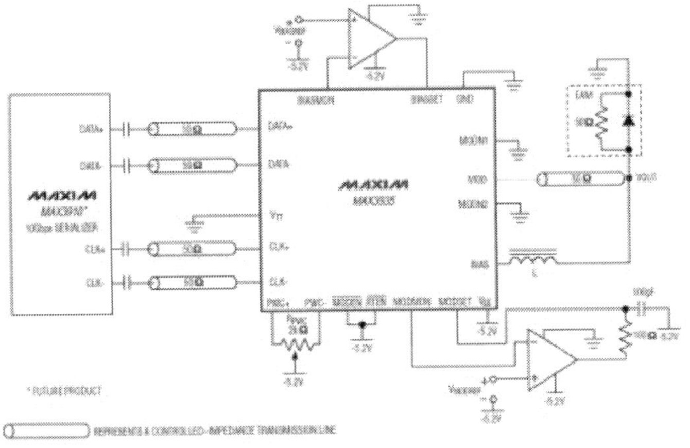

Figure 18.18: *Sample Driver Circuit for EML Laser 10 Gbps*

The EA modulator is essentially a photodiode that has an absorption band close to the wavelength of the laser. When the EAM is reverse-biased and an electric field is created in the EAM, its absorption band is shifted to the laser wavelength. The incident photons are absorbed, and an electrical current is produced. Thus, the effective impedance of the EAM is dependent on drive and bias levels. To fix the impedance match EAM, manufacturers effectively use a resistive match at the input of the EA section that creates about 6 dB of loss between the driver and the EAM. The parallel 50 Ohm resistor with the EAM in the above diagram is for descriptive purposes only. Since most EAMs at 10 Gbps and higher have coaxial RF input connectors, impedance matching between the driver, the PWB micro-strip, and the cable launch at 50 Ohms are the main critical parameters. This means the bias-T that brings the negative bias to the modulator is critical and needs to be a broadband RF choke. Other IC drivers have the bias-T internal to the IC, making the RF matching easier.

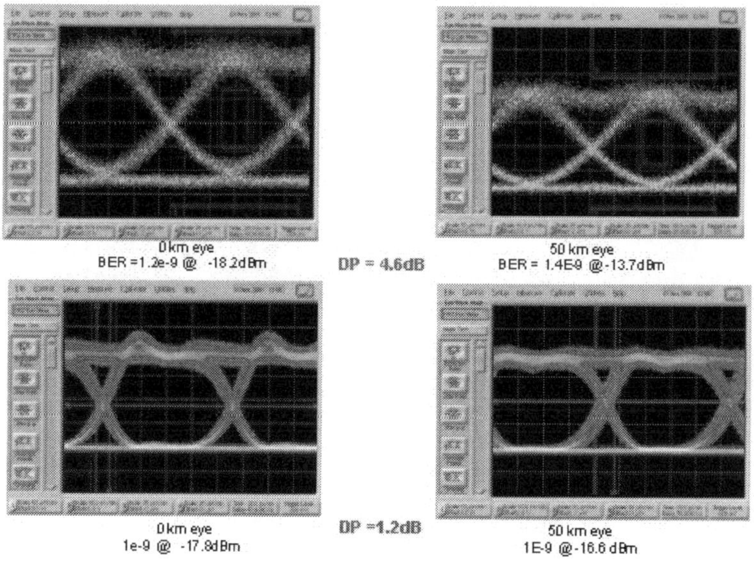

Figure 18.19: *EML Laser Performance at 10 Gbps*

A JDS Uniphase bulk-type EML was used to create the eye diagram in *Figure 18.19* and to illustrate how dispersion penalty can be improved by altering the driver parameters. The top two eye diagrams show how an acceptable eye diagram in a back-to-back measurement creates a degraded eye diagram after 50 km of standard single-mode (SM) fiber. The resultant dispersion penalty is 4.6 dB due to chirp and poor driver parameters. The two lower eye diagrams illustrate how adjusting the EA bias and driver output parameters can improve the eye diagram quality and dispersion penalty after 50 km of fiber. By reducing the pattern-dependent jitter, decreasing the rise and fall times out of the driver, and adjusting the bias of the EA section, a dispersion penalty of 1.2 dB can be achieved.

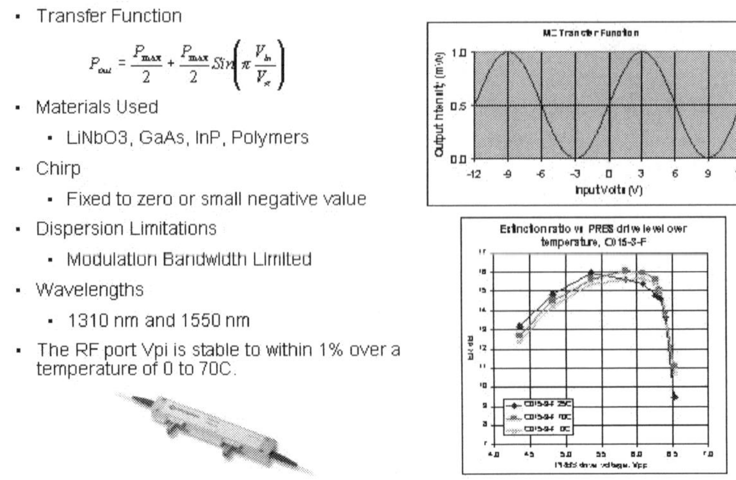

Figure 18.20: *Mach-Zehnder Modulator Fundamentals*

MZ modulators create the highest-quality optical eye diagram. MZs modulate the light by interference, so they have a sinusoidal transfer function that repeats itself periodically. The period of the sine transfer function is referred to as 2 Vpi. In the above transfer function plot, the Vpi voltage is 6V. There transfer function is very repeatable and stable over time and temperature, if the correct bias is maintained. The only downfall of MZs is that they have a bias point that varies over time and temperature and that must be actively controlled in a closed loop. The above plots illustrate the excellent extinction ratios that can be achieved with an MZ modulator and how stable it is over temperature. The Vpi of the tested modulator is close to 6V, which means that the best extinction ratio occurs when the output sing of the driver matches the MZ Vpi at 6 Vp-p. The plot also illustrates how over-driving the MZ modulator generates a very severe degradation in extinction ratio with only a slight increase above Vpi.

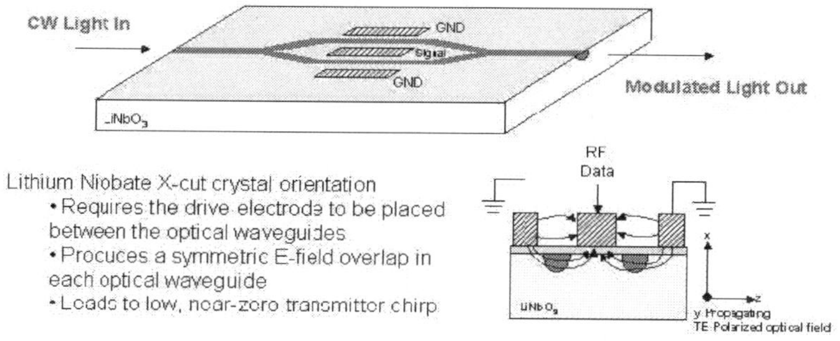

Figure 18.21: *Electrode Layout for X-Cut $LiNbO_3$ MZ Modulator*

The X-Cut lithium niobate modulators are commonly used due to their near-zero chirp performance. *Figure 18.21* illustrates how a Y-branch interferometric waveguide structure separates the continuous-wave CW light into two separate waveguides, then recombines the waveguides to create interference that provides the mechanism for amplitude modulation of continuous wave (CW) light. Lithium niobate modulators operate by the electro-optic effect, in which the applied electric field changes the refractive index, making light travel faster or slower in the two split waveguides of the Y-branch interferometer. The electro-optic is strongest along the Z axis, hence the CW laser output needs to have its electric field polarized along the Z axis of the lithium niobate crystal. By placing optical waveguides between the RF signal electrode and the ground electrodes that form the RF co-planar waveguide, a "push-pull" configuration is created that causes a symmetric E-field overlapping each optical waveguide leading to low, near-zero chirp.

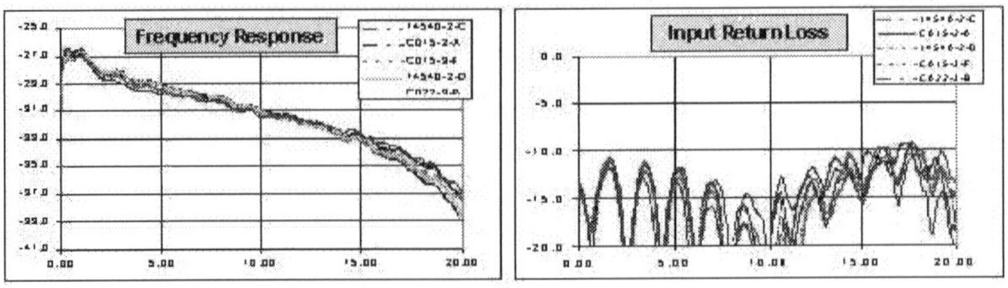

- Excellent Impedance Matching
  - Co-planar transmission waveguide on the $LiNbO_3$ crystal termnated in 50 Ohms
- Constant Bandwdth from unit-to-unit.
  - LiNbO3 modulator bandwidth maximized by phase matching the velocity of the RF wave guide with the optical wave guide.
- Important Characteristics Required in Input Eye Quality
  - Since the transfer function is sinusoidal, the eye diagram rise times, pattern dependent noise on the 1 and 0 levels are cleaned up by the modulator.

Figure 18.22: *Signal Integrity Concerns with MZ Modulators*

The frequency response in *Figure 18.22* illustrates how consistent the bandwidth is over several production units. And, the above input return loss measures the impedance mismatch reflections across the bandwidth of the device, revealing that from unit to unit the reflections are below –10dB out to 15 GHz. This illustrates how the LiNbO3 modulator has an extremely well-matched 50 Ohm termination that is very constant with drive levels. The bandwidth of the MZ is also very independent from bias levels and is very consistent from unit to unit. The sinusoidal shape of the transfer function transforms slow rise- and fall-time signals with noisy one and zero levels on the electrical input eye and produces sharper optical eye diagrams. However, a cleaner electrical input eye usually means less pattern-dependent jitter on the output optical eye diagrams. The one concern with having sloppy electrical eye diagrams at the input to the MZ is that the bias control loops of the MZ can be sensitive to eye quality. Hence, if the eye quality degrades over temperature (especially crossing ratio) the bias control loops of the MZ can severely distort the optical eye diagram or even become unstable.

10Gb/s Limiting Amplifier

- 50 Ohm input and output impedance
- 7.5Vp-p Adjustable Output Swing
- 250mVp-p to 1.5Vp-p input levels

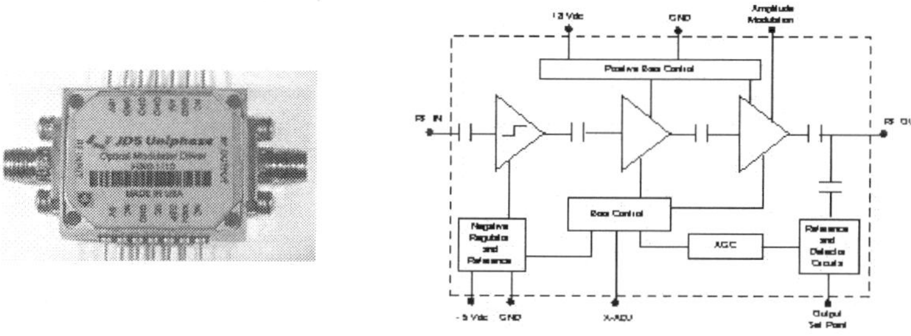

Figure 18.23: *Sample Driver Circuits for MZ Modulators*

An ideal driver for an MZ modulator would be a limiting amplifier that compensates for different or changing input levels and maintains the same output level over temperature. The limiting amplifier would also have the additional advantage of sharpening up the input waveform from a 10 Gbps serializer. *Figure 18.23* illustrates the H302 driver amplifier designed by JDS Uniphase to provide up to 7.5 Vp-p swings, which is required for some LiNbO3 MZ modulators.

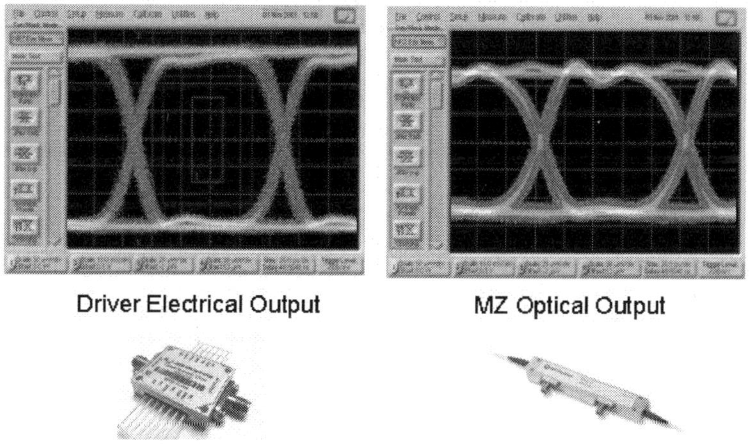

Figure 18.24: *Mach-Zehnder Performance at 10 Gbps*

The above eye diagrams illustrate the superb eye quality out of the hybrid MZ driver and how the JDS Uniphase MZ modulator even improves the eye diagram further by sharpening the rise and fall times and by removing the ringing on the one level. The output-level swing of the driver can be adjusted to maximize extinction ratio and rise times out of the MZ modulator. Improving rise times and pattern-dependent jitter will decrease the dispersion penalty. At 10 Gbps, an X-cut lithium niobate modulator is dispersion-limited to about 60 km for less than a 1 dB dispersion penalty over standard single-mode fiber in the 1,550 nm range.

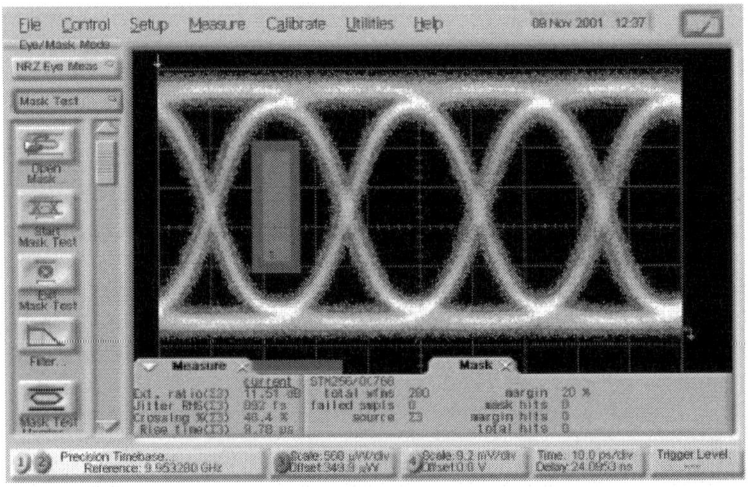

Figure 18.25: *Mach-Zehnder Performance at 40 Gbps*

At 40 Gbps external modulation is the only way to achieve acceptable eye quality. This means that EAM or MZ modulators will most likely be used for all fiber-optic transmitters at 40 Gbps and higher. *Figure 18.25* illustrates recent developments in using a JDS Uniphase X-cut LiNbO3 modulator in a fiber-optic transmitter. At 40 Gbps, the X-cut lithium niobate modulator is dispersion-limited to about 3.7 km for less than a 1 dB dispersion penalty over standard single-mode fiber in the 1,550 nm range, assuming a dispersion coefficient of 17 ps/(nm*km). This means that the dispersion limited distance decreased by a factor of 16, with only a factor of four increase in bit rate when going from 10

Gbps to 40 Gbps. One can use the equation DL < 105/B2 to determine the dispersion allowed for a specific bit rate of NRZ data, assuming zero chirp and a 1 dB dispersion penalty. D is the dispersion coefficient of the fiber in ps/(nm*km), L is the length of fiber in km, and B is the bit rate in Gbps. This illustrates how, at increasing data rates, the fiber dispersion becomes the significant limitation that needs to be solved.

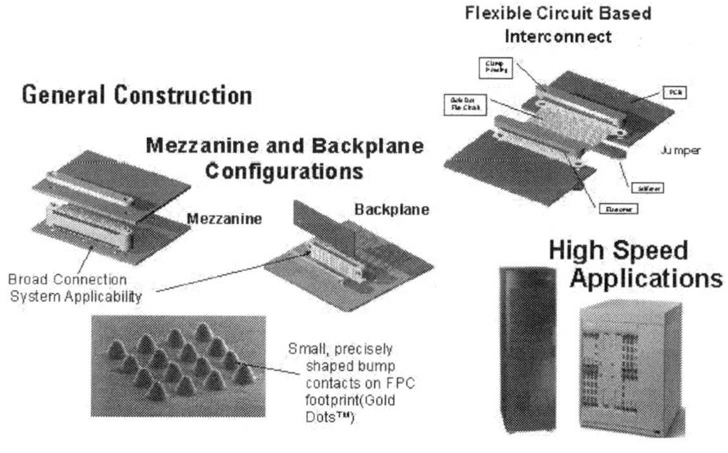

Figure 18.26: *Gold Dot Interconnect Profile*

Printed circuit boards (PCBs) are often implemented as the interconnect between the laser driver circuit and the laser diode. Considering the requirements on signal integrity characteristics and the ability to manufacture a variety of configurations, flexible circuit interconnects offer advantages over traditional PCBs. The gold dot flexible circuit has superior electrical performance in harsh environments. The small, precisely formed bump bonds provide the physical transition to the PCB and provides less impedance mismatch and change in contact resistance through environmental extremes. The bump bonds can either mate to PCBs in an elastomer-based clamping system for a temporary connection or a chip-on-dot for a permanent connection. The flexible circuit can be manufactured in a stripline or microstrip ground configuration. Additionally, the circuit may be used as a jumper, a mezzanine, or a backplane connection. Design geometries available yield configurations and footprints optimized for applications in high-speed switching units, routers, mobile computers, and cell phones. Optical transmitter driver applications are a new and exciting technology that is studied in this paper.

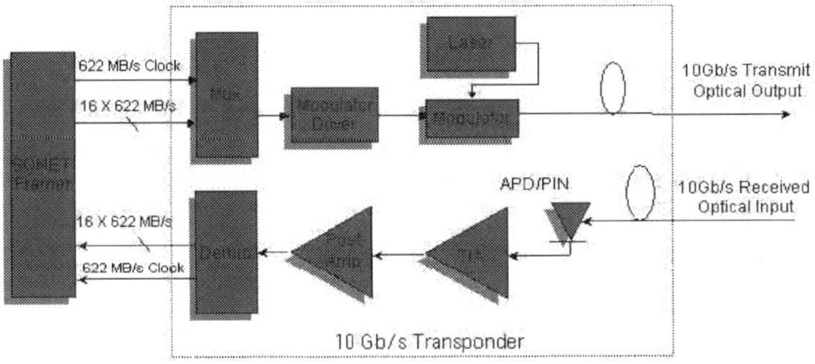

10 Gb/s Fiber Optic Transponder Block Diagram

Figure 18.27: *Laser Transmitter Interconnect Solutions*

Fiber-optic transmitters are commonly integrated into transponders that contain both a fiber-optic transmitter and receiver. The fiber-optic transponder usually contains a serializer (Mux) and deserializer (Demux). The Mux transforms 16 channels of a lower data rate into a single channel of data at 16 times faster than the lower data rate. The Demux returns the 16 channels of the lower data rate from the high-speed single channel, along with a recovered clock that is in phase with the incoming data stream.

The interconnect between the transponder and the synchronous optical network (SONET) framer contains differential controlled impedance lines on all channels and clocks. Hence, high-density, high-speed interconnects with controlled impedances and equal line lengths are necessary in order not to introduce line skew and other undesirable characteristics into the optical system. At high data rates such as a 40 Gbps transponder that would have 16 channels of 2.5 Gbps, these issues become even more critical. Flexible interconnects could be an ideal way to maintain the differential controlled impedances and equal line lengths and solve other technical issues of connecting the transponder to the SONET framer.

Furthermore, there are interconnect requirements between each of the critical blocks inside the transponder. Here signal integrity is of utmost concern, especially between the optical devices and the interface ICs that drive or receive the high-speed data. Due to packaging restraints and the ever-increasing desire to reduce the overall size of the transponder, flexible interconnects provide a way to solve the technical issues of connecting the high-speed components.

For example, the driver circuit interconnects require not only signal traces on high densities, but also that these signals not be distorted while being sent from one component to another (e.g., a laser driver diode to a laser). The optical eye diagram performance of the laser is dependent on the electrical characteristics of the transmitted signals between components in a laser driver circuit. High-density signal lines, both single-ended and differential must carry signals in excess of 2.5 Gbps. All of these considerations must be taken into account when routing high-density, high-speed interconnects from one optical device to another. Though environmental conditions are not extreme, operating temperatures in the range of –40 to 80°C are not uncommon and may be only one of the many environmental extremes to be considered when choosing an interconnect for a fiber-optic transponder.

Parameter	Performance
Impedance	50 Ohms Single Ended 100 Ohms Differential
Cross Talk NEXT	<5% Single Ended <.5% Differential
Attenuation(5 GHz)	<.5dB/inch
Propagation Delay	<200 ps/inch
Data Rate	>2.5 Gbps

Figure 18.28: *Interconnect Electrical Performance*

A flexible circuit interconnect is able to provide single ended and differential signaling from one device or component to another. Crosstalk, attenuation, and propagation delay have been measured on a host of materials and signal densities. In general, this performance of these flexible circuits can be predicted with simulation tools. The general electrical performance of a gold dot flexible circuit interconnects has been characterized in the time domain and frequency domain. Performance characteristics are highly dependent on material properties, flexible circuit stack-up, signal trace dimensions, and the manufacturing process.

Electrical Requirements

Impedance	50 Ohms
Differential Impedance	100 Ohms
NEXT	<5%
Differential NEXT	<.5%
Data Rate	> 10 Gbps
Propagation Delay	<170 ps/inch

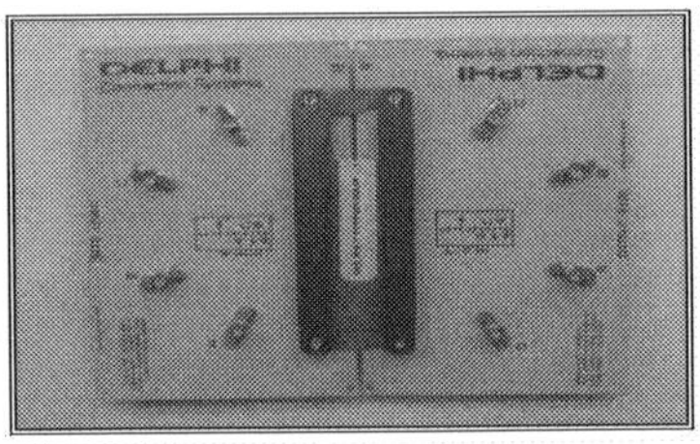

300 Way Differential Gold Dot Jumper

Figure 18.29: *Proposed Interconnect for Modulating Semiconductor Lasers*

A gold dot jumper interconnect configuration was evaluated as the interconnect between the laser driver circuit and the laser diode. With the gold dot flexible circuit supported by two PCBs, inserting this device in series with the driver circuit would quickly and effectively determine how this type of interconnect would affect the performance of the laser driver circuit. The gold dot flexible circuit supported on the PCBs mounted with SMAs would allow the signals to be sent and received through the gold dot interconnect and laser driver circuit. The elastomer backed clamping system provides the necessary pressure to the gold dot to the PCB pad interface. Four SMAs mounted on each of the PCBs allow four signal lines of the 300 available signal lines in the circuit to be characterized. The desired electrical performance for the laser driver circuit was determined prior to designing the gold dot circuit and test boards with the key characteristics being a single-ended impedance of 50 Ohms and a data rate in excess of 10 Gbps.

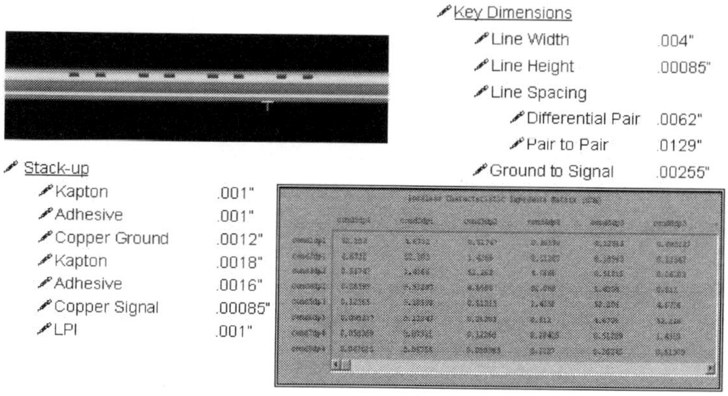

Figure 18.30: *Laser Transmitter Flexible Interconnect Design*

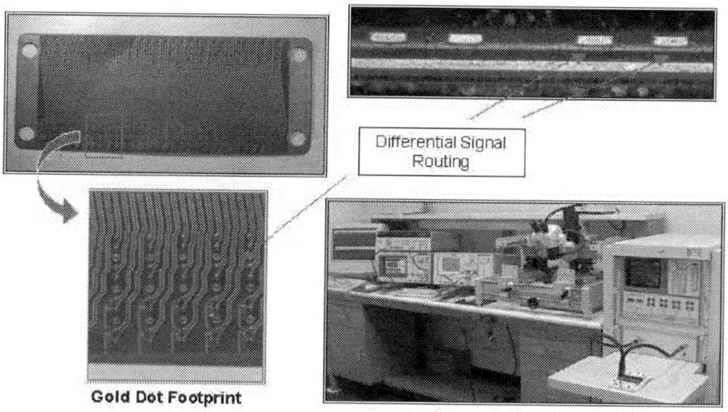

Figure 18.31: *Manufacturing and Testing of Gold Dot Flexible Circuits*

For controlled impedance circuits at data rates in excess of 2.5 Gbps, manufacturing tolerances on line width and spacing have a strong effect on desired electrical performance. The gold dot flexible circuit interconnect requires more than 30 wet processes, which all affect signal trace width and spacing as well as reliability in environmental extremes. Tolerances on trace width of +/- .0002 inches equate to a +/- 2.2 percent single-ended impedance variation. With the stringent process controls in place, line widths, and spacings of less than .003" are achievable well within the requirement of less than +/- 10 percent impedance variation.

Once the gold dot circuit is manufactured, the required performance must be confirmed in the test and measurement laboratory. Confirming signal integrity characteristics on a high-density, high-speed interconnect requires the measurement of time and frequency domain characteristics such as impedance, crosstalk, insertion loss, and return loss. Eye diagrams to confirm the data rate are also necessary. The test equipment often includes a time domain reflectometer (TDR), vector network analyzer (VNA), and bit-error rate test (BERT). Stimuli and responses of the gold dot interconnect under test require a test fixture or a probe station. In this specific case, two PCBs with SMA connectors provide the stimuli and responses to be launched and detected by the measurement equipment. Electrical performance is measured and a comparison to the predicted performance is made.

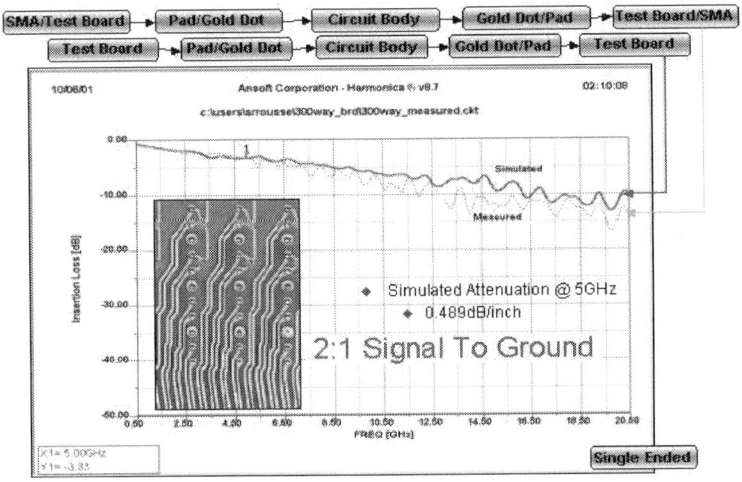

Figure 18.32: *Electrical Test Data Insertion Loss*

To predict the insertion loss of the circuit while the device was being manufactured additional simulation tools were applied. Ansoft Harmonica was able to quickly predict the insertion loss of a model composed of the gold dot flexible circuit and the two PCBs. The SMAs on the PCBs were not included in the model. The prediction of insertion loss and the actual measurement on an Agilent 8510C VNA illustrate close correlation up to approximately 10 GHz. Deviation of the measurement from the simulation can be attributed to the omission of the SMAs and vias from the simulation model.

Parameter	Simulation	Measured	Goal
Single Ended Impedance	52.1 Ohms	53 Ohms	50 +/-10% Ohms
Differential Impedance	95.2 Ohms	98 Ohms	100 +/- 10% Ohms
Attenuation (5GHz)	<.44 dB/inch	<.44 dB/inch	<.5 dB/inch
Propagation Delay	152 ps/inch	158 ps/inch	170 ps/inch
Single Ended NEXT	<4.5%	<4.5%	<5%
Differential NEXT	<.3%	<.3%	<.5%
Data Rate	10 Gbps	10 Gbps	10 Gbps

Figure 18.33: *Performance of Flexible Circuit Interconnect*

The electrical performance of the gold dot flexible using the printed boards as a part of the device under test (DUT) was measured in both the time and frequency domain. Eye diagrams at both 2.5 and 10 Gbps were also performed while applying the OC-48 and OC-192 masks. The single-ended and differential impedance were within the predetermined requirements for the laser driver circuit. Though the measurements of attenuation and propagation delay were within the goal performance, these characteristics will be improved in the actual flexible circuit for the laser driver circuit. Adhesives of lower dielectric constant and an impedance of 50 +/- 2 Ohms would allow signals of higher data rates to be sent through the gold dot flexible circuit interconnect. Additionally, the use of printed circuit boards and SMAs contribute to degrading the rise time of a signal and add further attenuation, which would not be present in the actual gold dot circuit used in a packaged laser driver circuit and EML.

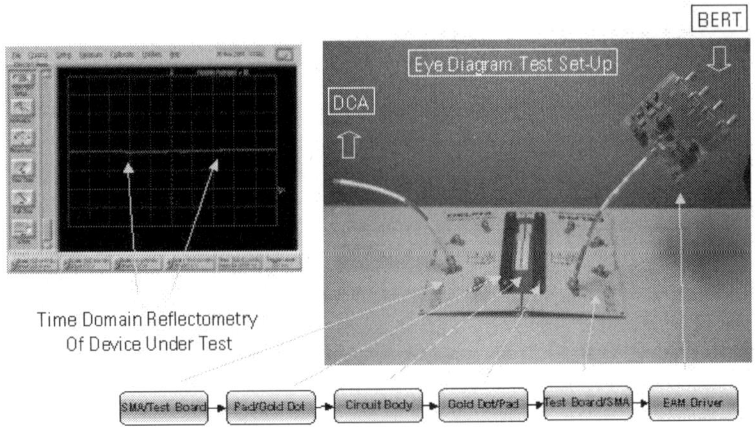

Figure 18.34: *Electro-Absorption Modulator Driver Test Setup*

The test system used to analyze the eye diagrams of the test board/flex circuit is shown in *Figure 18.34*. A BERT capable of generating a 10 Gbps pseudo-random binary sequence (PRBS) was used to drive the EAM. The resultant data stream output was fed into a digital communications analyzer (DCA) with a 65 GHz bandwidth electrical module. The TDR measurements were made using a differential TDR module in the DCA as a source rather than the BERT. The TDR measurement shows a very well-controlled impedance environment throughout the DUT. However, this is not very interesting to analyze, so we will expand the vertical scale to exaggerate the reflections.

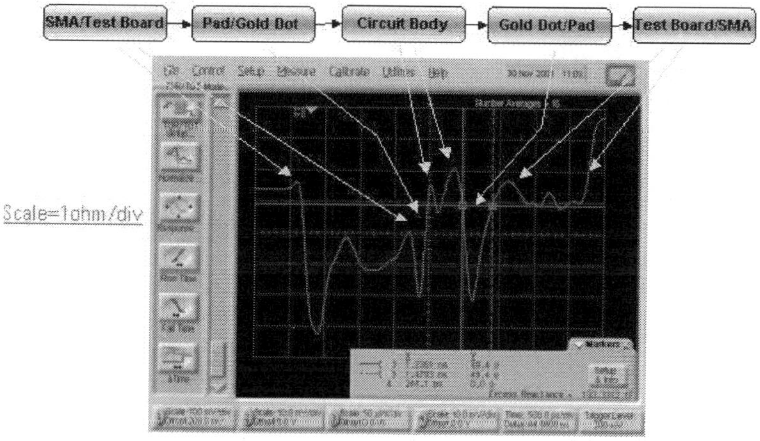

Figure 18.35: Time Domain Reflectometry Details

The impedance profile of the flex circuit/test board assembly can be measured by using time domain reflectometry. The TDR test equipment launches a 200 mV amplitude, 35 ps rise-time step into the DUT. When this step encounters impedance discontinuities, reflections are sent back into the test equipment module. The reflection coefficient (rho) is then calculated and the impedance value is extracted. In the flex/test board assembly, the first section shows a large negative reflection indicating excess capacitance due to the SMA launch into the test board. The next section is a slightly lower impedance of the left half of the test board (remember the vertical scale is expanded to 1 ohm/division). The next capacitive dip is the excess capacitance of the pad/gold dot interface. The TDR step then encounters the relatively flat impedance profile of the circuit body itself. After transitioning through another gold dot/pad interface and right half test board, the TDR step exits the SMA connector and reflects off an open circuit (indicated by a large positive reflection).

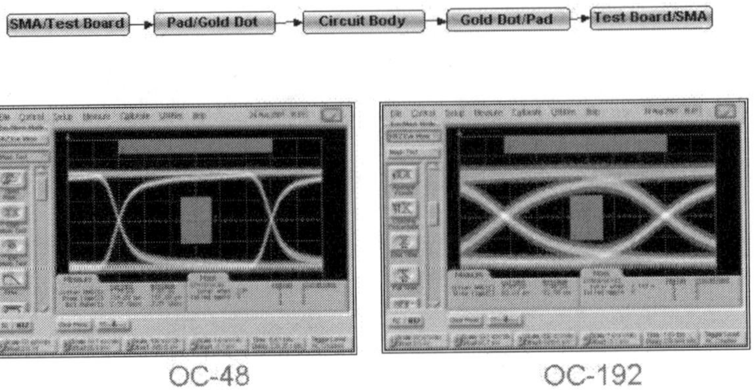

Figure 18.36: *Eye Diagram Analysis of Flex Circuit*

Eye diagram analysis was performed on the flex circuit/fixture assembly in order to determine level of compliance to SONET standards. By driving the assembly with the pattern generator of a BERT, it was clear that it passed the OC-48 standard at the 2.5 Gbps data rate. With less margin, but still passing, the assembly was compliant to the OC-192 standard at 10 Gbps.

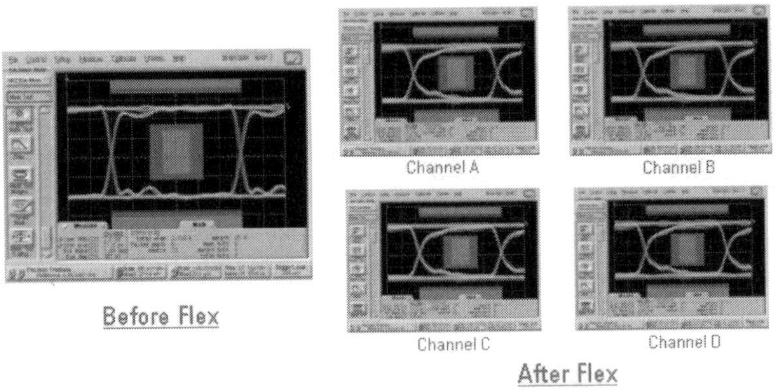

Figure 18.37: *EAM Driver Test Results (2.5 Gbps)*

The EAM driver passes OC-48 compliance testing with more than a 25 percent margin on the test mask.

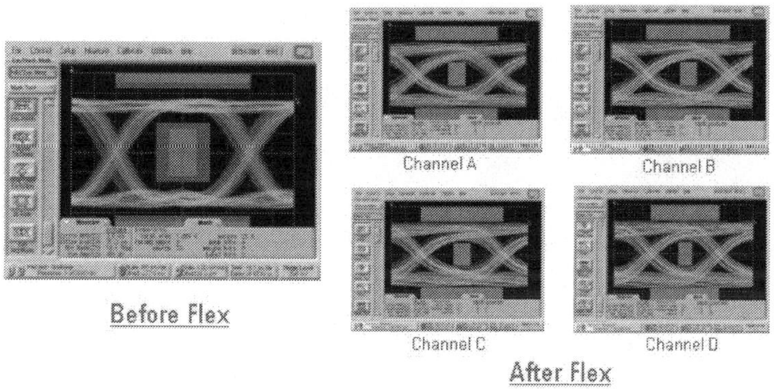

Figure 18.38: *EAM Driver Test Results (10 Gbps)*

The EAM driver passes OC-192 compliance testing. An interesting observation indicates deterministic jitter is present in this DUT. When viewing the cross section of any one of the eye diagrams, you can see a certain structure within the cross-section, somewhat of a "banding." This is caused by a short word pattern being output from the BERT (a shorter word repeats itself more often than a longer word in a given length of time). This creates more stress on the DUT that is susceptible to pattern dependencies.

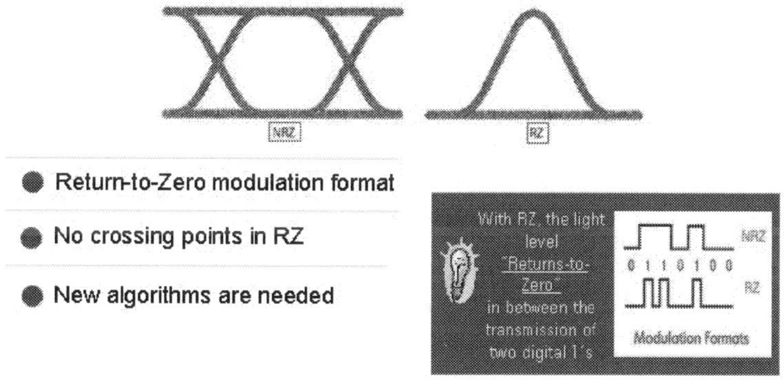

Figure 18.39: *Latest Measurement Trend: 40 Gbps Return-to-Zero Waveform Analysis—What Is It?*

The traditional non-return-to-zero (NRZ) signals that we are familiar with have a very nice characteristic to them: they have what are called crossing points. These are the areas where the transitions from one to zero and from zero to one intersect and form an "X." The crossing points are the reference point for NRZ waveform analysis. Bit rate, jitter, eye opening, etc., are all based on where the crossing points are located. As you can see, the RZ waveform does not have crossing points. This means that new measurement algorithms need to be created in order to characterize the RZ modulation format. New measurement concepts need to be invented, also. Contrast ratio and eye opening factor are measurements needed to fully characterize RZ waveforms.

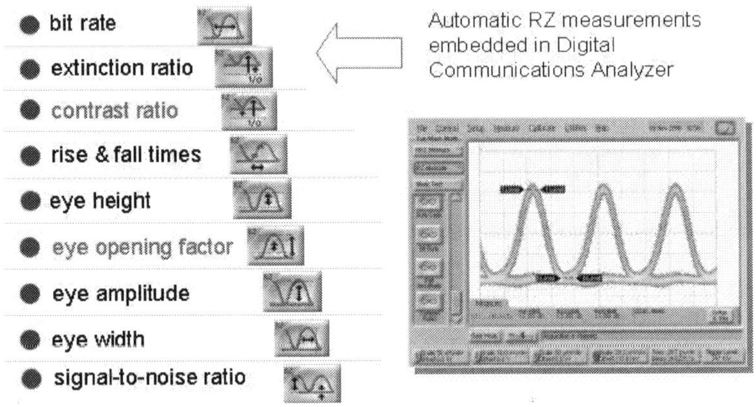

Figure 18.40: *New RZ Measurements*

The standard eye diagram measurements are still used for RZ characterization. The new measurements are contrast ratio and eye opening factor. The contrast ratio measurement is a ratio of the one level at the center of the eye diagram compared to the one level found midway between the eye diagram peaks. This indicates how well the logic 1 levels return to the logic zero level. This is sometimes referred to as modulation depth. Contrast ratio indicates how well a laser transmitter turns off between consecutive ones. Eye opening factor is similar to eye height, except eye opening measures the actual eye opening relative to an ideal, noise-free eye. While the eye height measurement uses 3 sigma for noise contribution, the eye opening factor measurement uses 1 sigma.

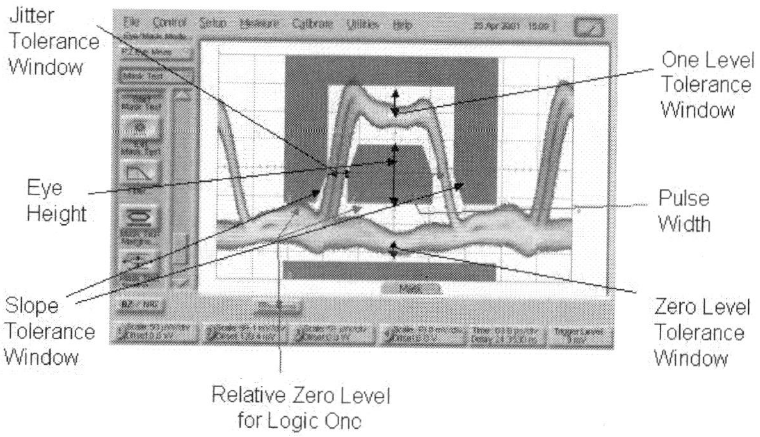

Figure 18.41: *40G RZ Mask Testing*

Standard not-return-to-zero (NRZ) masks are readily available in the commercial measurement and testing products, while no standard RZ mask has been established so far. One way to yield a RZ mask is to generate customized masks via creating coordinates for all desired polygons, a process that is tedious and error-prone. Mask changes must be done by editing the file containing the mask description. Another way to process mask changes is to add a generic mask editor function, which allows coordinate manipulation directly from the screen, via touch area, markers, dialog boxes, or some combination. However, the generic mask editor is cumbersome and not user-friendly. To overcome these obstacles, a hybrid method of RZ mask creation has been developed that is much more appealing and physically sound. This hybrid method reveals a generic RZ/NRZ mask and an efficient way to alter the mask by means of a limited set of physically meaningful mask parameters. This RZ mask creation allows rapid customized symmetric mask creation without sacrificing accuracy and flexibility.

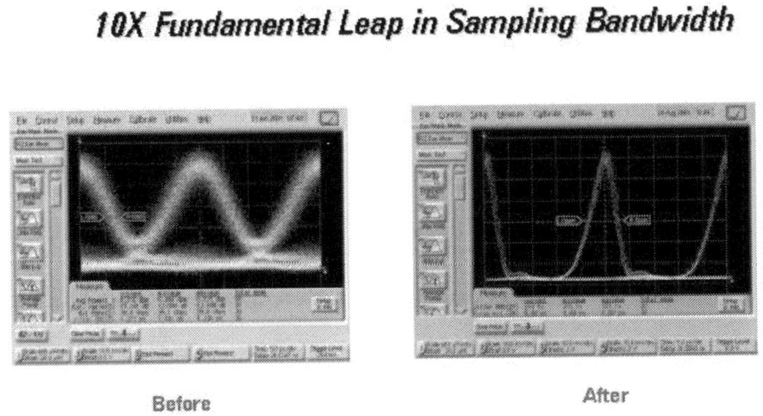

Figure 18.42: *Future Measurement Trends: Optical Sampling Technology*

In order to accurately characterize the modulation properties of 40 Gbps laser transmitters, it is necessary to obtain an order of magnitude increase of bandwidth in the test system. This goal is not easily achievable by conventional sampling circuits because they are limited by the semiconductor material switching speeds. Consequently, a new sampling method must be utilized. This sampling method must withstand the rapid pace of component development and the ultra-high data rates that they produce (up to 160 Gbps). The technology of choice for this challenging application is optical sampling. Agilent Labs Japan and Agilent Labs Palo Alto have collaborated to develop an optical sampling technique that allows more than 500 GHz of bandwidth. Optical experimentation utilizing nonlinear crystalline materials has yielded astonishing results. Eye diagrams as never observed before are now routinely generated on a laboratory bench in a very–high-throughput environment. Optical sampling technology will no doubt be utilized while characterizing the physical layer of the future 40–160 Gbps networks.

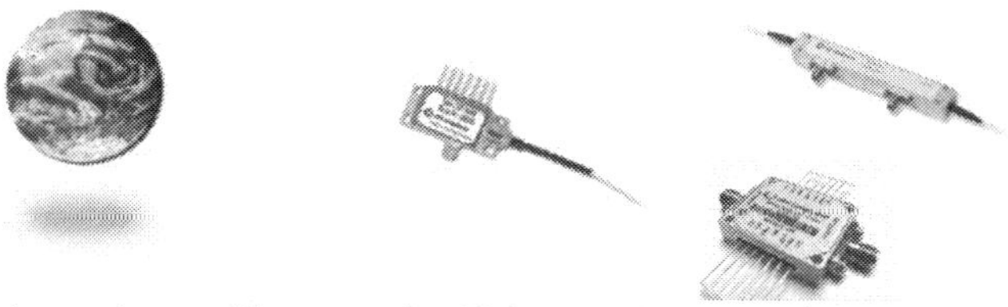

- Laser transmitters require high speed electrical drivers
- Signal integrity is a challenge
- Good engineering design will minimize problems
- Proper measurement and simulation tools are a must
- 160 Gb/s serial prototype systems complete Q2 of 2002

Figure 18.43: Summary

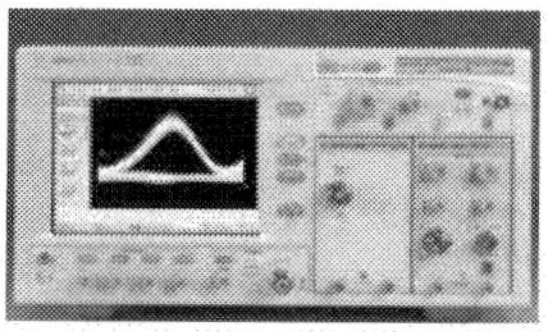

Agilent 86100A Digital Communications Analyzer
Agilent 86107A Precision Timebase Module
Agilent 86116A 55G Optical/65G Electrical Module
Agilent 54754A Differential TDR Module
www.agilent.com/comms/dca

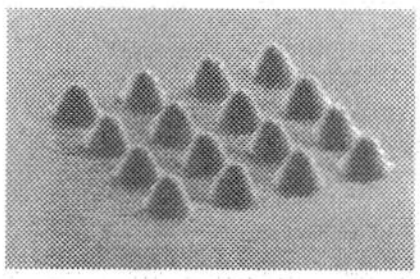

Delphi Gold Dot Interconnects
www.delphiauto.com

JDS Optical Components
www.jdsuniphase.com

Figure 18.44: Resources

Authors

Stephen Reddy, Senior Design Engineer, Transmission Subsystems Group, JDS Uniphase

Laurie Taira, Senior Project Engineer, Research and Development, Delphi Connection Systems

Mike Resso, Signal Integrity Application Scientist, Component Test Division, Agilent Technologies

Part V

Analysis of New Technologies

Chapter 19

The Role of Dielectric Constant and Dissipation Factor Measurements in Multi-Gigabit Systems

19.1 Abstract

In all high-speed serial links above about 2 Gbps, the interconnect usually limits performance. Though the trend is for lower loss and lower dielectric constant for signal paths and higher loss and higher dielectric constant for the power path, many values are acceptable. The most important ingredient to a successful design is having an accurate and stable value. The stack-up design of a board can be adjusted for a wide range of dielectric constants, if it is known early in the design phase. Transmitter and receiver equalization in the serializer/deserializer (SERDES) transceivers can overcome excessive material losses if their coefficients are optimized, based on accurate values of the insertion and return loss of the differential channel.

If a system has been designed with a specific dielectric constant and dissipation factor as the target, there may be performance problems in the field due to production variation and environmental variation in the dielectric properties. These factors are driving the need for a robust, routine technique to characterize the dielectric constant and dissipation factor of laminate materials used in the fabrication of printed circuit boards (PCBs), both for the signal paths and the power and ground paths.

This paper reviews a new implementation of the general approach of resonant cavity, precision measurements of laminate materials. This new approach requires minimal sample preparation and can give 1 percent accuracy in the dielectric constant and values accurate to within 0.0005 in the dissipation factor. A two-port vector network analyzer (VNA) is used to measure the insertion and return loss of a cavity structure without the sample and then with the sample added.

The before and after measurements, along with the information about the geometry of the cavity, are combined with sophisticated analysis software to extract the real and imaginary dielectric constant at the resonance frequency of the cavity. With four cavities, the important signal integrity range of 1 to 10 GHz can be spanned. This technique has the advantage over other, similar methods by being simple, routine, and automated.

19.2 Introduction

Telecommunication, computer, and even consumer products are pushing the limits of performance using conventional materials and conventional design rules. The low-hanging fruit has been harvested.

It is no longer cost-effective to over-design a product by adding extra margin to reduce the risk of performance problems. Engineers must utilize advanced design tools to create sophisticated network equipment that can transmit serial channel data at 10 Gbps and above and simultaneously achieve cost and schedule targets.

This means accurate prediction of system performance is increasingly important. The input to every design tool are the materials' electrical properties such as dielectric constant and dissipation factor. Likewise, the final system performance achieved depends on the actual, as-manufactured material properties of the interconnects. An important aspect of meeting signal integrity performance goals is carefully controlling the design and the material properties. These principles apply across the entire spectrum of interconnects, including backplanes, line cards, memory cards, motherboards, connectors, integrated circuit (IC) packages, and cables.

Accurate knowledge of the dielectric constant and dissipation factor of all the interconnect materials in the selection phase and the manufacturing phase are essential ingredients to a modern product design flow that balances the lowest cost for acceptable performance.

19.3 Dielectric Properties of Laminates

The insulating material keeping the signal conductor a controlled, precision distance from the return conductor in a transmission line has only two electrical properties that influence the signal integrity of the interconnect: a dielectric constant and dissipation factor [1]. These are often considered as the two components of the complex dielectric constant of a material, which can be written as

$$\varepsilon(\omega) = \varepsilon'(\omega) - i\varepsilon''(\omega)$$

The real part of the dielectric constant is what is traditionally called the dielectric constant, while the imaginary part is related to the dissipation factor. When the complex dielectric constant is plotted in the complex plane, as shown in *Figure 19.1*, the angle the complex dielectric constant vector makes to the real axis is traditionally labeled with the Greek letter delta (δ), and is called the loss angle.

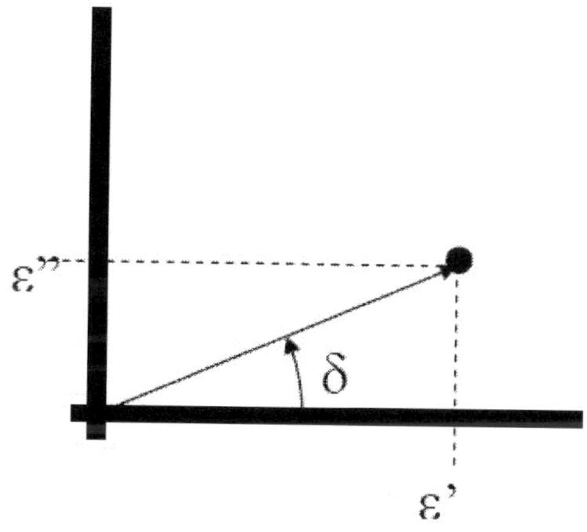

Figure 19.1: *Complex Dielectric Constant Plotted in the Complex Plane*

It is an unfortunate coincidence that the same Greek letter was chosen for the loss angle as is also used to represent the skin depth in a conductor. Even though both terms relate to losses in transmission lines, there is no connection at all between them. They refer to completely different effects.

The tangent of the loss angle, is the ratio of the imaginary part of the complex dielectric constant to the real part of the complex dielectric constant. This tangent is also called the dissipation factor.

$$\text{dissipation factor} = \tan(\delta) = \frac{\varepsilon''}{\varepsilon'}$$

Since the complex dielectric constant may be frequency-dependent, the real part and the dissipation factor may be frequency-dependent as well. Usually the dielectric constant is referred to as Dk and the dissipation factor as Df. The use of English letters, rather than Greek letters, minimizes the possible problems from publishing software tools not having the correct Greek letter fonts. All the important electrical properties of the insulating materials used in signal integrity applications are contained in the dielectric constant and dissipation factor, and how they vary with frequency.

19.4 Impact of Dielectric Materials in Signal Integration

The ideal dielectric for signal paths between the signal conductor and the return conductor is air, with a dielectric constant of 1 and a dissipation factor of 0. All materials used with interconnects strive to come close.

In high-speed digital products, the dielectric constant will influence both the time delay of a signal and the characteristic impedance of the transmission line. The dissipation factor will affect the rise time of the transmitted signal and the generation of inter-symbol interference (ISI). *Figure 19.2* is an example of the eye diagram from a 20-inch interconnect through a backplane comparing a laminate with a dissipation factor of 0.025 and 0.01.

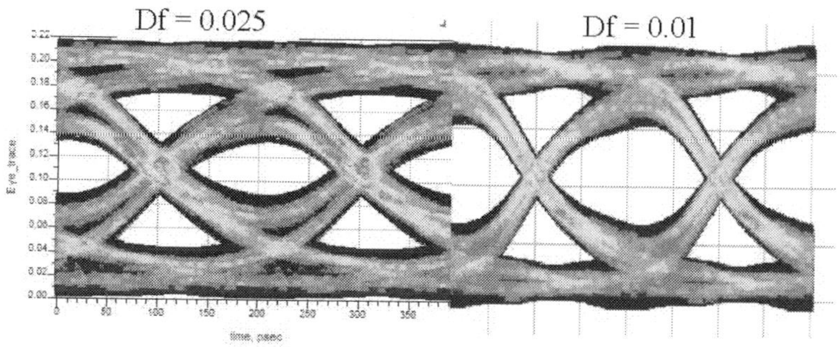

Figure 19.2: *Eye Diagrams of 5 Gbps PRBS through 20 Inches of Interconnect, with Two Dissipation Factor Materials*

In RF and mixed-signal products, the dielectric constant and dissipation factor can influence the resonant frequency and Q of embedded passive filters or patch antennas. All things being equal, a lower dielectric constant and lower dissipation factor will enable better performance. A lower dielectric constant will result in a shorter time delay of each transmission line, lower crosstalk between adjacent traces, and a thinner total board dimension, which means shorter via stubs. A lower dissipation factor will result in less rise-time degradation and a more open eye diagram.

In the real world, all things are not equal, and a lower dielectric constant or lower dissipation factor comes at a premium price. This means a balancing act between performance and price. A far more cost-effective quality than low dielectric constant is stable and known dielectric constant. Likewise, while low dissipation factor is valuable, with a stable and known value, TX and RX equalization can be implemented to compensate for a large dissipation factor.

Commercially available 2D and 3D field solvers are accurate to better than 1 percent in predicting characteristic impedance and attenuation. However, the fundamental input to all field solvers are the material properties of the laminate: the dielectric constant and dissipation factor.

With an accurate value of the dielectric constant, precise time delay and characteristic impedance targets of all the interconnects can be achieved by good design rules. For example, the time delay and characteristic impedance of an interconnect are each related to the square root of the dielectric constant. A 2 percent uncertainty in the dielectric constant will result in a 1 percent uncertainty in the time delay or the characteristic impedance.

If the typical variation in the characteristic impedance of traces across a set of boards is typically 10 percent, it might be asked, why is a variation in characteristic impedance from dielectric constant uncertainty of 1 percent important? It is all about yield [2].

There are many factors contributing to the variation in characteristic impedance of a trace. Some of them are related to process stability, and some of them are related to design accuracy. If the distribution of trace impedances is well centered between the upper and lower control limits and the limits are close to the actual distribution, then small variations in a contributing factor (e.g., uncertainty in the dielectric constant) will shift the distribution so that some of the parts are out of spec. This is illustrated in *Figure 19.3*.

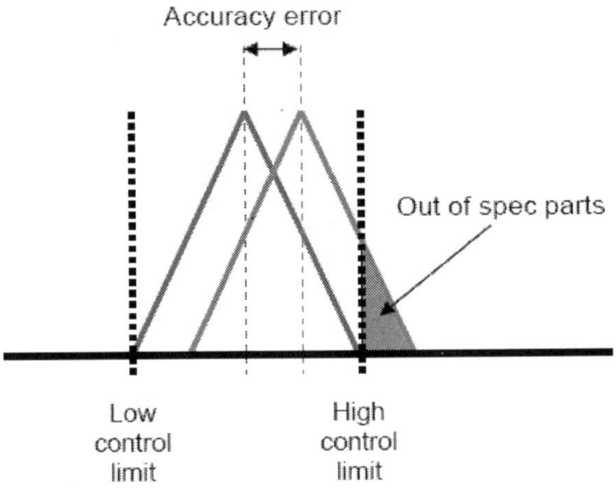

Figure 19.3: *Impact on Yield from Accuracy of Dielectric Constant*

An important element of yield improvement, or hitting target specs, is translated into the uncertainty in the dielectric constant. There are two factors contributing to the uncertainty: the measurement uncertainty and the actual variation of the material's dielectric constant. A similar argument can be made for dissipation factor.

Accurate and reproducible measurement of the dielectric constant and dissipation factor of laminate materials is an essential ingredient to achieve a high yield.

19.5 Measurement Methods

There are four general classes of measurement techniques for material properties, each with a different set of tradeoffs in terms of sample prep, measurement frequency range, ease of use, and measurement quality.

These techniques are illustrated in *Figure 19.4.*

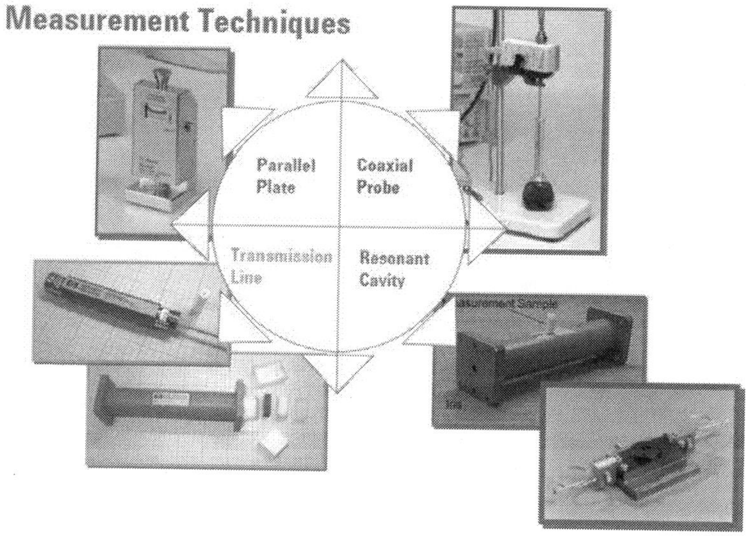

Figure 19.4: *Four Measurement Techniques for Dielectric Constant and Dissipation Factor*

In the parallel plate method, an unclad sheet of the material is placed between two flat parallel plates and the complex impedance is measured. By using a guard electrode around the edge of the planes, fringe field effects can be minimized. The impedance can be interpreted in terms of a capacitance, from which Dk can be extracted, and the conductance, from which a Df can be extracted.

The frequency range that can be measured is limited to where the sample behaves like a lumped element, before distributed effects arise. This is typically up to the frequency where the longest lateral dimension is comparable to about 1/20 a wavelength. For a sample 2 inches in diameter, the highest frequency range is about 100 MHz.

The sample prep is trivial, as all that is required is an unclad sheet. However, air between the sample and the plates will cause an anomalously low value for the dielectric constant. This technique also uses a simple LCR meter to perform the measurement and is easy to understand.

In general, many laminate materials show some frequency dependence, especially at low frequency below 50 MHz, while, above 500 MHz, the dielectric properties are relatively constant with frequency. This technique measures the low-frequency dielectric properties and, if done at 1 MHz, will give a value of Dk that can be 2 to 5 percent higher than at 1 GHz.

A coaxial probe takes advantage of the fringe fields at the end of a polished 50 Ohm coaxial line. The fringe fields penetrate into the material to be tested and their properties affect the capacitance and its Q. The change in the fringe field capacitance between the tip in air and embedded in the sample can be measured from the 1-port S-parameters and the Dk and Df can be extracted.

The useful range of this measurement can be from the low MHz range up to more than 20 GHz; however, the sample must be thick enough so that all the fields are encircled by the material. In addition, any air gap at the interface will introduce an artifact in the material properties measurements. This technique is really ideal for liquids where these limitations are not important.

The third general technique is the transmission line method, often referred to as the transmission reflection (TR) method. This is the most common method for general-purpose laminate materials. A transmission line is constructed with the some of the path containing the dielectric material. A two-port measurement is performed and the insertion and return loss interpreted in terms of the dielectric constant and dissipation factor of the sample.

There are really two types of TR techniques: free space and uniform transmission line. In the free-space version, an unclad sheet is placed between two antennas and the reflection and transmission through the sample is measured. This technique is analogous to an optical measurement and is suitable to high frequencies well above 20 GHz.

In the uniform transmission-line approach, a transmission line is constructed with the sample, as a co-planar waveguide (CPW), a microstrip, or a stripline. The Dk and Df are extracted from the measurement of the insertion and return loss. If plates are used to clamp on an unclad sample, air at the interface can often introduce an artifact. The IPC-TM-650 2.5.5.5 is based on the clamped stripline method.

One significant advantage of this technique is to enable a test structure to be embedded in the fabricated board for in situ measurements of Dk and Df. Separating dielectric loss from conductor loss can be a challenge, so using wide signal paths is common. To minimize the artifacts of the launches, different length lines can be compared, and the connector effects can be de-embedded from a topology-based model [3]. This technique can be used for frequencies in the 1 MHz to 20 GHz range.

The last method is the resonant cavity technique. This involves constructing a high-Q resonant cavity and measuring the perturbation on this cavity after a dielectric sample is inserted. The resonant frequency is shifted due to the dielectric constant of the sample and the fill factor. The width of the peak is spread out due to the dissipation factor of the material. From the cavity geometry and sample geometry, the Dk and Df of the sample can be extracted. This technique is inherently only for fixed frequencies but can be very accurate, especially for low-loss materials.

Cavity resonance is commonly used to measure the dielectric

properties of power and ground laminates [4]. The power and ground planes make up the cavity, and the resonant frequencies are measured using the two-port, low-impedance technique. From the geometry and resonant frequencies, the Dk and Df can be extracted. Fringe fields from the edges and spreading inductance from the probe locations into the planes often limit the accuracy to no better than 3 percent. It is often difficult to separate the conductor losses from the dielectric losses, even for moderate-loss materials. The IPC-TM-650 2.5.5.6 method is based on this approach.

Floating microstrip transmission line structures printed in the copper cladding of a laminate sample have also been made up as resonant cavities. Short, weakly coupled feed lines couple into the ends of the floating transmission line to measure the resonant frequencies and peak widths. As with other clad methods, it is sometimes difficult to extract the dissipation factor from the conductor loss for low-loss materials.

These various techniques are summarized in *Table 19.1*.

	Capacitance	Coax probe	TR/uniform transmission line	TR/free space	Resonant cavity/printed sample	Resonant cavity/unclad
Sample prep	Unclad sheet	Thick, unclad sheet	Printed CPW, microstrip or stripline in clad laminate	Unclad sheet	Clad sheet: rectangular or transmission line patterns	Unclad sheet
Frequency range	Low frequency: 5 Hz to 100 MHz	Broad band: 1 MHz to > 20 GHz	Broad band: from 1 MHz to 20 GHz	Broad band: 10 GHz to 100 GHz	Selected frequencies from 100 MHz to 5 GHz	Specific frequency values from 100 MHz to 20 GHz
Strengths	Simple sample prep Easy to understand	Simple sample prep Best for liquids	Provides in situ measurement Mimics application environment	Best for highest frequency Simple sample prep	Simple sample prep Simple extraction of Dk, Df	Simple sample prep Accurate values of Dk, Df even for low loss materials Rapid measurement and analysis
Weakness	Intrinsically low frequency Difficult ot get low loss Artifact if air is between the sample and plates	Sample must be thick enough Artifact if air is between the sample and probe	Requires patterning of clad laminate Difficult to separate conductor loss and dielectric loss for low loss laminates Challenge to extract accurate values of Dk and Df from S-Parameter measurements	Complex measurement techniques Challenge to extract accurate values of Dk and Df Not very accurate for low loss materials	Difficult to separate Df from conductor losses even for moderate loss materials Requires subtle analysis to extract accurate values of Dk and Df	Only for selected frequency points E field in XY plane, not Z axis Requires sophisticated software analysis to extract accurate Dk, Df values

Table 19.1: *Summary of the Features for the Four Generic Measurement Methods*

19.6 Split-Post Dielectric Resonator Method

Up until now, the resonant cavity technique using unclad samples was limited to high-end research labs due to the complexity of converting the perturbed resonant frequency shift and line width increase into the dielectric constant and dissipation factor of the material. Recently, Agilent Technologies introduced a new series of split-post dielectric resonator (SPDR) cavities and an analysis package, the 85071E option 300, which performs these measurements and calculations with a single click and extracts Dk values to better than 0.7 percent accuracy and Df values as low as 2 x 10^{-5}.

Of course, the influence of the dielectric sample on the cavity properties depends on the geometry of the cavity. There are two commonly used cavity configurations: an SPDR and a split cylinder dielectric resonator (SCDR). The sample prep and principles of operation are identical: it is in the analysis software and the frequency ranges where they differ.

A typical SPDR cavity is shown in *Figure 19.5*. The resonant cavity is the space between the two electrode posts. The frequency is determined by the lateral dimension of the post. A resonant frequency around 1 GHz has a post diameter of about 4 inches.

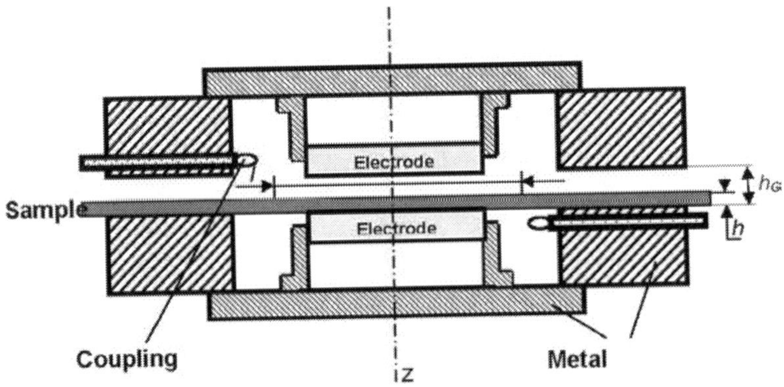

Figure 19.5: *Cross-Section of an SPDR Cavity*

One side of the cavity is excited by a small loop antenna driven by port 1 of the VNA. A second pick-up loop on the other side of the cavity connects to port 2 of the VNA and measures the transmission through the cavity. *Figure 19.6* is an example of an SPDR with a resonant frequency of about 1.095 GHz.

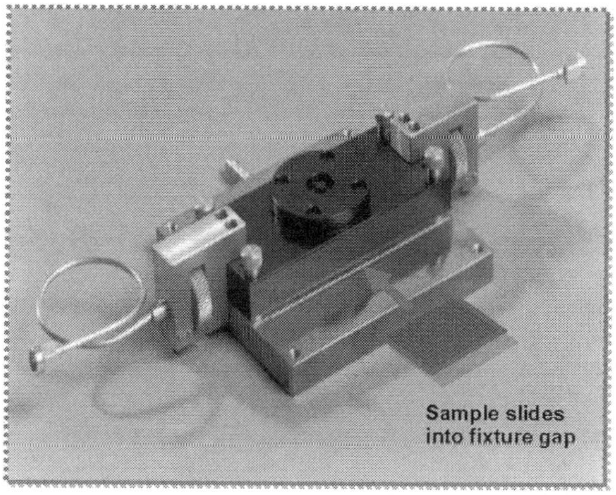

Figure 19.6: *Example of a SPDR Designed for Laminate Materials Measurement with a Resonant Frequency of 1.095 GHz.*

The transmitted signal, S21, is most sensitive to the cavity resonance. *Figure 19.7* shows the measured S21 for a 1.095 GHz SPDR cavity.

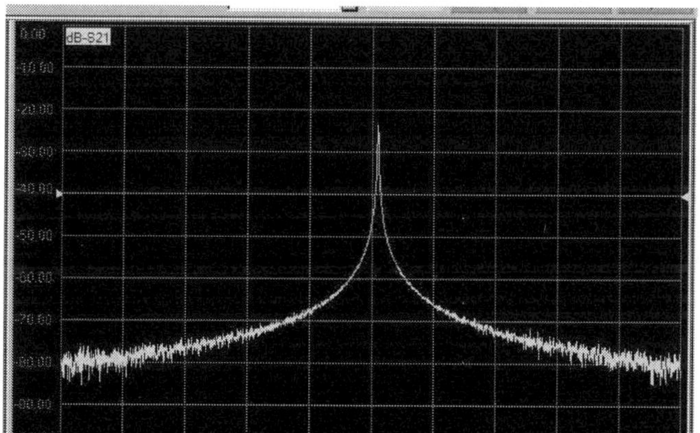

Figure 19.7: *Measured Insertion Loss of 1.095 GHz SPDR Empty Cavity. Horizontal Scale Is 1.095 GHz Center with 5 MHz per Division. The Vertical Scale Is the Insertion Loss in dB*

When an unclad, uniform-thickness, dielectric sample is inserted into the cavity, the resonant frequency shifts to lower frequency. An example of the insertion loss with a sample of FR4 inserted in the cavity is shown in *Figure 19.8*.

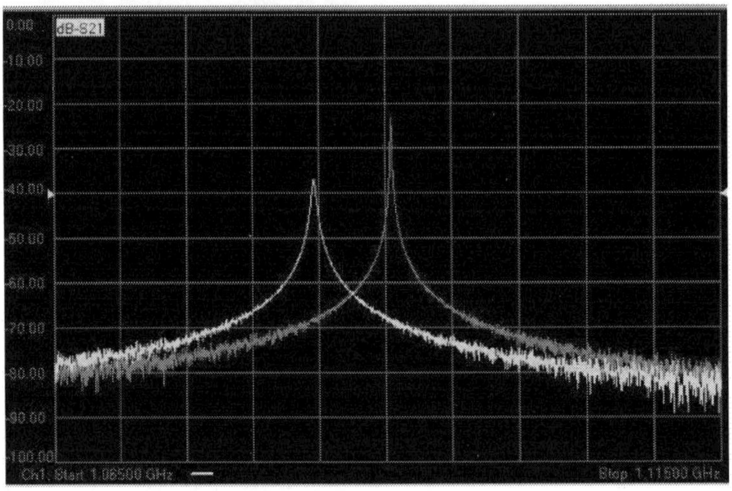

Figure 19.8: *Measured Insertion Loss of an SPDR with and without FR4 Sample in the Cavity*

From the two sets of measurements, the Dk and Df values can be determined at the resonant frequency of about 1.095 GHz. In this example, Dk = 4.45 and Df = 0.019.

The Raleigh-Ritz method is used to calculate the dielectric constant and dissipation factor [5]. The dielectric constant is related to the shift in resonant frequency and can be calculated from:

$$Dk = 1 + \frac{f_0 - f_s}{h\, f_0\, K(Dk, H)}$$

where
Dk = the dielectric constant of the sample
f_0 = the resonant frequency when empty
f_s = the resonant frequency when filled with the sample

h = the sample thickness
K(Dk, h) = a special function based on the specific SPDR cavity

The dissipation factor is extracted from the measurement of the -3 dB bandwidth of the cavity, from

$$Df = \frac{1}{p_{es}}\left(\frac{1}{Q} - \frac{1}{Q_{DR}} - \frac{1}{Q_c}\right)$$

where

p_{es} = the energy filling factor and is computed for each specific SPDR cavity
Q = the Q of the filled SPDR
Q_{DR} = the Q of the empty SPDR cavity
Q_c = the Q of the filled SPDR due to just the conductor loss

In this new implementation of the SPDR method, the calculations for each SPDR cavity for the K function and the energy filling factor have already been done, and custom software performs the measurements and the calculation. This approach makes the value of extracted Dk and Df insensitive to the sample thickness, as long as it fits between the posts of the cavity.

Figure 19.9 shows the measured dielectric constant of multiple layers of a polyester film. The total spacing in the cavity was 2 mm. From 0 to 0.6 mm thick sample, there was no impact on the extracted dielectric constant or dissipation factor.

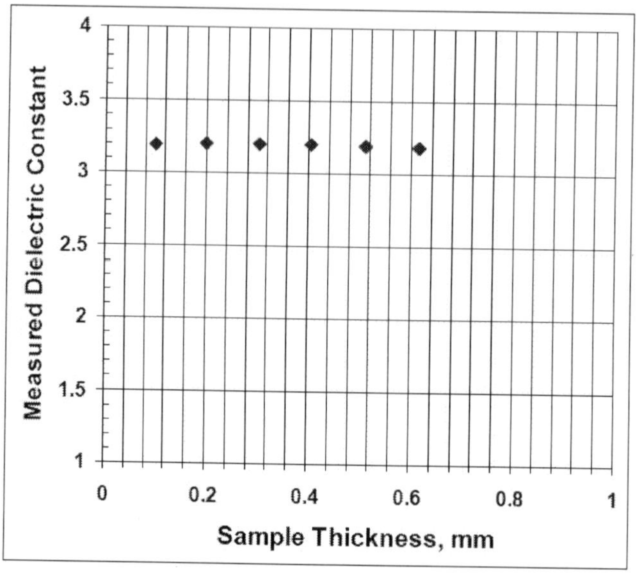

Figure 19.9: *Measured Dielectric Constant at 5 GHz for Polyester Samples of Different Thicknesses Showing Insensitivity of Extracted Dk on Sample Thickness*

To perform a measurement, the user is required to enter the sample thickness and push one button when the cavity is empty and one button when the sample is placed in the cavity. The value of Dk and Df at the resonant frequency is then displayed. An example of the setup screen is shown in *Figure 19.10*.

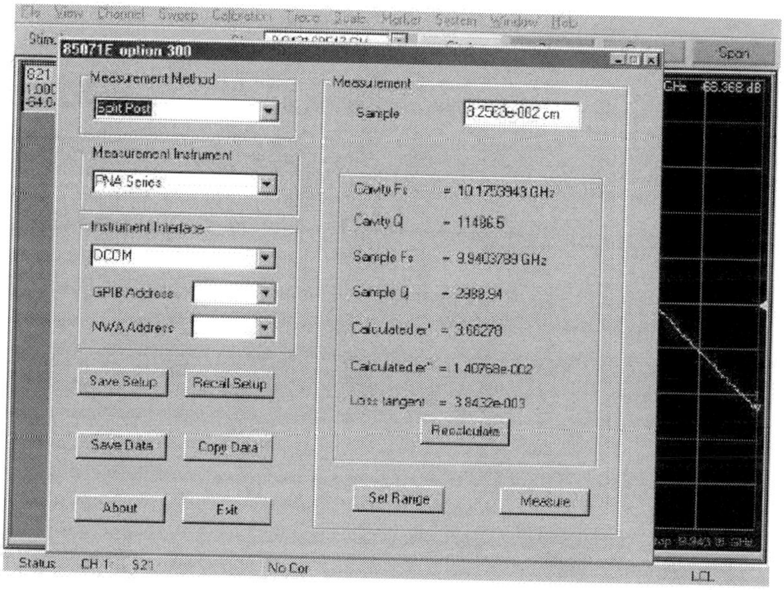

Figure 19.10: *User Interface Screen for Entering Sample Thickness and Initiating the Measurement*

There is one important limitation with this technique. The resonant mode of the cavity has the E field parallel to the surface of the electrodes of the posts. In all interconnect applications, the direction of the E field in the dielectric is normal to the plane of the laminate, along the Z axis. For homogenous, uniform materials, the dielectric constant is a scalar and its value is independent of the direction of the electric field. However, for inhomogeous materials such as fiber-reinforced resins, there may be a difference in the absolute value of the dielectric constant measured in the plane of the sheet and normal to the surface. This difference can be as much as 10 percent in some materials. Care should be taken in interpreting and comparing the dielectric constant measured with the SPDR method and other methods such as the clamped stripline. The dissipation factor is less sensitive to the field direction.

19.7 Conclusion

A new implementation of a SPDR technique is now available, which dramatically simplifies the measurement of in plane dielectric constant and dissipation factor properties of laminate materials. The difficulties of extracting the material properties from changes in the resonate frequency and Q are eliminated by using precision cavities and automating the calculations with custom software. This simple technique requires a laminate film be inserted into the cavity and, in less than 10 seconds, a measurement is automatically performed. This technique could be adopted in production environments where simple, fast, reproducible, and accurate measurements can contribute to incoming QC validation and process control. This will catch problems in incoming materials before they impact product and increase production yields.

Acknowledgments

The authors gratefully acknowledge useful discussions with Russ Hornung of Arlon, Al Horn of Rogers, and Bob Nurmi of Taconic.

References

[1] E. Bogatin, *Signal Integrity – Simplified*, Prentice Hall, 2004.

[2] E. Bogatin, "Controlling Controlled Impedance Boards," *Printed Circuit Design & Manufacture Magazine*, p. 20, May 1, 2006.

[3] E. Bogatin, "Materials Requirements for High Performance Digital Systems", *Circuitree Magazine*, August 2003.

[4] N. Biunno and I. Novak, "Frequency Domain Analysis and Electrical Properties Test Method for PCB Dielectric Core Materials," DesignCon East, 2003.

[5] "Split Post Dielectric Resonators for Dielectric Measurements of Substrates," Agilent Technologies application note, 5989-5384EN, 2006.

Authors

Eric Bogatin, Signal Integrity Evangelist, Bogatin Enterprises

Shelley Begley, Team Leader, Agilent Technologies

Mike Resso, Signal Integrity Application Scientist, Component Test Division, Agilent Technologies

Chapter 20

Designing Scalable 10G Backplane Interconnect Systems Utilizing Advanced Verification Technologies

20.1 Abstract

The design and implementation of high-speed backplanes requires substantial effort both in pre-prototype modeling and in post-prototype testing and measuring. Extant methods for modeling backplane signal paths have become very sophisticated and time-consuming. Correspondingly, current test methods for design verification have also relied on direct measurement techniques, which are often useful for only a single test condition requiring multiple test runs. This paper presents techniques for design that significantly reduce modeling requirements for the design of high-speed backplanes in conjunction with advanced testing techniques that provide maximum channel characterization with the minimum amount of time.

20.2 Approach

Companies requiring 10 Gbps (and higher) backplane solutions all face the same challenges of cost, power, scalability, and integration. The current standard design approach of embedding backplane serializer/deserializer (SERDES) input/output (I/O) within application-specific integrated circuits (ASICs) to save cost and extend performance has been effective for third generation (3G) and for some 6G backplanes. At 10G, issues of signal loss due to material, crosstalk, and power consumption make it, at best, risky and more costly to pursue the same design and testing approaches as previous generations.

Instead of relying entirely on electronic enhancements based on SERDES technology, an improvement of channel capability is proposed for the high-speed signals. Multiple benefits are created by improving the channel for high-speed signals. First, materials that reduce the insertion loss, and consequently reduce I/O power requirements, can be selected. Signal path structures can be optimized to minimize signal distortion, thereby further reducing I/O power consumption due to relaxed signal recovery timing requirements. Crosstalk, a major component of signal integrity, can also be significantly minimized, contributing to a lowering of overall signal-to-noise considerations in the signal receiver.

Improved channel construction holds the promise of much higher bit rates than a particular design point. For example, a channel that has been improved for use at 10 Gbps could easily be considered for use at 12, 15, or even 20 Gbps given the proper I/O electronics. The current methodology for analyzing a channel and its performance is a resource-intensive activity, given the cost and time associated with complexity of test setup and analysis. Projecting the performance of a channel at multiple bit rates is time-consuming and arduous when one considers the non-linearities associated with traditional channel construction operating at different frequencies. The idea of a singular design point has also permeated the back end of product design wherein testing and verification has been limited to the intended operating point (e.g., 10 Gbps).

Given the foregoing, it is clear that an updated test and measuring methodology is needed to take advantage of improved signal channel capability. Performance data collected at a single operating point, while useful, is certainly not efficient when there is need to extrapolate product performance into the future. Thus the approach taken for this project is to expand the test and measurement of the signal channel to include a full characterization of its properties over a 10x range of operating frequencies. The resulting performance data, in the form of S-parameters, is then used to construct performance characteristics from different operating points through the use of accurate modeling transformations. Significant savings in testing are gained since accurate test results (time domain reflectometry [TDR] plots and eye patterns) from different operating points are synthesized conveniently without having to continuously return to a test lab.

20.3 Current Design Impediments and Approaches

Powerful evolutionary forces in product design conspire to keep incremental improvement approaches in play for as long as possible to forestall the expense of new technology adoption. Backplane and chassis construction certainly adhere to these conditions and no doubt will continue into the foreseeable future.

A traditional backplane implementation is shown in *Figure 20.1*.

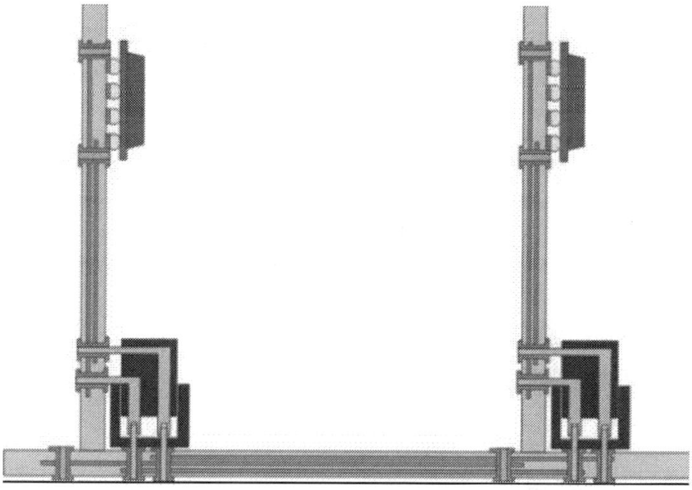

Figure 20.1: *Traditional Backplane Construction.*

For many backplane generations, the issues of impedance, loss, signal stubs, lumped parasitics, and cost have been in the forefront of design considerations. As bit rates have risen over time, the deleterious effects of the aforementioned elements on signal quality have increased significantly, and so also has I/O circuit design complexity. To combat the increase in signal degradation issues, electrical engineers, along with electro-mechanical engineers, have waged independent battles. The electrical engineers have taken advantage of Moore's law by using the ever-increasing supply of cheap transistors on a die to serve yeoman's duty in I/O circuits. On the electro-mechanical side of the battle, the designers of connectors and IC packages have incrementally parlayed new material and processes into interconnection elements that, while substantially traditional, perform better at higher speeds.

One set of electromechanical impediments associated with backplane printed circuit boards (PCBs) are summarized in *Figure 20.2.*

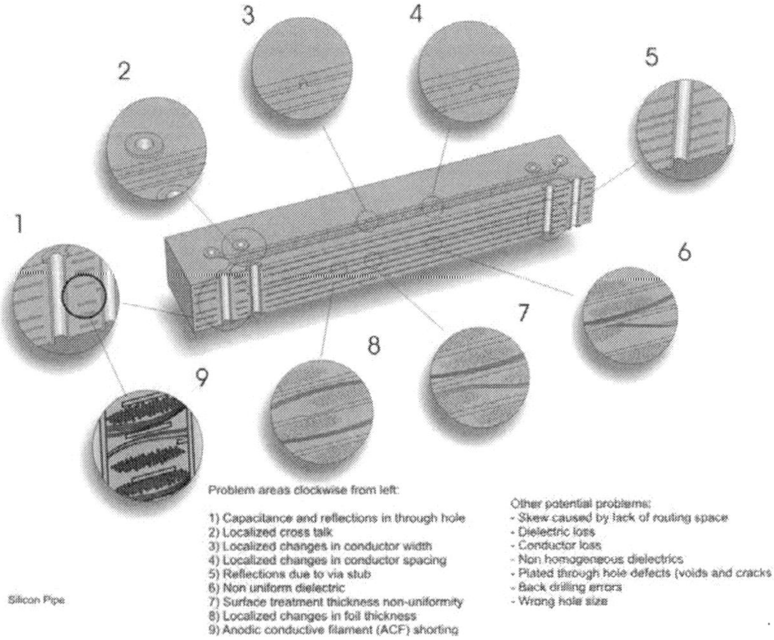

Figure 20.2: *PCB Problems*

These PCB and package structures translate directly into signal quality impediments. In particular, the device-to-package solder bump and package-to-board solder ball interfaces are high-impedance structures that create impedance compensation difficulties. In addition, signal layer transitions in both the package and board, needed to route the signal from device to device, create significant low- and high-impedance changes in the signal path.

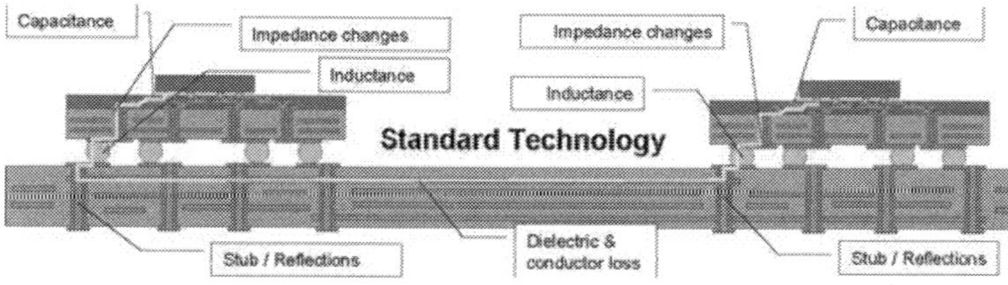

Figure 20.3: *PCB Signal Integrity Impediments*

It is not only the structure of PCBs but also the type of material that determines the performance capability of a backplane system as illustrated in *Figure 20.4*.

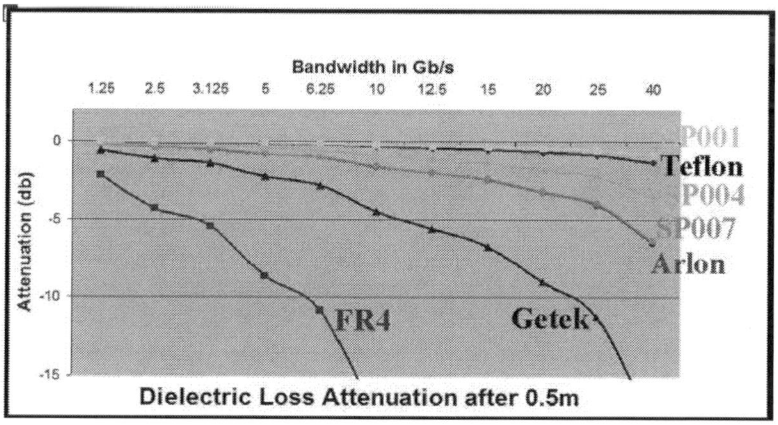

Figure 20.4: *Loss Performance for Backplane Materials*

The chart in *Figure 20.4* was generated by plotting the dielectric performance of a variety of commercially available PCB laminate materials. There are many beneficial attributes of FR4, which continue to make it a favorite choice among system designers. Foremost among these attributes are cost and manufacturing familiarity. Unfortunately, the performance of FR4 falls off dramatically as the data rate approaches 10 Gbps.

It is easy to see how the interconnection elements necessary to implement a 10G backplane create many challenges for the delivery of signals. To overcome signal quality issues, electrical engineers have implemented enhanced I/O electronics.

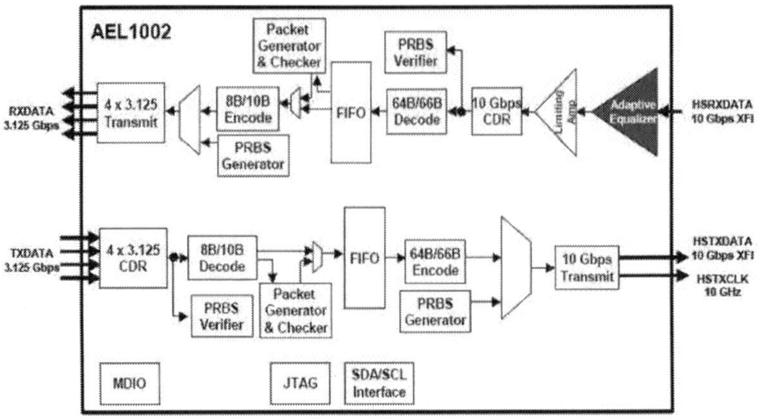

Figure 20.5: *SERDES for 10G Signal*

Figure 20.5 illustrates a simplified block diagram of a SERDES designed to transmit and receive 10G signals. The SERDES provides the necessary signal processing to overcome the signal degradation that occurs through the signal channel. In the transmitter, the data stream is encoded using 64B/66B code words to reduce the effects of inter-symbol interference (ISI). In the receiver, a clock/data recovery (CDR) circuit provides the necessary separation of clock from the data, while the 64B/66B decoder reconstitutes the original data stream. When the protocol allows, sometimes it is beneficial to encode the digital data with redundant bits of data with intent of adding error detection capability. This way, it is possible to detect certain type of common errors. Although such a scheme adds overhead to the data transmission and usually increases the raw rate of data bits being transmitted, error correction coding can be a very effective way to achieve a lower bit-error rate (BER). Highlighted in blue in *Figure 20.5* is an implementation of an equalizer to reduce some of the effects of the bandwidth non-linearity of the interconnection channel.

Signal conditioning techniques can be used to compensate the signal degradation due to the channel interconnections. One is to pre-condition the signal at the transmitter. This can be implemented with passive networks or active circuits using linear transversal filters. A linear transversal filter with a predetermined bit time interval is relatively easy to implement, especially with current mode drivers. For example, most XAUI drivers implement a variable-amplitude, one-bit–delayed tap that is opposite to the main signal bit. Such pre-emphasis has a net effect of reducing the low-frequency component of a transmitted signal. Such a scheme effectively reduces the effect of long PCB traces on high speed signals, which attenuates high-frequency components due to skin effect and dielectric loss. The technique can be extended to multiple bits of signal pre-emphasis, and the spacing can be extended to sub-bit times as well.

Another place to implement signal conditioning is at the receiver, as shown in *Figure 20.5*. The same linear transversal filter for transmitter pre-emphasis can be implemented at the receiver before the signal slicer. The advantage of having the filter at the receiver is that the receiver can adaptively adjust the filter coefficients without cooperation from the transmitter, while the transmitter cannot do so without the prior knowledge of the channel or cooperation from the receiver. In addition to a linear transversal filter, the receiver can also use a nonlinear decision feedback (DFB) equalizer, which has certain advantages over the linear filter.

There are also other equalizer implementation techniques. A common approach for equalizing cable or PCB loss is a split-path equalizer, in which the signal is split into two paths with different frequency responses. The output signal is a weighted sum of the two signal paths. By adjusting the frequency responses and the weighting ratio, one can achieve a robust equalizer with relatively simple circuitry.

21.4 AE 1002 Equalization

The optimal choice of signal conditioning technique strongly depends on the channel as well as the protocol used. In general, the cleaner the signal channel is, the less signal conditioning is required.

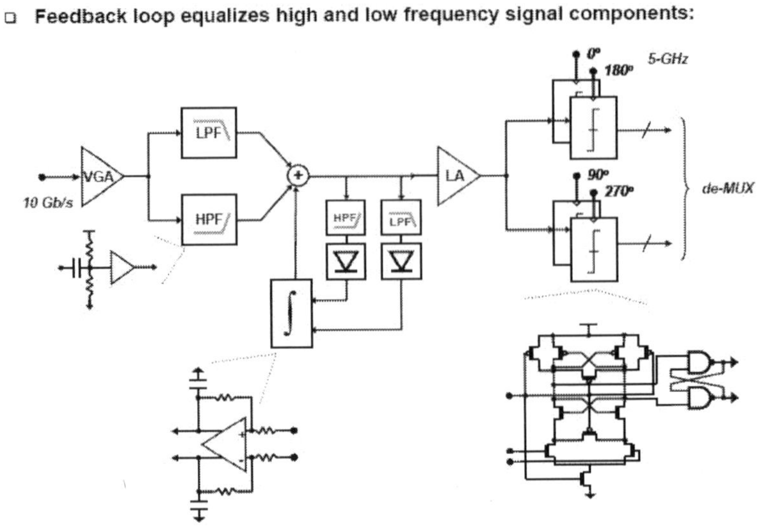

Figure 20.6: *AE1002 Adaptive Equalizer*

The details of the AE1002 adaptive receiver-side equalizer are shown in *Figure 20.6*. The input signal is first amplified with a variable gain amplifier and split into two signal paths—one with low-pass–frequency response and one with high-pass response. By adjusting the ratio between the two signal paths, the circuit can compensate for the frequency-dependent loss introduced by PCB traces. The feedback is carried out with two peak detectors that sense the signal amplitudes at both high- and low-frequency portions of the combined signal. The goal is to have a relatively flat frequency response before the input slicers (right-hand side).

21.5 Improving the Channel

As previously noted, powerful status quo forces in product design continue to keep incremental approaches in play for as long as possible to avoid the expense of new interconnection technologies. The electrical engineering design community has successfully addressed the problem so far through their proficiency at creating SERDES technology to address challenges related to interconnection issues. However, the ability of FR4 material to conduct high-speed signals drops off rapidly as the signaling rate climbs above 6 Gbps (*Figure 20.4*). To address this and other signal degradation issues, an alternate channel can be constructed that significantly improves signal transmission characteristics.

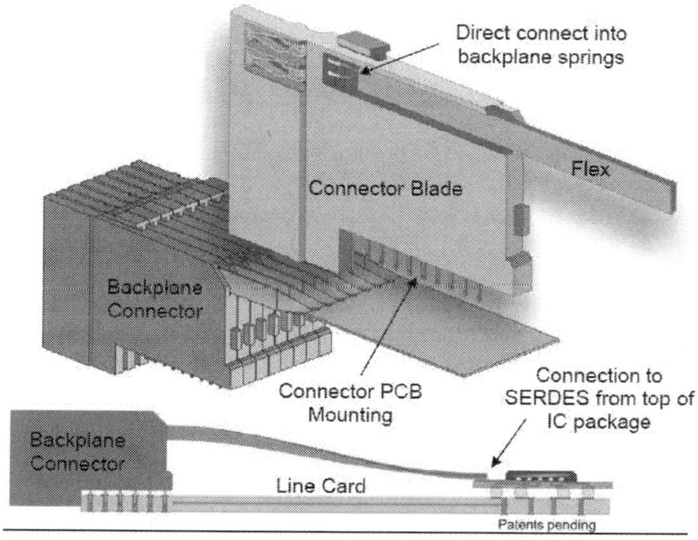

Figure 20.7: *Improving Signal Path for High-Speed Signals*

Figure 20.7 illustrates how a high-speed signal path can be constructed for the separate transmission of high-speed signals from an IC to a backplane connector. As shown, flexible material with significantly better dielectric properties than FR4 (dielectric constant of ~4.4) provides reduced channel insertion loss. For this project, a polyimide material with a dielectric constant of 3.4 was used. In addition, the channel can be easily constructed with no thru-hole in the signal path thereby providing fewer impedance discontinuities. Thru-holes are also a significant source of signal crosstalk.

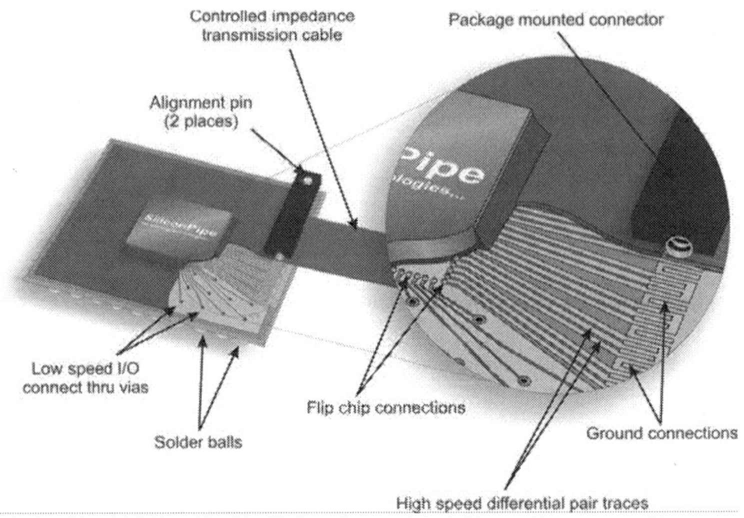

Figure 20.8: *High-Speed Signal Transmission with No Thru-Holes*

Figure 20.8 illustrates how high-speed signals can exit an IC package without the need for thru-holes. In this particular example, an IC connects into differential pair microstrip signal paths where they are routed to the IC package edge. At the edge of the IC package, a direct contact connector system with a controlled impedance flex circuit is utilized to carry the signal further into the system.

The equivalent of a 30-inch 10G backplane was built using this approach. Components from Aeluros, ERNI, and Sanmina-SCI (10G SERDES, backplane connector, and backplane) were used in the system.

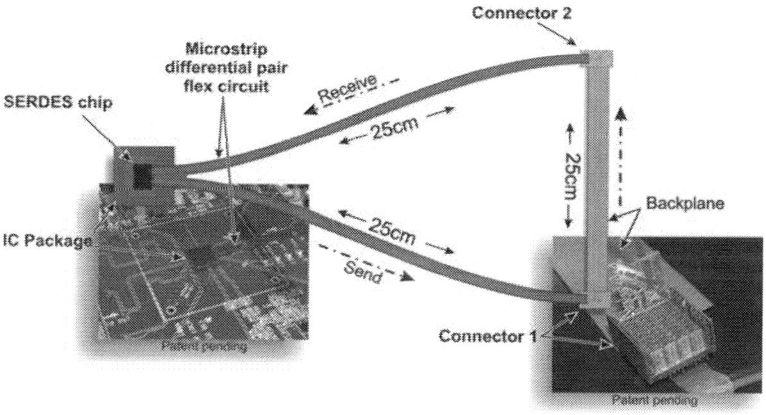

Figure 20.9: *10G Backplane Interconnection System Utilizing Improved Signal Channel*

A single 10G SERDES (Aeluros AEL1002) was used to provide both the transmitter and receiver for the test. A controlled impedance polyimide flex circuit (fabricated by Altaflex, Santa Clara, CA) was designed to mount onto the AEL1002 IC substrate and provide a direct attach for both transmit and receive differential pairs. The assembly was performed by Nxgen Electronics, San Diego, CA. The flex circuit was designed in a split fashion to allow for the receive pair to be separated from the transmit pair at the opposite end of the flex cable. This provided an ability to insert the flex cables at each end of a 10-inch backplane. The 10-inch backplane was manufactured by Sanmina/SCI with FR4, except for the microstrip signal layers, which were constructed with Rogers material (dielectric constant ~3). At the opposite ends of the flex, standard GBX connector blades (supplied by ERNI) were attached by bypassing the existing signal paths in the ERNI blades (*Figure 20.9*). The complete IC/Flex assembly was reflowed onto an Aeluros 10G evaluation board.

Figure 20.10: *SERDES Flexible Circuit IC Attachment*

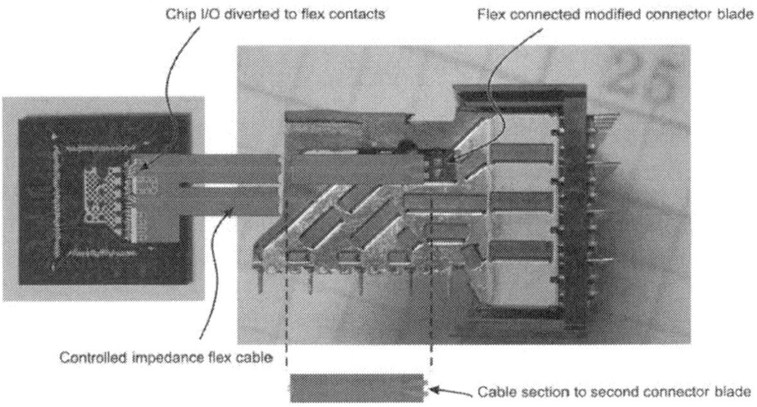

Figure 20.11: *GBX Flex Blade Attachment*

The AEL1002 10G SERDES is not designed as a backplane SERDES. The AEL1002 is designed specifically for XFI (a serial 10 Gbps electrical interface application), in which the chip is used as a bridge chip between an ASIC with an XAUI interface and an XFP optical module. In such applications, the distance between the SERDES chip and the XFP module can be up to 12 inches. The XFP module is an optical module that can be hot-plugged into a system with an XFP connector. The combined effect of the XFP connector and the potential long PCB trace will degrade the 10 Gbps electrical signal such that signal conditioning is required. The AEL1002 is designed with an adaptive receiver side equalizer to mitigate the variable degrees of PCB loss for 12-inch FR4 signal traces.

The AEL1002, in addition to the required functionality of providing 10G signal processing, also provides many test features that help reduce device testing time and are useful for system performance analysis. Extensive pseudo-random bit sequence (PRBS) and 10 gigabit Ethernet packet generation and checking are included. On the parametric side, the transmitter incorporates programmable amplitude adjustment and the receiver timing is adjustable for checking the timing margins of the signal link.

In this demonstration application, the backplane test environment consisted of a total signal path length of 30 inches with two industry-standard backplane connectors.

21.6 Initial Functional Testing

No attempt was made by the design team to determine in advance, through modeling, the capability of the test system. We anticipated a significant improvement in insertion loss and signal quality at a qualitative level from the use of controlled-impedance flex circuits and the elimination of thru-holes in the signal path. Owing to the test features of the AEL1002, it was possible to implement the test system and determine general performance capability without the use of signal path test equipment. Instead, the use of a laptop computer provided the operating test results through the test features built into the AEL1002.

During initial system bring-up, operating performance was observed at 10 Gbps with 800 mVp-p driver signal strength through the Aeluros evaluation board software running on a laptop. Through the Aeluros evaluation board software, the test team was quickly able to adjust the operating conditions of the system and determined that the system operated with 60 percent receiver timing margin even at 100 mVp-p driver strength. Unfortunately, the driver signal strength could not be adjusted down further to explore the boundaries of the system's operating margin. The system performance surpassed expectations of the design team, given no prior analysis had been performed.

21.7 Full System Analysis

While the operating results exceeded performance at 10G data rates by a wide margin, it was unknown as to why the system performed as well as it did. A thorough analysis was needed to understand and corroborate overall performance.

A full measurement of the 30-inch backplane system was performed to assess the system's overall performance and data rate scalability for all bit rates, not just 10G. Unlike the traditional approach of using a TDR oscilloscope to measure the performance of the channel at a set rise time corresponding to, for example, 10G, the team chose a new measurement methodology, which made it possible to evaluate the performance of the system at any bit rate between 100 Mbps and 25 Gbps, even months after the test fixtures have been removed.

Measurements were provided by GigaTest Labs and Agilent Technologies using a four-port network analyzer and micro-probing equipment. Since a complete differential channel characterization was required for validation of the interconnection channel, the N1930A Agilent physical layer test system (PLTS) was used as the calibration, measurement, and analysis tool. A single connection of four cables (two input and two output) yields all 16 differential and mixed-mode S-parameters, including differential insertion loss (SDD21). This measurement system allows extremely accurate, repeatable S-parameter data to be taken on a wide range of differential interconnect devices. For this measurement, four-port single-ended and two-port mixed-mode S-parameters were taken on the 30" channel in 25 MHz increments through 25 GHz. The resulting dataset of S-parameters for the channel allowed for the accurate synthesis of single-ended as well as differential and common-mode channel performance (TDR plots and eye diagrams).

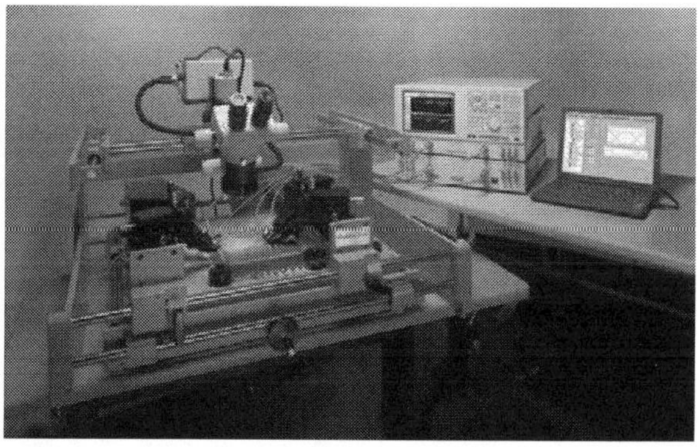

Figure 20.12: *High-Precision/High-Frequency Probe Station (GigaTest Probe Station)*

Equipment used:

- GTL-4060 probe station
- 26 GHz coaxial cables (3.5 mm SMA–compatible connectors)
- Agilent Technologies 8364B vector network analyzer (50 GHz) with four-port test set
- GGB 40A-GS/SG/GSG-450-DP Probes with CS-11 calibration substrate
- Agilent Advanced Design System software (ADS 2003C)
- Agilent PLTS

Vector network analyzer settings used:

- Start frequency: 25 MHz
- Stop frequency: 25 GHz
- Frequency steps: 25 MHz
- Input power: 0 dBm (equivalent to 0.6 Vp-p into 50Ω)
- Calibration kit: CK11450, SOLT style

Measurement of the channel required the use of a flex cable assembly without the AEL1002 SERDES IC. Instead of the AEL1002 driving to and receiving from the flex, the IC was replaced by probes connected into the test equipment.

Generating accurate channel measurements of the interconnection channel demanded the use of accurately positioned, high-performance microprobes. *Figure 20.13* shows a photograph of the end of the flex cable designed to connect into the SERDES. The SERDES is not attached. Instead, differential probes, with grounds, are positioned to make contact to the flex in the same area where wire bonds from the SERDES would be positioned.

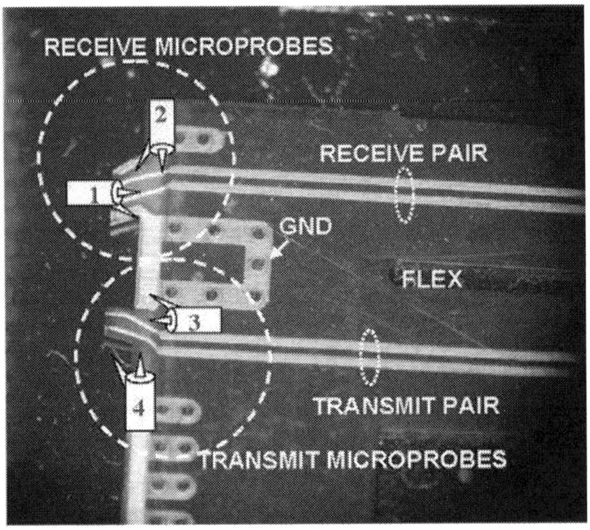

Figure 20.13: *Flex Cable Probe Locations*

The overall view of the flex assembly and backplane as positioned on the probe station is shown in *Figure 20.14*.

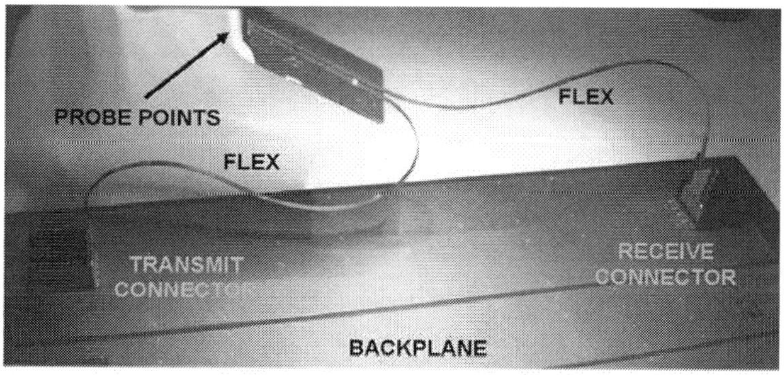

Figure 20.14: *Flex and Backplane Setup for Probe GTL Probe Station*

The full interconnect assembly can be seen in *Figure 20.14* being readied for the micro-probes. S-parameter data was taken for the system between 25 MHz and 25 GHz in 25 MHz increments. This data was captured using Agilent's PLTS four-port characterization software.

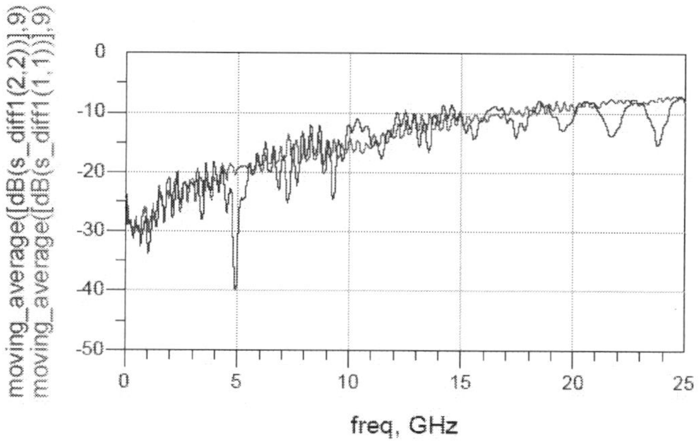

Figure 20.15: *VNA Differential Reflection Test Results*

The SDD11, differential return loss data is shown in *Figure 20.15*. This data shows the return loss to be better than -10 dB up to about 15 GHz. The notch at 5 GHz is the result of series capacitors (DC blocking) in the flex approximately two inches from the AEL1002.

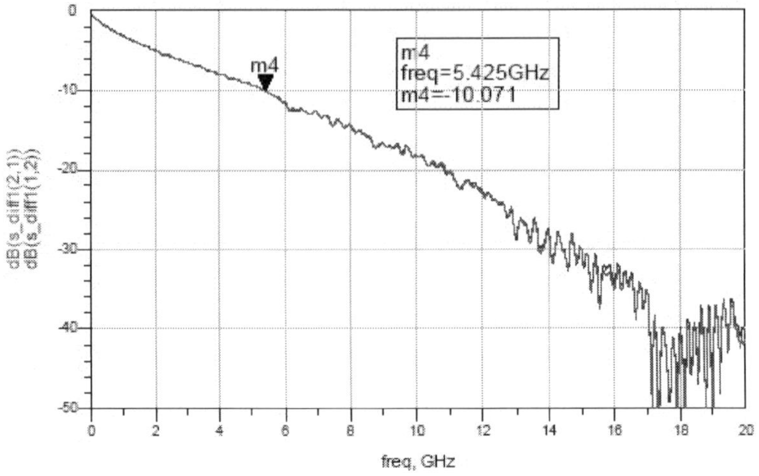

Figure 20.16: VNA Differential Attenuation Test Results

SDD21, differential signal attenuation, is shown in *Figure 20.16*. Note the excellent linearity of the frequency response up to 16 GHz. This linearity improves the effectiveness of frequency compensation such as the Aeluros equalizer.

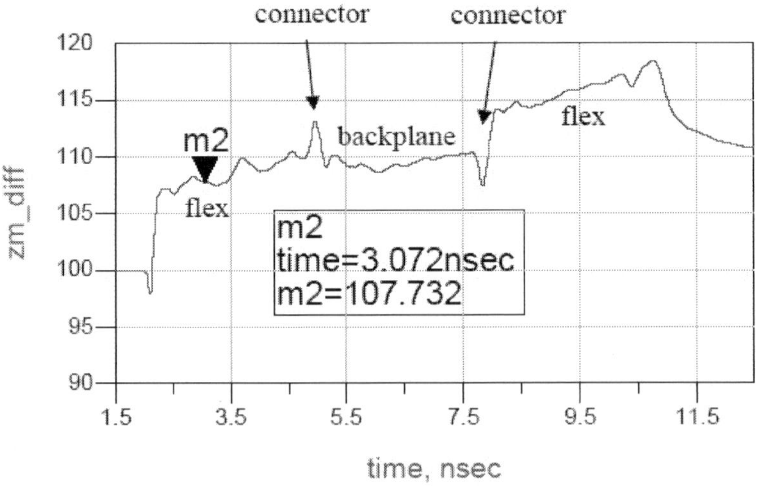

Figure 20.17: Differential Impedance TDR Plot Generated from S-Parameters

Differential TDR plot generated from the S-parameters shows the impedance in the interconnect channel. The impedance through the flex is approximately within the 100 Ohm +/- 10 percent specification.

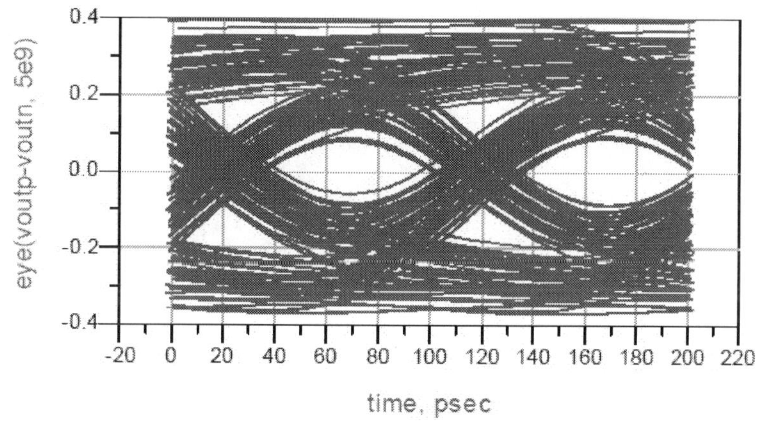

Figure 20.18: *10G Eye with Single Pole Pre-Emphasis (10Ω, 20 pF) Generated from S-Parameters*

A powerful advantage to the S-parameter test methodology is the ability for test engineers to generate "what if" scenarios for different design points. *Figure 20.18* is an eye diagram generated for the test system channel at 10 Gbps using a driver rise time of 25 ps, wire bond parasitics of 0.7 nH and 0.13 pF, and a single-pole 10Ω, 20pF pre-emphasis filter. The driver, wire bond, and filter model can each be independently manipulated to ascertain overall system performance without resorting to test bench setup changes.

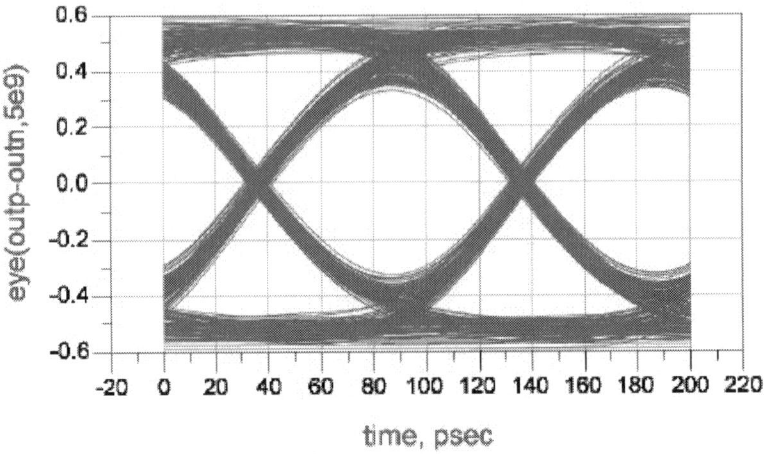

Figure 20.19: *Simulated 10G Eye Diagram after AEL1002 Adaptive Filter*

Figure 20.19 illustrates what the 10G eye diagram looks like within the AEL1002 at the input to the signal slicers. This eye diagram was generated from the same S-parameter data that generated the eye diagram in *Figure 20.18*. Clearly, the adaptive ability of the AEL1002 provides for much better overall performance and opens the door for increased bit-rate operation.

Although the receiver equalizer is designed for XFI channels with 12 inches of FR4 trace and a high-quality edge connector, it has no problem in equalizing this backplane link with 30 inches of traces and two backplane connectors. This is the benefit of a clean channel construction. The simulated timing degradation due to the channel and the adaptive equalizer is around 10 ps. Using the built-in timing margin capability, the system timing margin is measured to be around 60 percent unit interval (UI). This is consistent with a 70–75 percent UI timing margin measured with a clean back-to-back connection. In addition, since the channel has a very clean and well-behaved insertion loss up to 20 GHz, it is expected that with a similar but extended receiver equalizer, it should be able to handle 20 Gbps without difficulty.

20.8 Summary

The goal of building a 10G backplane interconnection system that minimized signal loss and distortion was achieved. The performance of the system was observed with actual signal transmissions in addition to being verified through a thorough channel test methodology using S-parameters generated by the industry leading test equipment. Overall, the performance of the system was above expectation for the team involved with the project. The test methodology of using S-parameters was validated and resulted in much less engineering test effort to determine channel performance utilizing various design parameters (e.g., emphasis, frequency).

The project did not provide an assessment of several aspects of the new interconnection approach insofar as multiple signal paths (crosstalk), manufacturing process, manufacturing costs, design and test costs, or overall usability in future backplane implementations. The authors believe that each of these issues, while important and relevant to the overall decision to adopt a new interconnection approach, will be shown to be acceptable and in line with modern product design objectives.

Several intangible and unforeseen benefits presented themselves as a result of pursuing the project. The most celebrated of the benefits among the group was the savings in engineering time and cost needed to design high-speed systems. The trend in electronic product design has been to solicit and use increasingly sophisticated design tools and experts in order to implement seemingly complex high-speed designs. Unlike product designs of 20 years ago when signal integrity was tantamount to a Schottky clamping diode on an input, today's digital designers are beholden to by a phalanx of tools and analysts making sure that the off–IC signals arrive at their destinations properly. With a cleaner channel, the time and cost of such analysis is eliminated or significantly reduced, thereby speeding up product design and lowering product risk.

The second benefit learned by the team was that I/O power for high-speed signals can be significantly reduced and consequently relax the overall system power requirements. Upward of 20 percent of an IC's power consumption can be attributed to processing needed to handle high-speed I/O. In particular, phase-locked loops (PLLs) designed for clock/data separation can consume large amounts of power to achieve fine timing accuracy. So when it was learned that an eight-fold reduction in driver voltage was possible for high-speed signals, the system-wide implications of overall power savings came to light.

A compelling conclusion as a result of pursuing this project is the predicted performance of a backplane solution below 10 Gbps. With the S-parameter data available up to 25 GHz, and a correlated result of the system at 10 Gbps, an eye diagram for 5 Gbps is presented in *Figure 20.20*.

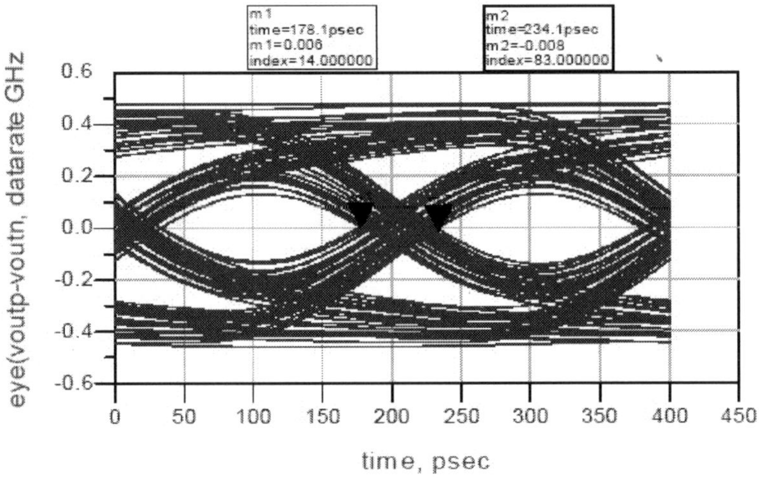

Figure 20.20: *Eye Diagram for 5G Backplane.*

The 5 Gbps eye diagram in *Figure 20.20* tells us that interconnection systems based on clean signal channels can be constructed with simplified I/O electronics. The eye diagram in *Figure 20.20* has no transmitter pre-compensation or receiver adaptive equalization included! Given the openness of the received eye pattern, one can conclude that simple digital I/O techniques are possible well beyond today's design guidelines.

Contributing Companies

The project was made possible through the combined personnel, equipment, and manufacturing contributions of the following companies:

Backplane—Sanmina-SCI (www.sanminasci.com).
SERDES and evaluation system—Aeluros (www.aeluros.com)
Flex cable—AltaFlex (www.altaflex.com)
Micro-probing and test analysis—Gigatest (www.gigatest.com)
Test equipment and test analysis—Agilent (www.home.agilent.com)
Assembly—NxtGen Electronics (www.nxgenelect.com)
Connectors—ERNI (www.erni.com)
System—SiliconPipe (www.sipipe.com)

Authors

Kevin Grundy, Chief Executive Officer, SiliconPipe

Haw-Jyh Liaw, Director, Systems Engineering, Aeluros

Gary Otonari, Engineering Project Manager, GigaTest Labs

Mike Resso, Signal Integrity Application Scientist, Component Test Division, Agilent Technologies

Chapter 21

Investigating Microvia Technology for 10 Gbps and Higher Telecommunications Systems

21.1 Abstract

Backplane technology is the foundation for today's telecommunication systems. The evolution of various backplane architectures has pushed system bandwidth from hundreds of megabits per second to terabits per second. In the pursuit of the ultimate data throughput, the physical layer (PHY) structures in the backplane take on critical roles. Pin density of connectors, via stubs, and routing of traces create challenges for controlling the excess reactance throughout the channel. However, with the use of advanced microvia technology and surface mount connectors, digital designers can now extend the barrier of telecommunication systems. This paper will show design techniques for implementing and evaluating such PHY structures.

21.2 Introduction

Today's telecommunication platforms depend on high-speed serial data transmission. Leading-edge digital designers push the performance limit of what is possible to achieve on copper. The proliferation of serial links beyond 10 Gbps has exposed signal integrity issues not typically encountered in the standard digital design laboratory. Optimization of signal integrity by focusing on the PHY structures within these high-speed channels can produce astonishing results. The fundamental insight of how signals propagate can be clearly understood with the proper design tools and methodologies. Network switches and routers have recently employed advanced backplane technology to break the terabit barrier. This accomplishment is in part due to sophisticated design techniques within the PHY components. A major portion of this design cycle is geared toward modeling, simulation, and measurement validation. Reflections, crosstalk, impedance mismatch, and loss can be visualized and these complex phenomenon become intuitive through the use of design tools that allow both time and frequency domain analysis.

Real estate constraints of high-speed digital systems necessitate the use of microvia technology to allow more components to be placed on a single circuit board. With more companies using microvia technology, the process has been able to advance rapidly from controlled-depth drilling to more advanced laser ablation techniques. Printed circuit board (PCB) manufacturers are tasked with developing processes for microvias that meet the aspect ratio requirements of today's multilayer backplanes. Implementing microvias opens the door for SMT board-to-backplane connectors and the overall system performance improvements inherent in these connector systems.

21.3 Telecom System Physical Layer Overview

Typical 10 Gbps Telecom System

During the late 1990s and early 2000s, the focus of network OEMs was the delivery of high-performance, technology-leading communication systems to meet the demands for ever-greater telecom bandwidths. Chassis and backplane design were key differentiators for market-leading manufacturers. The communications industry today is evolving toward modularity in a manner very similar to the server world transition in the early '90s. Telecom systems such as those shown in *Figure 21.1* typically achieve high-speed data transport through a switch fabric interface that can be used as a secondary communications channel in parallel with the base interface. In most high-speed networking applications, the base interface will be used to carry communications between the control-plane processors on each line card. This PHY copper interface creates many challenges for signal integrity engineers designing, developing, and testing network elements. One of the most challenging and interesting areas for high-speed design is in backplane applications. Performance of routers and switches are fundamentally limited by the bottleneck created by the backplane components; therefore, this is an area rich for technology breakthroughs and innovation.

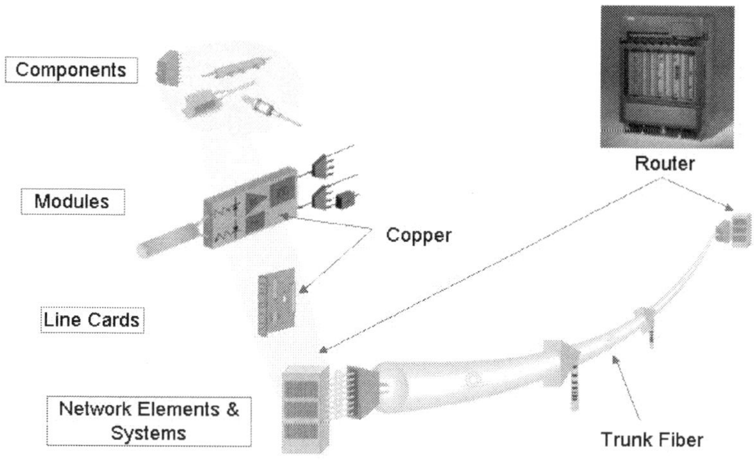

Figure 21.1

Backplanes Are a Critical Link

Today's standards efforts for a 10 Gbps Ethernet backplane are under development in the 802.3ap task force. The goal is to use an ordinary copper backplane, as shown in *Figure 21.2*, to send 10 Gbps Ethernet signals between line cards using no optics. A standard would give systems designers a head start, allowing them to choose among several PHY chips that would be pre-wired for the standard. Exciting work is being done with 10 Gbps serial and novel signaling schemes such as binary signaling or PAM4 that can help achieve this high-speed data transmission. However, the ultimate limit of serial rates will be most likely dictated by the signal integrity of the PHY backplane. Achieving a controlled impedance environment throughout the complete backplane channel from chip to chip will demand careful and meticulous design methodologies. The backplane connector plays a critical part in this channel.

Figure 21.2

Backplane Connectors Are Advanced

Designing a surface-mount (SMT) board-to-backplane connector has numerous requirements. First, the interface must withstand the mechanical conditions faced by standard board applications and be very rugged. Second, the connector must be able to transmit data at speeds exceeding 10 Gbps. Recent designs of SMT backplane connectors have evolved from press-fit connector technology, including many of the same mechanical features such as the 1.5 mm x 2.5 mm pin grid. The two main differences in the connector designs revolve around the use of SMT signal leads and the C-shaped pin-in-paste ground shield pin. The exploded view shown in *Figure 21.3* shows the details of construction of a 10 Gbps connector.

A well-designed high-speed board-to-backplane connector integrates the mechanical, chemical, and electrical properties of the device seamlessly. Orientation of differential pairs, spacing of contacts, and selection of component materials all play key roles in the overall performance. It is a challenge to find the proper combination of these design criterions without impacting the signal integrity of the connector. A great deal of time and effort goes into the design and modeling of these types of connectors before the first piece of steel is cut.

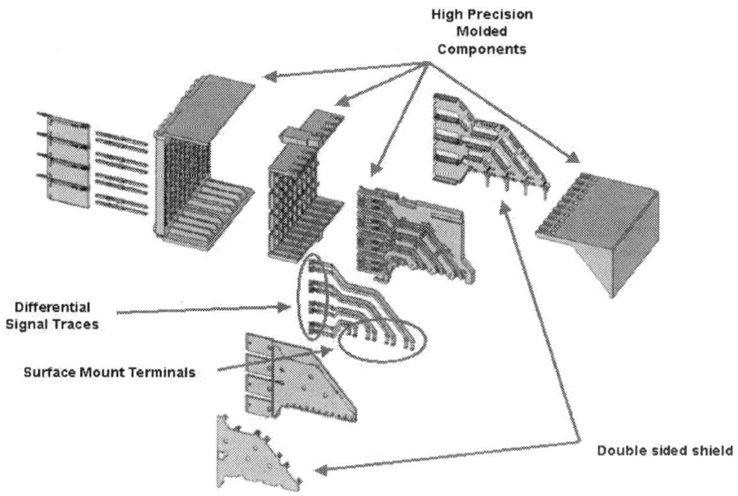

Figure 21.3

Via Stubs Create Capacitive Loads

It is essential to reduce the amount of via stub to successfully transmit data at 10 Gbps. Connectors that require interfaces to plated thru-holes (PTHs) are susceptible to capacitive loading that is inherent to the geometry of these commonly used board attachments. To overcome this challenge, the most critical lines would need to be routed closest to the bottom surface of the PCB, as shown in *Figure 21.4*, or the via barrels back-drilled to reduce the via stub. This may lead to longer design times and more layers to achieve the desired signal integrity performance.

Many board designers implementing press-fit connectors try to reduce the resonant behavior of a PTH by routing signals near the bottom layer of the PCB or back-drilling critical lines to reduce the via stub. With SMT connectors, there is no need for back-drilling, since the connector is mounted on the top surface of the PCB and the signals are attached with blind or buried vias. This type of connection scheme allows the system bottleneck to move from the connector to the PCB material.

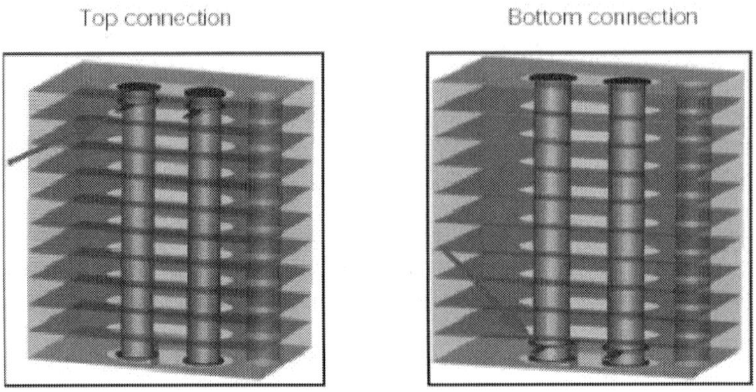

Figure 21.4

Connector to Board Interface

With an SMT termination, the reflective behavior associated with a PTH is reduced or even eliminated, since there is very little dangling stub. The interface-to–SMT devices, whether it is resistors, silicon integrated circuits (ICs), or connectors, must be made on the outer surfaces of the PCB. It would be impossible to route all of the signal lines of a high-density, high speed, differential connector on the outer surface alone, so an alternate approach must be taken. Additionally, these high-speed lines will need to interact and connect with inner routing layers to achieve all of the desired functionality of the system. Different via structures can be used in conjunction with the backplane connector to interface the connector to the inner board traces. Graphical examples of both PTH and SMT are shown in *Figure 21.5*.

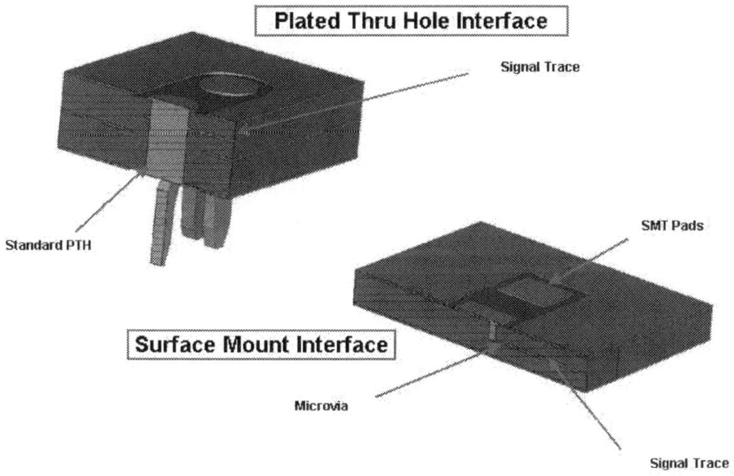

Figure 21.5

Microvia with SMT Improves Signal

Figure 21.5 shows a typical PTH and microvia structure within a mixed-board stack-up. The same type of SMT lead from the connector is attached to the PCB, but we can use two techniques to bring the signals from the connector to the PCB traces. With the illustration on the left, a PTH connects the SMT pad to the trace near the bottom surface of the PCB. Since there is no pin inserted into this PTH, the via diameter can be shrunk to a size that reduces the capacitive effect while still meeting the aspect ratio requirements of the board vendor. Using this smaller via allows for added signal performance with respect to a standard PTH while creating a cost savings over more expensive via alternatives. Additionally, using a full plated thru barrel allows signals to be accessed at any layer in the PCB stack, although accessing signal traces that are close to the surface will introduce stubbing effects into the signal path. In the example on the right, a small microvia is used to connect the SMT pad to an inner board trace. This via can be made even smaller in diameter, in relation to the PTH, as it is often formed by more precise methods than the mechanical drilling process used to create PTHs. Selectively stacking microvias to reach a desired layer allows the board designer to achieve optimal signal performance by eliminating the electrical stub.

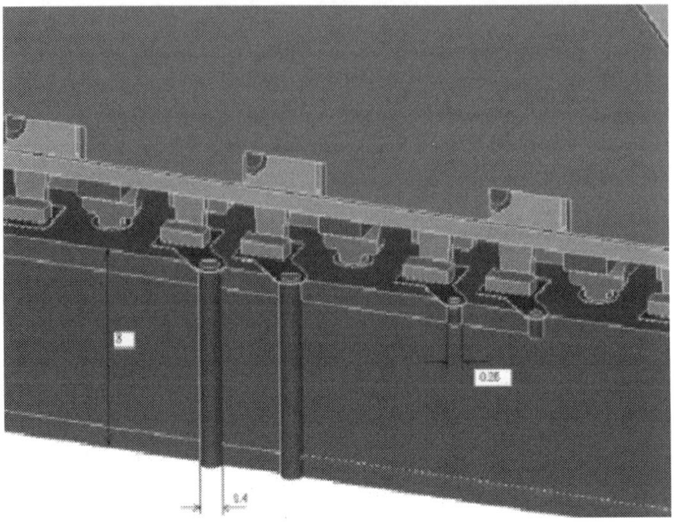

Figure 21.6

22.4 Signal Integrity and Differential Signaling Backplane Data

Rates Are Increasing

The proliferation of many new high-speed digital standards depicted in *Figure 21.7* push the envelope of what is possible on copper. The data must be transmitted with very few bit errors to maintain system reliability. Unfortunately, the signal integrity suffers when the risetime of the data transition from a one to a zero becomes faster. This faster rise time emphasizes poor design technique of any PHY component in the system, including every stripline, microstrip, cable, and connector. Frequency-dependent effects are now commonplace across most high-speed digital designs and knowledge of transmission line theory is now a requirement for leading-edge design. To complicate matters further, the majority of these standards use differential circuit topology. A paradigm shift in measurement technology is under way to achieve the goals of the advanced differential interconnect.

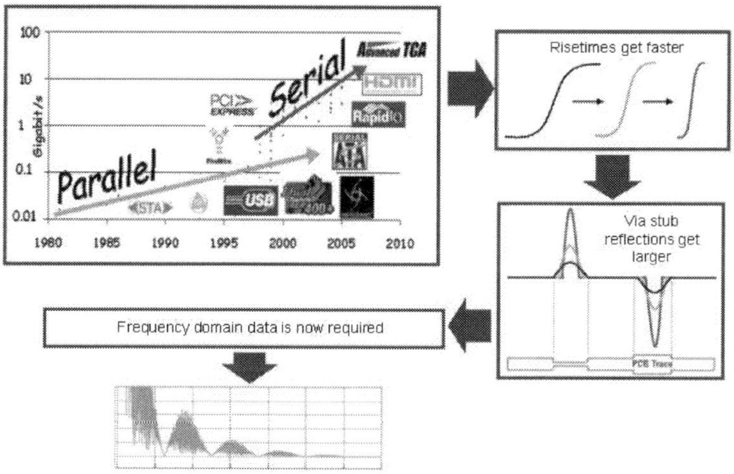

Figure 21.7

Transmission Lines Are Differential

Using *Figure 21.8* as a guide for understanding current flow, imagine two transmission lines that are driven by single-ended signals that are exactly out of phase (we call this differential driving). As the signal propagates down the differential pair, there is a voltage pattern between each signal line and the reference plane below. In addition, there is a signal between the two signal lines. This is called the difference signal or differential signal.

Differential impedance is simply the impedance the difference signal sees that is driven between the two signal lines in the differential pair. The impedance the difference signal sees is the ratio of the signal voltage (difference voltage) to the current in the line. The difference voltage is twice the voltage of the edges driven into each line. The current into each line is related to the impedance of each individual line in the pair. There is an additional current between the signal lines that is due to the coupling between the traces themselves. This is in general a small amount, but it cannot be neglected.

If there were no coupling between transmission lines, the impedance of a line, as defined by the ratio of the voltage across the paths and the current through them, would be dependent on just the line parameters of the one line. However, as soon as coupling is introduced, the voltage on one line may be dependent on the current in an adjacent line. To include these effects, the concept of impedance or characteristic impedance must be expanded to allow for one trace interacting with another. This is handled by expanding the impedance into an impedance matrix. Matrix math is very useful when quantifying the performance of differential transmission lines, as will be evident in the next discussion that describes another type of matrix called the mixed-mode S-parameter matrix.

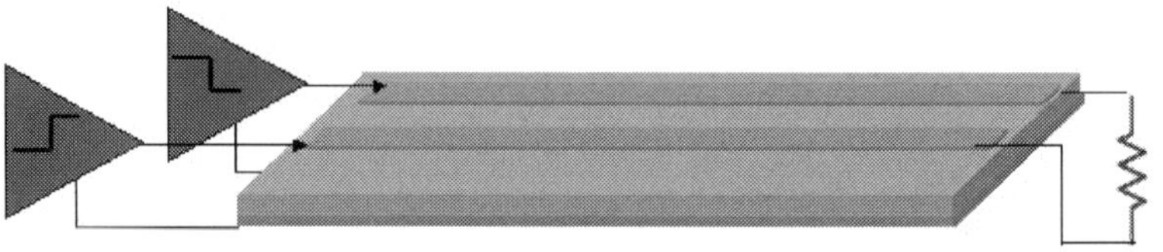

Figure 21.8

Single-Ended Parameters

To lay a foundation for understanding how to characterize a PHY device in a 10 Gbps telecom system, a brief discussion of multiport measurements is in order. The four-port device shown in *Figure 21.9* is an example of what a real-world structure might look like if we had two adjacent PCB traces that are operating in a single-ended fashion. Assume that these two traces are located in relative proximity to each other on a backplane and some small amount of coupling might be present. Since these are two separate single-ended lines in this example, this coupling is an undesirable effect, and we call it crosstalk. The matrix on the left shows the 16 single-ended S-parameters that are associated with these two lines. The matrix on the right shows the 16 single-ended time domain parameters associated with these two lines. Each parameter on the left can be mapped directly into its corresponding parameter on the right through an inverse fast Fourier transform (IFFT). Likewise, the right hand parameters can be mapped into the left-hand parameters by a fast Fourier transform (FFT). If these two traces were routed very close together as a differential pair, then the coupling would be a desirable effect, and it would enable good common-mode rejection that provides EMI benefits.

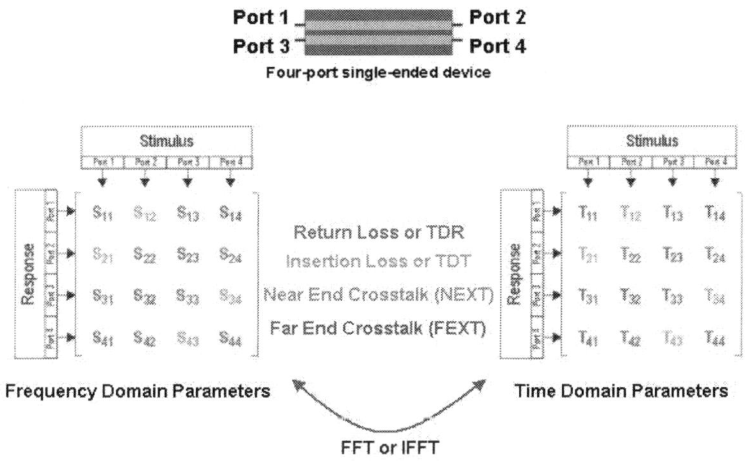

Figure 21.9

Single-Ended to Differential S-Parameters

Once the single-ended S-parameters have been measured, it is desirable to transform these to balanced S-parameters to characterize differential devices. This mathematical transformation is possible because a special condition exists when the device under test (DUT) is a linear and passive structure. Linear passive structures include PCB traces, backplanes, cables, connectors, IC packages, and other interconnects. Utilizing linear superposition theory, all of the elements in the single-ended S-parameter matrix on the left of *Figure 21.10* are processed and mapped into the differential S-parameter matrix on the right. Much insight into the performance of the differential device can be achieved through the study of this differential S-parameter matrix, including electromagnetic interference (EMI) susceptibility and EMI emissions.

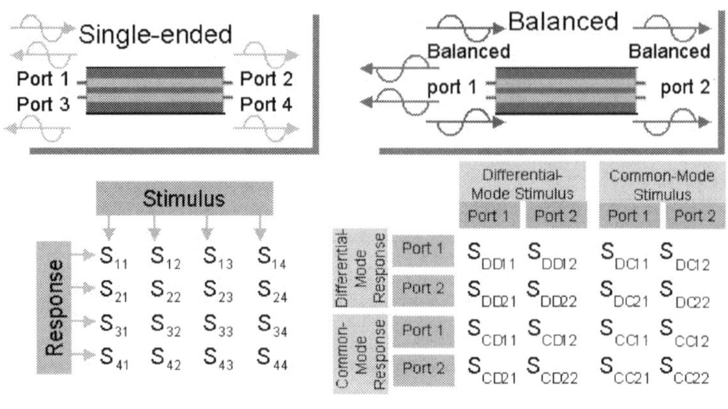

Figure 21.10

Differential S-Parameters

Interpreting the large amount of data in the 16 element differential S-parameter matrix is not trivial, so it is helpful to analyze one quadrant at a time. The first quadrant in the upper left of *Figure 21.11* is defined as the four parameters describing the differential stimulus and differential response characteristics of the DUT. This is the actual mode of operation for most high-speed differential interconnects, so it is typically the most useful quadrant that is analyzed first. It includes input differential return loss (SDD11), forward differential insertion loss (SDD21), output differential return loss (SDD22), and reverse differential insertion loss (SDD12). Note the format of the parameter notation SXYab, where S stands for scattering parameter (S-parameter), X is the response mode (differential or common), Y is the stimulus mode (differential or common), a is the output port and b is the input port. This is typical nomenclature for frequency domain S-parameters. The matrix representing the 16 time domain parameters will have similar notation, except the "S" will be replaced by a "T" (i.e., TDD11). The fourth quadrant is located in the lower right and describes the performance characteristics of the common signal propagating through the DUT. If the device is designed properly, there should be minimal mode conversion and the fourth-quadrant data is of little concern. However, if any mode conversion is present due to design flaws, then the fourth quadrant will describe how this common signal behaves. The second and third quadrants are located in the upper right and lower left of

Figure 21.11, respectively. These are also referred to as the mixed-mode quadrants. This is because they fully characterize any mode conversion occurring in the DUT, whether it is common-to-differential conversion (EMI susceptibility) or differential-to-common conversion (EMI radiation). Understanding the magnitude and location of mode conversion is very helpful when trying to optimize the design of interconnects for gigabit data throughput.

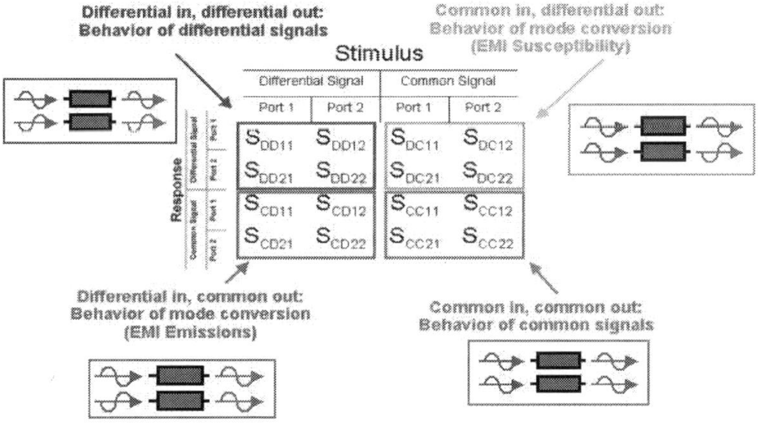

Figure 21.11

21.5 Four Port Microvia Measurements

Measurement Setup

The test equipment used in this experiment consisted of a four-port performance network analyzer (PNA) and four-channel TDR running PLTS software. Both instruments were simultaneously on the *GPIB* bus and were used to validate measurements between each other. See *Figure 21.12.*

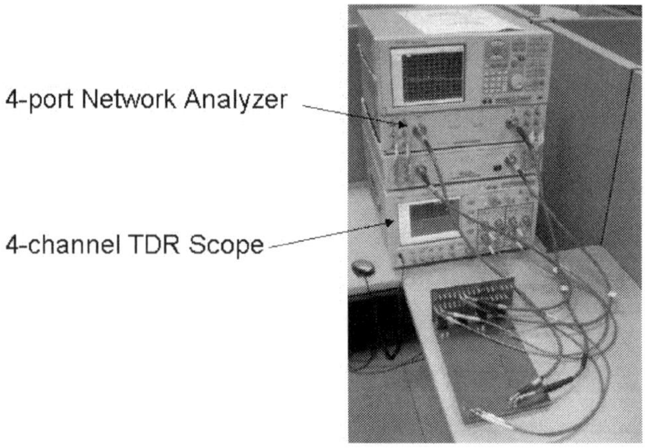

Figure 21.12

Frequency Domain Analysis

Now that we have a good understanding about four-port S-parameters, we will interpret the actual data in *Figure 21.13*. The more intuitive parameter to review first is typically differential insertion loss or SDD21. This is the frequency response seen by the differential signal as it propagates through the device. At lower frequencies (DC to 10 GHz), both vias perform nearly identical. However, the microvia structure clearly shows less attenuation of higher frequencies when compared to the standard via. This indicates a channel structure that allows higher frequencies to pass without significant degradation. This will inevitably result in an eye diagram that is more open, as will be shown shortly. The standard via, on the other hand, shows higher frequencies being attenuated more than the microvia. The second set of curves is perhaps less intuitive but equally important to analyze. The differential return loss (SDD11) indicates the magnitude of reflections occurring at various frequencies within each structure. Again, the low-frequency response is very similar for both vias. However, the magnitude of reflections in the standard via is higher than the microvia from 12 GHz to 20 GHz. Reflections are due to a poorly controlled impedance environment and the spacing between the nulls is related to the spacing of the resonant cavity within the structure. In the case of the standard via, this is related to the length of the via stub.

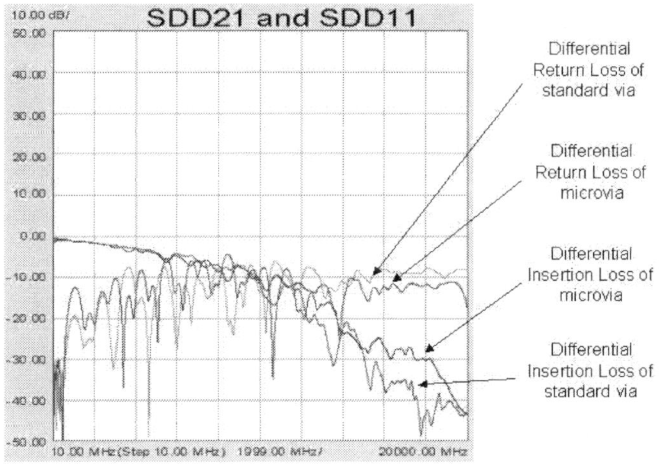

Figure 21.13

Differential Eye Diagram Analysis

The eye diagrams in *Figure 21.14* are synthesized from the four-port S-parameters. This method of creating eye diagrams correlates well with the standard method of compliance testing with a pattern generator and a sampling scope with standard masks. As can be seen, the eye diagrams for the microvia are clearly more open than the standard via, even at 20 Gbps.

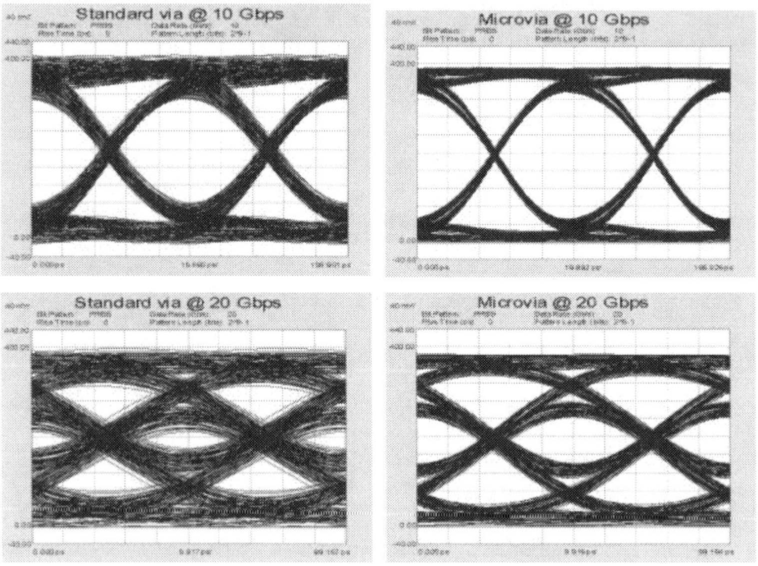

Figure 21.14

21.6 Microvia Construction

Laser Drilling for Microvia Forming

With laser drilling, a similar process to mechanical drilling is used, except the holes are formed by the ablation of material by the laser. When accessing layers beyond layer 1, a number of techniques can be used. Different laser technologies are often used to ensure that the correct features are formed. UV-YAG lasers will cut through metal layers but will not damage the organic material of the PCB. CO2 lasers will only cut through the organic material and stop when they reach a metal layer. Using these types of lasers allows precise forming of the via down to the desired layer in the PCB. Again, this physical forming requires additional real estate. CO2 lasers operate using wavelengths in the 9 to 11 μm range, which limits their ability to primarily cutting dielectrics. As a result of using CO2 lasers, a hole in the outer copper foil is needed and larger inner-layer pads are used to compensate for any image registration issues. UV lasers have a distinct advantage in the creation of microvias, as they operate at wavelengths less than 400 nm. At these wavelengths, the laser can be used to ablate a wide range of materials from metals and organic materials to glasses and inorganic materials. UV-YAG lasers are particularly advantageous given their ability to rapidly and precisely cut through multiple copper layers, as shown in *Figure 21.15*. The more precise the laser, the smaller the inner pad dimensions are required, which will further reduce routing density. As a rule of thumb, aspect ratios less than 1:1 should be maintained due to plating limitations. The sidewalls of the microvia are tapered slightly to help in the plating process and the thermal characteristics of the microvia. The ease of implementation of laser drilling into standard PCB manufacturing lines has made this methodology the most widely accepted microvia formation technique. Of boards using microvia technology, more than 90 percent of them were processed using laser drilling.

Figure 21.15

Sequential Stack-Up

Since the introduction of microvia technology there has been a rapid evolution in complexity from the single-layer laser-drilled to stacked "inline" vias, to even more complex structures. In some cases more than one via technology has been used to create a multi-layer PCB that combines superior signal integrity and efficient use of board space. Two examples are shown in *Figure 21.16*. The applications will dictate which configuration is best, as in the case of mobile devices. These devices are leading-edge, so they require using the most advanced silicon, which may be packaged in BGA configuration. Mobile devices also have to be small and cost-effective, as they tend to become outdated very quickly and must adjust rapidly to changes in the market. For these type of applications, the simple-laser drilled PCB will often provide the needed board space and desired signal performance in the smallest real estate.

For higher-density applications that require more interconnection to inner routing layers, the stacked (or inline) via approach may fit the bill. Any of these approaches can be used in conjunction with standard PTH technologies to link older legacy devices (press-fit or PTH solder devices) to newer high-density components.

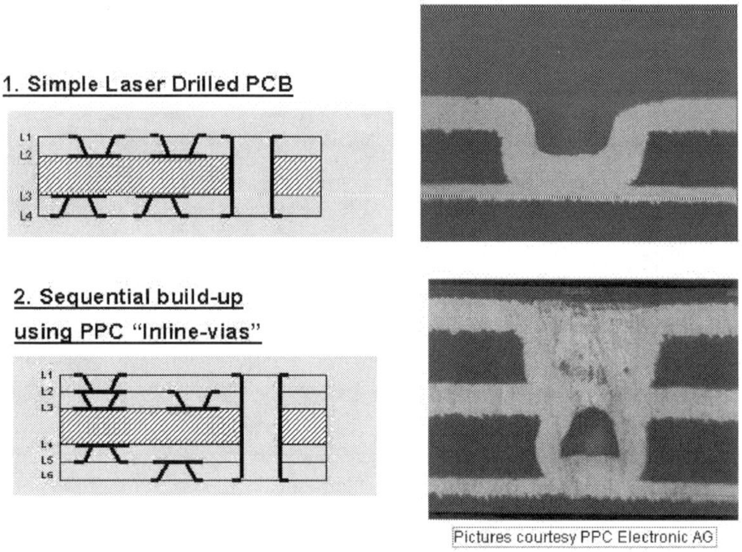

Figure 21.16

High Density Microvia Applications for Telecom

The most complex microvia applications are reserved for ultra-high–density telecom and datacom switches and routers. An example of one of these structures is shown in *Figure 21.17*. These systems require PCBs that have laser-drilled vias, stacked vias,

PTHs, and even buried vias. Sophisticated BGA ICs and field programmable gate arrays (FPGAs) are often used to perform the advanced functions of these networking devices and must have circuits routed in the most efficient manner. A sequential build-up approach to the fabrication of the PCB can reduce the via congestion by routing signals to various inner layers while still using the minimal board footprint. The benefit is seen in the reduced layer count for a microvia board in comparison to a PCB designed with standard thru vias. The use of microvias will also reduce the amount of time that is needed to route these complex devices. This is because the auto-routing functions of the CAD layout software can easily determine efficient routing schemes for microvias.

3. Sequential build-up in combination with buried vias

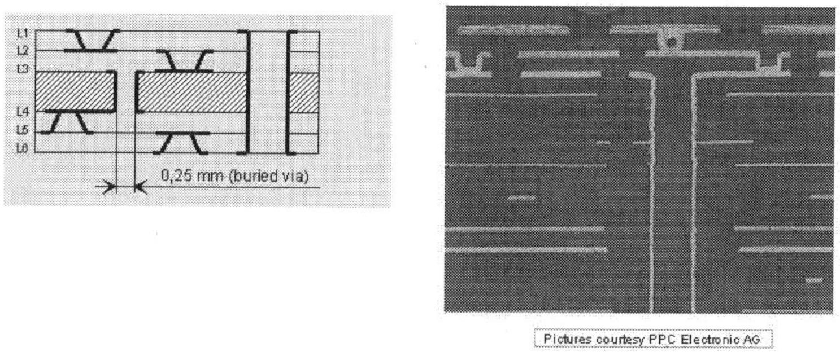

Figure 21.17

Microvia Manufacturing Process Comparison

Today's microvias can be created using a variety of processes, but these processes can be divided into two main categories: physical forming and chemical forming. With physical forming, the two most common approaches are mechanical and laser drilling. Chemical processes exist that allow board manufacturers to reduce the via size to its smallest size and also use very little board space. Two common forms of chemical via forming are photo forming and plasma etching. A tabularized summary of these processes is shown in *Figure 21.18*.

Mechanical drilling requires very little investment by the PCB suppliers since the equipment that is used to process standard PTHs can be used to create controlled-depth vias. The limitations of this approach are in the microvias' diameter and depth-control accuracy of the drill press. Additionally, mechanical drilling requires drilling through multiple layers of the PCB so additional real estate is needed. The mechanical drilling approach is like inverting the back-drilling approach for standard PTHs.

Factor	*YAG - Laser*	*CO2 - Laser*	*Mechanical drilling*	*Plasma*
Via min. diameter	1-2 mil	12 mil	8 mil	3 mil
Via max. diameter	7 mils	14 mils	none	none
Via aspect ratio	<= 1:1	<= 1:1	<= 1 :1	<= 0.5 :1
Consistant dielectric thickness	Not so important	important	Not so important	Very important
Ablation rate (5 mil vias 1.5 mil dielectric)	4500 vias/min.	8500 vias/min.	Limited by sequential drilling	All vias etched simultaniously
Via clean after ablation	No	Yes	No	Yes
Special preparation prior to plating	No	No	No	Cu overhang removal
Proprietary process	No	No	No	Dyconex, APS
Type of Equipment	UV Laser	Pulsed CO2 Laser	Standard drilling machine	Plasma gas in Vacuum

Figure 21.18

Different Microvia Routing Patterns

As circuit densities push higher and higher, it is essential to find alternative methods for interconnecting devices and alleviating via congestion. With microvia technology, the designer is capable of utilizing via in pad technology to reduce the physical space requirements of routing high-density components. This via in pad approach shown in *Figure 21.19* does not use any additional space to route the signals than the mechanical outline of the SMT pad. This is huge space savings on high-density components such as BGA devices, but it is limited to via diameters less than 4–5 mils. Larger diameter vias placed in the SMT pad could show voiding the component is soldered.

Microvias also open up the opposite side of the PCB for additional components and circuitry, since the vias do not extend through the

PCB. This extra space can be used to reduce the overall number of PCB layers or add functionality that would not be possible with conventional via approaches (see *Figure 21.19*).

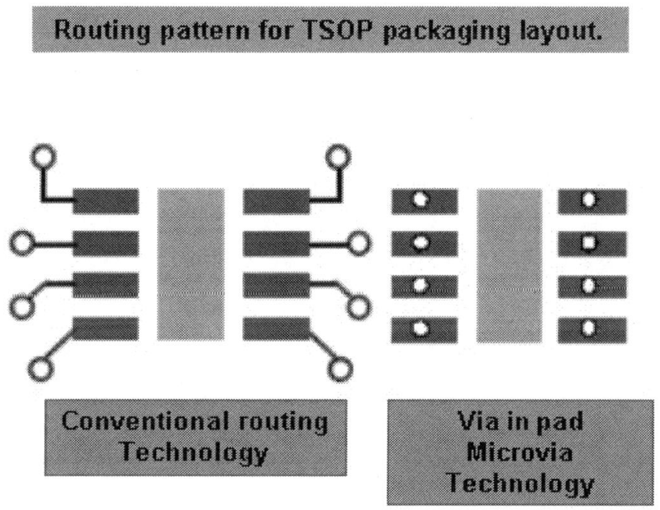

Figure 21.19

21.7 Modeling and Simulation Case Study

Three Configurations Modeled with Advanced Design Software

This section addresses the particular advantages of microvias in conjunction with SMT connectors with respect to the electrical behavior. The connector-to-board interface of the ERNI ERmet 0XT connector is taken as an experimental vehicle for this investigation. The sketch in *Figure 21.20* shows the connector-to-board interface of the connector receptacle, which is soldered to a typical board with 16 layers and an overall thickness of 4 mm. Using this realistic assumption, the following three typical configurations will be compared.

Looking at the entries in *Figure 21.20*, it can be identified that the selected signal layer is different in Setup 3. The reason is that a connection to a signal layer very close to the bottom of a PCB is typically a non-critical case because the remaining stub is very small and the diameter of the via and the antipad can be optimized to achieve good matching behavior. The critical case is the wiring from a connector to one of the upper signal layers in the PCB. Using conventional press-fit or SMT technology would lead to a via stub with a very large length. The very high capacitive load of this stub would result in a significant impedance mismatch that would reduces the overall signal quality. This is the reason why this "worst" case is not taken into account and the two solutions that are currently used to overcome this limitation are compared. One of these options is to apply back drilling to these critical vias, and the second one is to implement microvia technology.

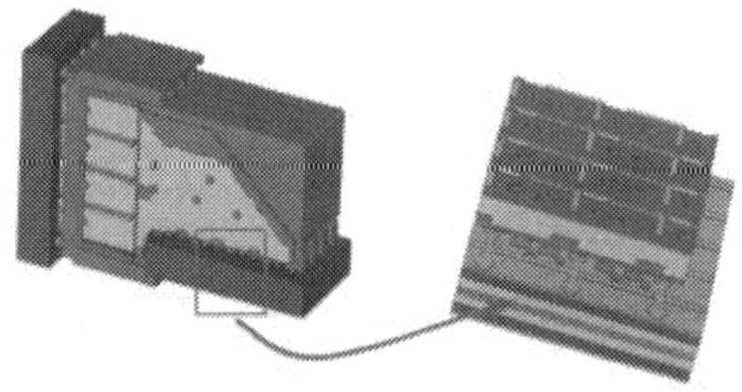

	Setup 1	Setup 2	Setup 3
Zoomed cross section	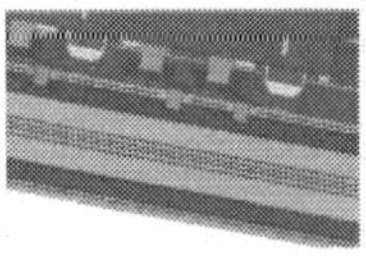	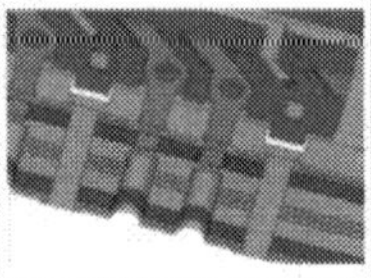	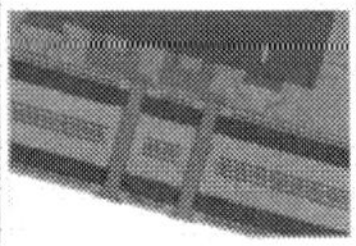
Connector pad	SMT	Pressfit	SMT
Via type	Microvia	Throughole with backdrill	Through hole
Via diameter	0.35	0.5	0.3

Figure 21.20

Characteristic Impedance and Crosstalk

The comparison is done by a simulation in the time domain using a full three-dimensional (3D) high-frequency field solver, where a step signal with a rise time of 50 ps (20 to 80 percent) is applied to the DUT. The differential signal is fed into the remaining part of the connector. Such a waveform is typical for a 10 Gbps serial transmission. As an outcome of the simulation experiment, the characteristic impedance Z0 of the system can be evaluated together with near-end crosstalk (NEXT) and far-end crosstalk

(FEXT). Looking at the impedance mismatch in *Figure 21.21*, it is evident that the microvia approach showed the smallest impact on the impedance profile compared to the other two setups. It can be seen that the crosstalk in the relatively short vias of the setups with microvias and back-drilled press-fit connections is much lower compared to that in the long thru-hole vias. In both cases (e.g., FEXT, NEXT), the microvia approach showed a better crosstalk behavior.

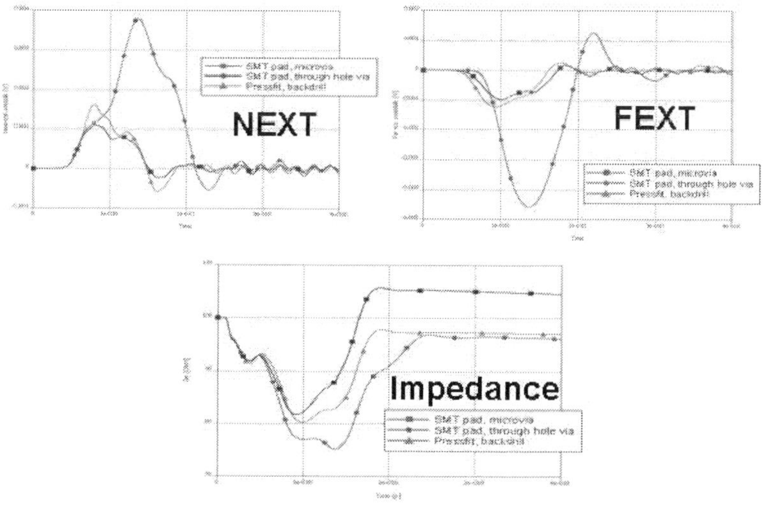

Figure 21.21

System Simulation Setup

This section discusses a comparison of the electrical performance of a serial link in two backplane systems using either microvias in conjunction with a SMT connector or thru-hole vias with a press-fit connector. In a first step, both alternative scenarios are evaluated in a circuit simulator. The original goal of this investigation is to demonstrate the influence of several design parameters, including the following:

- Distance between daughtercards
- Connector-to-board interface
- Width of transmission lines

- PCB material
- Metalization layer in the PCB
- Termination of transmission lines
- Type of transmission lines
- Statistical variations

For this special application, two typical cases will be highlighted that demonstrate the superior behavior of microvia technology and SMT connectors, compared to that of traditional press-fit connectors and thru-hole vias. The general system scenario in *Figure 21.22* is used for this simulation study.

The signal path is a point-to-point connection between two ICs on two daughtercards. The daughtercards are plugged into a backplane PCB. The connections are done using the ERNI ERmet ZD connector (press-fit pins) or the ERNI ERmet 0XT connector with SMT interface. The characteristic impedance of differential striplines is Z0diff = 100 W, and the data rate applied is 10 Gbps using a non-return-to-zero (NRZ) 8B10B code. The table in *Figure 21.22* also lists the different design parameters of the two experimental test vehicles.

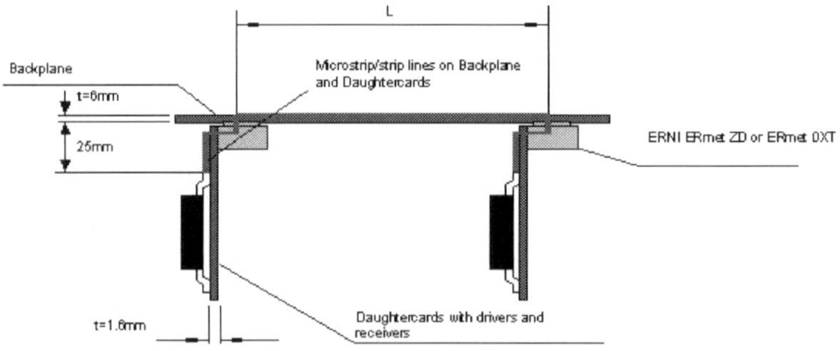

Design Parameter	Values for system with	
	micro vias / SMT connector	through hole vias / pressfit connector
PCB Material	Rogers	Rogers
Trace length backplane [mm]	380	190
Length of stub [mm]	-	0.5
Thickness daughtercard [mm]	3	3
Thickness backplane [mm]	3	6
Diameter via [mm]	0.35	0.7

Figure 21.22

Figure 21.23

Time Domain Results – Simulation versus Measured

The diagrams in *Figure 21.24* show both results from the two system setups from the measurement and the circuit simulations. The left column shows a very smooth eye diagram in both simulation and measurement of the system with microvia connections. In contrast, the right column contains the corresponding diagrams of the system with press-fit connectors and thru-hole vias. Here, we can see a heavily distorted eye with a reduced eye height compared to the results of the microvia system. The overall system behavior in the case of microvia interfaces is much smoother compared to the strong reflections in the case of PTH connections. The eye diagrams for the microvia are more open, indicating better performance at higher data rates. In both cases, a very good agreement between the simulated and measured behavior of the overall system behavior can be seen in these diagrams.

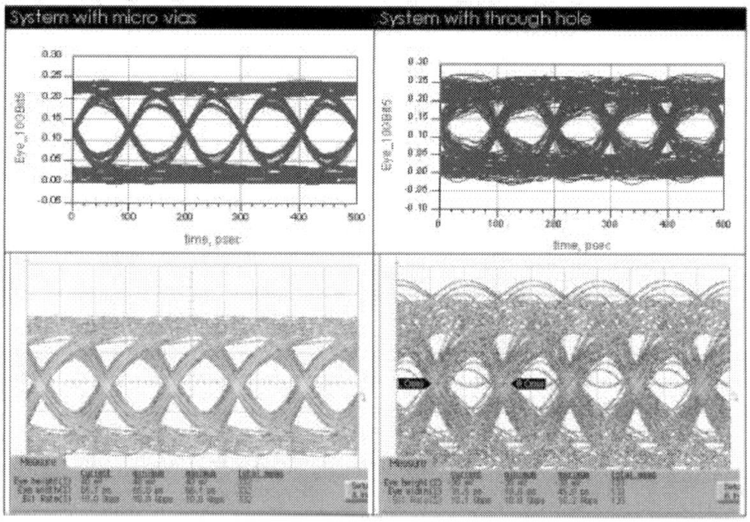

Figure 21.24

21.8 Summary and Conclusion

With data rates approaching 10 Gbps, small geometry changes in components and attachment features will become more evident. If care is taken in both the design of the SMT connector and the board-to-connector interface, very-high–speed serial links can be achieved. Understanding, testing, and designing these features to be as quiet as possible is essential for optimizing the complete signal path. This paper has examined the frequency domain effects that limit the performance of backplane structures and used intuitive time domain analysis such as the eye diagram and differential impedance profile. A design case study using 3D field solver simulations of various microvia technologies and implementations allowed for a thorough investigation of attachment performance prior to board layout. A complete signal path has been created in a simulation environment by using models exported from the 3D field solver in conjunction with other known models of connectors and traces. A four-port measurement tool environment has been used to characterize differential insertion loss, differential return loss, and differential eye diagrams in order to optimize the signal integrity of the microvia structures.

Authors

Mike Resso, Signal Integrity Application Scientist, Component Test Division, Agilent Technologies

Thomas Gneiting, Founder, AdMOS Advanced Modeling Solutions

Roland Mödinger, Senior Engineer, ERNI Electroapparate GmbH

Jason Roe, Application Engineer, ERNI Electroapparate GmbH

Chapter 22

ATE Interconnect Performance to 43 Gbps Using Advanced PCB Materials

22.1 Abstract

Printed circuit board (PCB) materials directly influence attenuation and near-end crosstalk (NEXT)/far-end crosstalk (FEXT) signal integrity of an automated test equipment (ATE) loadboard design. Balancing performance, cost, and ease of fabrication requires a quantitative understanding of the impact that the dielectric material has on the performance of a multi-gigabit loadboard signal path. An understanding of how the material will perform when used to fabricate 20+/-layer count ATE boards with thicknesses of more than 200 mm is required. This paper provides an analysis of various loadboard PCB materials for a nominal loadboard test fixture design with 25 cm (10 inches) of path length and data rates up to 43 Gbps.

22.2 Introduction

The ongoing explosion in device pin count and input/output (I/O) data rates that the semiconductor industry is going through creates significant challenges for test engineers working to characterize and verify these devices. The complexity of modern I/O cells and semiconductor manufacturing processes combined with short development and debug cycles pushes semiconductor manufacturers leading this wave to utilize ATE for thorough and precise device characterization. Previous methods of customized bench instrumentation are running into problems with the complexity and the need for short development cycles. The ATE system makes it possible not only to characterize multiple I/O cells running concurrently, but also to gather statistical data over several device lots.

The application of ATE systems to device characterization does have its challenges with respect to the signal integrity of the multi-gigabit I/Os. *Figure 22.1* shows that there is a significant distance between the ATE pin electronics and the device under test (DUT), which can easily degrade the signals of interest.

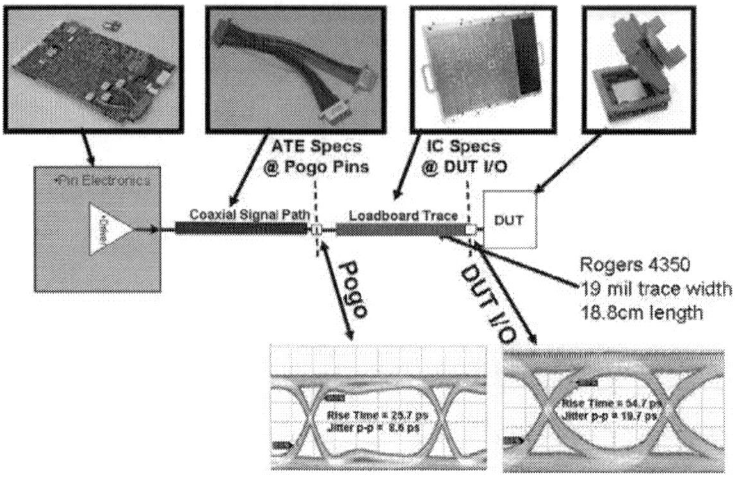

Figure 22.1: *The ATE Signal Integrity Challenge*

Each of the items in *Figure 22.1* must be optimized and characterized so that the measurement instrumentation on the ATE system accurately measures the real performance of the DUT [1, 2]. One of the most problematic items is the signal loss in the ATE test fixture also known as device interface board (DIB) or DUT loadboard. Due to the high pin count of modern SOCs with hundreds of multi-gigabit I/O cells, ATE test fixtures can be very large with signal traces in the 24–50 cm length range (depending on the ATE platform and application). *Figure 22.2* shows an ATE test fixture for a device with multiple I/O cells in the 6 Gbps range docked to an ATE system.

Figure 22.2: *Example of an ATE Test Fixture for Multi-Gigabit Graphics Device (Left), Using Coaxial Cables to Try to Avoid the PCB Signal Trace Loss (Right)*

The loss in PCBs can be divided into different factors as shown in *Figure 22.3*. The two dominant factors for an ATE test fixture running at multi-gigabit data rates are skin effect and dielectric losses due to the length of the signal traces involved. The larger of these two is typically the skin effect, which can be minimized by increasing the trace width of the controlled impedance transmission line [3]. Increasing the trace width has the disadvantage of also requiring a larger dielectric thickness to maintain the controlled impedance, and this is not always an option for high-layer-count ATE test fixtures that are already at the maximum height for the PCB fabrication process. The other option is to improve the dielectric loss by using specialized PCB materials with lower-loss tangent values and lower dielectric constants. Lowering the dielectric constant of the PCB material also has the advantage of increasing the trace width of the controlled impedance transmission line (lowering skin effect losses) for the same height in dielectric materials. This double benefit of improving losses by lowering the dielectric constant of the PCB material makes it worthwhile to investigate this further.

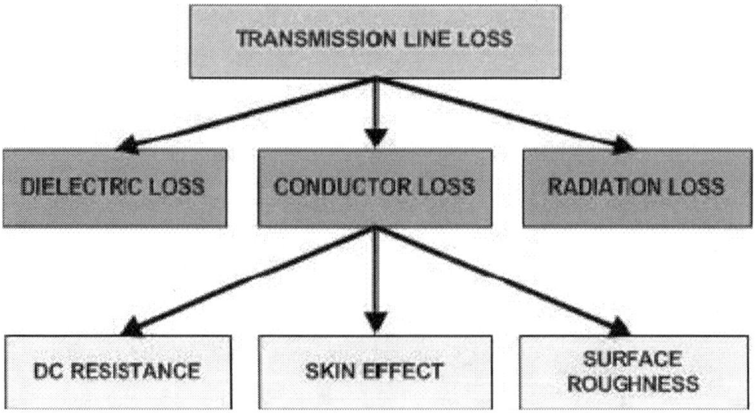

Figure 22.3: *Components of the Loss from a PCB Signal Trace.*

In this paper, we will concentrate on the dielectric loss for ATE test fixtures by comparing several dielectric materials available in the industry, including new advancements in Teflon-based dielectric materials for multilayer applications. We will start by an introduction to dielectric materials and the challenges they present followed by the latest in low dielectric constant materials for multilayer PCBs. We will then present several experimental results at 10, 20, and 43 Gbps with an ATE test fixture and show how equalization addresses the remaining test fixture loss challenge after an appropriate choice of dielectric material and trace geometry. The paper will conclude with an evaluation of the benefits of a lower dielectric constant PCB material for a real PCB stack-up of a high-density ATE test fixture. The results will clearly show the ability of PCB materials to extend the data-rate range of ATE test fixtures beyond 10 Gbps and improve the performance of existing high-density designs.

22.3 Dielectric Materials for ATE Test Fixtures

ATE test fixture PCBs, including probe cards for wafer sort or test fixtures with sockets for packaged semiconductor device testing, tend to be thick (200–300 mil) multilayer PCBs that can include as many as 56 layers of circuitry. The size can be as large as 22 inches by 17 inches (55 cm by 43 cm). This is similar to large-format telecom backpanels with regard to length, width, and thickness, and they face the same signal integrity challenges that these characteristics present. The multiple layers of an ATE test fixture accommodate a variety of routing needs, including power planes for low impedance, separation of analog and digital circuitry, and high-speed controlled impedance. A multilayer PCB is typically limited to around 250 mm in thickness due to manufacturing restrictions. As PCB thicknesses increase beyond 250 mm the PCB manufacturer is faced with the challenge of layer-to-layer registration and the plating difficulties of higher aspect ratio plated thru-holes while trying to maintain the same feature tolerances as a thinner board. These problems are exacerbated as designers try to squeeze more and more transmission lines onto a PCB by using smaller drilled holes, smaller pads, and smaller spaces between signals.

Adding additional signals to a design without increasing the total PCB thickness requires that a designer either reduce the thickness of the individual dielectric layers or increase the density of routing on an individual layer by reducing the width of the existing copper traces, the diameters of the drilled holes, and the corresponding pads. Squeezing additional transmission lines onto the same dielectric layers ultimately leads to a more complex PCB to design and manufacture with tighter tolerances on layer-to-layer registration and drilling accuracy. The drive for densification has led to some creative fabrication techniques such as flip drilling on a PCB, where half the hole is drilled from the top and the other half from the bottom in order to maintain the required drill registration of a via.

Reducing the layer thickness while using the same PCB material or increasing the routing density on a layer to accommodate more signals both have the adverse effect of reducing trace width and increasing signal losses. An alternate solution is to reduce the layer thickness and at the same time reduce the dielectric constant of the dielectric material so that the trace width can be maintained for a given impedance. If a design is converted from a PCB material having a dielectric constant of 3.5 to a dielectric constant of 2.8, the resulting PCB will be 20–30 percent thinner, allowing more room for additional layers while maintaining the existing trace widths and avoiding the increased copper losses of a thinner trace.

The selection of the appropriate material for an ATE test fixture requires the evaluation of cost, performance, and manufacturability to determine what is best for a given application [4, 5]. A variety of dielectric materials are available to test engineers designing ATE test fixtures for high-speed digital applications. *Table 1* shows a comparison of dielectric materials, including a rough cost comparison with FR4. Note that in a high-speed digital test fixture, the cost that the dielectric material represents can vary from approximately 25 to 40 percent of the total PCB cost, depending on the design specifics and the number of boards being manufactured.

Material	**εR**	**Tanδ (1 GHz)**	**Tanδ (10 GHz)**	**Cost**
FR4	4.4	0.018	N/A	1
Nelco 4000-13 SI	3.4	0.008	0.008	1.5
Arlon 25FR and 25N	3.38	N/A	0.0025	1.75
Rogers 4003	3.38	0.0027	0.0027	2
Rogers 4350	3.5	0.0031	0.0037	2
Teflon Glass	2.4	N/A	0.0014	2
Teflon Ceramic Filled	2.98	0.004	0.0025	2
Speedboard C	2.6	0.004	0.004	2
FastRise 27	2.7	0.0020	0.0020	2
TSM29	2.94	0.0012	0.0014	2

Table 22.1: *List of Typical Dielectric Materials Used for Multi-Gigabit ATE Test Fixtures (FR4 Included for Comparison)*

The electrical performance of the laminate material will depend on three key variables: the dielectric material, the reinforcement (e.g., style of fiberglass), and the copper surface roughness. The dielectric material has a loss factor called the "loss tangent" or tanδ value, which relates to the polarization of the atoms when subjected to changing electrical and magnetic fields. The ideal material does not interact with a propagating electrical signal, and no energy is lost as heat. FR4 is an epoxy thermosetting material that has a lot of non-reacted polarizable atoms following lamination. FR4 is typically reinforced by fiberglass and has a fiberglass content from 40-60 wt%. Nelco 4000-13SI differs from standard FR4 in the following ways: the dielectric material is a blend of epoxy and cyanate ester (cyanate ester is an organic material that, when cured, yields a lower loss than epoxy but is typically blended with epoxy because the pure cyanate ester is quite brittle), and it uses the Nittobo's NE fiberglass imported from Japan. Fiberglass normally contains various metal oxides such as silicon dioxide, calcium oxide, and magnesium oxide. The NE glass uses less calcium and additional boron such that the fiberglass has a reduced dielectric constant (4.4 versus 6.4 of the E glass used in FR4) and a reduced dissipation factor (0.0035 versus 0.0067 at 10 GHz). The 25FR, 25N, 4003, and 4350 are all dielectric materials based on polybutadiene, chemically not very different from natural rubber. The 25FR and the 4350 are flame-retardant versions of silica-filled rubber, whereas the 4003 and 25N has no flame retardant. The silica-filled rubber is typically reinforced with a standard E-glass type 1080 or 106 fiberglass. Simple polytetrafluoroethylene (PTFE) laminates consist of Teflon and woven fiberglass. These materials tend to suffer from PCB fabrication problems of poor drilling and too much material movement, which can cause significant problems on a high–layer-count ATE test fixture. Ceramic-filled PTFE composites are a dramatic improvement relative to simple PTFE–fiberglass composites. PTFE laminates with a high loading of micro-dispersed ceramic yield a much higher-quality drilled hole and have less z-axis expansion. For these reasons, ceramic-filled PTFE laminates are a staple of military designs. TSM29 is a ceramic-filled PTFE composite containing a very low glass content (9 wt%).

Speedboard C is based on a PTFE film that has been stretched in the x and y directions, causing the PTFE to fibrillate and form a porous spider web–like structure. The non-reinforced Web is then impregnated with ceramic-filled organic thermosetting resins. Its biggest advantage is that it has a high degree of flow to fill the gaps between copper features. However, this high degree of flow can also be a disadvantage for high–layer-count ATE test fixtures where the thick PCB requires registration over many layers and predictable movement of the core materials and prepregs is required.

The choice of reinforcement has a lot of implications on the electrical and mechanical properties of the composite. Various authors have published papers in the last few years describing the effects of fiberglass on delay differences or skew between same-length traces on a PCB [6, 7]. Fiberglass in the woven state (*Figure 22.4*) can have some degree of weave distortion before impregnation.

Figure 22.4: *Fiber Glass in Woven State*

The basic challenge with fiberglass is that its dielectric constant of 6.4 is a poor match to the 3.2–3.8 permitivity for an epoxy or 2.1 for PTFE. The result is that the propagation velocity along a copper transmission line will speed up or slow down depending on how it is routed over the underlying fiberglass. In a differential environment, the variation in signal speed between the coupled lines leads to skew, which can overwhelm the other sources of loss depending on frequency. One author suggests using a tremendous amount of fiberglass such that there are no windows. This will eliminate the dramatic impedance fluctuations when measured with a TDR, but that solution suggests using composites with a high density of very lossy fiberglass. For lossy dielectric materials such as FR4, a high fiberglass content can actually lower the overall dissipation factor. For high-performance materials such as PTFE, introducing a high content of lossy fiberglass is a very unattractive option. Flat fiberglass weaves are fiberglass structures where the fill yarns have been spread out to look flat and close the windows. However, the warp yarns are preserved as tight rods, so it is a partial solution, and moreover, one is still left with lossy fiberglass.

Elimination of fiberglass reduces the intra-pair electrical skew associated with fiberglass, and it also eliminates a very lossy component (0.0067 tanδ at 10 GHz). Discussion of using nonreinforced materials has to be divided into two topics, as a prepreg material and as a core material. Manufacturing non-reinforced materials puts a greater burden on the laminate supplier because non-reinforced materials have little mechanical strength. Careful attention must be paid to Web handling equipment so as not to stretch, wrinkle, or distort the composites. At the PCB fabricator the non-reinforced core material is difficult to handle. Depending on the thickness of the nonreinforced core, handling non-woven material from some vendors can be the equivalent of handling chewing gum. The PCB fabricator has the added complexity of print and etching a delicate laminate and maintaining registration during lamination. The non-reinforced prepreg, on the other hand, is relatively easy for the fabricator to handle, as it only has one process step where it is interleaved between core materials prior to lamination. The authors at this time have only investigated the combination of a non-reinforced prepreg with reinforced core material.

The choice of copper for a laminate is a balance between copper adhesion for mechanical strength and copper conductive losses for electrical performance. *Figure 22.5* shows the microstrip insertion loss of a laminate measured with various copper types.

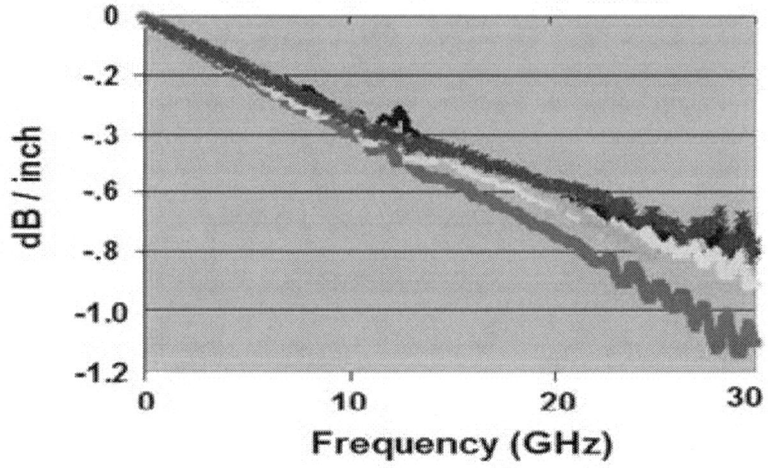

Figure 22.5: *Insertion Loss of a Laminate Measured with Various Copper Types for a Dielectric Thickness of 16 mm (DT) and a 37 mm Trace Width (TW)*

The worst performing copper has a surface roughness of 5.6 microns (Rz). Generally speaking, the lighter the copper weight, the less surface roughness. Reverse-treated half-ounce copper, for example, is a better choice than reverse-treated 1 ounce. *Figure 22.6* shows a photomicrograph of the surface roughness of the copper used on the Rogers 4350B series of laminates.

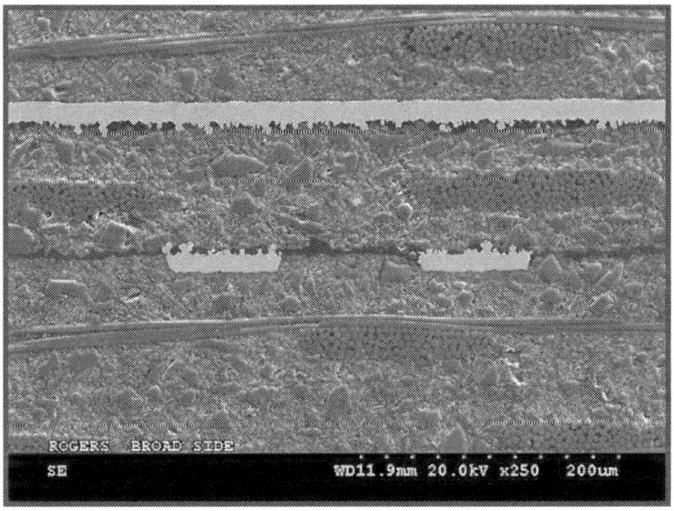

Figure 22.6: *Photomicrograph of the Copper Surface Roughness Used in Rogers 4350 Series of Laminates*

Outer-layer copper adhesion is important to prevent the surface traces from losing adhesion to the laminate. On the inner layers, designers have to consider the copper losses due to the rough copper treatment from the material supplier on one side and, on the reverse side, the additional inner-layer copper treatment that a fabricator will add to copper to insure layer-to-layer adhesion, known in the art as a metal oxide or alternative oxide treatment.

The variety of materials with their different chemical and structural compositions, along with manufacturability and cost tradeoffs, provides for significant customization depending on the application, and no one material seems to meet all of the ATE test fixture applications. R4350 and Nelco 4000-13SI are two commonly used materials that have been an enabling technology for high-density, high–layer-count, multi-gigabit I/O ATE test fixtures. However, as I/O counts and their speeds increase, even these high-end materials are struggling to keep up with the requirements of the ATE test fixture and new materials are being investigated.

22.4 The Taconic Fast-Rise Dielectric Materials

To address the needs of the high-speed and high-frequency world of electronics, one of the new PCB materials being developed that is of interest for ATE test fixtures is a new Taconic FastRise 27 Teflon based prepreg material. The FastRise 27 prepreg is a nonreinforced 2.7 DK prepreg designed to eliminate intra-pair skew between coupled differential traces. One potential disadvantage of using a nonreinforced prepreg is the possibility of excessive flow causing layer-to-layer misregistration. FastRise 27 consists of a film that is coated with a low-temperature thermosetting adhesive for multilayer lamination, as described in [8]. *Figure 22.7* left shows a photomicrograph of FastRise 27 between two black FR4 inner-layer cores for contrast. The white continuous film is incapable of flow and therefore maintains a relatively flat plane during lamination. A low-temperature thermosetting adhesive is coated onto the surface of the film to flow and fill the artwork and bond the inner-layer cores together. Because most of the mass is a non-flowing, non-melting film, the composite maintains good registration over many layers.

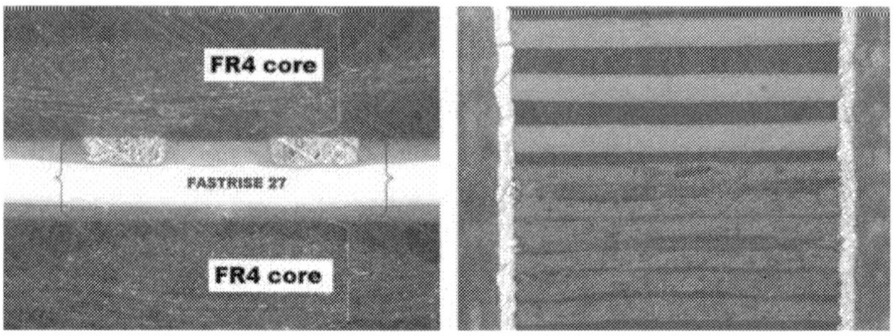

Figure 22.7: *FastRise Dielectric between Two FR4 Inner Layers (Left) and Cross-Section of a 200 mm Thick PCB between Two Plated Thru-Holes (Right)*

Figure 22.7 (right) is a cross-section of a portion of a 200 mm thick PCB between two plated thru-holes where multiple plies of the FastRise 27 have been combined with a standard ceramic-filled PTFE core. The obvious goal is to create a homogenous dielectric environment regardless of where circuit traces are located. The planar FastRise consists of a film having a dielectric constant of 2.6 and an adhesive material having a dielectric constant of 2.8 such that the film and adhesives are well matched with regards to permittivity. The primary benefits to the designer are a low 2.7 dielectric constant, allowing the designer to reduce dielectric thicknesses while maintaining trace widths; a very homogeneous dielectric material, eliminating skew variations; and the lowest-loss thermosetting prepreg commercially available.

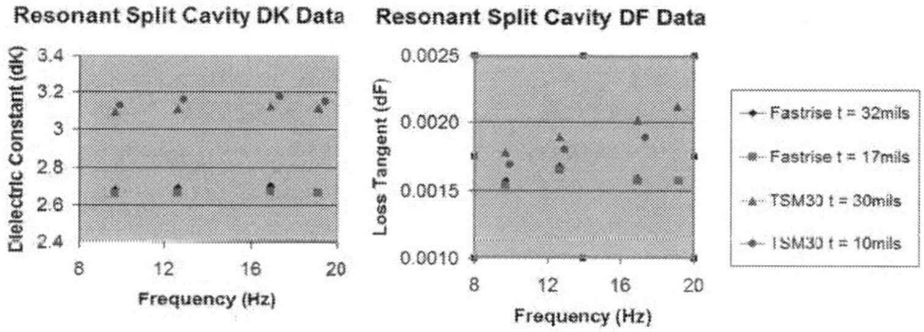

Figure 22.8: *Dielectric Constant and Loss Tangent for FastRise 27 prepreg and TSM 30 Core PTFE–Based PCB Materials. The TSM30 Is a Similar Composition to the TSM29 Used in the ATE UAB Test Fixture, Just a Slightly Higher Dielectric Constant of ~3.0 versus ~2.9 (Data Was Measured with Agilent Technologies' 85072A Split-Cavity Resonator Using IPC TM-650 2.5.5.13 Test Method)*

Independent measurements of this material using Agilent Technologies 85072A split-cavity resonator test method [9] yield data in agreement with the manufacturer's data sheets for the FastRise 27 and one of the TSM30 ceramic-filled PTFE core. The data shows minimal variations in the 8 to 20 GHz region and that the material properties do vary slightly with material thickness shown in *Figure 22.8*. Dielectric material measurements are critical for maintaining tight control on the product supplied. Final losses seen in an ATE test fixture are composed of copper losses and the final laminated material combination and are best measured in-situ on the real test fixture or with a test coupon fabricated on the same panel.

22.5 Experimental Results

To compare the performance of different dielectric materials for an ATE application, it is important to use a test fixture that represents a typical ATE application test fixture. Verigy developed the PinScale HX universal access board (UAB) that contains stripline traces on four inner PCB layers connecting on the ATE side to pogo vias at the location for the Verigy 12.8 Gbps PinScale HX card and on the other side to a surface-mounted SMA connector. The SMA connector has an SMT signal pin that connects at the top of the PCB for improved electrical performance and four ground legs that go through the PCB for mechanical strength. The signal traces on the UAB board are 25 cm (10 inches) long with a 19 mm trace width. A typical length for a medium ATE application on the Verigy V93000 platform is 25 cm (10 inches). Note that the pogo via and the via at the SMA connector have also been optimized for maximum performance through the correct placement of ground vias and back-drilling techniques [1, 10]. *Figure 22.9* shows a picture of the UAB test fixture and the time domain measurement setup.

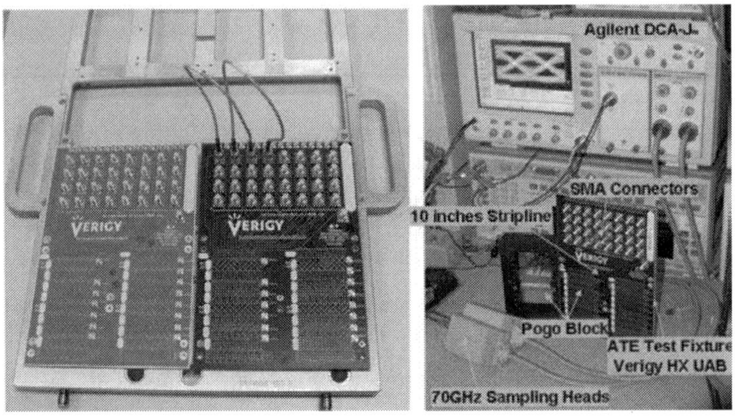

Figure 22.9: *The Verigy UAB ATE Test Fixture Implemented in Two Dielectric Materials (Right) and the Time Domain Measurement Setup with a 43 Gbps Data Source at Agilent Technologies*

Figure 22.10 shows a comparison of the insertion loss per inch for different dielectric materials obtained through measurements and simulations. The first important point is that the simulated results are more optimistic than the measured data. This is expected since the simulations were based on the specifications from the dielectric manufacturers that do not reflect manufacturing effects and the combined properties of the core and prepreg. Also, the model does not take into account the copper roughness that might be different from material to material due to manufacturing requirements.

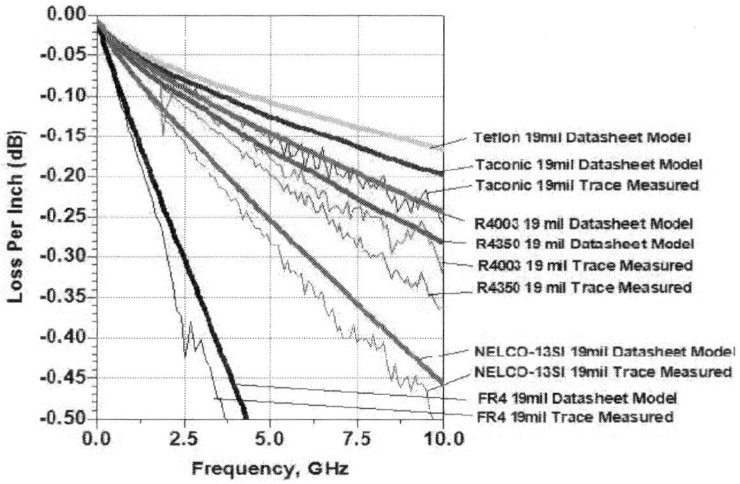

Figure 22.10: *Comparison of Simulated and Measured Insertion Loss for Different Dielectric Materials Using a 19 mm Wide Stripline*

This dependence on the fabricated laminate for the trace loss can also vary among fabrication vendors, since the details of surface treatments and lamination processes are often proprietary and not directly described on the PCB fabrication drawing. This should always be checked when changing fabrication vendors on high-speed digital designs. The data for two particular vendors using R4350 agreed quite well for the material losses but showed significant differences in the performance of the via transitions, which affects losses above 10 GHz (see *Figure 22.11*). Comparing the stripline loss performance from multiple layers on the same board from one vendor indicates that the via transition has some variations at higher frequencies, but these variations are not as large as that seen between vendor A and vendor B (see *Figure 22.11*). This demonstrates how critical the connector and via topology are for transitioning into and out of an ATE test-fixture PCB if one wants to achieve the full benefit of lower-loss materials at higher data rates.

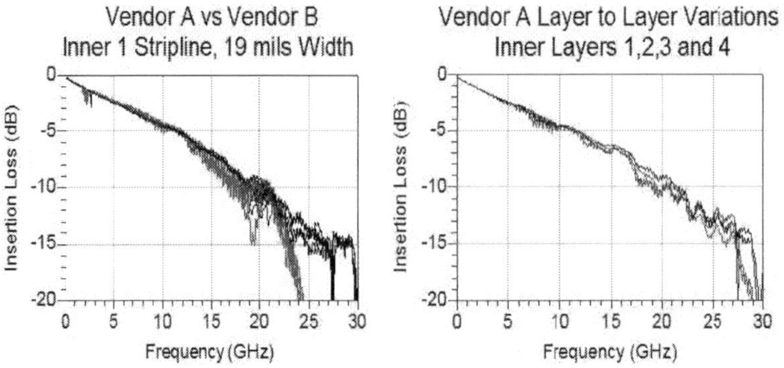

Figure 22.11: *Measured Data for the ATE HX Universal Access Board Comparing the 10-Inch Stripline Loss for Two Vendors (Left) and for Four Layers in One Vendor's Board (Right)*

One can also look at the quality of the transmission-line impedance as a function of distance along the trace using time domain analysis, as shown in *Figure 22.12*. The data for four board materials from three vendors shows that 50 Ohm impedances with +/-5% tolerances are reasonable to expect for the higher-performance materials.

To compare the time domain performance on the UAB test fixture manufactured with the Rogers 4350 and Taconic FastRise dielectric, a 6.5 ps 10–90 rise-time driver was used as a time domain stimulus source for data eye measurements up to 43 Gbps. The data eyes were measured using an Agilent Technologies DCA-J with a precision time base and a 70 GHz remote sampling head. *Figure 22.13* shows the performance of the stimulus source measured at its output with a PRBS31 data pattern at 10 Gbps and 43 Gbps.

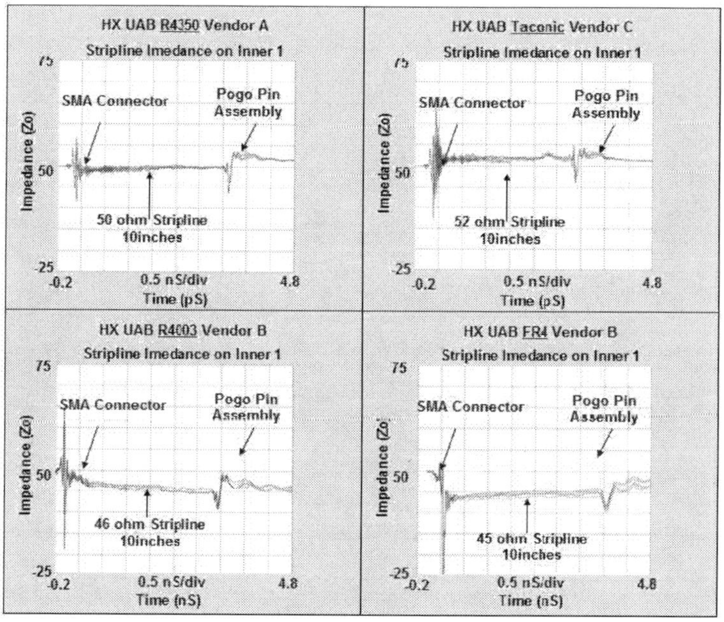

Figure 22.12: *Repeatability of Trace Impedance on Inner Layer 1 for Six Signal Lines; Comparison of Four PCB Materials from Three Fabrication Vendors*

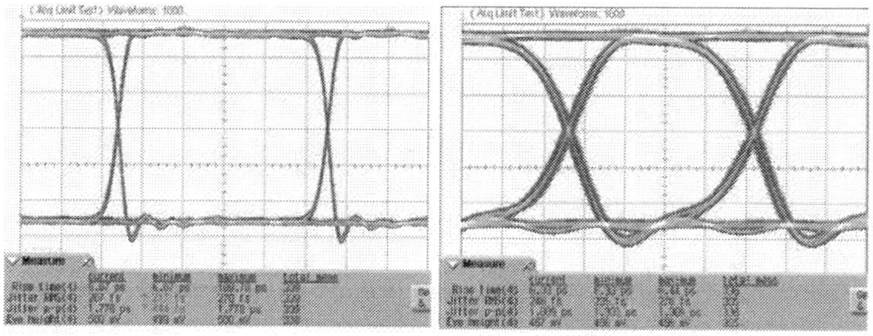

Figure 22.13: *Stimulus Source Performance (Left: 10 Gbps; Right: 43 Gbps) with a PRBS31 Data Pattern*

Figure 22.14 shows the measured data eyes obtained with the UAB board in Rogers 4350/4450B and Taconic FastRise27/TSM29 at 10, 20, 30, and 43 Gbps. Note that in both boards, the trace width is the same (to keep skin effect losses the same) and all measurements do include an ATE pogo assembly with 5 cm coaxial cable on the pogo side. From the measured data, it is possible to see that the Taconic-based board does have a higher performance, with the difference being more significant at 20 and 30 Gbps.

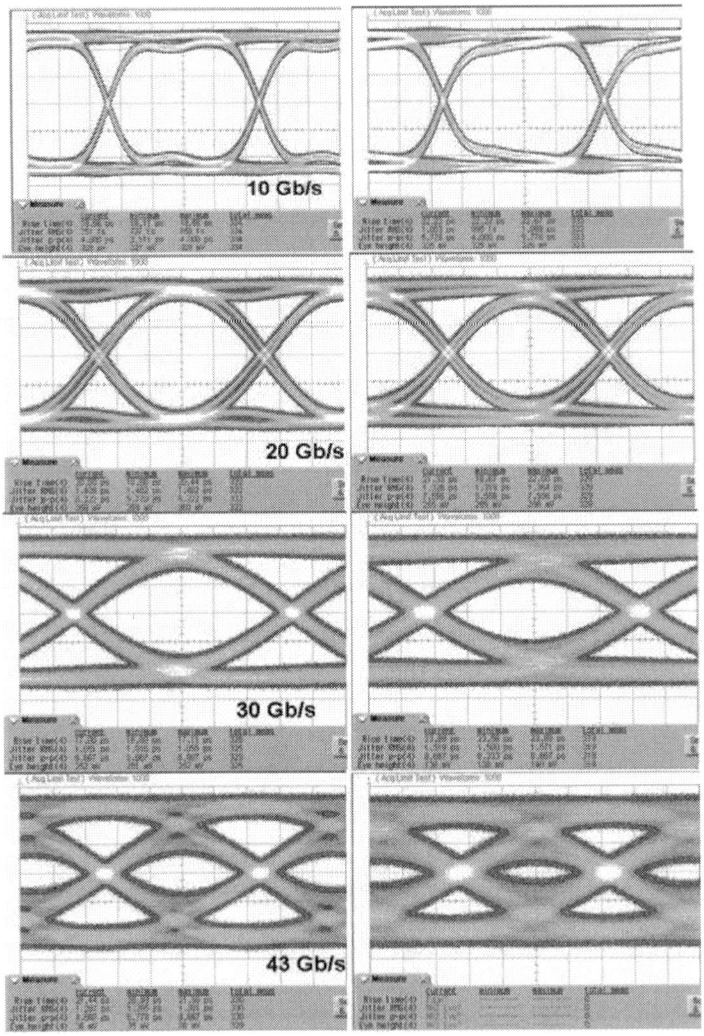

Figure 22.14: *Comparison between Rogers 4350 (Right) and Taconic FastRise (Left) for a PRBS 31 Data Pattern*

This can again be seen in *Figure 22.14*, where the measured data eye height comparison is displayed in a graph.

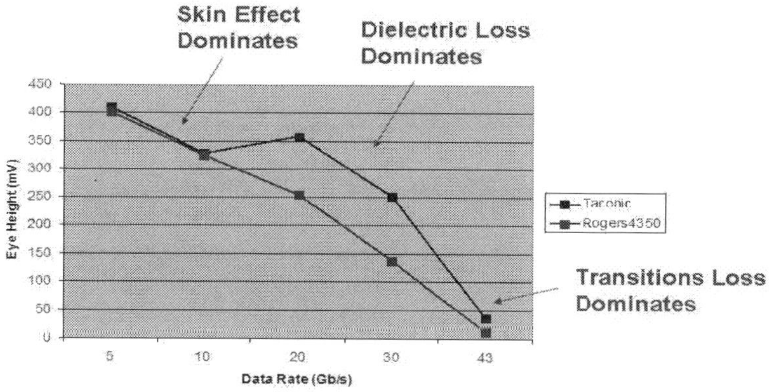

Figure 22.15: *Comparison of the Data Eye Height at Different Data Rates between the UAB Test Fixture in Rogers 4350 and Taconic Fast Rise for a PRBS31 Data Pattern*

Figure 22.15 shows that for lower data rates, the loss is mainly dominated by the skin effect. For higher data rates, the dielectric loss becomes a more significant contributor to performance and the dielectric loss difference between the Taconic and Rogers materials becomes a factor. At 43 Gbps, the transitions on the board become the main performance bottleneck and the eye amplitude is dropping rapidly. Remember that all measurements include the ATE pogo assembly and the SMA connector. Here the Taconic material still has a measurable eye opening, while the R4350 is almost closed.

22.6 Equalization to the Rescue

To compare the performance of different dielectric materials for an ATE application, it is important to use a test fixture that represents a typical ATE application test fixture. Verigy developed the PinScale HX universal access board (UAB) that contains stripline traces on four inner PCB layers connecting on the ATE side to pogo vias at the location for the Verigy 12.8 Gbps PinScale HX card and on the other side to a surface-mounted SMA connector.

The SMA connector has an SMT signal pin that connects at the top of the PCB for improved electrical performance and four ground legs that go through the PCB for mechanical strength. The signal traces on the UAB board are 25 cm (10 inches) long with a 19 mm trace width. A typical length for a medium ATE application on the Verigy V93000 platform is 25 cm (10 inches). Note that the pogo via and the via at the SMA connector have also been optimized for maximum performance through the correct placement of ground vias and back-drilling techniques [1, 10]. *Figure 22.16* shows a picture of the UAB test fixture and the time domain measurement setup.

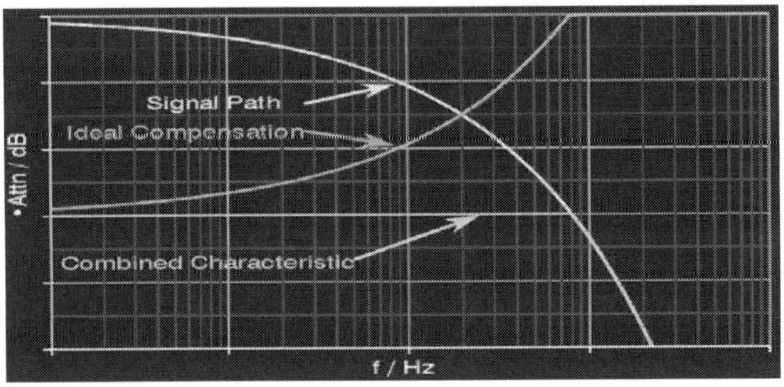

Figure 22.16: *Basics of Signal Path Loss Equalization through a Passive Equalization Filter*

The objective of equalization is to remove the frequency-dependent loss that is inherent to any real PCB. To demonstrate the possible improvements that equalization can provide, we used the Verigy V93000 PinScale HX ATE pin electronics card that includes integrated equalization that is able to achieve a 12.8 Gbps data rate. The objective is to compare the improvements provided by equalization in conjunction with different materials (the equalizer on the Verigy PinScale HX card is a passive equalizer with a fixed response tailored for a typical text fixture). *Figure 12.17* shows the measured power spectrum at the driver output on a special bench setup. One curve is at the end of the test fixture with equalization, and another curve is at the end of the test fixture without equalization for the Rogers 4350 dielectric materials. From *Figure 12.17,* it is possible to observe that the equalization compensates for the low-pass effect that the test-fixture creates on the data signal spectrum. In the presented case, the equalizer is able to get the data spectrum closer to the original one measured at the driver output.

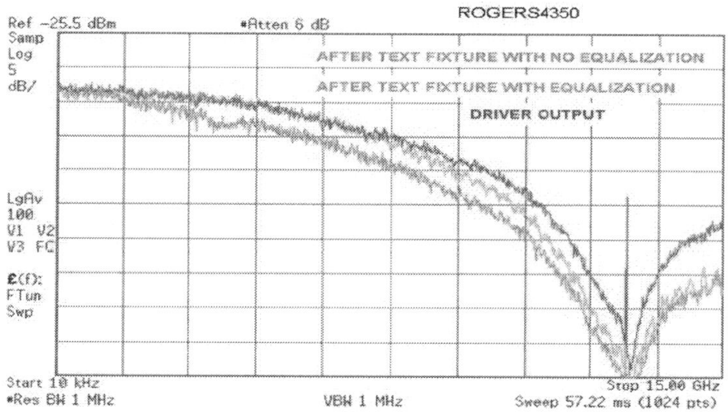

Figure 22.17: *Power Spectrum of a 12.8 Gbps PRBS31 Data Signal at the Driver Output and Measured after 10-inches of a 19 mm Stripline in Rogers 4350 Dielectric Material with and without Equalization*

This same phenomenon can also be seen in *Figure 22.18*, which shows the measured data eyes at 12.8 Gbps with and without equalization. Cleary for data rates above 10 Gbps, including 43 Gbps, the same approach has to be taken and equalization must be used to compensate for the trace loss.

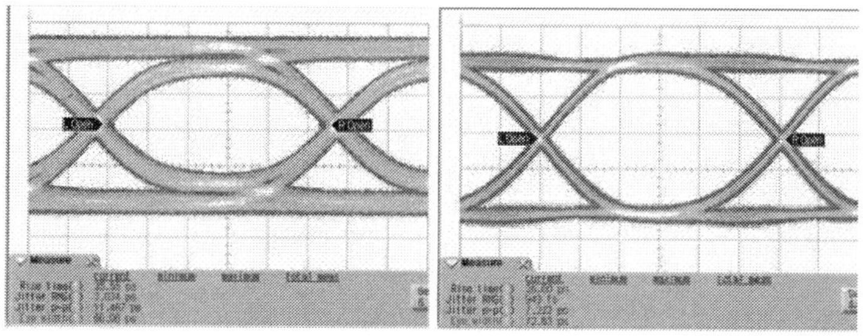

Figure 22.18: *Comparison of the Data Eye at the Output of the ATE Test Fixture with and without Equalization Using the Rogers 4350 Dielectric Materials with a PRBS 31 Data Pattern at 12.8 Gbps*

22.7 NEXT/FEXT Crosstalk Variations with PCB Materials

The selection of dielectric material not only impacts the signal performance in terms of its effect on the signal trace loss, but also can have an impact on crosstalk in an ATE test fixture. This is especially true given the fast rise times that are a result of the higher data rates of modern I/O cells. Ten Gbps telecommunication systems can have 30 ps rise times, which is equivalent to a distance of 171 mm for FR4 at a dielectric constant of 4.4 and 210 mm for a Taconic material with an average dielectric constant of 2.9. The shorter the length of the rise-time edge, the more sensitive it will be to the feature variations on the PCB. The sensitivity is due to the fact that the voltage will vary dramatically across this distance, and any feature on the order of the length of the rise-time edge or larger will require impedance matching to prevent reflections. The benefit of the lower dielectric constant to reduce reflections may be minimal but is in the right direction for the added benefit of lowering the dielectric constant of the PCB material. Typically the worst crosstalk offenders in ATE test fixtures are the pogo vias [14] and the vias to the socket. *Figure 22.19* shows the NEXT and FEXT results for the UAB boards manufactured in three materials (NELCO 4000-13 SI, Rogers 4350, and Taconic TSM 29 with FastRise 27 prepreg).

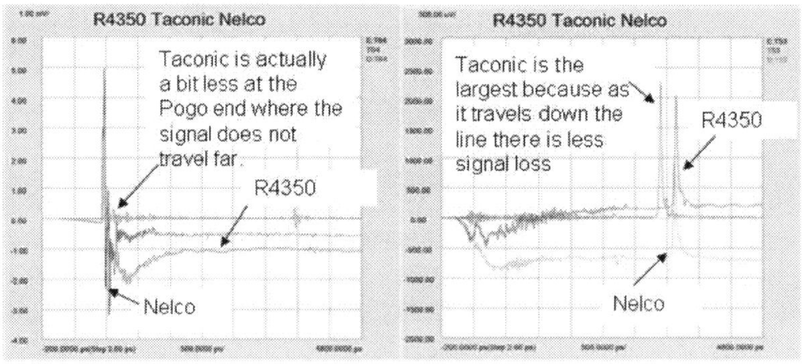

Figure 22.19: *FEXT at the SMA End (Left) and NEXT at the Pogo Block (Right)*

Lowering the dielectric constant of the material and therefore reducing the material losses actually has a mixed result on the

NEXT and FEXT. In the case of the Verigy HX UAB test fixture, the crosstalk is coming from the pogo via field. As this crosstalk signal travels down the line to the other end for the FEXT case, the material with the highest losses has the best result by attenuating this signal. In the case of the NEXT, which has a smaller distance to travel, there is a slight reduction in the crosstalk for the lower dielectric material, which can be attributed to the longer rise-time edge and less sensitivity to the impedance discontinuities.

22.8 Dielectric Influence on Complex ATE Test-Fixture Stack-Up Decisions

The previous sections have addressed the performance of different dielectrics at different data rates and compared the advantages of lower dielectric constant low-loss Teflon-based material. The important point to note on the previous section results is that the trace width was kept constant (19 mm) on all measurements so that the skin-effect loss was the same. For test engineers developing test fixtures for complex ATE applications such as microprocessors, the number of layers needed for signal routing and power prevents the use of these large trace widths that take full advantage of lower dielectric losses.

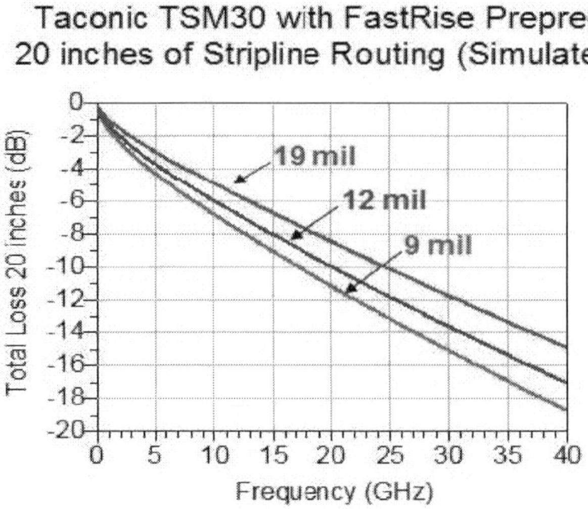

Figure 22.20: *Simulated Data for Trace Loss When the Trace Width Is Decreased on a 20-Inch Trace. Loss Model Based on Measured Data for the 19 mm Trace Width*

Evaluating the lower–dielectric-constant Taconic materials for ATE applications where narrow (less than 10 mm) trace widths are a necessity still shows a significant benefit. The dielectric plays a key role in the fact that the dielectric constant εR determines the height of the dielectric material needed for obtaining 50 Ohm common-mode impedance (100 Ohm differential when coupling is minimal) of the transmission line for a given trace width. Materials with a lower εR have less capacitance and thus require the reference ground planes to be closer to the signal trace. This increase in trace width can be seen in *Figure 22.21*, where the stack-up for an ATE test fixture for a microprocessor application is shown. The layers INR1 to INR4 are used to route the high-speed signals that in this application can run at data rates above 5 Gbps with a maximum trace length of 16 inches. The stack-up thickness does not allow the test engineer to use a larger trace width (e.g., 19 mm) on layers INR1 to INR4 since this would make the PCB height 287 mm and not feasible to manufacture.

The figure shows three dielectric materials (NELCO 4000-13 SI, Rogers 4350, and Taconic FastRise) and the maximum trace width one could obtain on the INR1 to INR4 layers with the restriction that the stack-up height needs to be below 250 mm. The important point to notice is that Rogers 4350 does not provide an advantage in regards to Nelco and the maximum trace width. However Taconic material does provide an increase of more than 20 percent on the trace width compared to R4350, which is significant for skin-effect loss. This type of reasoning needs to be applied by the test engineer when evaluating which dielectric material to use on a test fixture. Also note that depending on the application requirements, the designer can instead keep the trace width constant and reduce the dielectric spacing, enabling the use of more signal layers.

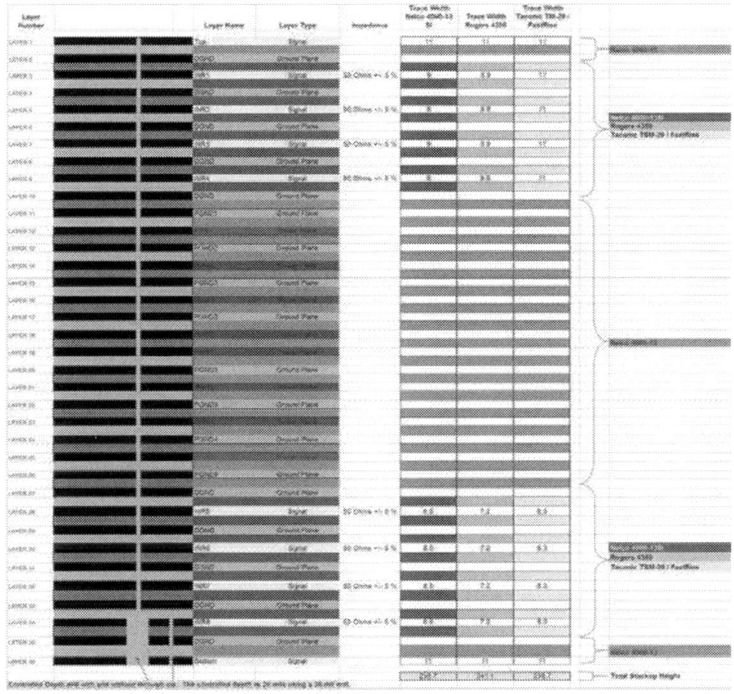

Figure 21: Example of an ATE test fixture stack-up for a microprocessor application using three different materials (Nelco 4000-13 SI, Rogers 4530 and Taconic FastRise). The layers INR1 and INR4 are the high-speed layers with worst case trace lengths of 16 inch.

Figure 22.21: *Example of an ATE Test Fixture Stack-Up for a Microprocessor Application Using Three Materials (Nelco 4000-13 SI, Rogers 4530, and Taconic FastRise). The Layers INR1 and INR4 Are the High-Speed Layers with Worst-Case Trace Lengths of 16 Inches*

22.9 Conclusion

This paper has shown that the selection of high-performance PCB laminate materials can enable ATE test fixtures using pogo pin–type interfaces for use in applications at 43 Gbps. Proper design of the interconnects, trace geometry, and dielectric material allow for an open 43 Gbps data eye to be measured even after a typical ATE test fixture trace of 25 cm. Although the data eye performance is not enough for a test and measurement application, the development of appropriate equalization techniques can compensate for this increasing loss with frequency and improve the data eye to meet the needs of an ATE application. It is important to note that equalization does not provide a perfect solution and will have some limits due to the amount of power that the ATE system can give up to increase the equalization strength (increasing pin counts on ATE systems create significant power management challenges). Selecting a lower-loss dielectric material will minimize the amount of correction required by the equalizer, leaving more power available for the transmitted signals.

The results presented in this paper comparing several dielectric materials with a typical ATE test fixture clearly show that available high-end multilayer dielectric materials provide similar performance at 10 Gbps, where losses are still similar in magnitude and are dominated by the skin-effect losses. At higher data rates, the lower losses of the Taconic materials did show an improved performance when compared to existing high-performance materials such as Rogers 4350 and Nelco 4000-13SI. Comparison of crosstalk performance shows that FEXT from the pogo via transition can actually be higher with a low-loss material where there is less attenuation of the crosstalk, while NEXT can actually improve due to a reduction in reflections at the pogo via as the wavelength or rise-time edge is lengthened by the lower εR value. A lower-dielectric constant material also provides an indirect improvement in performance at lower data rates for certain high-density ATE applications. When a high-density ATE test fixture is limited in layer count due to the maximum stack-up height that can be fabricated, then a lower εR value reduces the dielectric height that is required for the high-speed signal layers of a given trace width. This benefit allows the designer to either add more signal layers to improve signal routing or to widen the trace width of the controlled impedance lines to reduce the skin-effect losses.

It is important to make clear that the full benefits of a lower-loss dielectric material can easily be masked by the losses of the via transitions. In the case of the ATE test fixtures measured for this paper, it was found that the pogo via and connector via transitions can start degrading performance above 10 Gbps and are the limiting factor at 43 Gbps. So the next challenge to moving the data rates ever higher on an ATE test fixture will be in the area of via transition optimization and tighter fabrication tolerances for repeatable electrical performance.

Acknowledgments

We would like to acknowledge Altanova for layout and assembly of the measured ATE test fixtures and specifically Don Sionne at Altanova for his contributions to the cost comparisons of PCB materials. We would like to thank Wendell Anderson and Shelly Baglia of Agilent Technologies for their assistance in measuring PCB material properties. We would like to thank Morgan Culver of Agilent Technologies for the use of the fast–rise-time 43 Gbps bench test system for the time domain measurements.

References

[1] Jose Moreira, Heidi Barnes, et al., "PCB Loadboard Design Challenges for Multi-Gigabit Devices in Automated Test Applications," IEC DesignCon, 2006.

[2] Masashi Shimanouchi, "New Paradigm for Signal Paths in ATE Pin Electronics are Needed for Serialcom Device Testing," Proceedings of the International Test Conference, 2002.

[3] Jeff Loyer, "Proper PCB Stackup and Trace Geometry Design are Key Elements in the Fight to Lower Losses," Printed Circuit Design & Manufacture, 2006.

[4] J. Moreira, H. Barnes, W. Burns, D. Sionne, C. Gutierrez, and F. Azeem, "Influence of Dielectric Materials on ATE Test Fixtures for High-Speed Digital Applications," Proceedings of the Sixth International Kharkov

Conference on the Physics and Engineering of Microwave, Millimeter and Submillimeter Waves, 2007.

[5] Martin W. Jawitz and Michael J. Jawitz, "Materials for Rigid and Flexible Printed Wiring Boards," CRC Press, 2006.

[6] Jeff Loyer, Richard Kunze, and Xiaoning Ye, "Fiber Weave Effect: Practical Impact Analysis and Mitigation Strategies," IEC DesignCon, 2007.

[7] Scott McMorrow and Chris Head, "The Impact of PCB Laminate Weave on the Electrical Performance of Differential Signaling at Multi-Gigabit Data Rates," IEC DesignCon East 2005.

[8] Thomas F. McCarthy and David L. Wynants Sr., "Low signal loss bonding ply for multilayer circuit boards," US patent 6,500,529.

[9] "Basics of Measuring the Dielectric Properties of Materials," Agilent Application Note, PN 5989-2589EN.

[10] H. Barnes, J. Moreira, H. Ossoinig, M. Wollitzer, T. Schmidt, and M. Tsai, "Development of a Pogo Via Design for Multi-Gigabit Interfaces on Automated Test Equipment," Proceedings of the 2006 Asia-Pacific Microwave Conference.

[11] J. Moreira, H. Barnes, H. Kaga, M. Comai, B. Roth, and M. Culver, "Beyond 10Gb/s? Challenges of Characterizing Future I/O Interfaces with Automated Test Equipment," First IEEE International Workshop on Automated Test Equipment: VISION ATE 2020.

[12] Wolfram Humann, "Compensation of Transmission Line Loss for Gbit/s Test on ATEs," Proceedings of the International Test Conference 2002.

[13] Jose Moreira et al., "Passive Equalization of Test Fixture for High-Speed Digital Measurements with Automated Test Equipment," Proceedings of the 2006 International Design and Test Workshop.

[14] B. Szendrenyi, H. Barnes, J. Moreira, M. Wollitzer, T. Schmid, and M. Tsai, "Addressing the Broadband Crosstalk Challenges of Pogo Pin Type Interfaces for High-Density High-Speed Digital Applications," 2007 IEEE International Microwave Symposium.

Authors

Heidi Barnes, Senior Application Consultant, Verigy

Jose Moreira, Senior Application Consultant, Verigy

Tom McCarthy, Vice President, Taconic

William Burns, Senior Applications Engineer, Altanova Corporation

Crescencio Gutierrez, Engineering and Research and Development Manager, Harbor Electronics

Mike Resso, Signal Integrity Application Scientist, Agilent Technologies

Part VI

Future Directions

Chapter 23

Design and Test Challenges Facing Next-Generation 20 Gbps Interconnects

23.1 Abstract

The increased emphasis on high-performance serial data lines as the primary means of moving data through datacom and telecommunication systems presents new challenges for digital design engineers. Backplanes, cabling between boards and systems, circuit boards, and high-speed chip packages require an ever-increasing sophistication in the tools, methods, and techniques used in product design, characterization, and production test.

Establishing an accurate measurement at 20 Gbps and higher requires a test environment that can precisely measure and extract data at microwave frequencies. This paper uses a novel test fixture design to demonstrate solutions to the calibration, de-embedding, and measurement techniques necessary to achieve accurate test results.

23.2 The Need to Measure Signal Integrity

Communication link architecture plays a major role in the advancement of telecom systems. These changes also present new engineering challenges. As parallel bus structures give way to fast serial bus structures, the signal transition from a logical zero to a logical one becomes much shorter. This faster rise time results in larger reflections from impedance discontinuities in the physical layer. Backplanes, line cards, cables, and even integrated circuit (IC) packages will fail to provide adequate bit-error-ratio performance to transmit data properly when reflections increase in magnitude. This creates a signal integrity bottleneck that needs to be removed.

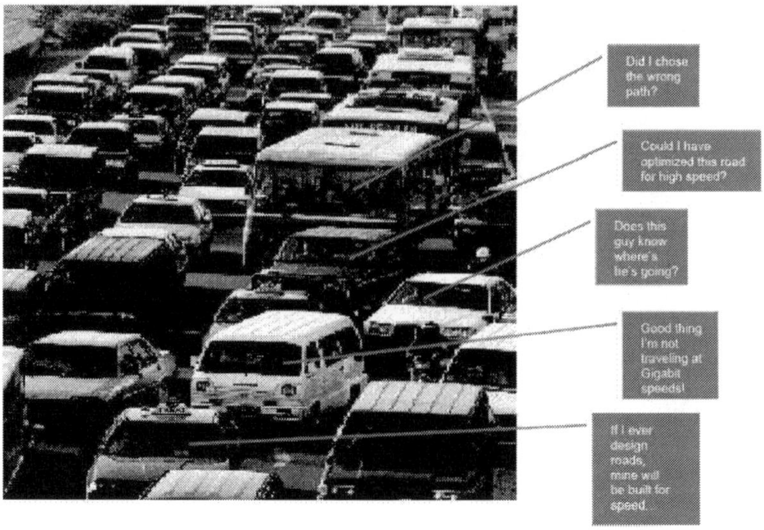

Figure 23.1: *Cars and Roads*

A popular analogy used in signal integrity testing is called "cars and roads" (see *Figure 23.1*). The cars are the electrons propagated through the transmission channel, and the roads are the transmission media such as FR4. No matter how fast of a Ferrari you may have, if the road has huge potholes, you will not be traveling very fast. In a similar fashion, the electrons in a 10 Gbps data packet will not transmit error-free if your channel is not designed with a controlled impedance environment. Removing physical layer reflections (potholes) is critical to the success of any telecom or computing application. The majority of this paper will address how to characterize and optimize the roads.

This paper is presented in three parts: Signal Integrity Analysis, High-Performance Test Fixtures, and Test Fixtures: A User's Perspective.

23.3 Signal Integrity Analysis

Most signal integrity problems can be divided into two major categories: active signal integrity and passive signal integrity. The active signal integrity problems arise from active devices that are generating their own signals. A good example of this is a serializer-deserializer (SERDES) chipset transmitting and receiving data between boards. The passive signal integrity problems revolve around structures that do not generate their own signal, but merely propagate a signal emanating from another source (e.g., simple printed circuit board [PCB], cable).

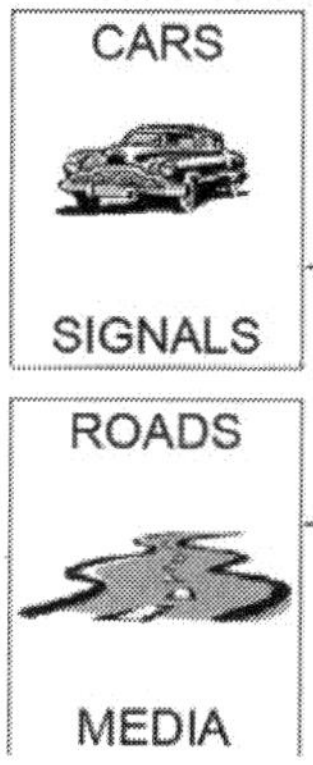

LAYER	ANALYSIS	TOOLS
Data Layer	Timing Measurements, State Measurements, Basic Protocols	Logic Timing Analyzers, Logic State Analyzers, Bus Protocol Analyzers, Bit Error Rate Tester
Signal Layer	Wave-shape Measurements, Eye Diagrams, Bathtub Curves, Jitter	Real-Time Scope, Sampling Scope, Bit Error Rate Tester
Clock Distribution	Clock Jitter, Data Jitter & Jitter Analysis	Sampling Scope, Real-Time Scope Spectrum Analyzer, Signal Source Analyzer
Signal Path Integrity	Signal Path Measurements Cross-Talk, Loss, Reflection, Model Extraction,	Physical Layer Test System with Vector Network Analyzer or TDR
Power Distribution Integrity	Capacitive & Inductive loading with/without Power Supply Bias, Model Extraction	Power Distribution Measurement System with Vector Network Analyzer
Materials Measurements	Dielectric Constant Measurements at Frequency Points	Materials Measurement System with Resonant Cavity

Figure 23.2: *Passive and Active Signal Integrity*

As outlined in the *Figure 23.2*, there are different test and measurement tools for active signal integrity and passive signal integrity problems. There are specific Optical Internetworking Forum (OIF) standardized sections or layers defined for both active and passive signaling components and systems. The signal path physical layer plays a major role in the performance capabilities in today's telecom systems. Even though most typical PCBs do not receive much attention from the great design houses in the technology community, it is well known that a poorly designed backplane can break a system quickly. At best, a poorly designed cable or backplane will limit the lifetime of a switch or router when data rates increase according to Moore's Law.

PCB manufacturing techniques are tuned for high volume, especially those made with woven-glass dielectric material similar to FR4. From a signal integrity standpoint, FR4 has a relatively good performance at lower data rates. However, when data rates exceed 10 Gbps, the problems increase dramatically. As engineers, we must mitigate these problems. Some of the most common PCB problems are shown in *Figure 23.3*. Looking at inset 1, we see the excess capacitance of a thru-hole affecting the differential transmission line characteristics. Inset 2 depicts localized crosstalk between vias in proximity to a differential transmission line. Insets 3 and 4 show localized changes in conductor trace width and spacing, respectively. Inset 5 represents the notorious via stub that creates capacitive reflections and loads down transmission line impedance. Insets 6, 7, and 8 describe problems associated with dielectric lamination layers, including non-uniform dielectric, surface treatment thickness non-uniformity, and localized changes in foil thickness, respectively.

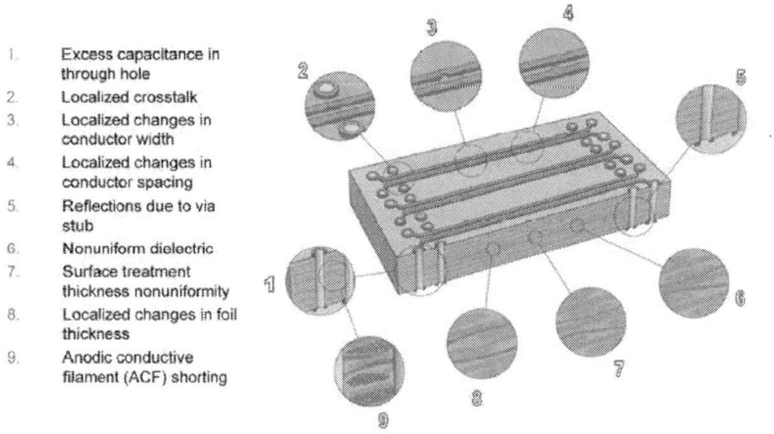

Figure 23.3: *High-Speed PCB Signal Integrity Challenges*

In inset 9, we see a problem called anodic conductive filament (ACF) shorting. This is a problem discovered by Bell Labs that is created by gaps in the lamination layers of epoxy resin and fiberglass. Since a differential transmission line has positive and negative voltages imposed upon it simultaneously by definition, any lamination problems will create an unwanted migration of copper ions between positive and negative terminals. Eventually, plating out copper on one via creates an impedance discontinuity that can reduce bandwidth and attenuate high frequencies.

Improving Signal Integrity Measurements

Over the years, many approaches have been developed for removing the effects of the test fixture from the measurement, which fall into two fundamental categories: direct measurement (pre-measurement process) and de-embedding (post-measurement process). An approximation of ease of use and accuracy of these two techniques is shown in *Figure 23.4.* Direct measurement requires specialized calibration standards that are inserted into the test fixture and measured. The accuracy of the device measurement relies on the quality of these physical standards. De-embedding uses a model (typically a Touchstone or citifile) of the test fixture and mathematically removes the fixture characteristics from the overall measurement. This fixture de-embedding procedure can produce very accurate results for the non-coaxial device under test (DUT) without complex non-coaxial calibration standards.

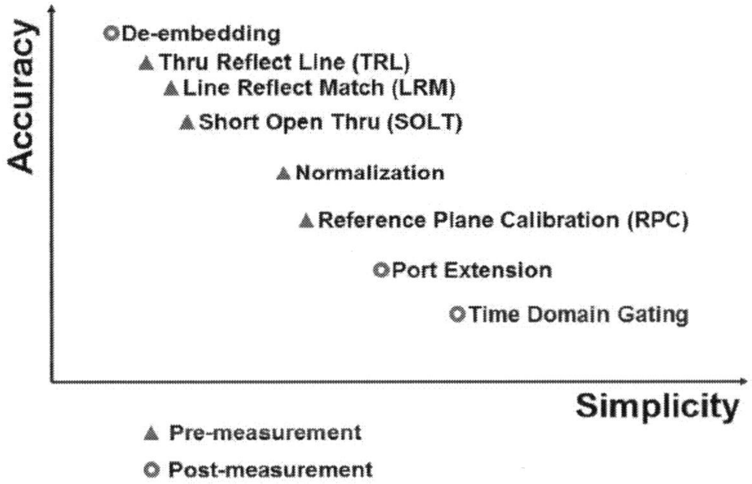

Figure 23.4: *Pre- and Post-Measurement Process Measurement Error Correction Techniques. De-Embedding Is the Most Accurate, but Also the Most Difficult if the De-Embed File Is Not Provided with the Test Fixture*

The process of de-embedding a test fixture from the DUT measurement can be performed using scattering transfer parameter (T-parameter) matrices or in this case, the de-embedded measurements can be post-processed from the measurements made on the test fixture and DUT together. Also, modern EDA tools have the ability to directly de-embed the test fixture from the vector network analyzer (VNA) measurements using a negation component model in the simulation (see *Figure 23.5*).

A brief description of the user interface used for reference plane adjustment will help explain the process used for this design case study. *Figure 23.5* shows a step-by-step method for assigning a user-defined reference plane. The first step in using the port reference adjustment is choosing the adjustment method and its options. There are four adjustments: two-port de-embedding, four-port de-embedding, port rotation/extension, and port reference impedance. Each adjustment has its own options that are displayed when the adjustment is selected from the list. Step two is to select the appropriate de-embed data file that represents the fixture to be removed. This can be a Touchstone or citifile data format.

Alternately, the citifile format could have been used as well. Step three is to assign the de-embed file to the appropriate port of the test system. The last step is to apply the reference plane adjustment by clicking on the "apply" button in the bottom portion of the dialogue box.

> *Note: Normally, the easiest way to obtain the de-embed file is to insist the fixture or probe supplier provide a Touchstone format file (S-parameter file) with the device when it is purchased. However, if the supplier has not done its homework, then it may not have this immediately available, and other methodologies must be employed to create the de-embed file*[1].

---

[1] A good discussion of Touchstone files can be found in the Cascade Microtech application note at www.home.agilent.com/upload/cmc_upload/All/ONWAFER.pdf

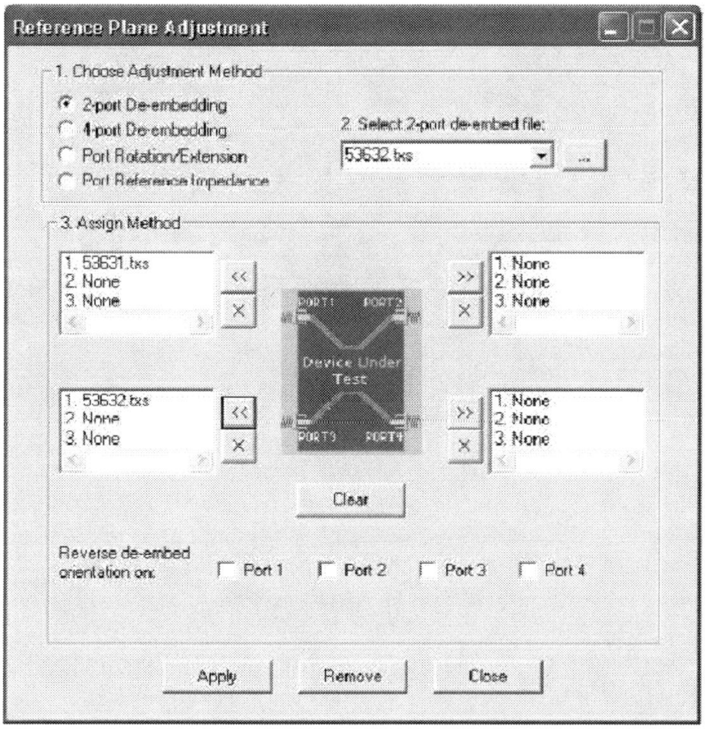

Figure 23.5: *De-Embedding the Test Fixture*

Typical signal integrity laboratories have oscilloscopes, time domain reflectometers (TDRs), logic analyzers, and VNAs. The leading-edge SI labs have four-port VNA–based physical layer test systems (PLTSs) with either 20 or 50 GHz of bandwidth.

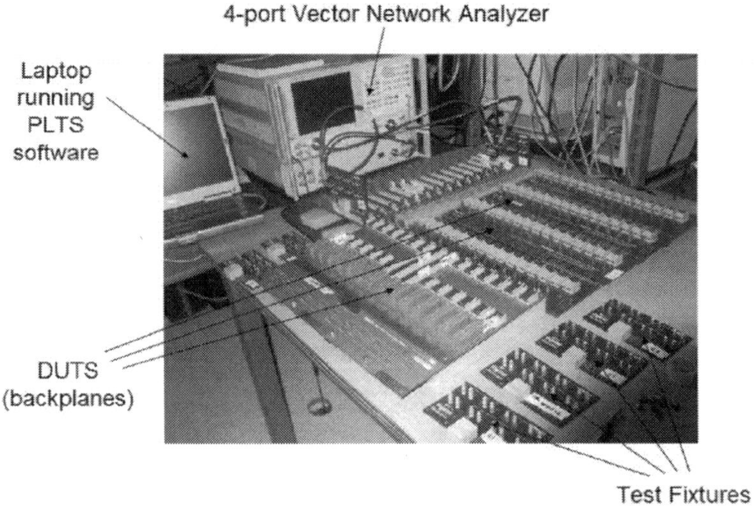

Figure 23.6: *University of New Hampshire's Signal Integrity Lab (Photo Courtesy of Bob Noseworthy)*

This particular lab *(Figure 23.6)* is at the leading edge because of its work in standards committees. This lab is located at the University of New Hampshire's Interoperability Laboratory (IOL). Many companies send their new prototype backplane designs to the IOL for evaluation. The extended attachment unit interface (XAUI) backplane transmits serial data at 3.125 Gbps, while the advanced telecommunications architecture (ATCA) backplane transmits data at 10 Gbps. The test fixtures in *Figure 23.6* are used to measure the performance of high-speed backplanes. Using a similar methodology, the authors will discuss measurements made with test fixtures and a high-speed cable.

One new trend in physical layer testing is the use of 12-port VNAs to measure multiple aggressor differential crosstalk. The four-port VNA has been the workhorse of today's leading signal integrity labs worldwide, but it is important to understand this new design tool.

Multiple Aggressor Differential Crosstalk

In order to fully characterize the interaction between one differential channel and the two differential channels on either side of it, a 12-port measurement data set is required. A typical block diagram of 12-port VNA is shown in *Figure 23.7*. This test system alleviates the need for multiple measurements, multiple calibrations, and multiple cable connects and disconnects. Also, it is an efficient way to save valuable time in the lab. Of course, it is possible to use four-port measurements to construct the equivalent 12-port measurement, but it takes many more measurements and therefore much more time. The manual concatenation of smaller Touchstone files will need to be done if the number of device ports exceeds the number of ports on the test system. However, be forewarned that this is a painstakingly lengthy process that is inherently error-prone. In fact, this will require the merging of 15 four-port measurements.

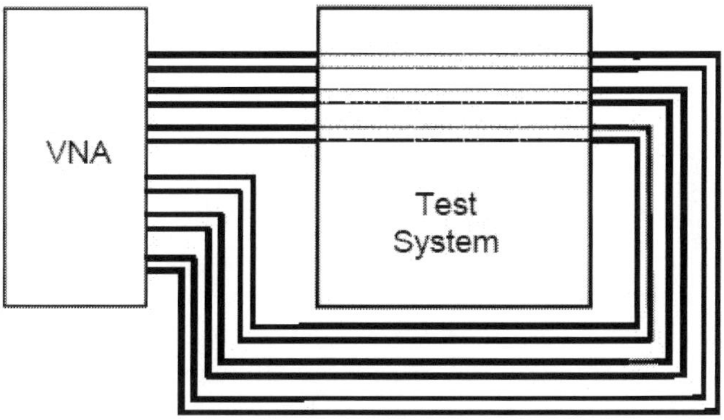

Figure 23.7: *Testing Multiple Aggressor Differential Crosstalk with a 12-Port VNA*

Depending on the number of ports in the test system, the subsequent data file produced at the end of the measurement sequence has different information. The most popular file type imported into simulators is the Touchstone file format using the .snp format, where n equals the number of ports in the test system. If the multiport test system is a full crossbar architecture that allows all ports to be sourced and received, then a fully populated [n x n] matrix of S-parameter elements is achieved. For proper modeling in Touchstone format, each element in the [n x n] matrix must be present. As previously mentioned, if the DUT includes more ports than the test system, then multiple measurement configurations must be made (disconnecting and re-connecting cables and terminations). *Table 23.1* compares the calibration and measurement time and effort for testing a 12-port DUT using test systems with various numbers of ports.

Multiport VNA Comparison				
	12-port VNA	8-port VNA	4-port VNA	2-port VNA
Touchstone data file	.s12p	.s8p	.s4p	.s2p
Measurements	1	6	15	66
Cable connect/disconnect	12	48	60	132
Termination connect/disconnect	none	48	60	132
Cal time w/o Ecal (minutes)	23	15	7	3
Cal time w/Ecal (minutes)	8	6	2	1
Measurement time (minutes)	3+10	4+53	4+73	5+220

Table 23.1: *Multiport VNA Calibration and Measurement Time Comparison*

One differential channel can be completely described with a four-port S-parameter matrix. The 16 elements of this [4x4] matrix contain information about the behavior of that differential channel (single-ended signals, differential signals, and common signals). This is incredibly important information, but it is limited. There is no information about the interaction of the differential channel with adjacent ones. By expanding the interaction to include three adjacent differential channels, the crosstalk performance can be included. The complete description of three adjacent differential channels can be described using 12 single-ended ports (see *Figure 23.8*). The conventional port assignment is with odd-numbered ports on the left input side and even-numbered ports on the right output side. Each individual signal line and its return path have a single-ended port on each side.

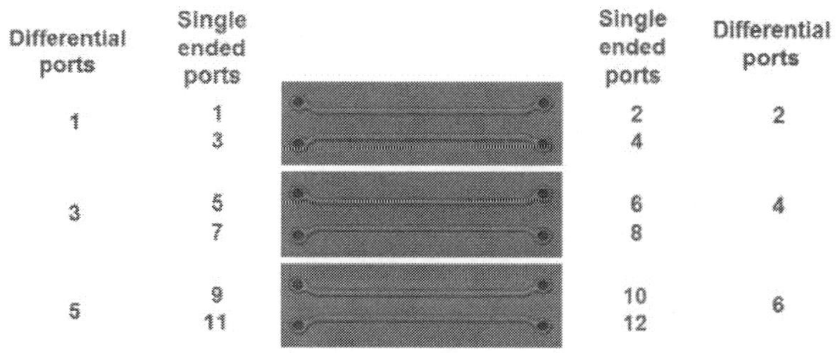

Figure 23.8: *The Conventional Definition of All Port Assignments in a 12-Port Device Helps Organize Data Analysis and Promotes Data Correlation between Various Signal Integrity Laboratories*

Each S-parameter is defined as the ratio of the sine wave voltage out of one port to the sine wave voltage into another port. With 12 ports, there are 144 matrix elements, but only 78 unique elements. This is a huge amount of information. Likewise, we can use differential signals and common signals to describe the interaction of sine waves with the differential channel. Each differential S-parameter is the ratio of a differential or common signal out of any port to a differential or common signal into any differential port.

The PLTS topology map for multiple differential channel analysis allows the user to name test ports with logical names (see *Figure 23.9*). This topology map is used extensively from the beginning of calibration throughout the final data analysis. The horizontal stripes in the middle of the dialog box represent a single-ended channel. The user can select either single-ended or differential channel analysis, depending upon the application and verification requirements.

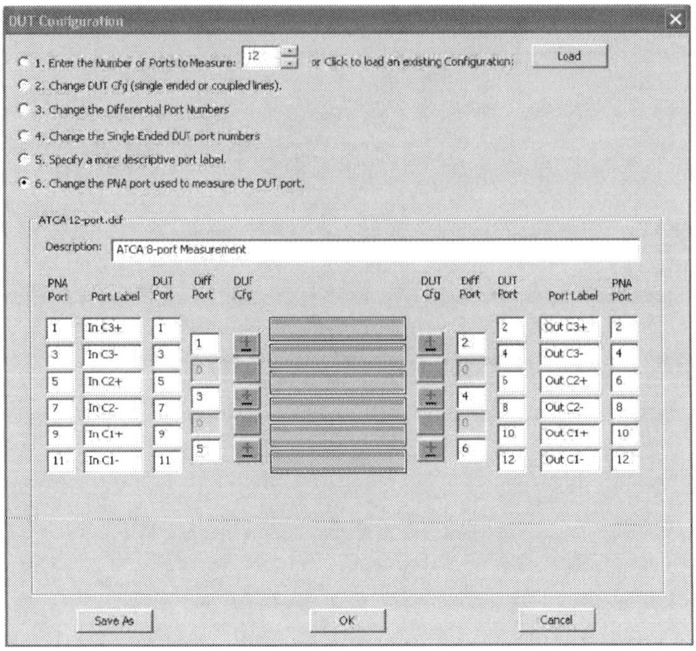

Figure 23.9: *The 12-Port PLTS Topology Map Allows an Intuitive User Interface for Managing the Huge Amount of Data in the [12x12] S-Parameter Matrix*

The ability to analyze the data in both time and frequency domain domains is very powerful. *Figure 23.10* shows one simulation (insertion loss) and six S-parameter elements of the 12 x 12 matrix. This may be an overwhelming amount of data to analyze all at once, so we will break it down into smaller sections. Basically, there are three types of measurements displayed in *Figure 23.10*: differential insertion loss, differential return loss, and differential crosstalk.

Looking at the differential insertion loss curves first, we see one simulation that is a fairly smooth curve with no ripple. While this lack of ripple is an ideal case and not realistic, the overall magnitude of the curve is a very good approximation to the measured data. This is the ultimate goal of simulation and modeling—correlating with real-world measurements.

In this example, the second set of curves is the differential return loss of two separate channels. The lower-frequency reflections start out at about -20 to -30 dB and then increase to slightly more than -10 dB. This is a typical return loss shape for a lossy differential transmission line. It should be noted that the upper envelope of the rippling curve should be the figure of merit, not the downward spikes. This is because the downward spikes represent nulls where reflection is at an absolute minimum. The channel performance will be dictated more by the overall worst-case reflection at a given frequency band, not at the nulls.

The third set of curves is the differential crosstalk. The magnitude of voltage coupled from a differential aggressor to a differential victim is shown for three adjacent channel pairs. The coupling at lower frequencies is quite small, starting at about -60 dB and then increasing with frequency to just over -40 dB. This -40 dB level represents 1 mV of crosstalk noise voltage, assuming a 100 mV drive signal. For high-speed digital systems performing at 20 Gbps, 1 percent crosstalk can easily kill the channel. Whether or not 1 percent crosstalk will kill a 10 Gbps channel depends on many factors, including data rate, design margin, and channel isolation tolerance. Most leading-edge SI labs today consider 1 percent crosstalk a yellow flag to be monitored as a potential cause of failure.

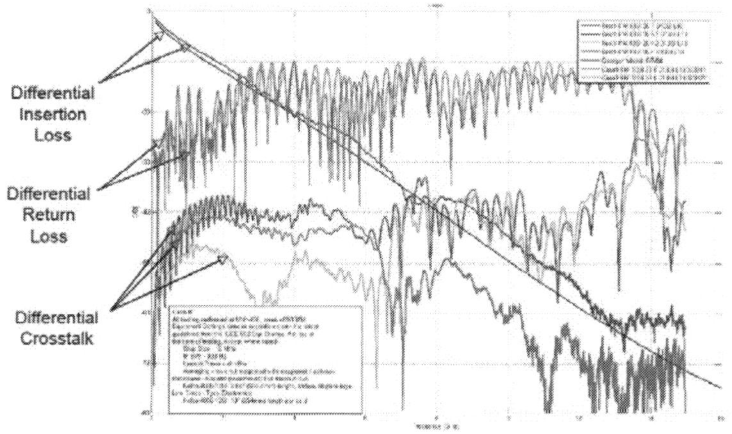

Figure 23.10: *12-Port VNA Measurements*

23.4 High-Performance Test Fixtures

The prior discussion demonstrates the increasing need for accurate measurements to better understand the sources of signal integrity problems in the electrical systems. Without this understanding, as will be discussed in the later part of this paper, the design engineer will be left with a system that has margin and performance problems that cannot be fixed.

Test fixtures are an integral part of the process of obtaining accurate and repeatable measurements. They adapt the connections of the device or cable under test to the connections on the measurement instrumentation, typically a female surface-mount assembly (SMA) on today's high-performance measurement systems. *Figure 23.11* shows two approaches to test fixture design for High-Definition Multimedia Interface™ (HDMI™)[2] used in newer digital televisions and set-top boxes. The fixture on the left is cable-based and the fixture on the right is board-based; these represent two approaches to fixture design.

Figure 23.11: *Comparison of Two Approaches to v1.3 HDMI Test Fixture Design*

The cable-based design interfaces from the HDMI plug to a small PCB that attaches to the coax cables and on to the SMAs. The board-based design requires the use of interface cables to adapt to standard SMA connectors. Both fixtures are v1.3 compliant and

[2] HDMI and High-Definition Multimedia Interface are trademarks or registered trademarks of HDMI Licensing LLC. www.HDMI.org.

have carefully matched traces as well as controlled impedance launches for the HDMI plug.

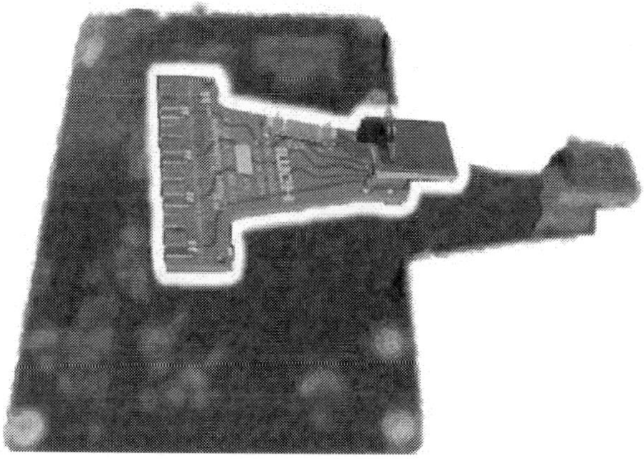

Figure 23.12: *Comparison of Size—Old and New Board-Based HDMI Fixtures*

Figure 23.12 shows the first- and second-generation HDMI fixtures stacked on top of each other so that an accurate comparison of the relative size can be made. The newer fixture is 2.4 inches from the tip of the plug to the tip of the connectors, approximately 50 percent of the length of the older fixture. Minimizing the trace lengths, trace skew, and impedance disruptions are some of the keys to a better test fixture design.

Test fixtures almost always induce some additional electrical discontinuities into the measurements being taken. At a minimum the test fixture has two connectors—one to the test instrument and one to the DUT. In the ideal world, all of these undesirable characteristics are de-embedded during the calibration and setup process so the test fixture does not interfere with the measurements. However, proper measurement science must be employed in order to completely de-embed all aspects of the fixture.

Historically, engineers used microprobes to directly connect the test instrument to the DUT. These microprobes worked well for "one-up" measurements but have fallen more to the wayside because of problems with robustness, reliability, and repeatability of microprobe-based measurements. Advances in connector technology and size have resulted in connectors that have excellent electrical characteristics to microwave speeds, allowing plenty of guardband for testing at today's multi-gigabit speeds. *Figure 23.13* shows the header pins used on the older 2002 HDMI test fixture. To obtain the bandwidth beyond the 10.2 Gbps necessary to test the latest HDMI v1.3 speeds, the new fixture (in the middle) adapts the HDMI receptacle to miniature GPPO™ connectors[3]. On the right, the test fixture is shown with cables that adapt from the GPPO connectors to the standard SMA connector to connect to the test instrument. The use of SMA connectors on the test fixture would have dramatically increased the size of the test fixture, and while that would have removed the GPPO connector from the circuit, the tradeoff in electrical performance and fixture size would not have been worthwhile.

Figure 23.13: *Comparison of Launch Connectors—HDMI Fixtures*

Mechanical Issues

Mechanical robustness sets the foundation for reliable and repeatable measurements. The test fixture must be able to withstand the physical rigors of the repeated connections that will occur in an engineering analysis environment such as the "wiggling" when connecting a cable to the receptacle on the test fixture. The use of test adapters in engineering test labs, for production tests, and for quality assurance tests all influence different aspects of the mechanical requirements.

One key area of mechanical and electrical overlap, which is not the focus of this paper, is the impact on electrical performance of the

[3] GPPO is a trademark of Corning Gilbert. www.corning.com/corninggilbert/products__services/microwave/gppo.asp

mechanical wear induced by repeated insertions into the DUT receptacle. Much has already been written and discussed about this issue, and the limitations/requirements this places on test adapter use in a high-contact environment such as a manufacturing test. The measurements in this paper assume a relatively new and unworn receptacle.

Another overlap of mechanical and electrical design goals occurs at the SMA end of the test adapter. The user must be careful to not over-torque the SMA connection to minimize the stress on the PCB and on the cable assembly. Too much torque when tightening the connection and too much twist on the cable assembly may affect the measurement reliability and accuracy and will eventually damage the PCB and/or cable assembly. Efficere has addressed this concern with a unique mechanical modification of the SMA to minimize the twisting that comes from torquing the bolts holding a vertical-launch SMA to the PCB.

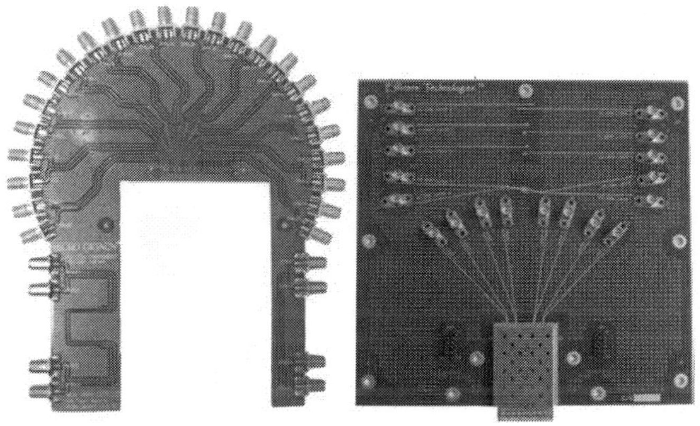

Figure 23.14: *Two Fixtures Showing Different Mechanical Attributes.*

Figure 23.14 shows two examples of approaches taken with the mechanical aspects of high-performance test fixture design. Note that both fixtures have a radial design relative to the relationship of the SMAs to the receptacle. The board on the left uses edge-launch SMAs, while the board on the right uses vertical-launch SMAs. Electrical performance of these two fixtures is discussed and shown in *Figure 23.17*.

In addition to these mechanical requirements, there are form and fit requirements in some applications that place specific constraints on the size and shape of the test adapter. These constraints can make it difficult to achieve all of the electrical requirements, especially for plug-based test adapters. Adapters for testing cable assemblies tend to not have these physical limitations.

Setup Issues

An additional consideration when using a test fixture is any additional hardware that becomes part of your test setup. The length of cables back to the test instrumentation, any added barrel connectors, the switch boxes to an intermediate fixture for volume or manufacturing tests, and in some cases even the spatial orientation of the test fixture relative to other electrical signals and surfaces may affect the measurement results. Every potential signal and field disruption needs to be either calibrated out or factored in to the final electrical measurement.

This is particularly important when using an intermediate test adapter in manufacturing. These intermediate adapters sit between the test fixture and the final product to be tested. Often they allow the test fixture to be switched between multiple production test setups to optimize throughput or they are used to reduce the wear and tear on the test fixture receptacle on by providing a "throw-away" fixture that can take the wear and then be replaced without creating that same wear on the test fixture.

Calibration and De-Embedding

Earlier in this paper there was a discussion about calibration and de-embedding. While this is a good foundation to discuss test fixture characterization in the scope of this paper, it is by no means an exhaustive study. Reference ii provides some background for pursuing detailed investigation into advanced error-correction techniques. (See the end of the paper).

It is very important that the designer of the test fixture provide a set of calibration structures that accurately reflect the structures used for the data paths. There are many methods being used for calibration, including four or five variants of thru-reflect-line (TRL), short-open-load-thru (SOLT), several variants of line-reflect-match (LRM), gating time domain gating, and normalization. The designer of the fixture will often provide only one of these solutions for a given fixture.

The InfiniBand 25 Gbps test fixture described later in this section uses a SOLT calibration philosophy. The traces provided include a "2X thru" trace that includes two vertical-launch SMAs, four vias, and the landing for the InfiniBand receptacle (see *Figure 23.15*).

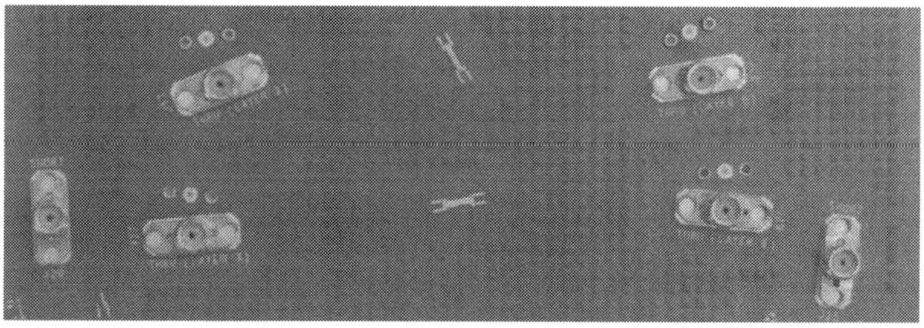

Figure 23.15: *SOLT Calibration Showing InfiniBand 4X Launch Pad*

One thing that all of the calibration approaches benefit equally from is having calibration adjacent to the data paths on the same physical board. This assures that any variations in manufacturing processing will affect the calibration structures the same as the data path structures. There has been some interesting work done in the area of PCB fiberglass yarn weave that suggests that care needs to be taken when designing the orientation of the calibration structure on PCB fixtures.[ii]

A reminder that de-embedding is a function performed by the test instrument (or by the software tool post-processing the measurement data) that effectively moves the measurement point reference plane from the end of the test cable to another point on the test fixture, typically to the receptacle launch/PCB interface. Usually, because the receptacle does not have high electrical signal integrity, the receptacle is not de-embedded.

Buy or Build?

It is the decision of each engineer to understand the importance of the test fixture in achieving accurate measurements in his or her own test environment. Traditionally test fixtures have been designed in-house either directly with in-house staff or using a consultant, or they have been designed and supplied by connector or cable manufacturers. A recent third choice of test fixture supply is an independent supplier that is focused on providing the highest-quality fixture no matter what deficiencies it shows in the product tested. The fixtures shown in *Figure 23.16*, the fixture on the left being supplied by a connector manufacturer and the fixture on the right being supplied by a leading independent test fixture supplier, were compared in *Figure 23.17* to show the difference in performance of similar fixtures. The independent fixture offers considerably more bandwidth with insertion loss of -3 dB at 8.2 GHz and return loss of -10 dB to more than 25 GHz.

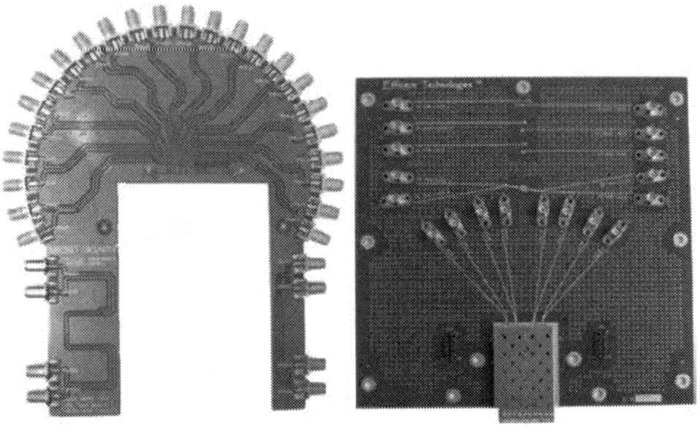

Figure 23.16: *These Fixtures, First Shown in Figure 23.14, Have Nearly Identical Calibration Traces That Were Used to Gather the Data Shown in Figure 23.17*

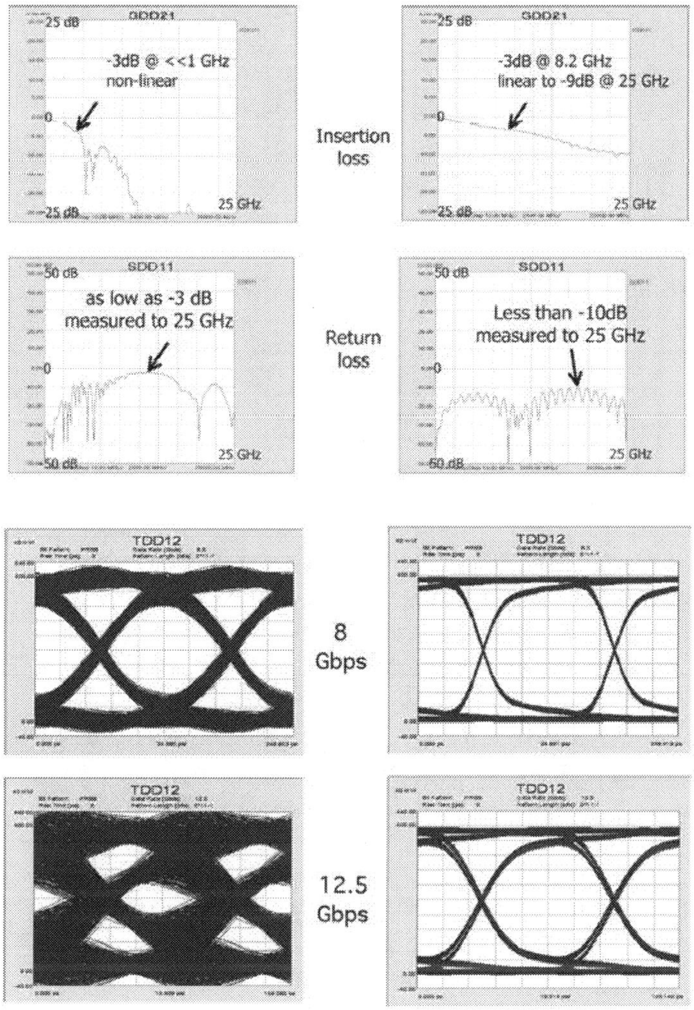

Figure 23.17: *Fixture on Left Supplied by a Connector Manufacturer Compared to Fixture on Right Supplied by an Independent Fixture Supplier Shows a Significant Increase in Bandwidth*

Test fixtures, if designed properly, can offer a mechanical robustness and an electrically sound environment that enables the design engineer to obtain accurate, reliable, and repeatable measurements.

25 Gbps Test Fixture

It is possible to make the test fixture perform better so that it is not a significant contributing factor in the measurements being taken. This section of this paper discusses various issues surrounding the design of a unique test fixture that has overcome the mechanical and electrical performance issues that have plagued high-speed InfiniBand test fixtures. The 4X InfiniBand fixture shown in *Figure 23.18* is able to perform at speeds to 25 Gbps per channel due to some unique design elements.

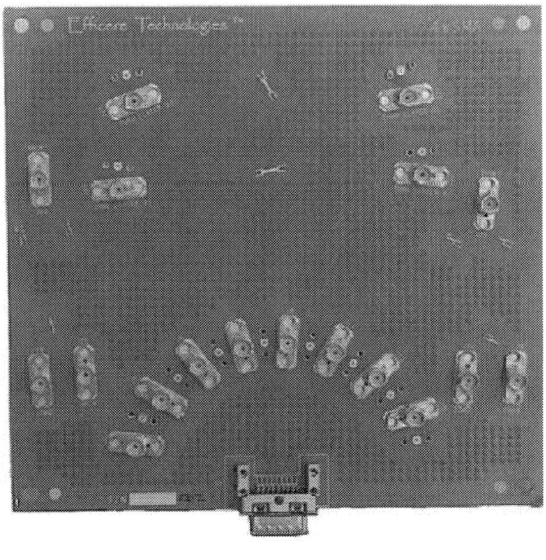

Figure 23.18: *25 Gbps InfiniBand 4X Test Adapter*

Electrical Design Goals

The electrical performance of the fixture does not take a back seat to the mechanical requirements. The electrical design goal of a test fixture is to provide an environment that electrically observes the DUT, not the test fixture. The fixture needs to have bandwidth guardband to facilitate measurements well beyond the specification of the product under test. Impedance matching at all potential points of disruption, predictability, and repeatability from trace to trace and measurement to measurement are key design and performance issues.

To test and characterize InfiniBand cables to the current double data rate (DDR) speeds of 5 Gbps, the fixture was specified to an insertion-loss bandwidth of five times the fundamental frequency of 2.5 GHz, which equates to -3dB at 12.5 GHz. The return loss was specified to three times the fundamental frequency or -20 dB at 7.5 GHz. The initial prototypes of the board exceeded the requirements for testing DDR 5 Gbps InfiniBand.

These performance improvements were to be achieved using traditional wet-etch PCB manufacturing processes and with industry-standard design tools.

Design Methodology and Results

In order to achieve these high-performance levels, special consideration must be paid to the various points of signal disruption starting at the SMA launch into the board and then going to the vias that propagate the source signal into the inner board layers, the inner layer traces, and the transitions back to the outer layers, and then finally launching into the 4X receptacle.

The design process starts with a systemic model from SMA to receptacle so that the critical signal disconnects and disruptions can be fully understood and simulated. In addition, this baseline model allows the impact of specific design approaches and structure implementations to be simulated and optimized. Where necessary, both electrical and mechanical models in two and three dimensions were used to capture all aspects of the fixture into the simulations.

An inter-layer interconnect structure is used to facilitate and optimize the internal connections on the fixture. Some additional insight can be gained today by analyzing two areas of particular challenge: the SMA launch and the receptacle. In both cases, the use of third-party–supplied components limits the flexibility in making changes within those components.

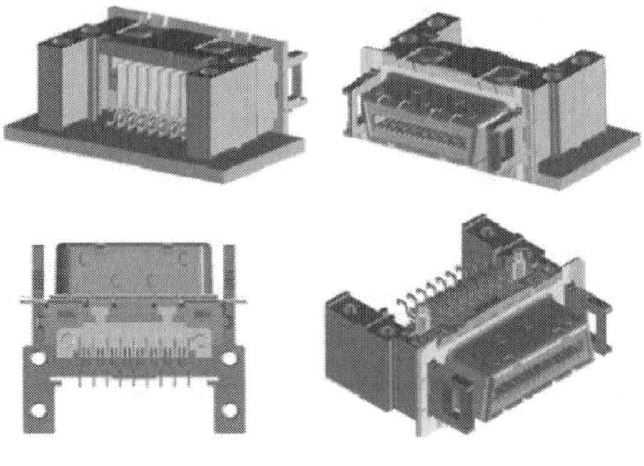

Figure 23.19: *InfiniBand Receptacles*
(Photo Courtesy of Molex Corporation)

The receptacle in *Figure 23.19* shows a typical high-speed InfiniBand connector that is widely used today. Optimizing the complete signal channel is always a way for clever design engineers to improve their products and using standard components in a novel way, thereby differentiating systems. Optimizing the connector/PCB interface to create a smooth launch with minimal reflections is the major design challenge. The receptacle requires certain pad geometries to enable efficient manufacturing assembly. It also performs a combined electrical and mechanical translation from the two-dimensional PCB surface to the three-dimensional dual-row female receptacle. The dramatic impact of the receptacle on measurements taken with test fixture can be seen by comparing *Figure 23.20* to *Figure 23.22*.

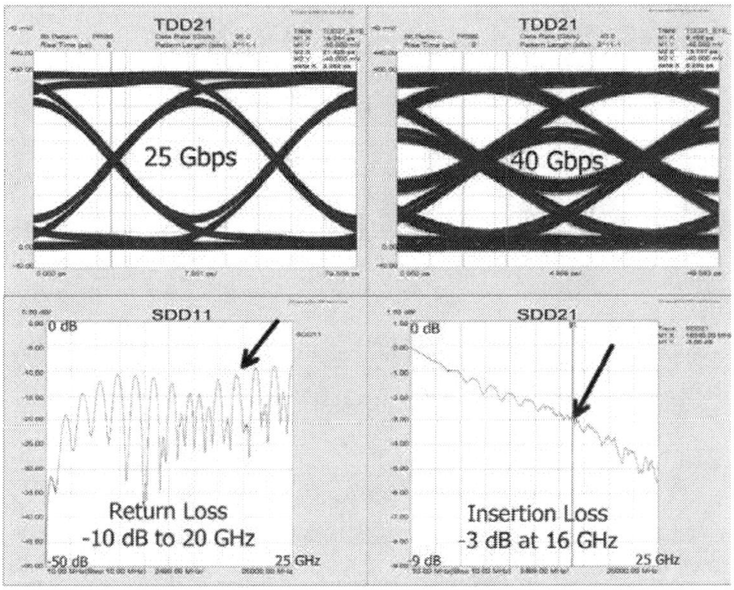

Figure 23.20: *2x Thru Traces—No Receptacles*

Figure 23.20 shows the test results using the 2x thru calibration trace that includes two SMAs, four vias, identical traces as used in the signal paths, and the receptacle launch pad structure—no receptacle. The test fixture achieves more than 40 Gbps with an acceptable eye, a return loss of -10 dB to 20 GHz, and an insertion loss of -3 dB at 16 GHz. This fixture is currently under redesign to achieve the desired -20 dB of return loss.

To test the impact of the receptacle on the measurement results, the two test fixtures were connected to each other using a test interface board. The test interface board was used to avoid any further disruptions that might be incurred if an InfiniBand cable was connected between the two boards. The minimum distance from the contacts of one InfiniBand receptacle to the other when the boards were butted up together was measured to be 167 mm. The interface board, which essentially consists of thru traces from one fixture to the other, was carefully constructed to minimize any signal disruptions and to optimize signal integrity. The connected test fixtures are shown on the left side of *Figure 23.21*, and a drawing of the interface board is shown on the right.

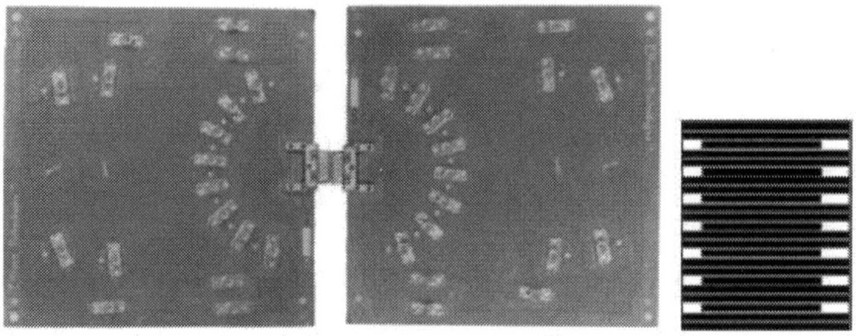

Figure 23.21: *Test Setup to Measure the Impact of the InfiniBand Receptacles on the Test Fixture—Mated Boards Shown on the Left with an Enlarged Drawing of the Adapter Board on the Right*

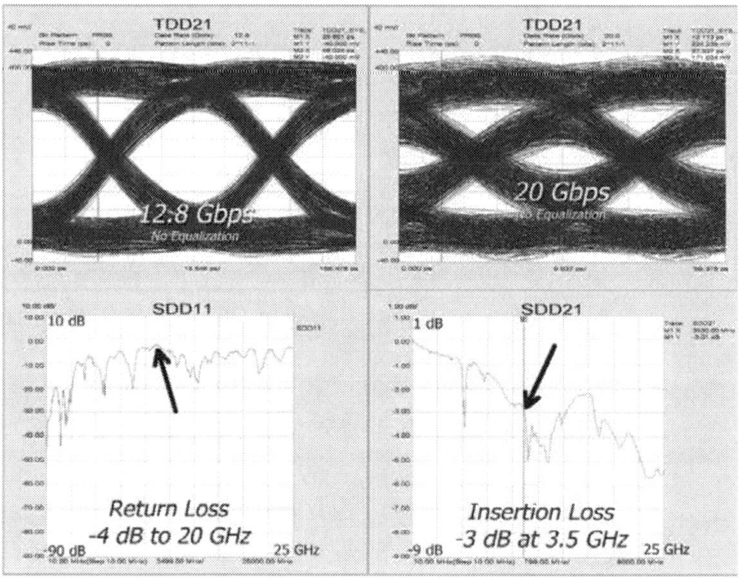

Figure 23.22: *Board-to-Board—Includes Two Receptacles*

Figure 23.22's notable electrical change from the 2X thru data in *Figure 23.20* is the addition of two receptacles into the signal path. The two receptacles reduce the bandwidth by some 50 percent with a data eye at 20 Gbps and achieves return loss of only -4 dB at 20 GHz and insertion loss of -3 dB at only 3.5 GHz.

The unique SMA launch directly into the via used in the 25 Gbps test fixture is shown in *Figure 23.23*. The SMA launches directly into the via to minimize additional disruptions.

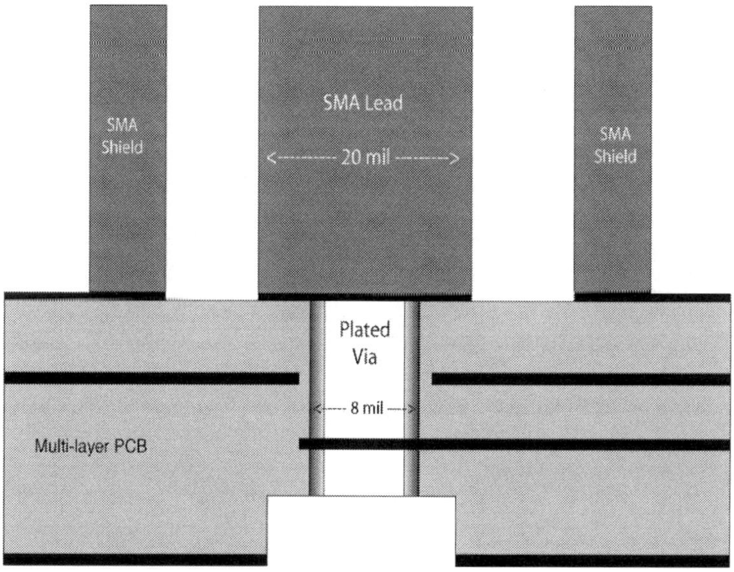

Figure 23.23: *Generalized View of the Efficere Combined Via/Launch Structure Minimizes Signal Disruptions at the SMA Launch Point*

Figure 23.24 shows TDR plots of two via structures interfacing to the launch. The plot on the left shows the results when the via is not fully optimized for the characteristics of the PCB stacking profile. The plot on the right shows the via after optimization.

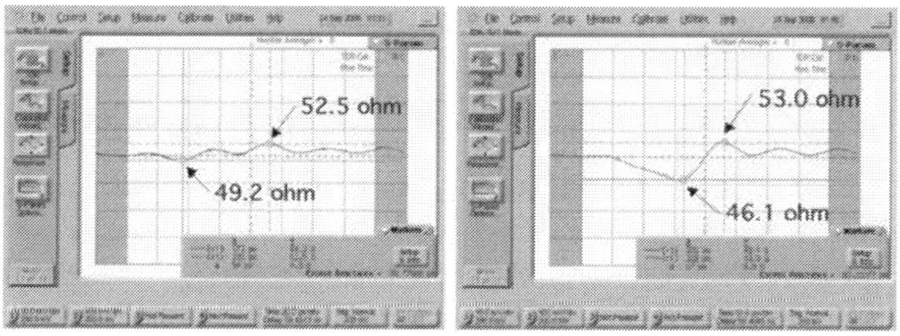

Figure 23.24: *Partially and Fully Optimized Via/Launch*

Figure 23.25 shows a TDR plot of the completed InfiniBand test fixture looking into the board from the SMA toward the InfiniBand receptacle. Note the two vias in the signal path. These vias are proprietary Electrically Invisible Via™[4] structures that virtually eliminate the signal disruption common in traditional via designs. For this particular board, the via shows a near-perfect 50 Ohm path that is actually better than the stripline traces that connect the via to other parts of the PCB. The initial via is 49.8 Ohms and the second via is 50.8 Ohms. Also note the very large impedance disruptions at the InfiniBand receptacle.

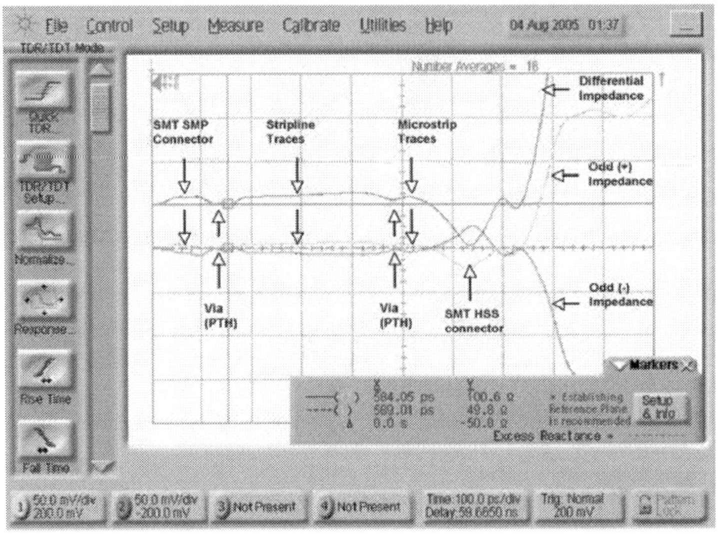

Figure 23.25: *Electrically Invisible Via™ Inter-Layer Connect Structure*

Substrate

It should be noted that even though this test fixture was built with standard wet-etch chemical processes, it was not built with a standard FR4 substrate. The custom stack-up of material used offers a better substrate than FR4 at a slight cost premium. In the case of test fixtures, which are low-volume and require high performance even at a slightly higher cost, the choice of material

[4] Invisible Via is a trademark of Efficere Technologies and refers to a patent-pending inter-layer interconnect structure.

was not a limiting factor. If some of these design structures and techniques were to be applied to a standard FR4 substrate, such as might be the case in a higher-volume, more cost-sensitive product, the limitations of that product would need to also be a part of the modeling and simulation.

Test Fixture Setup

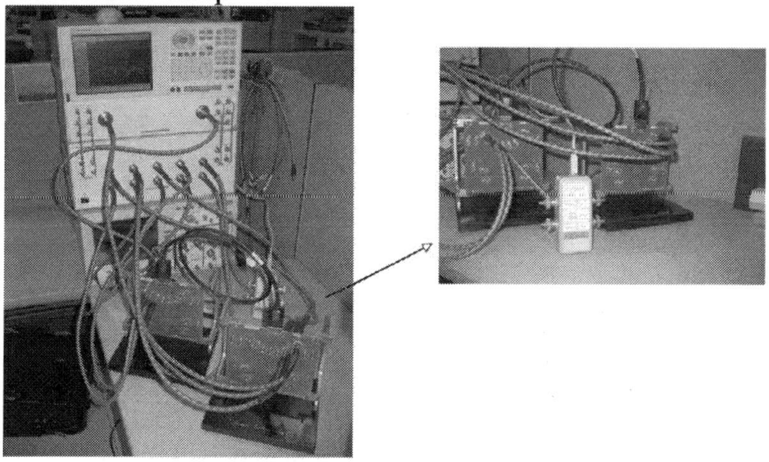

Figure 23.26: *12-Port Test Setup with 25 Gbps 4X–SMA Test Fixtures*

The test fixtures were tested on the new Agilent Technologies 12-port PLTS, consisting of three items: the E8362B VNA, the U3022AE10 10-port test set, and the N1930A PLTS software (see *Figure 23.26*). The fixtures were mounted on test stands to remove any potential effects of the table surface. A one-meter high-performance InfiniBand cable was used to connect the two test fixture boards.

The results of testing with the 12-port VNA are shown in *Figure 23.27* and *Figure 23.28*. It is convenient to perform post-measurement analysis with PLTS and look at the data displayed as the interaction of differential signals. Looking at the expanded differential near-end crosstalk (NEXT) plot in *Figure 23.28*, the two waveforms show differential crosstalk. The differential signal coupling between the first pair, with port 1 on the left side, and the second pair, with differential port 3 on the left side, is just SDD31 in the frequency domain or TDD31 in the time domain. This is the fraction of differential voltage noise picked up on the near end of the second pair, when a 1V differential signal is sent into the active line. The differential near-end noise between adjacent pairs, TDD31, is measured as less than 0.005 of the active signal, or less than 0.5 percent. This crosstalk coupling happens at the connectors, and there is virtually no differential near-end noise from the traces—they are just too far apart to generate any near-end noise between the differential pairs.

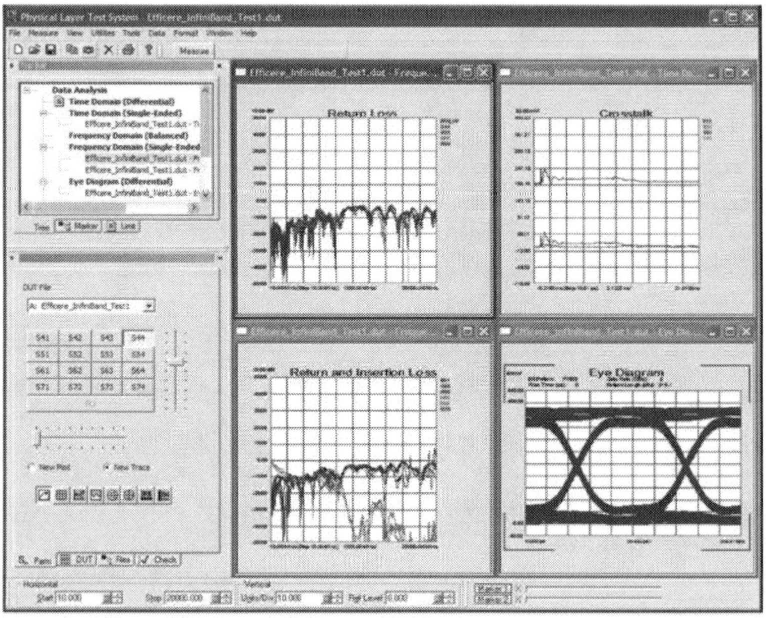

Figure 23.27: *12-Port Measurement Results (One-Meter Cable)*

Notice that the TDD31 noise is negative. This is because we define the polarity of the lines as p, n, p, n, p, n. As we might expect, most of the coupling between the differential ports 1 and 3 is from the two lines that are adjacent. The second line in the first pair has a negative polarity, while the first line in the second pair has a positive polarity. The negative signal in the second line of the first pair generates a negative signal in the first line of the second pair. This near-end noise is all about the coupling in the connector and the via field.

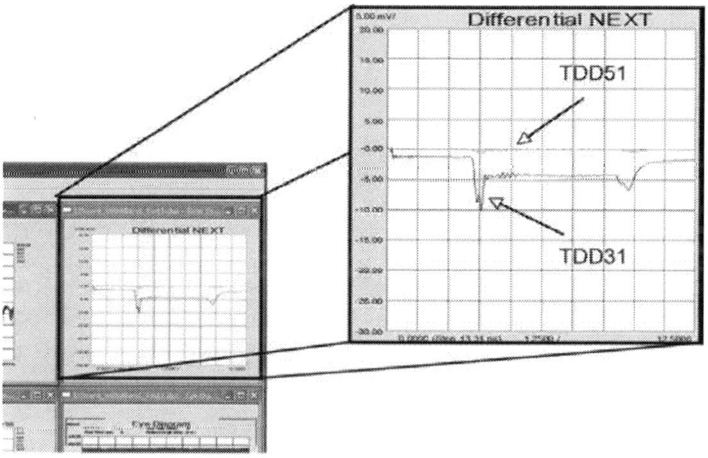

Figure 23.28: *NEXT Measured on a 12-Port VNA*

Displayed on the same scale in *Figure 23.28* is the differential near-end noise between the third pair and the first pair, TDD51. The near-end noise does not even register on this scale. This suggests that the near-end noise is only remotely significant between adjacent pairs. If it is small between adjacent pairs, it will be completely negligible between differential channels that are farther apart. However, when the connector or via field has not been designed well—for example, if the ground plane web between the thru-holes in the via field has been etched through and is disconnected—the ground bounce between connectors may increase considerably. Not having adjacent ground returns will increase the differential noise between adjacent via pairs. Even a great connector, in a board that has been improperly designed or fabricated, can have near-end noise problems and the TDD31 term and the TDD51 term will quickly identify this. This is a key tip to identify crosstalk problems.

The VNA was used to gather the data for the impedance profile shown in the expanded view of *Figure 23.29*. This profile, similar to the profile taken on the time domain–based TDR in *Figure 23.25*, shows a short length of test fixture transmission line followed by large inductive discontinuities of the InfiniBand receptacle. It also demonstrates the versatility of the frequency domain–based VNA in making measurements.

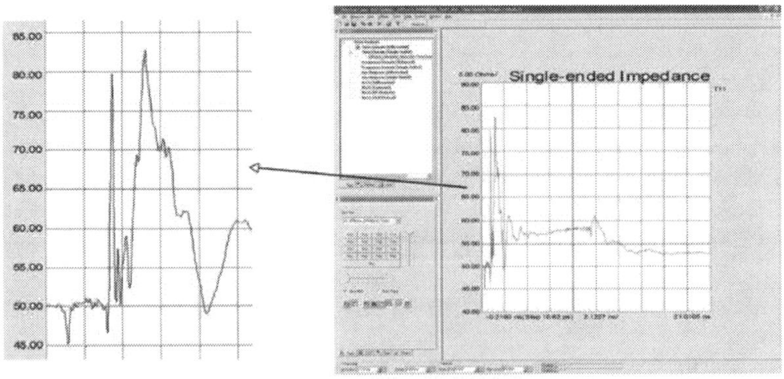

Figure 23.29: *Time Domain Impedance Profile from a 12-Port VNA*

The following section will discuss the ever-increasing demands being placed on the cables in high-performance switching, routing, and computing applications. These large systems, as well as a multitude of smaller high-performance systems, are pushing the envelope of cable length and bandwidth. We have discussed how high-performance test fixtures are designed and can be used to detect marginally performing cables that could cause a system to degrade or, in the worst case, stop performing entirely.

23.5 Test Fixtures: A User's Perspective

Implementation of larger and larger computing systems is forcing system developers to be constantly pushing the envelope of maximum cable length, even as the data transfer rates increase. These of course are conflicting goals, since increasing data rates tends to reduce cable lengths due to the higher attenuation as frequencies increase. Introduction of techniques to ameliorate this dilemma such as pre-emphasis (aka pre-distortion); equalization in cables, drivers, and/or receivers; and multilevel signaling have all been considered as acceptable methods of preserving signal integrity.

Historically, there has always been a crossover point after which one has to bite the proverbial bullet and switch from copper to fiber with its higher cost due to the optical transceivers involved. That boundary line, *Figure 23.30*, continues to be pushed out by both advanced error-correction measurements and active cable technology. Here is described a 25 Gbps fixture that will give us confidence in the results as technologists continue pushing that boundary. Let there be no mistake, there is still much high-speed digital performance left in copper.

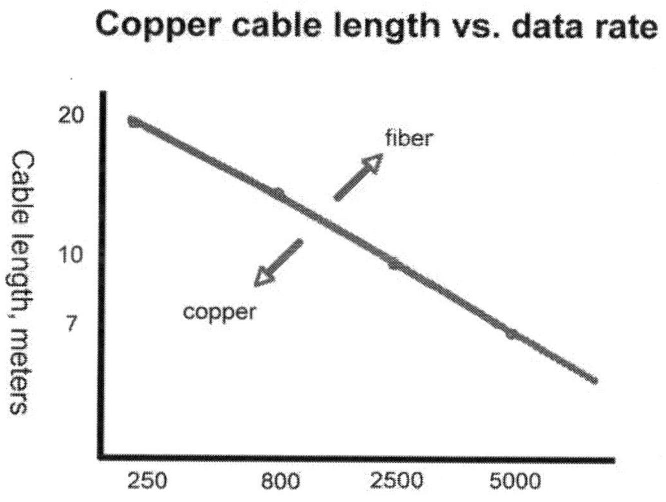

Figure 23.30: *Copper versus Fiber*

As long as one stays in the copper world, accurate and repeatable results for cable electrical measurements are critical to ensure that the relevant system specifications are indeed met. This will guarantee adequate system interoperability and provide for confidence when comparing lab test results. Accurate characterization and de-embedding of test fixture effects are required to provide confidence in the cable test results, a task made much easier with well-designed fixtures that minimize their effect on the measurement results.

One architecture being increasingly used in large computing applications is InfiniBand. Typical system installations are shown in *Figures 23.31, 23.32*, and *23.33*, showing the use of 96-port switches to connect processors and I/O devices. Although there are alternate paths between switches, one would only achieve optimum performance were all the links to be working with the zero or near-zero error rate possible only with cables that met the relevant electrical specifications. Good electrical fixtures are needed to verify the cables' conformance to the specification to make that a reality.

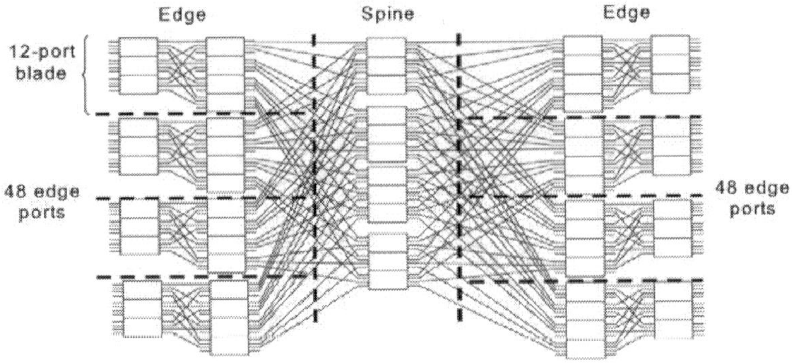

Figure 23.31: *Block Diagram of 96-Port Switch*

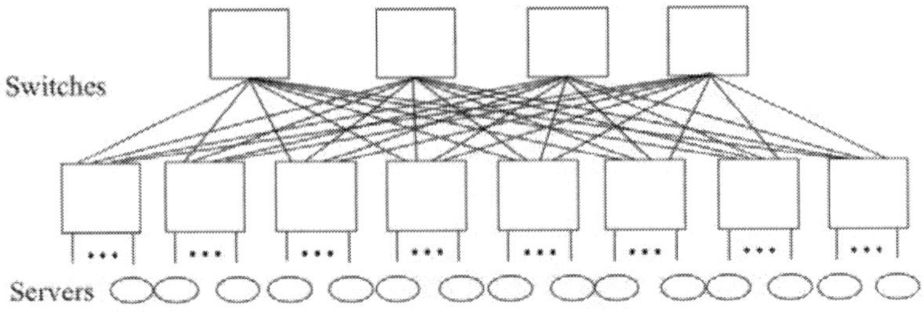

Figure 23.32: *System Interconnect*

Often the linear passive components in high-speed digital systems are not scrutinized for high-bandwidth performance. The active electrical transceivers have signal conditioning capabilities that can open the eye diagram by use of pre-emphasis and equalization. However, the limiting factor in how well these signal conditioning algorithms can work is dependent on the raw performance of the physical-layer passive components (e.g., cables, connectors, PCB traces). As a result, optimizing the cables in high-speed systems is an excellent way to gain more system performance. Most systems depend on reliable, very-low–error-rate cables to achieve their system performance objectives.

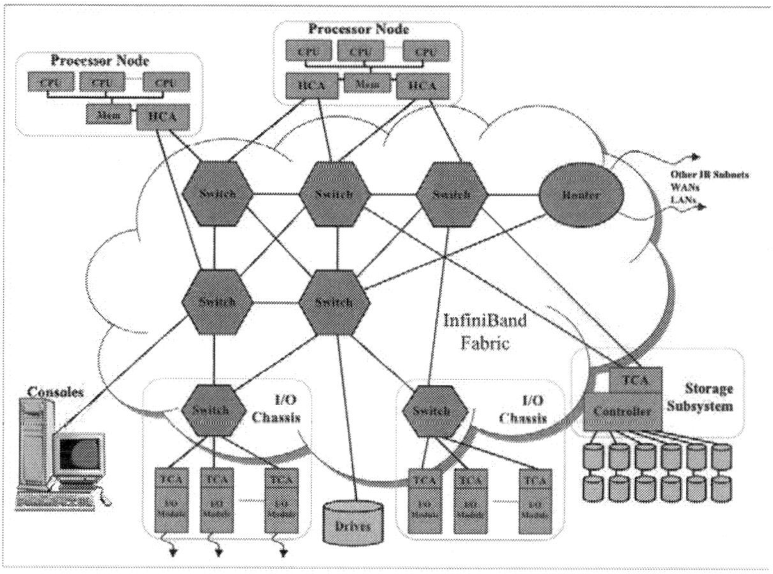

Figure 23.33: InfiniBand System Configuration

In the author's opinion, all industry-standard cables are definitely not created equal. This point is illustrated at industry "PlugFests," where various suppliers bring their cables for compliance testing. It is often the case that even when the cables have been tested at the supplier's facility, they do not pass the tests done at a PlugFest. Sometimes the failure is due to misinterpretation of the specification or the test requirements, and sometimes it is due to the design and manufacturing of the cable itself or its components. The reason that cable testing is becoming more important is that very large cluster systems are being built today that require huge numbers of cables to interconnect their processors.

One example of the importance of good cable signal integrity is IBM's Deep Computing Visualization initiative, which seeks to apply massive cluster computing capabilities to applications such as oil and gas exploration, genomic research, financial analysis, digital media content creation, medical imaging, and weather prediction. As an illustration of the importance of accurate cable measurements made possible through the use of high-performance fixtures, consider some large system installations.

One example of such a system is the Mare Nostrum European supercomputer[5], built in 2006 using off-the-shelf components in the Chapel Torre Girona at the Technical University of Catalunya Barcelona Supercomputing Center (BSC). This system, as shown in *Figure 23.34*, has 2,282 IBM blade servers housed in 163 chassis, 4,564 64-bit IBM Power PC processors, and 140 TB of IBM storage servers. At 94 TeraFLOPs' (trillion floating point operations') performance, it is currently the fifth-fastest supercomputer in the world.

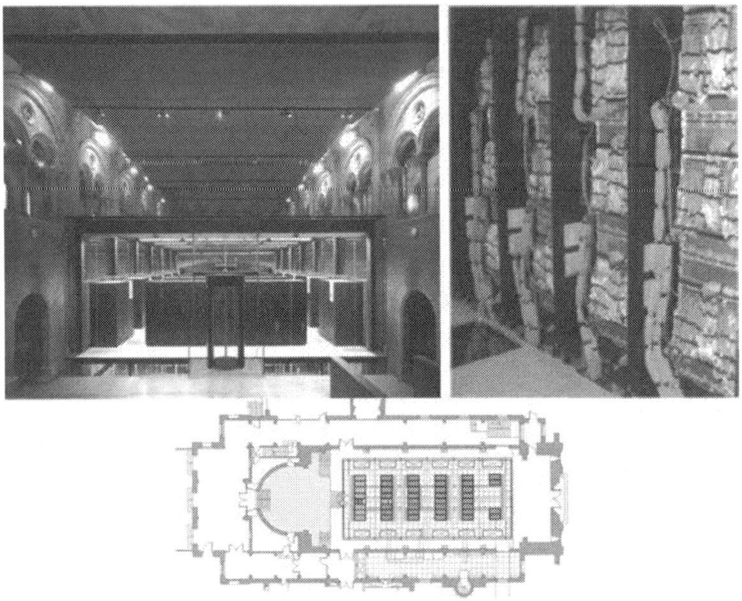

Figure 23.34: *Mare Nostrum (Barcelona), 94 TFLOPs*

Mare Nostrum System Architecture Overview:

- 10,240 processors in 31 racks (186 chassis)
- 120 m^2
- 94 TFLOPs
- 20 TB main memory
- 280 plus 90 TB disk storage in 20 nodes

[5] Mare Nostrum, www.bsc.es/plantillaA.php?cat_id=200.

- 2,560 switch ports
- Myrinet and Gigabit Ethernet interconnect architecture

The system has a dual purpose: to serve as a primary high-performance computing resource for the European e-science community and to demonstrate the many benefits of Linux on the POWER architecture in scale-out processing environments. Planned applications areas include computational sciences; earth sciences, including climate and atmospheric studies; and life sciences.

In order for world-class computing systems such as the Mare Nostrum to operate at peak capacity, the inter-processor cables must be optimized for signal integrity or bit errors will be introduced into the signal path. Such systems have redundant data paths between processors, but this fault-tolerant system design severely degrades computing performance.

Figure 23.35: *ASC Purple, which Has 17,000 Cables with a Total Length of 140 Miles*

Another large cluster application requiring excellent signal integrity cables is the ASC Purple supercomputer[6] installed at the Lawrence Livermore Labs' Terascale Simulation Facility, as shown in *Figure 23.36*. This machine has 94 TeraFLOPs of compute capacity and is composed of 269 racks, including 1,534 eight-way processors; 2 million GB of storage; four 16-way login nodes, each with eight 10

[6] ASC Purple, https://asc.llnl.gov/computing_resources/purple.

Gbps and two 1 Gbps network connections; 1,000 Serial ATA–and Fibre Channel–attached disk drives in 90 racks connected by 48 switch racks; and 17,000 cables totaling 140 miles in length. ASC Purple ranks fourth in the world in supercomputing performance. This system's primary application is nuclear weapons simulation, but many other applications in various fields can be imagined.

ASC Purple (CA) – System Architecture Overview:

- 12,304 processors in 269 racks with 94 TFLOPs of computing power
- 50 TB main memory
- 2,000 TB of storage spread across 1,000 drives in 90 racks
- 48 switch racks
- Proprietary interconnect similar to 12X InfiniBand cables
- 17,000 cables – 140 miles

The current great granddaddy of large system installations is the Blue Gene/L supercomputer[7] shown in *Figure 23.36* installed in 2005 at the Lawrence Livermore National Lab in California. As *Figure 23.37* indicates, it is made up of 72 racks of 2,048 processors each, interconnected by 4,352 cables. Every one of those cables has to be designed with a controlled impedance environment with minimal insertion loss for the system to run at its peak performance of 360 TeraFLOPs/second.

Figure 23.36: *BlueGene/L at Lawrence Livermore National Labs—147,456 Processors Connected by 4,352 Cables*

[7] BlueGene/L, https://asc.llnl.gov/computing_resources/bluegenel.

BlueGene/L Purple (CA) – System Architecture Overview:

- 147,456 processors in 72 racks
- 360 TFLOPs
- 32,768 TB of main memory
- Proprietary interconnect similar to 12X InfiniBand cables
- 4,352 cables

The BlueGene/L system currently holds the number-one spot on the TOP500[8] supercomputer performance-ranking list. Applications of the Blue Gene system include life sciences, financial modeling, hydrodynamics, quantum chemistry, molecular dynamics, astronomy and space research, materials science, and climate modeling.

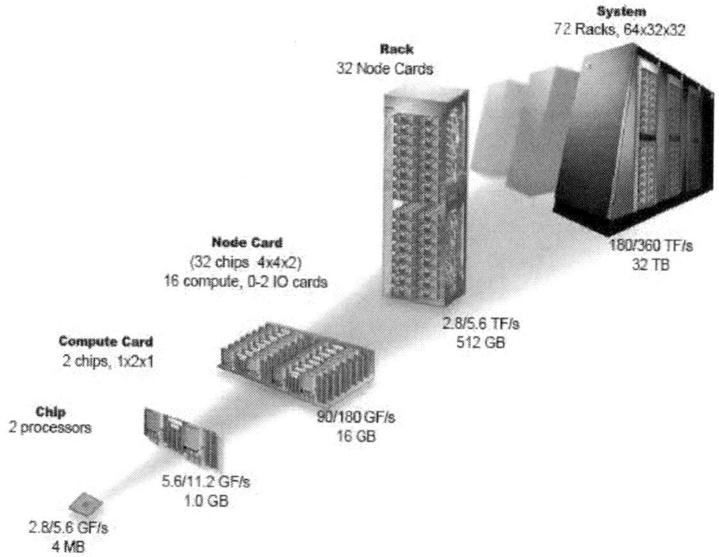

Figure 23.37: *BlueGene/L System Architecture*

The BlueGene supercomputer's 147,456 processors are interconnected in a 3D torus configuration as depicted in *Figure*

[8] The Top 500 Supercomputer list can be found at www.TOP500.org.

23.38. One group of four racks is shown in *Figure 23.39* with its interconnecting cables. One can imagine the mass of the 4,352 cables needed to interconnect 72 racks!

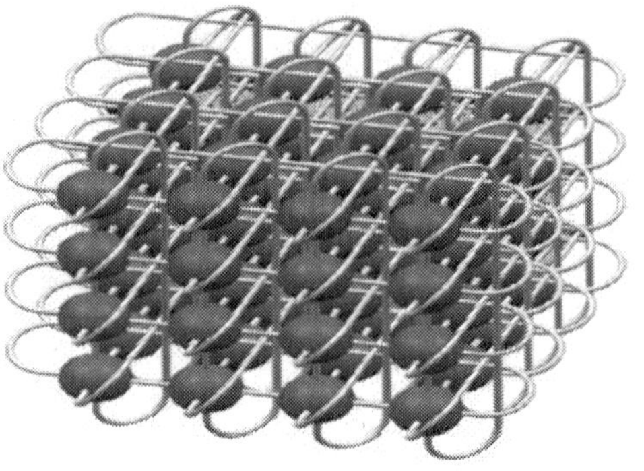

Figure 23.38: *BlueGene's 3D Torus Processor Configuration*

Figure 23.39: *BlueGene Cabling Showing Just a Few of the 4,352 Cables in the Complete System*

The concept of on-demand computing is fulfilled in the "fractional BlueGene" program in which a customer can rent just the amount of computing horsepower needed for the job at hand, rather than having to buy a whole machine whose capacity is only needed part of the time.

23.6 Conclusion

Silicon technology continues to drive the entire signal integrity ecosystem for electrical interconnects. As new and faster chipsets are developed and deployed, this pushes the performance level required for the chip package. As the chip package performance is enhanced, this pushes the performance of PCBs and high-speed backplanes. This ultimately requires better system performance from box to box and rack to rack. This whole cycle of signal integrity improvement can be broken if the cables are not designed properly. Good electrical interconnect design techniques are critical in order to keep up and provide the enabling pathways to the silicon engines scattered across boards, shelves, racks, campuses, and the world.

In this paper, we have discussed various aspects of interconnect design, including a new 12-port signal integrity analysis tool, two design case studies of a high-performance fixtures (InfiniBand and HDMI), and some of the critical computing applications that require leading-edge interconnects. The trends of increasing performance requirements and ever-higher data rates are not stopping anytime soon. The importance of fully characterizing linear passive devices such as cables is paramount to achieving world-class computing systems. By using the leading-edge tools and methodologies outlined in this paper, high-speed design engineers can continue developing leading-edge components and systems in the face of ever-increasing analog and high-frequency effects.

Acknowledgment

Dean Suhr of Efficere Technologies contributed to this report.

Reference

Duperron, Dunham, and Resso, "Practical Design and Implementation of Stripline TRL Calibration Fixtures for 10-Gigabit Interconnect Analysis," DesignCon Proceedings, 2006.

Authors

Jay Diepenbrock, Senior Technical Staff Member, Interconnect Qualification Engineering, IBM

Will Miller, Vice President, Engineering, Efficere Technologies

Mike Resso, Signal Integrity Application Scientist, Component Test Division, Agilent Technologies

Author Biographies

Heidi Barnes is a high-frequency device interface board designer for Agilent's 93000 semiconductor test system focusing on both digital and analog controlled impedance transition design for full path integrity. She initially joined Agilent Technologies in 1997 and was previously with its Microwave Technology Center working with thinfilm, thickfilm, PCB laminate chip and wire, and machined metal packaging technologies for DC to 20 GHz analog and digital applications. Heidi holds a B.S. degree in electrical engineering from the California Institute of Technology.

Shelley Begley currently leads a small team of both R&D and marketing at Agilent Technologies, which focuses on advancing dielectric measurement techniques. She has over 21 years experience in the network analyzer business, working in production engineering, electrical metrology, and product marketing. Shelley has written countless papers and has given numerous seminars around the world on the topic.

Orlando Bell is vice president of engineering at GigaTest Labs where he joined after completing his graduate school education. For the past decade, he has specialized in the characterization of high-frequency interconnects based on measurements and the design of measurement fixtures for the DC to 20 GHz range. In addition, he heads the device characterization effort at GigaTest Labs, a measurement and model extraction service for BJT, GaAs FET, pHEMT devices. He holds an MSEE degree from the University of Florida.

Osvaldo Buccafusca is a development scientist in the lightwave division of Agilent Technologies. He is responsible for the research and development of optical instrumentation which enables world-class measurements. His broad background includes ultrafast laser measurement techniques, VCSEL characterization, and manufacturing test optimization. His recent work has focused on a novel timebase architecture for digital communications analyzers. Osvaldo received his Ph.D. and M.S. degrees from Colorado State

University in 1993 and 1996, respectively, and his M.S. degree in physics from Buenos Aires University in Argentina in 1989.

William Burns is a senior applications engineer at Altanova Corporation. He is an expert in test fixture design for Verigy's 93000 and focuses on engineering and investigation projects. He has been working in the test fixture industry since 2004 and provided third tier support for Agilent's 84000 RFIC and MMIC testers and the RF subsystem of the 93000 SOC tester since 2000. William holds a B.S. degree in electrical and computer engineering from Brigham Young University and specialized in RF, MW, and communication systems.

Jack Carrel is a system IO specialist at Xilinx. He has over 25 years of experience in product development and design in the fields of instrumentation, test and measurement, and telecommunications. His background includes development of electro-optic modules, multi-gigabit transceiver boards, high-speed and high-resolution data acquisition systems for government and commercial applications. Most recently, Jack has been involved in product design using multi-gigabit transceivers with specific focus on PCB design issues. He received his B.S. degree in electrical engineering from the University of Oklahoma.

Jinhua Chen manages signal integrity group within EMC hardware engineering. He provides the high-speed design including interconnect, termination, clock distribution, and power plane decoupling for EMC hardware designs. Jinhua also evaluates future technologies such as SerDes and high-speed connectors, continually searches for new tools and technologies, and develops new design methodologies. Jinhua worked as an SI consultant for North East System Associates from 1996 to 2001 before joining EMC and, since 2003, he has been a DesignCon technical committee member. He has been awarded the outstanding paper award in DesignCon98 and in DesignCon East 2004. Jinhua Chen holds two U.S. patents and a Ph.D. degree in physics from the University of Massachusetts Lowell.

Antonio Ciccomancini is an application engineer at CST of America, Inc. In 2004, he received the CST University Publication Award for the use of the FIT technique in signal integrity applications and in August 2005 he joined CST. He is a member of IEEE, and he received his Ph.D. degree in electrical engineering from the University of L'Aquila, Italy in 2005. His dissertation focused on full-wave simulations and de-embedding techniques for the characterization of PCB discontinuities.

Michael Comai is a senior product engineer at Advanced Micro Devices where he has been working since 2001. He specializes in new methodologies for testing analog macros and IO PHYs as part of the strategic test technologies team. He received his B.S. degree in electrical engineering technology from Purdue University.

Steven Corey is principal engineer in the electro-optical product line at Tektronix, where he has contributed to the development of high-speed measurement and modeling technology. His interests include automation of model extraction from measurement and the conversion of academic research into mainstream design automation tools. From 1998-2005, he was principal engineer at TDA Systems, where he led research and development for the company's measurement-based modeling products, both in algorithm development and in steering product direction. Steven has been in the field of measurement-based model extraction for fourteen years and has published a number of relevant academic and trade magazine articles. He holds a Ph.D. degree from the University of Washington.

John D'Ambrosia is the manager of semiconductor relations for Tyco Electronics in Harrisburg, PA. He works with semiconductor vendors to explore the interaction between semiconductor devices and the company's interconnect solutions. This has led to his involvement in standards bodies defining solutions for backplane environments. He was an active participant in the development of XAUI, and served as the chair of the 10 Gigabit Ethernet Alliance XAUI Interoperability Group, which drove interoperability testing for XAUI solutions for the industry. John helped organize the High Speed Backplane Initiative (HSBI), and he also served as its secretary. He is now serving as the chair of the Optical

Internetworking Forum's (OIF) Market Awareness & Education Committee and is participating in the development of the OIF Common Electrical I/O (CEI) project. John received a B.S. in electrical engineering technology from Pennsylvania State University in 1989 and a M.S. in engineering management from the National Technology University in 1999.

Bill Dempsey is owner and president of Redwire Enterprises, LLC which is a design services company focused on high-frequency digital and RF design. His background includes design of military and commercial equipment for more than twenty years. He founded Red Wire Enterprises in 2001 to provide turn-key solutions for customers that include signal integrity, electromagnetic analysis, system design, and PCB layout. He is also an active member of the IEEE Microwave and EMC societies. Bill received a B.S. degree in electrical engineering from Texas A&M University in 1985.

Joseph C. (Jay) Diepenbrock is a senior technical staff member in the interconnect qualification engineering department in IBM's integrated supply chain organization. He works on the electrical testing and modeling of connectors and cables; this involves hardware testing and software simulation of various interconnect structures, lossy model generation and verification, and the development of electrical test procedures and software. Jay has been with IBM since 1976 and has worked in a number of areas including bipolar and CMOS IC design, analog and digital circuit design, backplane design and simulation, and network hardware and server product development. He has authored papers at various IEEE and other conferences, contributed to a number of industry standards including USB, EIA 364 electrical test methods, and PCI-Express specifications, and is currently leading the cable/connector sub-team of the Electromechanical Working Group of the InfiniBandsm Trade Association. Jay received the M. S. and Sc. B. degrees in electrical engineering from Syracuse University and Brown University, respectively.

Dave Dunham is the electrical engineering manager for the connector products division at Molex, Inc. He is responsible for high-speed development for backplanes and board-to-board and

I/O interconnect solutions. With over 15 connector patents, Dave and his team participate in next-generation connector designs and advocate SI testing and system/interconnect modeling. Dave has 34 years of experience in the design test and measurement on both coaxial and multi-pin connectors, with recent activities in the backplane connector development arena. Dave received his BSEE degree from University of New Mexico.

Vince Duperron is an electrical project engineer for the connector products division at Molex, Inc. where he designs connectors and test fixtures for a variety of products. Vince previously worked at Silicon Graphics, Inc. for five years during which he designed circuit boards and interconnect for three generations of super computers. Vince received a B.S. degree in electrical engineering from Michigan Technological University.

Thomas Gneiting is the founder of AdMOS Corporation which focuses on advanced modeling solutions. He is responsible for developing methods for the model parameter extraction of deep submicron MOS devices, as well as providing design support for high-frequency and high-speed communication systems. His company provides modeling services which cover state-of-the-art CMOS processes and such passive devices as high-speed connectors and IC packages. Thomas received his Dipl.-Ing.degree from the University of Applied Sciences in Esslingen, Germany and a Ph.D.from Brunel University of West London.

John Goldie is a member of the technical staff at National Semiconductor where he has been since 1988. He has since then worked on LVDS, BLVDS, CML, ECL, GTL, BTL, RS-xxx and other interface technologies. John has published a number of articles and routinely conducts seminars on Data Transmission topics. He holds a B.S. degree in electrical engineering, with a minor in design and industry, from San Francisco State University.

Kevin Grundy is a seasoned electronics industry executive who is chief executive officer of SiliconPipe. He has more than 25 years of experience in reducing complex systems into semiconductor circuits and has been credited for creating and managing an environment where high volume and high quality products are successfully brought to market. Kevin was previously with DirecTV Broadband where he was cofounder of its predecessor, Telocity, and was responsible for all engineering development and operations. He has held various engineering and managerial posts at several successful startups, most notably at NeXT Computer, where he led the design efforts for three generations of innovative desktop computers. Kevin has BSEE and MSEE degrees from California State Polytechnic University, Pomona.

Crescencio Gutierrez is the engineering and R&D manager at Harbor Electronics, Inc. He holds expertise in controlled impedance high-frequency multi-layer laminates for high density ATE "interface boards" for the semiconductor industry. Crescencio has over 18 years of experience in the PCB industry; he is directly involved in evaluating materials and techniques in increasing the high aspect ratio of multi-layer PCBs for high density 0.5mm pitch BGA's and evaluating materials for optimum impedance control at multigigabit data rates.

Abraham Islas is a senior product engineer at Advanced Micro Devices. He has been a part of the company since 2005. His professional specialty is on high-speed I/O bus characterization and debug. He received his M.S. degree in electrical engineering from Texas A&M University.

Haw-Jyh Liaw is director of system engineering at Aeluros, Inc. His professional interest includes simulation, measurement, and system-level design of high speed interconnects. He was previously with Rambus Inc. where he worked on memory interfaces and high-speed backplane links. Haw-Jyh received outstanding paper awards from 1995 ECTC and 1996 Design SuperCon. Mr. Liaw received his Ph.D. degree in electrical and computer engineering from the University of Illinois at Urbana-Champaign in 1996.

Jim Mayrand is an independent signal integrity consultant who operates offices in both San Francisco and Massachusetts. He has extensive backplane and ASIC design experience and has in-depth knowledge of both interconnect and IC device physics. Jim also has a novel perspective of signal integrity in tribute to his practical understanding of copper interconnect having worked his way through college as journeyman electrician. Prior to retiring from Hewlett-Packard as a hardware design engineer, he worked in the semiconductor product test and design engineering field. Jim holds four patents, and he received a M.S. degree from University of New Hampshire in electrical engineering.

Tom McCarthy is vice president at Taconic and focuses on the development and advancement of multilayer materials for high-speed digital applications. He was previously the company director of engineering for seven years and has co-authored 11 publications and 8 patents. He holds a Ph.D. in polymer science and engineering from the University of Massachusetts.

Will Miller is vice president of engineering at Efficere Technologies. He has over 24 years in design and simulation of very high-performance semiconductor automatic test equipment and advanced electro-optical instrumentation for aerospace and commercial applications. Will has worked at such companies as Integrated Measurement Systems and Credence, and he has developed manufacturing techniques for precision high-performance electronic signals on electronic substrates. He joined Efficere Technologies in 2005 to lead the company's engineering efforts in advanced test, memory module, power subsystem, and interconnect/cable/receptacle design.

Roland Moedinger joined the development department at ERNI Electroapparate GmbH where he is currently responsible for high-speed digital applications. He has held several technical positions where he has developed a specialty for custom applications and design utilizing both software tools and physical tools. Roland is a graduate of the University of Esslingen where he specialized in precision engineering.

José Moreira is a senior application consultant in the Center of Expertise of Verigy's semiconductor test solutions division in Böblingen, Germany. He initially joined Agilent Technologies (now Verigy) in 2001 and focuses on the challenges of testing high-speed digital devices particularly in the area of signal integrity and jitter testing. Jose holds a M.S. degree in electrical and computer engineering from the Technical University of Lisbon, Portugal.

Gary Otonari is the engineering project manager at GigaTest Labs. He initially joined the company in 1999 where he first served as vice president of business development and, for the last three years, he has managed activity in GigaTest Labs' signal integrity engineering business. His previous experiences have led him to work at Hughes Aircraft as a satellite communications RF payload design engineer and even to Incorporated and Agilent Technologies where he was for over 10 years in various positions. He is an expert in the use of high-frequency EDA software tools and signal integrity measurement techniques. He received his B.S. in electrical engineering from the University of California at Los Angeles.

Gautam Patel is a signal integrity engineer in the new product development group at Teradyne. He performs measurements, modeling, and simulation of backplane interconnects. In addition, Gautam's other functions include applications support, along with technical presentations and writing technical papers. He has an M.S. degree in electrical engineering from Northeastern University.

Stephen Reddy is a senior design engineer in the transmission subsystems group of JDS Uniphase based in Melbourne, FL. He is responsible for providing technical support for the design of 2.5 Gb/s, 10Gb/s, and 40Gb/s fiber optic transmitters and receivers. Stephen's recent projects include the development of a 40 Gb/s fiber optic transponders for telecommunication applications. He has over five years of experience in fiber optic transmitter and receiver design. Stephen received his M.S. and B.S. degrees in electrical engineering from the University of Central Florida, the former which was obtain through the Center for Research and Education in Optics and Lasers (CREOL).

Jason Roe is an application engineer with ERNI Electronics in Midlothian, VA. His is responsible for design and development of high-speed connectors. Prior to joining ERNI Electronics, Jason honed his engineering problem solving skills on telecommunications applications for Verizon. Jason received his BS degree in Electrical Engineering from Virginia Commonwealth University.

Anthony Sanders is principal engineer of Infineon Technology and is responsible for analog concept engineering within the Optical Networking Group, developing bipolar & CMOS 3/6/10/40Gbit line card products. His current research activities focus on the theoretical understanding of jitter phenomena based on observation from laboratory measurement, the modeling and optimization of optical and electrical serial links using equalization, and the limits of CMOS technology and its application to high-speed serial interfaces. Anthony has chaired the IEEE ad-hoc on XAUI jitter, various OIF jitter ad-hocs and is currently one of the editors of OIF's Common Electrical I/O (CEI) standard and author of the *StatEye*. He received his B.Eng. degree with honors in electrical and electronic engineering from the University of Nottingham, England in 1991.

Edward Sayre is owner and director of engineering of North East Systems Associates (NESA), which is a high-performance engineering and design firm for the computer and communication industries. Edward's company works with globally recognized semiconductor companies to provide interconnect reference designs for their new I/O products. Over fifty systems have been EMC engineered by NESA to pass FCC, Bellcore and CE compliance standards. Dr. Sayre pioneered time domain characterization of interconnects as well as proficient use of the better-known frequency domain instrumentation methods.

Robert Schaefer is a technical leader and R&D project manager for the signal integrity group of Agilent Technologies based in Santa Rosa, CA where his responsibilities include product planning, strategy, and development. Previously, he worked as a solution planner in one of the marketing solutions units for Agilent Technologies and, for more than 20 years, he worked in research and development as a designer and project manager. His design and management experience covers the breadth of GaAs IC and microcircuit design, RF and analog circuit design, instrument firmware and computer-aided test software, device modeling and design of modeling systems, and microwave and RF CAD products. Prior to joining Hewlett-Packard in 1976, Robert completed his BSEE and MSEE degrees from the University of Missouri at Rolla.

Hong Shi is a member of the technical staff in the packaging technology group of Altera Corporation. His current responsibilities include developing strategy for high density and high-speed IO packaging, simulating system level electrical performance of FPGA packages and creating chip-package-board interconnect co-design capability. His most recent activities include characterizing the simultaneous switching noise and power distribution network of high performance FPGAs. Before joining Altera, Hong was principal engineer and project leader for Agilent's first 40Gbps digital communications analyzer module. Hong has published over 30 technical papers and has twice received the Agilent "Spark of Insight" award for his contribution to the company. He obtained his Ph.D. in microwave opto-electronics field from CREOL College of Optics of University of Central Florida.

Dima Smolyansky currently works for Tektronix, Inc. where he works on business development for IConnect® TDR software products, which help with signal integrity and high-speed digital design work. He has spent his professional career in the instrumentation and measurement industry, working with high-speed time domain reflectometry oscilloscopes and frequency domain network analyzers. In the past several years, Dima has accumulated significant experience in the area of high-speed digital interconnect measurements and modeling. He has published a number of papers and taught short courses on interconnect analysis. Dima holds a MSEE degree from Oregon State University and the M.S. engineering degree from Kiev Polytechnic Institute.

Laurie Taira is a senior project engineer in the R&D department at Delphi Connection Systems. Her current responsibilities include simulation, design and test, and measurement of high density, and high speed flexible circuit interconnects. She is the task leader of the electrical simulation and measurement laboratory at Delphi Connection Systems and her laboratory was recently awarded the prestigious Center of Measurement Excellence from Delphi Corporate. Laurie was previously a microwave, fiber optic and telecom compliance engineer. She obtained her B.S. in physics from the University of California at Los Angeles and her M.S. in physics from California State University, Long Beach.

Francisco Tamayo-Broes is a product development engineer at Advanced Micro Devices where he has been since 2006 after first developing his career as a test engineer. His focus is on high-speed characterization and debug, as well as signal integrity analysis. He holds a B.S.E.E. degree.

Ming Tsai is a staff hardware development engineer in the product technology division for Xilinx. His responsibilities include signal integrity analysis for the highspeed signal transitions, high-speed loadboard design and characterization. He joined Xilinx in October 2004 and earned his Ph.D. degree from the electrical engineering department of University of California, Los Angeles in 1996.

Acronym Guide

AFE	analog front end
AI	artificial intelligence
ASIC	application-specific integrated circuit
ATCA	advanced telecommunications architecture
ATE	automated test equipment
BER	bit-error rate
BERT	bit-error rate tester
BGA	ball grid array
BIST	built-in self test
BU	bounded uncorrected
BUJ	bounded-uncorrelated jitter
CDF	cumulative distribution function
CDR	clock and data recovery
CDRC	clock and data recovery circuit
CJTPAT	compliant jitter tolerance pattern
CML	current-mode logic
CMOS	complementary metal-oxide semiconductor
CMRR	common mode rejection ratio
CMU	clock multiplication unit
CPRI	Common Public Radio Interface
CR	clock recovery
DAC	digital-to-analog converter
DCD	duty cycle distortion
DDJ	data-dependent jitter
DDN	data-dependent noise

DFE	decision-feedback equalizer
DFT	design for testability
DJ	deterministic jitter
DN	deterministic noise
DRSL	differential Rambus signaling level
DUT	device under test
EMC	electromagnetic compatibility
EMI	electromagnetic interference
EPA	edge placement accuracy
EPD	eye-pattern diagram
ESL	equivalent series inductance
ESR	equivalent series resistance
FC	Fibre Channel
FCC	Federal Communications Commission
FDTD	finite difference time domain
FEC	forward error correction
FEXT	far-end crosstalk
FFT	fast Fourier transformation
FIR	finite impulse response
FPGA	field programmable gate array
FT	Fourier transformation
GbE	Gigabit Ethernet
GJ	Gaussian jitter
GTL	Gunning transistor logic
HDMI	high definition multimedia interface
HPJ	high-probability jitter
HSS	high-speed serial

HSSL	high-speed serial link
IC	integrated circuit
IFFT	inverse fast Fourier transform
IIR	infinite impulsive response
ISI	intersymbol interference
ITRS	International Technology Roadmap for Semiconductors
LICA	low-inductance capacitor array
LTI	linear time invariant
LRL	line-reflect-line
LRM	load-reflect-match
MG	multiple Gaussian
MoM	method of moment
MTS	mega transfers per second
NEXT	near-end crosstalk
NF	noise figure
NRZ	non-return to zero
OBSAI	Open Base Station Architecture Initiative
ODR	octal data rate
OIF	Optical Internetworking Forum
OS	over-sampling
PAM	pulse amplitude modulation
PBGA	plastic ball grid array
PC	personal computer
PCI-X	Peripheral Component Interconnect eXtended
PCB	printed circuit board
PDF	probability density function
PDN	power distribution network

PEEC	partial element equivalent circuit
PHY	physical layer
PI	phase interpolator
PJ	periodic jitter
PLI	programming language interface
PLL	phase-locked loop
PLTS	physical layer test system
PMD	polarization mode dispersion
PRBS	pseudo-random bit stream
PRF	parallel resonant frequency
PSD	power spectrum density
PSRR	power supply noise rejection ratio
PTH	plated through hole
PVT	process, voltage, and temperature
PWB	printed wiring board
RC	resistive and capacitive
RJ	random jitter
RN	random noise
RSL	Rambus signaling level
Rx	receiver
SAS	Serial-Attached SCSI (Small Computer System Interface)
SATA	Serial Advanced Technology Attachment
SCA	static crosstalk analysis
SDI	Serial Digital Interface
SerDes	serializer/deserializer
SG	single Gaussian
SI	signal integrity

SoC	system on chip
SRF	self resonant frequency
SRIO	Serial Rapid Input Output
SSC	spread spectrum clock
SSN	simultaneous switching noise
SOLT	short-open-load-thru
SSO	simultaneous switching output
STA	static timing analysis
TDR	time-domain reflectrometer
TDT	time-domain transmission
TIA	time-interval analyzer
TJ	total jitter
TMM	transmission matrix method
TN	total noise
TRL	thru-reflect-line
TRM	thru-reflect-match
Tx	transmitter
UDSM	ultra-deep sub-micron
UI	unit interval
VCS	Verilog compiler simulator
VNA	vector network analyzer
VRM	voltage regulator module
XAUI	extended attachment unit interface

Made in United States
Orlando, FL
05 November 2026

90230136R00442